Properties of Real Numbers

Commutative

$$a + b = b + a$$
$$ab = ba$$

Associative

$$(a + b) + c = a + (b + c)$$
$$(ab)c = a(bc)$$

Distributive

$$a(b + c) = ab + ac$$

Identities

$$a + 0 = 0 + a = a$$

$$a \cdot 1 = 1 \cdot a = a$$

Inverses

$$a + (-a) = (-a) + a = 0$$

$$a \cdot \frac{1}{a} = \frac{1}{a} \cdot a = 1$$

Equations and Inequalities

If $a = b$, then $a + c = b + c$

If $a = b$, then $ac = bc$

If $a < b$, then $a + c < b + c$

If $a < b$ and $c > 0$, then $ac < bc$

If $a < b$ and $c < 0$, then $ac > bc$

Sets of Real Numbers

Real numbers: R is the set of all numbers corresponding to a point on the number line.

Rational numbers: $Q = \left\{ \dfrac{p}{q} \middle| p \text{ and } q \text{ are integers, } q \neq 0 \right\}$

Irrational numbers: Q' is the set of real numbers that are not rational.

Integers: $Z = \{ \ldots -3, -2, 1, 0, 1, 2, 3, \ldots \}$

Natural or counting numbers: $N = \{1, 2, 3, \ldots \}$

Factoring Polynomials

$$ax + ay = a(x + y)$$
$$a^2 - b^2 = (a + b)(a - b)$$
$$a^3 + b^3 = (a + b)(a^2 - ab + b^2)$$
$$a^3 - b^3 = (a - b)(a^2 + ab + b^2)$$
$$a^2 + 2ab + b^2 = (a + b)(a + b)$$
$$a^2 - 2ab + b^2 = (a - b)(a - b)$$

INTERMEDIATE ALGEBRA

INTERMEDIATE

ALGEBRA

Donald Hutchison
Clackamas Community College

Louis Hoelzle
Bucks County Community College

James Streeter
Late Professor of Mathematics
Clackamas Community College

McGRAW-HILL, INC.

New York   St. Louis   San Francisco   Auckland   Bogotá   Caracas
Lisbon   London   Madrid   Mexico City   Milan   Montreal   New Delhi
San Juan   Singapore   Sydney   Tokyo   Toronto

INTERMEDIATE ALGEBRA

This book is printed on acid-free paper.

2 3 4 5 6 7 8 9 0   DOW   DOW   9 0 9 8 7 6 5

ISBN 0-07-062602-2

This book was set in Times Roman by York Graphic Services, Inc.
The editors were Michael Johnson, Karen M. Minette, and Jack Maisel;
the designer was Joan Greenfield;
the production supervisor was Paula Keller.
R. R. Donnelley & Sons Company was printer and binder.

Photo Credits for Chapter-Opening Photos
 1   Comstock/Comstock Inc.
 2   Comstock/Laura Elliot
 3   Comstock/Boyd Norton
 4   Comstock/Comstock Inc.
 5   Comstock/Russ Kinne
 6   Comstock/Jack Clark
 7   Tony Stone/Larry Ulrich
 8   Comstock/Denver Bryan
 9   Tony Stone/Poulides/Thatcher
10   Comstock/Marvin Koner
11   Comstock/Hartman-DeWitt
12   Comstock/Comstock Inc.

Library of Congress Cataloging-in-Publication Data

Hutchison, Donald, (date).
    Intermediate algebra / Donald Hutchison, Louis Hoelzle, James
Streeter.
       p.   cm.
    Related work: Intermediate algebra / James Streeter.
    Includes index.
    ISBN 0-07-062602-2
    1. Algebra.   I. Hoelzle, Louis F.   II. Streeter, James (James A.)
III. Streeter, James (James A.) Intermediate algebra.   IV. Title.
QA154.2.S7819   1995
512.9—dc20   94-7766

DONALD HUTCHISON is active in several professional organizations. He is a member of the ACM committee that has undertaken the writing of computer curriculum for the two-year college. Since 1990 he has chaired the Technology in Mathematics Education committee for AMATYC.

Don spent his first 10 years of teaching working with disadvantaged students. He taught in an intercity elementary school and an intercity high school. He also worked for two years at Wassaic State School in New York and two years at the Portland Habilitation Center. He worked with both physically and mentally disadvantaged students in these two settings.

In 1982 he was hired by Jim Streeter to teach at Clackamas Community College. It was here that he discovered the two things that, along with his family, serve as a focus for his life. Jim introduced him to the joy of writing (with the first edition of *Beginning Algebra*) and Jack Scrivener converted him to a born-again environmentalist. In 1989 Don became Chair of the Mathematics department at Clackamas Community College.

LOUIS HOELZLE has been teaching at Bucks County Community College for 23 years. He has taught the entire range of courses from Arithmetic to Calculus. This gives him the perspective of the current and future needs of developmental students.

Over the past 30 years Lou has also taught Physics courses at four year colleges. This gives him the perspective of the practical applications of mathematics. In addition, Lou has done extensive reviewing of manuscripts and writing of several solutions manuals for major texts. In these he has focused on writing for the student.

Lou is also active in professional organizations and has served on the Placement and Assessment Committee for AMATYC since 1989. In 1989, Lou became Chair of the Mathematics Department at Bucks County Community College.

While a graduate student at the University of Washington, JAMES STREETER paid for his education as a math tutor. In 1968 he moved on to Clackamas Community College to become their first mathematics chair. Jim recognized that he faced a very different population than the one for whom he had tutored at UW. It was here that he began to formulate the ideas that would eventually become the basis for this series.

CONTENTS

Nine | Other Types of Equations, Functions, and Inequalities 461

Ten | Graphs of Quadratic Functions and Conic Sections 511

Eleven | Exponential and Logarithmic Functions 551

Intermediate Algebra is designed for a one-term course. The focus of the text is to make students better problem-solvers. Our emphasis as writers is to communicate this to our readers. By directing our attention to readability, we encourage the participation of students in the learning process. Each topic is presented in a straightforward fashion with numerous examples to clarify the concept being developed. All the features are designed to encourage, facilitate, and motivate problem solving among the students.

Functions and Graphing The vast majority of students are visual learners, so we introduce functions and graphing as early as Chapter 3. We then integrate the use of functions and graphing throughout the text. This approach allows for a visual interpretation of mathematical concepts. By introducing such topics early in the book, students become familiar with, and also more comfortable with, concepts that will be critical to their success in future mathematics courses.

The benefits of the functional approach are particularly apparent in our discussion of quadratic functions. By looking at each quadratic in the form $f(x) = a(x - h)^2 + k$, students are inspired to view each function geometrically as a translation and stretch of $g(x) = x^2$.

Technology Many problems are included that require the use of a scientific calculator or graphing tool ⬚. In addition, there are discussions about how technology

can enhance the study of mathematics through **exploration, visualization,** and **geometric interpretation.**

These discussions usually appear as margin notes, which are designed to make a connection between the algebra and the more intuitive graphic representation while maintaining the integrity of the algebra.

Applications The problem sets in an algebra text should relate enough to student experience to stimulate interest in the topic. One of our goals has been to create a set of applications that will be relevant to the students' lives. As a consequence, students will be more ready to expend the time and effort involved in learning techniques needed to solve the problems.

Each chapter opens with an **environmental essay** that connects mathematics to the real world. The essays can be used to encourage class discussions and collaborative learning. In addition there are exercises in each chapter that specifically relate to the environmental essay for that chapter. These are easily identified by a tree logo ✿ that appears next to the exercise number. Our goal has been to produce a set of noncontrived applied problems that students can solve with the skills that they have just learned. The graphing calculator is a useful tool for solving most of **these** applications.

Exercises In response to the NCTM standards and AMATYC and MAA guidelines, several types of exercises involving **writing, discussion, critical thinking, exploration,** and **technology** appear in the book.

- **Check Yourself exercises,** which follow each example, are designed to actively involve the student in the learning process. Answers are provided at the end of each section for immediate feedback.

- **Build Your Skills exercises** allow the student to practice and master the basic skills of each section.

- **Transcribe Your Skills exercises** require written or verbal answers that aid in interpretation, conceptualization, and comprehension. They also provide opportunities for individual students to have varying (but correct) answers to the same question. This can lead to important class and small-group discussions.

- **Think about These exercises** require the student to extend or generalize from the concept just learned. These exercises promote and utilize students' critical skills.

- **Skillscan exercises** draw problems from previous sections of the text. They are designed to aid the student in the process of reviewing concepts that will be applied in the next section.

- **Scientific and Graphing calculator exercises** give students the opportunity to use a tool to help them visualize the solution to a problem that cannot be easily solved algebraically.

  These exercises are also designed to teach students how to use the calculator effectively as a tool for approximating solutions that can be found easily only by algebraic methods.

- **Chapter Summaries and Summary Exercises** give the student an opportunity to practice and review at the end of each chapter.

- **Self-Tests** at the end of each chapter give students guidance in preparing for in-class tests.

- **Cumulative Review Exercises** are designed to give the student further opportunity for building skills and gaining confidence.

Pedagogy Each concept is illustrated with an example that is introduced by a transitional sentence or paragraph. The transitional material allows the student to see the need for a new example. This prepares the student to focus on the key element of each example.

  Every example in the text is followed immediately by a parallel problem **(Check Yourself).** Answers to these problems are provided at the end of the section for immediate feedback. This procedure keeps the student continually involved in the learning process. Examples frequently contain *annotations* within the solution that help the students understand the more complicated algebraic steps.

Screens are used in examples in which the student is asked to simplify an expression. By the use of such screening, what the student is expected to simplify is made immediately clear.

Margin notes serve several purposes. They give related historical information, they explain word origins, they remind students of related material, and they caution students against making common mistakes.

Student Supplements

- A **Student's Solutions Manual** is available through the college bookstore. It contains worked-out solutions and answers to the odd-numbered end-of-section exercises.

- **Student Activities and Manual** contains individual and group activities that help illustrate and reinforce algebraic topics.

- **Mathworks** is a self-paced interactive tutorial specifically linked to the text. The Mathworks logo ▦ appears next to each text section for which the tutorial can be used. It reinforces selected topics and provides unlimited opportunities for the student to review concepts and to practice problem solving. It requires virtually *no* computer training and is available for IBM, IBM compatible, and Macintosh computers.

- Course **Videotapes** are available for use from instructors. The videotape logo ▦ appears next to each section for which they can be used.

- A **Graphing Calculator Enhancement Manual** presents an integrated approach that utilizes calculator-based graphing to enhance understanding and development. It includes calculator exercises and examples as well as appendixes on how to use the most popular brands of calculators.

Instructor's Supplements

- A **Teacher's Annotated Answer Manual** contains all the exercises in the text along with answers.

- An **Instructor's Resource Manual** contains multiple-choice placement tests for two levels, three forms of multiple-choice and open-ended chapter tests, two forms of multiple-choice and open-ended cumulative tests, two forms of multiple-choice and open-ended final tests, and an answer section. It also contains hints and suggestions on how to utilize the *Student Activities Manual.*

- The **Professor's Assistant** is a unique computerized test generator available to instructors. It is a system that allows the instructor to create tests using algorithmically generated test questions and those from a standard testbank. This testing system enables the instructor to choose questions either manually or randomly by section, question type, difficulty level, and other criteria. It is available for IBM, IBM compatible, and Macintosh computers.

- A **Printed and Bound Testbank** is also available. It is a hard-copy listing of the questions found in the standard testbank.

For further information about these supplements, please contact the local college division sales representative.

Acknowledgments The authors would like to thank the following people for their contributions to the development of *Intermediate Algebra.*

B. P. Bockstege, Broward Community College

William Dunn, Las Positas College

Barbara Gale, Prince George's Community College

Frank Gunnip, Oakland Community College

Carol Hay, Northern Essex Community College

Lou Ann Mahaney, Tarrant County Junior College

Julia Monte, Daytona Beach Community College

Linda Murphy, Northern Essex Community College

Nancy Nickerson, Northern Essex Community College

Barbara Sausen, Fresno City College

Diane Tesar, South Suburban College of Cook County

Joyce Vetter, Indiana University-Purdue University at Ft. Wayne

McGraw-Hill and the authors would also like to thank the following numerous people who have reviewed the worktext version of this text and therefore contributed to the evolution of this hardcover edition.

Robert Calabrese, Miami-Dade Community College

Rita P. Hussing, College of Mount Saint Joseph

Glenn E. Johnston, Morehead State University

Margaret A. Karpinski, Holy Family College

Philip Montgomery, University of Kansas

Carol O'Loughlin, Northern Essex College

Murray B. Peterson, College of Marin

Richard D. Semmler, Northern Virginia Community College

Mark Serebransky, Camden County College

Edith A. Silver, Mercer County College

Margaret Stevenson, Massasoit Community College

Ara B. Sullenberger, Tarrant County Junior College

Thank you to Roseann Foglio of Gloucester County College for checking the accuracy of examples, exercises, and answers. Thanks also go to Nathan Kirkman and Micol Hutchison for their work on the index. Very special thanks go to the editorial and production staffs at McGraw-Hill, Inc., for their dedication to producing an outstanding text and supplements package. These people include Michael Johnson, Mathematics Editor; Karen M. Minette, Associate Mathematics Editor; Jack Maisel, Editing Supervisor; and Paula Keller, Production Supervisor.

Donald Hutchison
Louis Hoelzle

You are about to begin a course in intermediate algebra. We have made every attempt to provide a textbook that will help you understand what algebra is about and how to effectively use it. In this text we assume that you have taken a previous course in beginning algebra. However, we understand that your level of skills in the topics of beginning algebra depends on what you gained from a previous course and on how long you have been away from algebra. Therefore, we have provided a thorough review of all the basic concepts that you will need as a foundation for this course.

There are some specific features in this textbook that will aid you in your studies. Here are some of our suggestions on how to use those features and how, in general, to be successful in your study of mathematics.

1. In a lecture class, make sure that you take the time to read the appropriate section of the text *before* your instructor's lecture on the subject. You may not understand everything that is developed in the text before the lecture, but reading in advance will greatly help you understand the concepts that are being presented in the lecture setting. Take careful notes on the examples that your instructor presents during class.

   Attend all class sessions. You should not depend on other students' notes. The only way that you will know exactly what your instructor is expecting of you is to be there yourself.

   An important note: a set of Skillscan exercises concludes every section exercise set. These refer to concepts learned earlier that will be applied in the next section. It is helpful to review the exercises before you begin to read a new chapter section.

2. After class, use the text to work through examples that are similar to those that were presented during the lecture. Try to understand each of the steps that are shown in the text. Examples are followed by Check Yourself exercises. Algebra is best learned by a student's being involved in the process, which is the purpose of the Check Yourself exercises. Always have a pencil and paper in hand, work the problems that are presented, and check your results immediately. If you have difficulty with any of the problems, go back and carefully review the previous examples. Make sure that you understand what you are doing and why. A good test of whether you understand a concept is whether you can explain that concept to one of your fellow students. Try working together.

3. At the end of each chapter section, you will find a set of section exercises. Work these exercises on a regular (preferably daily) basis.

   Our experience with students has shown that consistent study of the concepts is one of the most important keys to success in studying mathematics.

Again, learning algebra requires becoming involved. As is the case with learning any other skill, the main ingredient is practice.

You will find the answers for the odd-numbered exercises in the back of the book. If you have difficulties with any of the exercises, by all means do not become discouraged. You cannot expect to completely understand all the topics that are presented on your first reading. Go back and review the appropriate parts of the chapter section. If your questions are not completely cleared up, ask your instructor or an available tutor for further assistance.

4. To aid your retention of the topics you have studied in a chapter, chapter summaries are provided. In these summaries, you will find all the important terms, definitions, and algorithms along with examples illustrating the techniques that were developed in the chapter. Following each summary are exercises for further practice. These exercises are keyed to the appropriate chapter sections so that you will know exactly where to turn if you are still having problems with the material.

5. After you have successfully completed the summary exercises for a chapter, the Self-Test that concludes each chapter will help you check your mastery of the material and also provide you with an actual practice test to use in preparation for in-class testing.

6. Finally, another key element of success in studying mathematics lies in the process of cumulative review. This will give you a means of assimilating related concepts as you progress through the course. We have provided a set of Cumulative Review Tests throughout the textbook. These will help you in reviewing not only the concepts of the chapter that you have just completed, but also those of previous chapters. You may want to use these in preparation for any midterm or final examinations. If it appears that you have forgotten some of the topics that are being tested, go back and review the sections where the idea was initially explained or study the appropriate chapter summary.

We hope that you will find the suggestions above helpful as you work through this material, and we wish you the best of luck in this intermediate algebra course.

Donald Hutchison
Louis Hoelzle

INTERMEDIATE ALGEBRA

THE REAL NUMBERS: PROPERTIES AND OPERATIONS

INTRODUCTION *Mathematics students often ask, "Where will we ever use this?" This is an important question because students often find topics easier to understand if they can make a useful connection between the material studied and some real-world use.*

The environment is certainly an area of interest to many students today. Pollution, global warming, ozone depletion, nuclear power, overpopulation, and other environmental issues are topics that concern most students. It is these concerns that we hope to address with the essays and problems you will find in each chapter of this text.

Each essay will explore some topic concerning an environmental issue. The

essays will be followed up with application problems which show how the mathematics skills discussed in that chapter can be used to understand the environmental issue discussed there.

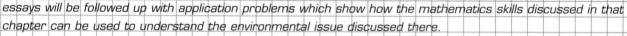

Looking at some of the numbers associated with environmental issues shows the need to have some mathematical understanding of these topics. Several thousand square kilometers of new deserts are being formed each year. Hundreds of thousands of square kilometers of forests are being cut every year. Half or more of the world's wetlands have been lost to development and pollution. Soil erosion carries away thousands of tons of topsoil each year.

When we see that each person in the United States is responsible for approximately 2 kilograms (kg) of garbage every day and that there are over 250 million people in the United States, a little arithmetic shows that the United States must dispose of 500,000 metric tons (t) of garbage every day.

Questions concerning this garbage might include, How much landfill space will be needed to contain this much waste every day? How long will our landfills last at this rate? Questions of how much, how many, or how long need some level of mathematics to answer. They are also some of the most important questions asked about the many different environmental issues we face today.

The essays and problems show the relationship of mathematics to the issues of the world. Students will see the usefulness of mathematics for understanding complex ideas and will have the opportunity to practice their mathematical skills to reach a deeper understanding of the world. ∎

1.1 Sets and the Real Number System

OBJECTIVES: 1. To use set notation
2. To recognize the basic subsets of the real number system

Algebra is a system consisting of a set together with operations that follow certain properties. The set of real numbers is the set we use most often. In this section, we will discuss the basic terminology connected with sets and the components of the real number system.

A *set* is a collection of objects. We often describe a set by using the *roster method,* enclosing the contents of the set in braces.

$\{1, 2, 3\}$ is a set containing the first three counting numbers.

$\{z, x, c, v, b, n, m\}$ is a set containing the letters found on the bottom row of a computer keyboard. We can also use *set builder notation* by enclosing a description of the contents of a set as follows:

This is read, "the set of all x such that x is a day of the week."

$\{x \mid x$ is a day of the week$\}$ is a set containing seven elements.

Frequently, we will name a set with a capital letter so that we may refer to it more easily. For example, we might say

$D = \{x \mid x$ is a day of the week$\}$

Any object or symbol that is contained in a set is called an *element,* or a *member,* of the set. The symbol $\in$ is used to indicate that an object is an element of a set. The symbol $\notin$ is used to indicate that an element is not a member of a set.

EXAMPLE 1 Set Membership

Complete each statement with the symbol $\in$ or $\notin$.

(*a*) If $A = \{1, 2, 3, 4\}$, 2 _____ A.

Here 2 is an element of the set $\{1, 2, 3, 4\}$, so we write $2 \in A$.

(*b*) If $M = \{x \mid x$ is a month of the year$\}$, Tuesday _____ M.

Tuesday is not a month of the year, so it is not an element of set M. This is written Tuesday $\notin M$.

Check Yourself 1

Complete each statement with the symbol $\in$ or $\notin$.

If $V = \{a, e, i, o, u\}$

1. a _____ V **2.** t _____ V

Two sets are *equal* if they contain exactly the same elements.

EXAMPLE 2 Set Equality

Are the sets $A = \{a, b, c, d\}$ and $B = \{d, c, b, a\}$ equal?

Solution Here the two sets consist of the same elements—the first four letters of the alphabet. So we can write

$$A = B$$

Check Yourself 2

In each of the following, label the given sets as equal or unequal.

1. $A = \{l, t, w, b\}$ $B = \{b, w, t, l\}$

2. $A = \{1, 4, 5, 6\}$ $B = \{5, 4, 1\}$

If the elements of a set can be ordered and we wish to indicate that a set continues as described, we use *ellipses*. These are three dots that mean "and so on."

$\{a, b, c, \ldots, z\}$ describes the set of letters in the English alphabet

$\{1, 2, 3, \ldots, 100\}$ describes the set of the first 100 counting numbers

$\{1, 3, 5, \ldots\}$ describes the set of odd counting numbers

> This is the set of positive odd numbers. They continue without end.

You will notice that the last set described above *ends* with the ellipses. This indicates that the elements continue without end. A set that has no end is said to be *infinite*. A set that has some specific number of elements is said to be *finite*. A set with no elements is said to be *empty*. The set $\{x \mid x$ is a month beginning with the letter $Q\}$ is an example of an empty set.

The empty set can be denoted either by the null symbol $\varnothing$ or by empty braces { }.

EXAMPLE 3 Recognizing Infinite Sets

Describe each set as finite or infinite.

(*a*) $\{a, b, c, d\}$

This set is finite since it contains a specific number of elements.

(*b*) $\{2, 4, 6, \ldots, 50\}$

This is a finite set since there are a specific number of elements in it.

(*c*) $\{3, 6, 9, \ldots\}$

This is an infinite set since it continues without end.

Check Yourself 3

Describe each set as finite or infinite.

1. $\{1, 3, 5, \ldots\}$

2. $\{1, 2, 3, \ldots, 99\}$

Recall that a prime number has *only* itself and 1 as factors.

3. $\{$all prime numbers$\}$

These are also referred to as the *counting numbers*.

Several numeric sets are used so commonly in mathematics that they have readily identifiable names. The rich history (or prehistory, in this case) of mathematics began with tallying, or counting. The set of numbers used for counting is called the set of *natural numbers* and is designated by the capital letter N. In set notation, we write

$$N = \{1, 2, 3, \ldots\}$$

An equally important mathematical set is the set of *integers*. This is how we describe the set of integers:

Note that every natural number is *also* an element of the set of integers.

$$Z = \{\ldots, -2, -1, 0, 1, 2, \ldots\}$$

First, we see that the set continues without end in *both* the positive and negative directions. Also note the choice of the letter Z to designate this particular set. This comes from the word ''Zahl,'' the German word for number.

A rational number has the form

$$\frac{p}{q}$$

where p and q are integers and q cannot be 0.

We also will be referring to the set of *rational numbers*. A rational number is one that can be written as the ratio of two integers. Since ratios can be considered as *quotients* (and also because R will be used to designate a different set of numbers), we denote the set of rational numbers by the letter Q.

Because we cannot list the rational numbers in any meaningful fashion, we define that set by *describing* the elements of that set instead:

Since any integer can be written as the ratio of two integers where $q \neq 0$, every integer is also a rational number.

$$Q = \left\{ \frac{p}{q} \,\middle|\, p, q \in Z, q \neq 0 \right\}$$

This is read ''the set of elements of the form p over q, such that p and q are integers and q is not equal to zero.''

There is another important characterization of the set of rational numbers. The decimal representation of any rational number is either a terminating decimal or a repeating decimal.

This is no surprise since

$$0.45 = \frac{45}{100}$$

the *ratio* of two integers.

So 0.45 and 0.825 name rational numbers.

Also 0.3333. . . and 0.272727. . . name rational numbers.

EXAMPLE 4 Elements of Sets of Rational Numbers

Complete each statement with $\in$ or $\notin$.

(*a*) 5 _____ Z

Since 5 is a positive integer, $5 \in Z$.

(b) $\dfrac{4}{3}$ _____ N

The set of natural numbers N contains only positive integers, and $\dfrac{4}{3}$ is not an integer, so $\dfrac{4}{3} \notin N$.

(c) -0.25 _____ Q

Since -0.25 can be expressed as the ratio of two integers $\left(\dfrac{-25}{100}\right)$, it is a rational number. Thus $-0.25 \in Q$.

(d) $0.\overline{16}$ _____ Q

Now $0.\overline{16}$ is a repeating decimal; it can be expressed as the ratio of two integers and is a rational number. So $0.\overline{16} \in Q$.

The line over the 16 indicates that the pattern repeats indefinitely, so $0.\overline{16} = 0.161616\ldots$

Check Yourself 4

Complete each statement with the symbol $\in$ or $\notin$.

1. 3 _____ Z

2. -3 _____ N

3. $\dfrac{2}{3}$ _____ Z

4. 0.25 _____ Q

5. $\dfrac{2}{3}$ _____ N

6. -5 _____ Z

7. $-\dfrac{3}{4}$ _____ Q

8. $0.\overline{35}$ _____ Q

Not every number can be expressed as the ratio of two integers. For example, it can be shown that the square root of 2 (denoted $\sqrt{2}$) cannot be written as the ratio of integers and is therefore *not* a rational number.

Constants such as $\sqrt{2}$, $\sqrt[3]{7}$, and π are called *irrational numbers*. None of their decimal representations will ever repeat. The set of irrational numbers can be designated Q'.

Now, if we combine the set of rational numbers with the set of irrational numbers, we call this new set the set of *real numbers* and use the letter R to designate the set. This is the set to which we refer most often in algebra.

The fact that $\sqrt{2}$ could not be represented as the ratio of two integers disturbed the Pythagoreans (400 B.C.), because it contradicted their belief that all lengths could be represented by ratios of integers.

The symbol Q' is read "Q prime."

EXAMPLE 5 Elements of Sets of Real Numbers

For the set $\left\{-3,\ 1.7,\ \sqrt{3},\ \pi,\ \dfrac{2}{3},\ 0.\overline{12},\ 0,\ 8\right\}$, which of the elements are (a) natural numbers, (b) integers, (c) rational numbers, (d) irrational numbers, and (e) real numbers?

(a) The only natural number is 8.

(b) The integers are -3, 0, and 8.

(c) The rational numbers are -3, 1.7, $\dfrac{2}{3}$, $0.\overline{12}$, 0, and 8.

(d) The irrational numbers are $\sqrt{3}$ and π.

(e) All the numbers in the set are real numbers.

Check Yourself 5

For the set $\left\{-4, 2.3, \sqrt{6}, -\pi, 0, \dfrac{3}{4}, 0.\overline{36}, 7\right\}$, list the elements that are

1. Irrational numbers   **2.** Real numbers   **3.** Natural numbers

4. Rational numbers   **5.** Integers

A convenient way to "picture" the set of real numbers is with a number line. This number line is constructed by drawing a straight line and then choosing a point to correspond to 0. This point is called the *origin* of the number line.

The standard convention is to allow positive numbers to increase to the right and negative numbers to decrease to the left. This is represented in the number line shown below.

Zero is neither positive nor negative.

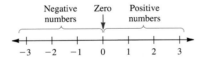

We associate numbers with points on the number line called the *coordinates* of those points. The set of all numbers that correspond to a point on the number line is the set of real numbers.

We often refer to the real number line for this reason.

Every real number will correspond to exactly one point on the line, and every point corresponds to exactly one real number.

EXAMPLE 6 Locating Points on the Number Line

Locate the point corresponding to each element of the set

$$\left\{-1, \sqrt{2}, \frac{5}{2}, -2.5, \pi\right\}$$

on the real number line.

An approximation for $\sqrt{2}$ is 1.414, and for π it is 3.14.

Solution

Check Yourself 6

Locate each element of set A on the number line, where

$$A = \left\{ -\pi, -2, \sqrt{3}, -\frac{3}{4}, 3.5 \right\}$$

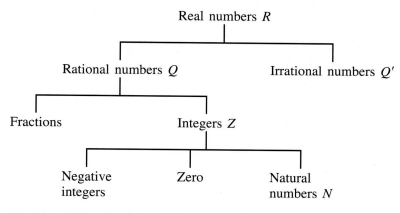

The following diagram summarizes the relationships between the various numeric sets that have been introduced in this section.

Real numbers R

Rational numbers Q Irrational numbers Q'

Fractions   Integers Z

Negative integers   Zero   Natural numbers N

CHECK YOURSELF ANSWERS

1. (1) a $\in$ V; (2) t $\notin$ V **2.** (1) $A = B$; (2) $A \neq B$ **3.** (1) Infinite; (2) finite;

(3) infinite   **4.** (1) $3 \in Z$; (2) $-3 \notin N$; (3) $\frac{2}{3} \notin Z$; (4) $0.25 \in Q$; (5) $\frac{2}{3} \notin N$;

(6) $-5 \in Z$; (7) $-\frac{3}{4} \in Q$; (8) $0.\overline{35} \in Q$ **5.** (1) $\sqrt{6}$, $-\pi$; (2) all; (3) 7; (4) -4,

2.3, 0, $\frac{3}{4}$, $0.\overline{36}$, and 7; (5) -4, 0, 7   **6.**

1.1 EXERCISES

Build Your Skills

Classify each set as finite or infinite.

1. $\{2, 4, 6, \ldots\}$

2. $\{1, 4, 9, 16, \ldots\}$

3. $\{1, 3, 5, \ldots, 49\}$

4. $\{2, 12, 22, \ldots, 72\}$

5. $\{x \mid x$ is a season of the year$\}$

6. $\{x \mid x$ is a state of the union$\}$

7. $\{x \mid x$ is a rational number$\}$

8. $\{x \mid x$ is an irrational number$\}$

Name all the numeric sets to which each of the following numbers belongs. The numeric sets consist of the natural numbers N, integers Z, rational numbers Q, irrational numbers Q', and real numbers R.

9. 3

10. 3.6

11. -2.4

12. -5

13. 0

14. $\sqrt{3}$

15. $-\sqrt{3}$

16. -3.1416

17. π

18. $\dfrac{2}{3}$

19. $\sqrt{9}$

20. $\sqrt{121}$

21. $-\dfrac{3}{5}$

22. $0.\overline{54}$

23. 2.45

24. 3π

Graph the points named by each element of the set.

25. $\{2, 4, 6\}$

26. $\{-3, 0, 3\}$

27. $\left\{-\dfrac{1}{2}, 2, \dfrac{7}{3}\right\}$

28. $\left\{\dfrac{1}{3}, \dfrac{5}{3}, -\dfrac{7}{3}\right\}$

29. $\{-\sqrt{2}, \sqrt{3}\}$

30. $\{-\sqrt{3}, -\sqrt{2}, \sqrt{5}\}$

Complete each statement with the symbol $\in$ or $\notin$.

31. 5 _____ Z

32. 2.4 _____ Z

33. -4 _____ N

34. $\dfrac{5}{4}$ _____ Z

35. $\sqrt{2}$ _____ Z

36. -6 _____ Z

37. $\dfrac{5}{7}$ _____ Q

38. $-\dfrac{22}{7}$ _____ Q

39. $\sqrt{2}$ _____ R

40. $\sqrt{2}$ _____ Q'

41. $\sqrt{36}$ _____ R

42. $\sqrt{-36}$ _____ R

43. 0 _____ N

44. $0.\overline{35}$ _____ Q'

For the set $\left\{-5, -\dfrac{4}{3}, 0, \dfrac{5}{2}, \sqrt{5}, 2\pi, 7.5, 0.\overline{23}, 6\right\}$, list the elements that are

45. Natural numbers

46. Rational numbers

47. Irrational numbers

48. Real numbers

49. Integers

Transcribe Your Skills

50. What is the difference between 0.21 and $0.\overline{21}$?

51. Why is it incorrect to write the empty set as $\{\varnothing\}$?

52. Can a number be both an integer and a rational number? Justify your answer.

53. Can a number be both a rational number and an irrational number? Explain.

Think About These

Determine whether each of the given statements is true or false. Remember, if a statement is true, it must always be true.

54. All even integers are rational numbers.

55. All odd integers are integers.

56. Every positive real number is a natural number.

57. Every natural number is an integer.

58. All integers are natural numbers.

59. Every real number is a rational number.

60. All integers are rational numbers.

61. Zero is not an integer.

62. No number can be both an integer and a rational number.

63. No number can be both a rational number and an irrational number.

64. Every real number is either a rational or an irrational number.

65. If $A = B$ and $4 \in B$, then $4 \in A$.

66. If $A = B$ and $4 \notin B$, then $4 \notin A$.

67. If $A \neq B$ and $4 \notin B$, then $4 \notin A$.

68. If $A \neq B$ and $4 \in B$, then $4 \notin A$.

Skillscan

Determine if the following are true or false.

a. $3(2 + 5) = 3 \cdot 2 + 3 \cdot 5$

b. $3 + (2 + 5) = (3 + 2) + 5$

c. $5 + (3 \cdot 4) = (5 + 3) \cdot (5 + 4)$

d. $15 - 4 + 9 = 15 - (4 + 9)$

Properties of the Real Numbers   1.2

OBJECTIVE: To recognize and apply the properties of the real numbers

TAPE IN1

In Section 1.1, we discussed the set of real numbers. In this section, we discuss the properties of the operations of addition and multiplication on the set of real numbers.

For the most part, you will find that these properties are familiar from your past experience in working with numbers. As just one example, you know that

$3 + 5 = 8$

But it is also true that

$5 + 3 = 8$

Your experience tells you that the *order* in which two numbers are added does not matter. This fact is called the *commutative property of addition* and is one of the properties we consider in this section. Let's proceed now with the formal statement of these properties.

PROPERTIES OF ADDITION ON REAL NUMBERS

For any real numbers *a*, *b*, and *c*:

CLOSURE PROPERTY

$a + b$ is a real number

In words, the sum of any two real numbers is always a real number.

COMMUTATIVE PROPERTY

$a + b = b + a$

In words, the *order* in which two numbers are added does not affect the sum.

ASSOCIATIVE PROPERTY

$a + (b + c) = (a + b) + c$

In words, the *grouping* of numbers in addition does not affect the sum.

The additive inverse is also called the *opposite* of a.

The $\boxed{+/-}$ key on your calculator is an additive-inverse key. Pushing this key gives you the opposite of the number in the display.

ADDITIVE IDENTITY There exists a unique real number **0** such that

$a + 0 = 0 + a = a$

In words, **0** is the additive identity. No number loses its identity after addition with **0**.

ADDITIVE INVERSE There exists a unique real number $-a$ such that

$a + (-a) = (-a) + a = 0$

In words, $-a$ is the *additive inverse* of a. The sum of a number and its additive inverse is **0**.

Our first example illustrates the use of the properties introduced above.

EXAMPLE 1 Properties of Addition

State the property used to justify each statement.

(*a*) $2 + (4 + 5) = (2 + 4) + 5$
(*b*) $-8 + 0 = -8$
(*c*) $(-8) + 5$ is a real number.
(*d*) $b + (-b) = 0$
(*e*) $62 + 8 = 8 + 62$

Solution

(*a*) Associative property of addition   The *grouping* has been changed.
(*b*) Additive identity
(*c*) Closure property of addition
(*d*) Additive inverse
(*e*) Commutative property of addition   The *order* has been changed.

Check Yourself 1

State the property (or properties) used to justify each statement.

1. $2x + y = y + 2x$

2. $3n + 0 = 3n$

3. $2x + (3x + 4y) = (2x + 3x) + 4y$

4. $7 + (-10)$ is a real number.

5. $5n + [7 + (-7)] = 5n + 0 = 5n$

Similarly, we have a set of properties for the operation of multiplication on the set of real numbers. In fact, the properties parallel those just stated for addition.

PROPERTIES OF MULTIPLICATION ON REAL NUMBERS

For any real numbers a, b, and c:

CLOSURE PROPERTY

$a \cdot b$ is a real number

In words, the product of any two real numbers is always a real number.

COMMUTATIVE PROPERTY

$a \cdot b = b \cdot a$

In words, the *order* in which two numbers are multiplied does not affect the product.

ASSOCIATIVE PROPERTY

$a(bc) = (ab)c$

In words, the *grouping* of numbers in multiplication does not affect the product.

MULTIPLICATIVE IDENTITY   There exists a unique real number **1** such that

$a \cdot 1 = 1 \cdot a = a$

In words, 1 is the multiplicative identity. No number loses its identity when it is multiplied by 1.

MULTIPLICATIVE INVERSE   For any real number a where $a \neq 0$, there exists a unique real number $\dfrac{1}{a}$ such that

$$a\left(\frac{1}{a}\right) = \left(\frac{1}{a}\right)a = 1$$

In words, $\dfrac{1}{a}$ is the multiplicative inverse of a. The product of a number and its multiplicative inverse is 1.

The multiplicative inverse of a is also called the *reciprocal* of a. This is the property that allows us to *define division* by any nonzero number.

CAUTION

$\dfrac{1}{0}$ is *not* defined.

Example 2 illustrates the use of the properties introduced above.

EXAMPLE 2 Properties of Multiplication

State the property used to justify each statement.

(*a*) $2(3b) = (2 \cdot 3)b$

(*b*) $\left(\dfrac{2}{3}\right)\left(\dfrac{3}{2}\right) = 1$

(*c*) $(-3)(-4)$ is a real number.

(*d*) $1(x) = x$

(*e*) $2(xy) = 2(yx)$

Solution

(*a*) Associative property of multiplication   The *grouping* has been changed.

(*b*) Multiplicative inverse   Since $\dfrac{3}{2}$ is the reciprocal of $\dfrac{2}{3}$

(*c*) Closure property of multiplication
(*d*) Multiplicative identity
(*e*) Commutative property of multiplication   Only the *order* has been changed.

Check Yourself 2

State the property used to justify each statement.

1. $(9)(-7)$ is a real number.

2. $2 \cdot x \cdot y \cdot x = 2 \cdot x \cdot x \cdot y$

3. $\left(\dfrac{1}{3} \cdot 3\right)xy = 1 \cdot xy$

4. $1 \cdot xy = xy$

5. $12(3ab) = (12 \cdot 3)ab$

In addition to the specific properties for addition and multiplication, we have one property that involves both operations.

Multiplication also distributes over addition in an expression such as

$(a + b)c$

and we can write

$(a + b)c = ac + bc$

or

$ac + bc = (a + b)c$

> ### DISTRIBUTIVE PROPERTY
> For any real numbers *a*, *b*, and *c*,
> $$a(b + c) = ab + ac$$
> In words, multiplication distributes *over* addition.

The following example illustrates the use of the distributive property.

EXAMPLE 3* Using the Distributive Property

Use the distributive property to simplify each expression.

(*a*) $4(3x + 7) = 4(3x) + 4(7)$ "Distribute" the multiplication by 4 over 3x and 7.
$$= 12x + 28$$

(*b*) $3(4a^2 + 5a) = 3(4a^2) + 3(5a)$
$$= 12a^2 + 15a$$

(*c*) $7(3x + 2y + 5) = 7(3x) + 7(2y) + 7(5)$
$$= 21x + 14y + 35$$

Even though the formal distributive property is stated for a sum with two terms, it is valid for a sum with any number of terms.

(*d*) $3y(y + 4) = (3y)y + (3y)(4) = 3y^2 + 12y$ $(3y)y = 3(y \cdot y)$ or $3y^2$

---

*In algebra texts, students are sometimes asked to simplify expressions and sometimes asked to solve equations. Throughout this text a screen will be used to indicate an expression that is being simplified.

Check Yourself 3

Use the distributive property to simplify each expression.

1. $5(4a + 5)$ **2.** $4(2x^2 + 5x)$

3. $6(4a + 3b + 7c)$ **4.** $5p(p + 5)$

Expressions such as 5, $3a^2$, $4b^3cd^2$, and x^2y are called *terms*. A *term* is a number or the product of a number and one or more variables raised to a power.

The numerical part of a term is called its *numerical coefficient,* or more simply its *coefficient.*

Like terms involve the *same variable* raised to the same power and can always be combined by use of the distributive property.

Let's see how that property is applied.

> The coefficients of the terms to the right are 5, 3, 4, and 1. When the numerical coefficient is omitted, it is understood to be 1.

EXAMPLE 4 Using the Distributive Property

Use the distributive property to simplify each expression.

> Here, *simplify* means "to combine like terms."

(a) $7x + 3x = (7 + 3)x = 10x$ Distributive property

(b) $5a^2 + 3a^2 = (5 + 3)a^2 = 8a^2$ Add the coefficients.

(c) $7x + 5 + 3x + 6$

    $= (7x + 3x) + (5 + 6)$ Commutative and associative properties

    $= (7 + 3)x + (5 + 6)$ Distributive property

    $= 10x + 11$

> We rearrange the terms so that like terms are grouped. Then we combine those terms by the distributive property.

Check Yourself 4

Use the distributive property to simplify each expression.

1. $5r + 6r$ **2.** $6x^3 + 4x^3$

3. $8w + 5 + 4w + 7$

The distributive property is frequently used in "mental arithmetic" to make problems easier.

EXAMPLE 5 The Distributive Property and "Mental Arithmetic"

Multiply.

(a) $25(98)$

$98 = 100 - 2$

so by the distributive property,

$25(98) = 25(100 - 2) = 25(100) - 25(2) = 2500 - 50 = 2450$

(b) $15(103)$

$15(103) = 15(100 + 3) = 15(100) + 15(3) = 1500 + 45 = 1545$

Check Yourself 5

Multiply.

1. $20(199)$ 　　　　　　　　　　　　　　　**2.** $12(1005)$

CHECK YOURSELF ANSWERS

1. (1) Commutative property of addition; (2) additive identity; (3) associative property of addition; (4) closure property of addition; (5) additive inverse and identity 　**2.** (1) Closure property of multiplication; (2) commutative property of multiplication; (3) multiplicative inverse; (4) multiplicative identity; (5) associative property of multiplication 　**3.** (1) $20a + 25$; (2) $8x^2 + 20x$; (3) $24a + 18b + 42c$; (4) $5p^2 + 25p$ 　**4.** (1) $11r$; (2) $10x^3$; (3) $12w + 12$ **5.** (1) 3980; (2) 12,060

1.2 EXERCISES

Build Your Skills

In each exercise, apply the commutative and associative properties to rewrite the expression. Then simplify the result.

1. $(a + 5) + 4$

2. $(w + 3) + 7$

3. $8 + (5 + m)$

4. $10 + (5 + p)$

5. $(2x + 7) + 11$

6. $(3m + 3) + 10$

7. $8 + (m + 7)$

8. $7 + (2y + 11)$

9. $(7 + b) + (-7)$

10. $-3 + (x + 3)$

11. $3(7x)$

12. $5(3a)$

13. $\frac{1}{3}(3x)$

14. $5n\left(\frac{1}{5}\right)$

15. $\left(\frac{2}{5}\right)\left(\frac{5}{2}\right)(x)$

16. $\left(\frac{2}{3}\right)(m)\left(\frac{3}{2}\right)$

In each exercise, apply the distributive property to rewrite the expression. Then simplify the result where possible.

17. $5(3x + 4)$

18. $3(4t + 2)$

19. $3w(w + 3)$

20. $5a(a + 7)$

21. $\frac{1}{2}(6x + 10)$

22. $\frac{1}{3}(9x + 15)$

23. $5(2x + 3y + 5)$

24. $7(2p + 7q + 4)$

In each exercise, apply the distributive property to simplify the expression.

25. $7a + 3a$

26. $10w + 3w$

27. $2x + 3x + x$

28. $a + 11a + 2a$

29. $\dfrac{1}{2}x + \dfrac{3}{2}x$

30. $\dfrac{2}{7}a + \dfrac{12}{7}a$

31. $\dfrac{1}{2}z + \dfrac{1}{3}z$

32. $\dfrac{2}{3}t + \dfrac{3}{4}t$

In each exercise, apply the appropriate properties to rewrite the expression. Then simplify the result.

33. $7y + (3 + 2y)$

34. $5m + (2 + 7m)$

35. $7x + (3x + 5)$

36. $8w + (5 + 6w)$

37. $2n + 7 + 3n + 9$

38. $3x + 1 + 9x + 8$

39. $2a + 3a + 4 + 5a$

40. $7y + 8 + 9y + 10$

41. $4 + 8x + (-4) + x$

42. $a + (-8) + 2a + 8$

43. $3 + 2(3x + 1) + 2x$

44. $5 + 3(2a + 2) + 4a$

45. $3w^2 + 2w(2 + w) + 3w$

46. $5m + 2m(m + 3) + 3m^2$

Use the distributive property to multiply each of the following "mentally."

47. $20(102)$

48. $30(103)$

49. $25(199)$

50. $50(198)$

51. $24(1003)$

52. $31(1002)$

State the property used to justify each of the following statements.

53. $3 + 6 = 6 + 3$

54. $5 \cdot 7$ is a real number

55. $3(x + 6) = 3x + 18$

56. $5(3x) = (5 \cdot 3)x$

57. $3 + (4 + 5) = (3 + 4) + 5$

58. $3 + (4 + 5) = (4 + 5) + 3$

59. $24a + 12 + 18a = 24a + 18a + 12$

60. $24a + (18a + 12) = (24a + 18a) + 12$

61. $(24a + 18a) + 12 = (24 + 18)a + 12$

62. $\left(\dfrac{2}{5}\right)\left(\dfrac{5}{2}\right) = 1$

63. $\dfrac{2}{5} + \left(-\dfrac{2}{5}\right) = 0$

64. $(x + 2)(x + 3) = x(x + 3) + 2(x + 3)$

Transcribe Your Skills

65. Can a real number be its own multiplicative inverse? Give an example, if possible.

66. Does every real number have a multiplicative inverse? If not, give an example of one that does not.

67. Is the set of odd integers closed with respect to addition? If not, give an example showing the set is not closed.

68. Is the set of integers closed with respect to division? If not, give an example.

Think About These

Determine whether each statement is true or false. If it is false, rewrite the right side of the equation to make it a true statement.

69. $2 + 4(x + 5) = 2 + 4x + 5$

70. $12 + 6x + 3 = 6(2 + x + 3)$

71. $5a + 6a = 11a$

72. $2x + (12 + 3x) = (2x + 3x) + 12$

73. $3(2x + 2) = 6x + 5$

74. $\dfrac{1}{7}x + \dfrac{2}{7}x = \dfrac{3}{14}x$

75. $2x + 3x + 1 = 5x + 1$

76. $4a + 3a + 2 = 9a$

77. $7x + (-7)x = 0$

78. $2x + (-2x) = x$

79. $2x + 5x = 7x^2$

80. $2m + m = 2m^2$

Skillscan (Section 1.1)

In each of the following, select the smaller number.

a. -3 and 5 **b.** -4 and -2 **c.** $\sqrt{2}$ and 1.5

d. -6 and 2 **e.** $\dfrac{8}{3}$ and 2.6 **f.** $\dfrac{1}{6}$ and $\dfrac{1}{4}$

1.3   Equality, Order, and Absolute Value

TAPE IN1

OBJECTIVES: 1. To recognize the properties of equality
2. To use the notation of inequalities
3. To use the absolute value notation

Thus far in this chapter we have considered the set of real numbers and introduced the properties of addition and multiplication on that set. We now look at relations on the set of real numbers.

The simplest relation is *equality*. The symbol for equality is an equals sign $=$, and we write

$$a = b$$

when we want to indicate that the two expressions, a and b, represent the same number.

Listed below are several important properties of the equality relation.

> The use of the $=$ symbol to represent equality is due to Robert Recorde (1510–1558), an English mathematician who first used it in print in 1557. According to Recorde, "noe 2 thynges can be moare equalle," referring to the two parallel line segments he used.

> Each property should seem "natural" given your previous understanding of the equality relation.

PROPERTIES OF EQUALITY

For any real numbers a, b, and c:

REFLEXIVE PROPERTY

$a = a$

SYMMETRIC PROPERTY

If $a = b$, then $b = a$.

TRANSITIVE PROPERTY

If $a = b$ and $b = c$, then $a = c$.

The following principle follows from the properties above.

SUBSTITUTION PRINCIPLE   If $a = b$, then a may be replaced by b, or b by a, in any statement, without changing the validity of that statement.

EXAMPLE 1   Properties of Equality

Which property of equality justifies each of the following statements?

(a) If $V = L \cdot W \cdot H$, then $L \cdot W \cdot H = V$.
(b) If $2(1 + 3) = 2 + 6$ and $2 + 6 = 8$, then $2(1 + 3) = 8$.
(c) $a + 4 = a + 4$
(d) If $x = y + 1$ and $2x = 4$, then $2(y + 1) = 4$.

Solution

> The form in (b) is, "If $a = b$ and $b = c$, then $a = c$."

> We have "replaced" x with $y + 1$.

(a) Symmetric property
(b) Transitive property
(c) Reflexive property
(d) Substitution principle

Check Yourself 1

Which property of equality justifies each of the following statements?

1. If $7x + 2 = y$, then $y = 7x + 2$.

2. If $3(y + 1) = 2x - 1$ and $x = y + 1$, then $3x = 2x - 1$.

3. $w + 5 = w + 5$

4. If $x + 1 = y$ and $y = 2$, then $x + 1 = 2$.

Let's now consider two other relations on the set of real numbers. These are the relations of order or inequality known as *less than* and *greater than*.

The set of real numbers is an ordered set. Given any two numbers, we can determine whether one number is less than, equal to, or greater than the other. Let's see how this is expressed symbolically.

We use the *inequality symbol* $<$ to represent ''is less than,'' and we write

$a < b$ This is read "a is less than b."

to indicate that a is less than b. The number line gives us a clear picture of the meaning of this statement. The point corresponding to a must lie *to the left* of the point corresponding to b.

> As was true with the equals sign, the inequality represents a *verb* phrase.

$$a < b$$

Similarly, the inequality symbol $>$ represents ''is greater than,'' and the statement

$a > b$ This is read "a is greater than b."

indicates that a is greater than b and means that the point corresponding to a on the number line lies *to the right* of the point corresponding to b.

> $a > b$
>
> and
>
> $b < a$
>
> are equivalent statements. The symbol "points to" the smaller quantity.

$$a > b$$

The following example illustrates the use of the inequality symbols.

EXAMPLE 2 The Inequality Symbol

Complete each statement by inserting $<$ or $>$ between the given numbers.

(*a*) 2 _____ 8

(*b*) 2.786 _____ 2.78

(*c*) -23 _____ -5

(*d*) $\sqrt{2}$ _____ 1.4

(*e*) $\dfrac{1}{2}$ _____ $\dfrac{1}{3}$

Solution

(a) $2 < 8$

(b) $2.786 > 2.78$

(c) $-23 < -5$

(d) $\sqrt{2} > 1.4$ Recall that 1.414 is an approximation for $\sqrt{2}$.

(e) $\dfrac{1}{2} > \dfrac{1}{3}$

Check Yourself 2

Complete each statement by using the proper verb phrase and inserting an inequality symbol.

1. $5 \underline{\hspace{1cm}} -2$

2. $3.14 \underline{\hspace{1cm}} \pi$

3. $-10 \underline{\hspace{1cm}} -15$

4. $\sqrt{15} \underline{\hspace{1cm}} 4$

5. $9.78 \underline{\hspace{1cm}} 9.87$

6. $-1.3 \underline{\hspace{1cm}} -\dfrac{4}{3}$

7. $\dfrac{1}{5} \underline{\hspace{1cm}} \dfrac{1}{4}$

Suppose we are given an inequality of the form

$$x > -1$$

The *solution set* for an inequality (as it is for an equation) is the set of all values for the variable that make the inequality a true statement. A convenient way to picture that solution set is by a graph on a number line. The following example illustrates.

EXAMPLE 3 Graphing Inequalities

Graph each of the following sets.

This set is read "the set of all x such that x is greater than -1."

(a) $\{x \mid x > -1\}$

Here we want to graph all real numbers greater than -1. This means we want to include all real numbers *to the right* of -1 on the number line.

The *open circle* at -1 means that the point corresponding to -1 is *not included* in the graph. Such a graph is called an *open half line*.

(b) $\{x \mid x < 4\}$

In this case we want to include all real numbers less than 4, that is, *to the left* of 4 on the number line.

Here the "left" arrowhead means the graph continues indefinitely to the left.

Check Yourself 3

Graph each of the following sets.

1. $\{x \mid x < 5\}$

2. $\{x \mid x > -3\}$

Two other symbols, $\leq$ and $\geq$, are also used in writing inequalities. In each case they combine the inequality symbols for less than and greater than with the symbol for equality. The following shows the use of these new symbols. The statement

$a \leq b$

is read "a is less than or equal to b." Similarly,

This combines the symbols $<$ and $=$ and means that either $a < b$ or $a = b$.

$a \geq b$

Either $a > b$ or $a = b$.

is read "a is greater than or equal to b." We consider the graph of inequalities involving these symbols in our next example.

EXAMPLE 4 Graphing Inequalities

Graph each of the following sets.

(a) $\{x \mid x \leq 3\}$

Now we want to graph all real numbers to the left of 3 but also *including* 3 because of the "is less than or equal to" symbol.

The closed circle at 3 means that the point corresponding to 3 is *included* in the graph. Such a graph is called a *closed half line*.

(b) $\left\{x \mid x \geq \dfrac{7}{2}\right\}$

Here we want all numbers to the right of $\dfrac{7}{2}$ *including* $\dfrac{7}{2}$.

Check Yourself 4

Graph each of the following sets.

1. $\{x \mid x \leq 7\}$

2. $\left\{x \mid x \geq -\dfrac{4}{3}\right\}$

Note You may very well encounter a different notation for indicating the graphs of inequalities. This involves the use of parentheses and brackets to represent open and closed half lines, respectively. For example, the graph of $\{x \mid x > 3\}$ can be drawn as

The left parenthesis (is used to indicate the *open* half line, extending to the right and *not including* 3.

and the graph of $\{x \mid x \leq -2\}$ as

The right bracket] is used to indicate the *closed* half line, extending to the left and *including* −2.

Our subsequent work with inequalities involves the use of a *double-inequality* statement such as

$$-3 < x < 4$$

We read this "x is greater than −3 and less than 4." This statement combines the two inequalities

The word "and" is implied in any double-inequality statement.

$$x > -3 \qquad and \qquad x < 4$$

or, more formally,

$$\{x \mid x > -3\} \cap \{x \mid x < 4\}$$

This set of numbers includes all real numbers between −3 and 4.

In our next example we look at the graphs of inequalities that have this form.

EXAMPLE 5 Graphing Double Inequalities

Graph each of the following sets.

(*a*) $\{x \mid -4 < x < 5\}$

For the solution set of this double inequality, we want all points that lie to the right of −4 ($x > -4$) and to the left of 5 ($x < 5$). This means that we should include all points that lie *between* −4 and 5.

The *open* circles indicate that the endpoints, −4 and 5, are *not included* in the graph. This is called an *open interval*.

(*b*) $\{x \mid 2 \leq x \leq 6\}$

Here we want all points between 2 and 6—in this case we also *include* the endpoints 2 and 6.

The *closed* circles mean that 2 and 6 are *included* in the graph. This is called a *closed interval*.

Check Yourself 5

Graph each of the following sets.

1. $\{x \mid -1 < x < 6\}$

2. $\{x \mid -2 \leq x < 8\}$

Once again, we refer to the number line to introduce our final topic of this section. If we locate the number 4 and its opposite, -4, on the number line, we see that both numbers correspond to points that are the same distance (4 units) from the origin.

When we are concerned not with the direction (left or right) of a number from the origin, but only with the distance from the origin, we refer to that number's absolute value.

An *absolute value* is the distance (on the number line) between the point named by that real number and the origin. We indicate the absolute value of a number with vertical bars as follows:

$$|5|$$

This is read "the absolute value of 5," and

$$|5| = 5$$

We read

$$|-3|$$

as "the absolute value of -3," and

$$|-3| = 3$$

Note that the absolute value of a number is always positive or zero.

In general, we can define the absolute value of any real number a as

$$|a| = \begin{cases} a & \text{if } a \text{ is positive} \\ 0 & \text{if } a \text{ is zero} \\ -a & \text{if } a \text{ is negative} \end{cases}$$

The third part of this definition may be hard to see. If a is *negative*, then its opposite, $-a$, must be *positive*, and we want a *positive* absolute value.

The use of the absolute value notation is illustrated in our final example.

EXAMPLE 6 Evaluating Absolute Value Expressions

Evaluate each of the following expressions.

(a) $|32|$ (b) $|-2.5|$ (c) $|\sqrt{2}|$

(d) $|-\sqrt{2}|$ (e) $-|-5|$ (f) $|-3| + |-7|$

(g) $\left|-\dfrac{5}{6}\right|$ (h) $\left|-\dfrac{3}{4}\right| + \left|\dfrac{5}{8}\right|$

Most graphing calculators have a key labeled abs. This can be used to evaluate the absolute value of an expression. To evaluate 6(b), enter abs (-2.5).

Solution

(a) $|32| = 32$

(b) $|-2.5| = 2.5$

(c) $|\sqrt{2}| = \sqrt{2}$

(d) $|-\sqrt{2}| = \sqrt{2}$

(e) $-|-5| = -5$ $|-5|$ is 5, so $-|-5|$ must be -5.

(f) $|-3| + |-7| = 3 + 7 = 10$

(g) $\left|-\dfrac{5}{6}\right| = \dfrac{5}{6}$

(h) $\left|-\dfrac{3}{4}\right| + \left|\dfrac{5}{8}\right| = \dfrac{3}{4} + \dfrac{5}{8} = \dfrac{11}{8}$

Check Yourself 6

Evaluate each of the following expressions.

1. $|121|$
2. $|-3.4|$
3. $|\sqrt{3}|$
4. $|-\sqrt{5}|$
5. $-|-8|$
6. $|-9| + |-2|$

CHECK YOURSELF ANSWERS

1. (1) Symmetric; (2) substitution; (3) reflexive; (4) transitive
2. (1) $5 > -2$; (2) $3.14 < \pi$; (3) $-10 > -15$; (4) $\sqrt{15} < 4$; (5) $9.78 < 9.87$; (6) $-1.3 > -\dfrac{4}{3}$; (7) $\dfrac{1}{5} < \dfrac{1}{4}$

3. (1) $\{x\,|\,x < 5\}$

(2) $\{x\,|\,x > -3\}$

4. (1) $\{x\,|\,x \le 7\}$

(2) $\left\{x\,\middle|\,x \ge -\dfrac{4}{3}\right\}$

5. (1) $\{x\,|-1 < x < 6\}$

(2) $\{x\,|-2 \le x < 8\}$

(This is a *half open* interval.)

6. (1) 121; (2) 3.4; (3) $\sqrt{3}$; (4) $\sqrt{5}$; (5) -8; (6) 11

1.3 EXERCISES

Build Your Skills

Which property of equality justifies each of the following statements?

1. If $2x + 1 = y$, then $y = 2x + 1$.

2. $m + 3n = m + 3n$

3. If $3x - 4 = y$ and $y = 2$, then $3x - 4 = 2$.

4. If $3x - 4 = y$ and $x = 2$, then $3(2) - 4 = y$.

5. If $a^2 + b^2 = c^2$ and $c^2 = 25$, then $a^2 + b^2 = 25$.

6. If $a^2 + b^2 = c^2$ and $b^2 = 16$, then $a^2 + 16 = c^2$.

7. $w + 17 = w + 17$

8. If $2x + y = 2z$, then $2z = 2x + y$.

9. If $3x + 2 = 4y$ and $y = x + 1$, then $3x + 2 = 4(x + 1)$.

10. If $x = 4y - 1$ and $4y - 1 = z$, then $x = z$.

Complete each of the following statements with an inequality symbol ($<$ or $>$) or equality symbol ($=$).

11. 7 _____ 5

12. -3 _____ 6

13. -2 _____ -1

14. -5 _____ -6

15. -3.8 _____ -3.9

16. -6.50 _____ -6.5

17. $-\dfrac{5}{4}$ _____ $-\dfrac{4}{3}$

18. -1.3 _____ $-\dfrac{4}{3}$

19. $\sqrt{2}$ _____ 1.41

20. $-\sqrt{3}$ _____ -1.6

21. 1.25 _____ $\dfrac{5}{4}$

22. $\dfrac{7}{3}$ _____ 2.33

23. $|-2|$ _____ -2

24. $|-2|$ _____ $|-1|$

25. $-|4|$ _____ $|-4|$

26. $|5|$ _____ $-|-5|$

Write each of the following inequalities in words.

27. $x > 5$

28. $x \geq 3$

29. $t \leq 3$

30. $y < -2$

31. $x \geq y$

32. $p < q$

33. $m < 0$

34. $x \geq 0$

35. $-2 < x < 5$

36. $-5 \leq x \leq -2$

Graph each of the following sets. Assume x represents a real number.

37. $\{x \mid x < 5\}$

38. $\{x \mid x \geq 3\}$

39. $\{x \mid x > -5\}$

40. $\{x \mid x \leq -1\}$

41. $\{x \mid -2 \geq x\}$

42. $\{x \mid 2 < x\}$

43. $\{x \mid x \geq 2\}$

44. $\{x \mid 0 \geq x\}$

45. $\{x \mid 2 < x < 3\}$

46. $\{x \mid -4 \leq x \leq 6\}$

Rewrite each of the following statements, using inequality symbols. Then graph the solution set for each inequality. Assume that x represents a real number.

47. x is less than 2.

48. x is more than -3.

49. x is at least -1.

50. x is no more than 6.

51. x is greater than 5.

52. x is at least 1.

53. x is no more than -1.

54. x is not less than -1.

55. x is between 2 and 5.

56. -5 is less than or equal to x, and x is less than -2.

Write each of the following expressions without the absolute value symbol.

57. $|4|$

58. $|-4|$

59. $|-3.5|$

60. $|2.5|$

61. $\left|\dfrac{7}{8}\right|$

62. $-\left|\dfrac{5}{6}\right|$

63. $-|1.5|$

64. $|-3.4|$

65. $-|-2|$

66. $-|-7|$

67. $|-4| + |-6|$

68. $|-3| + |-7|$

69. $-(|2| + |-3|)$

70. $-(|-5| + |-6|)$

Transcribe Your Skills

71. Give a geometric interpretation to the solution of the equation $|x| = 2$.

72. Under what circumstances is the statement $|x| = -x$ true?

73. Are there any conditions on x such that $x > \dfrac{1}{x}$?

74. Under what conditions on x is $x^2 < x$?

Think About These

In problems 75–86, label each statement true or false. If it is false, explain.

75. $|-3| \geq 3$

76. $|0| = 0$

77. $|-15| = |15|$

78. $|1 - 2| > |0 - 1|$

79. $|x - 1| = |1 - x|$

80. $|5 - 3| = |5| + |-3|$

81. $|-a| = a$. *Hint:* For this statement to be true, it must be true for all values of the variable.

82. $-|y| = -y$

83. The absolute value of any real number is positive or 0.

84. The absolute value of any real number is equal to the absolute value of its opposite.

85. Some real numbers have no absolute value.

86. There is only one real number which is equal to its own absolute value.

87. The Wilderness Society defines "classic" old-growth forests as "containing at least eight big trees per acre exceeding 300 years in age or measuring more than 40 inches [in] in diameter. . . ." Let N represent the number of big trees per acre, A the age in years, and D the diameter in inches. Write three inequalities that can be used to represent this definition of old-growth forests.

88. An old-growth Douglas fir is defined as being between 400 and 1000 years old. Write the age of an old-growth Douglas fir as a double inequality. (Include 400 and 1000.)

89. The typical color television set in U.S. homes uses between 75 and 1000 kilowatthours of electricity E per year (kWh/yr). Write this electricity usage as a double inequality. (Include 75 and 1000.)

90. The average person in the United States creates at least 2 kg of garbage every day. Write an inequality that expresses the number of kilograms of garbage G created by the average person each day in the United States.

Skillscan

Perform the indicated operations.

a. $8 + 7$ **b.** $4.3 + 2.6$ **c.** $\dfrac{1}{3} + \dfrac{1}{4}$ **d.** $(6)(5)$

e. $(3.2)(5.1)$ **f.** $\left(\dfrac{3}{5}\right)\left(\dfrac{2}{9}\right)$

1.4 Operations with Signed Numbers

TAPE IN1

OBJECTIVE: To perform the four arithmetic operations with signed numbers

Since the set of real numbers consists of positive and negative numbers, along with 0, it is sometimes called a set of *signed numbers*.

In this section we review the rules for the four basic operations—addition, subtraction, multiplication, and division—on the real numbers.

We start with addition. The justification for the addition rules we develop is most easily seen by returning to the number line. We have already established that real numbers involve a *distance* (absolute value) and a *direction* (to the right if the number is positive and to the left if the number is negative). With that background let's consider the sum

$2 + 3$

Using the number line, we can interpret this as follows: Start at the origin, and move 2 units to the right—the positive direction. Then move 3 more units to the right to find the sum. Graphically we have

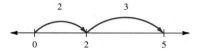

So

$2 + 3 = 5$

We can use the number line in a similar manner when the addition involves two negative numbers. For example, let's look at the sum

$-4 + (-7)$

Again we begin at the origin, but this time we move 4 units to the left—the negative direction. Then we move 7 more units to the left to find the sum.

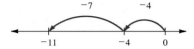

So

$-4 + (-7) = -11$

The situation is similar when we add a negative number and a positive number. For instance, look at the sum

$-2 + 5$

Graphically we start at the origin and move 2 units left. Then, to add 5, we move 5 units to the right.

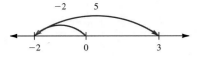

The numeral at the final position on the number line is the sum of the integers.

$-2 + 5 = 3$

Obviously we do not want to have to refer to the number line every time we want to find the sum of two numbers. The discussion above was simply intended to help you visualize the operation of addition and to notice some helpful patterns that will allow you to do the work mentally. For instance, consider the following four sums.

Same signs
$$\begin{cases} 3 + 6 = \quad 9 & \text{Positive sum} \\ -3 + (-6) = -9 & \text{Negative sum} \end{cases}$$

Opposite signs
$$\begin{cases} -3 + 6 = \quad 3 & \text{Positive sum} \\ 3 + (-6) = -3 & \text{Negative sum} \end{cases}$$

This should suggest that, in general, we have the following rules for the addition of real numbers.

> **RULES FOR ADDITION**
>
> 1. If the numbers have the *same sign*, add their absolute values. Give the sum the sign of the original numbers.
> 2. If the numbers have *opposite signs*, subtract their absolute values, the smaller from the larger. Give the sum the sign of the number with the larger absolute value.

Note that these rules are true, not just for integers, but for any two real numbers.

EXAMPLE 1 Adding Signed Numbers

Find each of the following sums.

The sum of two negative numbers is negative.

(a) $-20 + (-30) = -50$

Since -10.5 has the larger absolute value, the sum is negative.

(b) $3.5 + (-10.5) = -7$

Since 8.3 has the larger absolute value, the sum is positive.

(c) $-2.3 + 8.3 = 6$

(d) $-\dfrac{2}{3} + \left(-\dfrac{4}{3}\right) = -\dfrac{6}{3} = -2$

Check Yourself 1

Find each of the following sums.

1. $-12 + 8$

2. $-4.6 + (-6.4)$

3. $\dfrac{5}{6} + \left(-\dfrac{1}{3}\right)$

4. $14 + (-10)$

Having established our rules for addition, we could now define a set of rules for the subtraction of real numbers. However, it is far better to minimize the number of rules that must be learned, and this can be accomplished by defining subtraction in terms of addition.

> **SUBTRACTION**
>
> $a - b = a + (-b)$
>
> In words, subtracting b from a is the same as adding the opposite of b to a.

EXAMPLE 2 Subtracting Signed Numbers

Subtract, using the above definition.

(a) $10 - 12 = 10 + (-12)$ Add the opposite of 12, or -12, to 10.
$$= -2$$

(b) $-5 - 8 = -5 + (-8)$
$$= -13$$

(c) $6.2 - (-5.8) = 6.2 + 5.8$ Add the opposite of -5.8, or 5.8, to 6.2.
$$= 12$$

> The opposite of a negative number is *always* a positive number.

(d) $-8 - (-5) = -8 + 5$
$$= -3$$

Check Yourself 2

Subtract.

1. $-3 - 9$

2. $-2.1 - (-4.1)$

3. $\dfrac{5}{7} - \left(-\dfrac{2}{7}\right)$

4. $-13 - (-15)$

The following example illustrates how we evaluate expressions that combine the operations of addition and subtraction. Note that the definition of subtraction in terms of addition has allowed us to apply the associative property of addition.

EXAMPLE 3 Expressions with Signed Numbers

Evaluate each expression.

(a) $8 - 5 + 3 = 8 + (-5) + 3$
$$= 3 + 3$$
$$= 6$$

(b) $-7 - (-5) + 3 = -7 + 5 + 3$
$$= -2 + 3$$
$$= 1$$

> Technically we are using the associative property to "group" $8 + (-5)$ for this step. In practice, you can work from left to right when *only* addition and subtraction are involved in an expression.

Check Yourself 3

Evaluate each expression.

1. $10 - 7 - 3$

2. $5 - (-6) - 4$

Our work with signed numbers is particularly useful when we are trying to simplify an expression that involves terms with coefficients with various signs. We refer to this process as *combining like terms*. Recall again that like terms have the same variable raised to the same power.

EXAMPLE 4 Combining Like Terms

Simplify the following expression.

$$\begin{aligned}
2x + 6y - 5x - 4y &= 2x + 6y + (-5x) + (-4y) \\
&= [2x + (-5x)] + [6y + (-4y)] \\
&= [2 + (-5)]x + [6 + (-4)]y \\
&= -3x + 2y
\end{aligned}$$

As before, we use the commutative and associative properties to group the like terms in x and in y.

The distributive property. Recall that $(a + b)c = ac + bc$.

Check Yourself 4

Simplify the following expression.

$$5a - 4b + 3c + 2a + 6b - 7c$$

Note In practice, we would not write out the steps of the simplification shown above.

Once subtraction is defined in terms of addition, we can extend the distributive property as follows:

$$a(b - c) = ab - ac$$

We are now ready to proceed to the operations of multiplication and division. To develop rules for these operations, we build on our earlier work in this section. First, recall that multiplication is actually just a shorthand for repeated addition. For instance, we can think of

This shows that the product of two positive numbers must be positive.

$$4 \cdot 3 \quad \text{as} \quad 3 + 3 + 3 + 3 \quad \text{or} \quad 12$$

Similarly,

$$4(-3) = (-3) + (-3) + (-3) + (-3) = -12$$

In fact, the product of a positive number and a negative number can always be thought of as a sum of negative numbers. From our rules for addition, we know such a sum will always be negative, and we can state that the product of two real numbers with *opposite signs* will always be *negative*.

EXAMPLE 5 Multiplying Signed Numbers

Multiply.

The numbers have opposite signs, so the product is negative.

(*a*) $3(-4) = -12$

(*b*) $(-3)(4) = -12$

(*c*) $\left(\dfrac{3}{4}\right)\left(-\dfrac{2}{3}\right) = -\dfrac{1}{2}$

Check Yourself 5

Multiply.

1. $5(-6)$ **2.** $(-5)(6)$ **3.** $\left(-\dfrac{5}{6}\right)\left(\dfrac{3}{10}\right)$

We now know that the product of two positive numbers is positive, and that the product of a positive and a negative number is negative, but what about the product of two negative numbers?

Consider the product

$$(-3)(-4)$$

We can use the following argument to find such a product. First, we know that

$$4 + (-4) = 0$$ The additive-inverse property

$$-3[4 + (-4)] = -3 \cdot 0 = 0$$ Multiply by -3.

so

$$-3(4) + (-3)(-4) = 0$$ The distributive property

or

$$-12 + (-3)(-4) = 0$$

It follows that $(-3)(-4)$ must be the opposite of -12, which is 12. And we can state that

Since -12 and $(-3)(-4)$ add to 0

$$(-3)(-4) = 12$$

In fact, the argument made above can be applied to any product of negative numbers, and we can make the general statement that the product of two negative real numbers is always positive. To summarize, we have the following rules for multiplication.

RULES FOR MULTIPLICATION

1. The product of two numbers with the *same sign* is always positive.

2. The product of two numbers with *opposite signs* is always negative.

EXAMPLE 6 Multiplying Signed Numbers

Multiply.

(*a*) $(-3)(-3) = 9$

(*b*) $(-3)(-3)(-3) = (9)(-3)$
$$= -27$$

Note the application of the associative property.

$[(-3)(-3)](-3) = (9)(-3)$
$= -27$

We simplify inside the parentheses first:

$2 - 7 = -5$

Of course, the distributive property could also be applied.

(c) $-4(2 - 7) = (-4)(-5)$
$= 20$

(d) $-6\left(\dfrac{1}{3}\right) = -2$

(e) $(-5)^2 = (-5)(-5) = 25$

(f) $-5^2 = -(5)(5) = -25$? why is it neg-

Note that in Example 6*f* the exponent 2 applies only to 5.

Check Yourself 6

Multiply.

1. $(-5)(-5)$ **2.** $(-4)(-4)(-4)$ **3.** $-2(5 - 6)$

4. $-4\left(\dfrac{1}{4}\right)$ **5.** $(-4)^2$ **6.** -4^2 ?

In the argument above, you may have noticed our use of the fact that $-3 \cdot 0 = 0$. It is, of course, true that for any real number a,

$$a \cdot 0 = 0$$

This is sometimes referred to as the *multiplication property of zero* although technically the statement should be called a *theorem*, since it can be proved by use of the properties of real numbers introduced in Section 1.2.

We now want to complete our work in this section by considering the operation of division. Just as we defined subtraction in terms of addition, we define division in terms of multiplication. This avoids having to develop new rules for the operation of division.

DIVISION

$$\frac{a}{b} = a\left(\frac{1}{b}\right) \qquad b \neq 0$$

In words, if a and b are real numbers and $b \neq 0$, then the quotient of a and b can be expressed as the product of a and the multiplicative inverse (or reciprocal) of b.

Now, given this definition, consider a quotient such as

$$\frac{-20}{4}$$

By the above definition, this becomes

By the rules for multiplication, the quotient of a negative number and a positive number is negative.

$$\frac{-20}{4} = (-20)\left(\frac{1}{4}\right) = -5$$

Also

$$\frac{-30}{-6} = (-30)\left(-\frac{1}{6}\right) = 5$$

The quotient of two negative numbers is positive.

As our above examples suggest, the rules for signs in division exactly parallel those shown earlier for multiplication, and we can state the following.

RULES FOR DIVISION

1. **The quotient of two numbers with the *same sign* is always positive.**

2. **The quotient of two numbers with *opposite signs* is always negative.**

EXAMPLE 7   Dividing Signed Numbers

Divide as indicated.

(*a*) $\dfrac{-42}{6} = -7$

(*b*) $\dfrac{-80}{-8} = 10$

(*c*) $\dfrac{36}{-12} = -3$

Check Yourself 7

Divide.

1. $\dfrac{-75}{-5}$ **2.** $\dfrac{35}{-7}$ **3.** $\dfrac{-63}{9}$

We conclude this section by considering examples of evaluating arithmetic expressions when mixed operations are involved. Keep in mind that the rules for the order of operations must be carefully applied when different "levels" of operations appear in an expression.

The order of operations is

Parentheses

Exponents

Multiplication and Division

Addition and Subtraction

Some students memorize the sentence "Please Excuse My Dear Aunt Sally" to remember the order.

EXAMPLE 8   Using the Order of Operations

Evaluate each expression.

(*a*) $\dfrac{6(-5) - 10}{3 - (-2)} = \dfrac{-30 - 10}{3 - (-2)}$

The "fraction bar" serves as a grouping symbol, so simplify first in the numerator and then in the denominator. Multiply, then subtract.

$$= \frac{-40}{3 + 2}$$

$$= \frac{-40}{5} = -8 \qquad \text{Divide as the last step.}$$

(b) $\dfrac{4^2 + 3^2}{(-2)^2 - 3^2} = \dfrac{16 + 9}{4 - 9}$ Evaluate the powers, then add or subtract as indicated.

$$= \dfrac{25}{-5} = -5$$

(c) $-15 + 12 \div (-3) \div (-2) = -15 + (-4) \div (-2)$

$$= -15 + 2 \quad \text{Do all division first, from left to right.}$$

$$= -13$$

(d) $8 - 2^2 \cdot (-5) \div (-4 - 6) = 8 - 4 \cdot (-5) \div (-10)$

$$= 8 + 20 \div (-10)$$

$$= 8 + (-2)$$

$$= 6$$

Check Yourself 8

Evaluate each of the following expressions.

1. $\dfrac{(-8)(-7) + 4}{4^2 - 1}$

2. $\dfrac{(-6)^2 - (-9)^2}{(-5)^2 - 4^2}$

3. $-5 + 6 \cdot (-2) \div (-4)$

4. $10 - 2^3 \div 4 \cdot (-2 + 4)$

 A scientific (or graphing) calculator can be used to evaluate an expression.

> **Here are three rules that apply to evaluating expressions on a calculator.**
> 1. **Do not press the** $\boxed{=}$ **key until the entire expression has been entered.**
> 2. **Anything set aside by a grouping symbol must be enclosed in parentheses.**
> 3. **Parentheses can be nested (one inside another (like this)). Be certain that the number of left parentheses is the same as the number of right parentheses.**

EXAMPLE 9 Using a Calculator to Evaluate Expressions

Use a calculator to evaluate each expression.

(a) $2^{11} + 5^4 \cdot (-3)$

The normal sequence for entering this expression is

2 $\boxed{y^x}$ 11 $\boxed{+}$ 5 $\boxed{y^x}$ 4 $\boxed{\times}$ 3 $\boxed{+/-}$ $\boxed{=}$

The result is 173.

(b) $\dfrac{-3.5(3 \cdot 7 - 2 \cdot 8)^5}{7.25(15 - 11)^4}$

Remember that the entire numerator should be placed in one set of parentheses and the entire denominator in another.

Rounded to the nearest tenth, the answer is -5.9.

Check Yourself 9

Use a calculator to evaluate each expression. Where appropriate, round to the nearest tenth.

1. $126 - 2^{15} - 3^7(-15)$ $= 63$

2. $\dfrac{3.2(9 \cdot 12 - 8 \cdot 11)^4}{-2.5 \cdot 15 + 236 \cdot 659}$

The next example illustrates the evaluation of expressions involving variables that represent real numbers.

EXAMPLE 10 Evaluating Expressions by Substituting

Evaluate the following expressions if $a = -4$, $b = 2$, $c = -5$, and $d = 6$.

(*a*) $7a - 4c = 7(-4) - 4(-5)$
$= -28 + 20 = -8$

Always put negative numbers in parentheses when you are substituting.

(*b*) $b^2 - 4ac = 2^2 - 4(-4)(-5)$
$= 4 - 4(-4)(-5)$
$= 4 - 80$
$= -76$

(*c*) $\dfrac{3a - d}{a + c} = \dfrac{3(-4) - 6}{-4 + (-5)}$
$= \dfrac{-12 - 6}{-4 + (-5)}$
$= \dfrac{-18}{-9}$
$= 2$

Check Yourself 10

Evaluate each of the following expressions if $p = -4$, $q = 3$, and $r = 2$.

1. $2p^2 + q$

2. $p(q + r)$

3. $\dfrac{4p - 2r}{r^2}$

CHECK YOURSELF ANSWERS

1. (1) -4; (2) -11; (3) $\dfrac{1}{2}$; (4) 4   **2.** (1) -12; (2) 2; (3) 1; (4) 2   **3.** (1) 0;

(2) 7   **4.** $7a + 2b - 4c$ **5.** (1) -30; (2) -30; (3) $-\dfrac{1}{4}$ **6.** (1) 25;

(2) -64; (3) 2; (4) -1; (5) 16; (6) -16 **7.** (1) 15; (2) -5; (3) -7 **8.** (1) 4;

(2) -5; (3) -2; (4) 6   **9.** (1) 163; (2) 3.3   **10.** (1) 35; (2) -20; (3) -5.

1.4 EXERCISES

Build Your Skills

Perform the indicated addition or subtraction.

1. $5 + (-3)$

2. $8 + (-4)$

3. $5 - 8$

4. $2 - 12$

5. $-2 + 4$

6. $-5 + 7$

7. $3 + (-10)$

8. $2 + (-7)$

9. $-2 + (-3)$

10. $-8 + (-2)$

11. $-2 - 4$

12. $-5 - 9$

13. $2 - (-4)$

14. $5 - (-9)$

15. $-2 - (-4)$

16. $-5 - (-9)$

17. $3.8 + (-5.8)$

18. $-2.7 + (-3.3)$

19. $2.5 - (-3.5)$

20. $-4.7 - (-2.7)$

21. $-\dfrac{3}{4} + \left(-\dfrac{5}{4}\right)$

22. $\dfrac{19}{6} + \left(-\dfrac{1}{6}\right)$

23. $\dfrac{7}{8} - \left(-\dfrac{9}{8}\right)$

24. $-\dfrac{5}{4} - \left(-\dfrac{7}{4}\right)$

Perform the indicated operations.

25. $-5 + (-7) + 2$

26. $-4 + 3 + (-6)$

27. $2 + (-5) + (-7)$

28. $-6 + 3 + (-4)$

29. $-8 + (-5) + (-7)$

30. $-4 + (-7) + (-9)$

31. $5 + (-7) - 6$

32. $4 + (-8) - 3$

33. $5 - (-3) - 2$

34. $7 - (-2) + 5$

35. $-3 - (-5) + 7$

36. $-8 - (-3) - 5$

37. $(3 - 1) - 5$

38. $(9 - 16) - 7$

39. $-2 - (3 - 5)$

40. $-9 - (2 - 6)$

41. $5 - [8 - (6 - 3)]$

42. $9 - [8 - (4 - 7)]$

Rewrite each of the following as a sum of real numbers, evaluate the sum, and answer the question.

43. Lucy entered the elevator on the 34th floor. From that point the elevator went up 12 floors, down 27 floors, down 6 floors, and up 15 floors before she got off. On what floor did she get off the elevator?

44. A submarine dives to a depth of 500 feet (ft) below the ocean's surface. It then dives another 217 ft before climbing 140 ft. What is the depth of the submarine?

45. Jake bought a car and trailer together at an auction for $2275. He subsequently sold the car for $1550 and the trailer for $895. What was the net result of the transactions?

46. During one week, the Dow Jones average went up 12.55, up 33.61, down 51.68, down 33.91, and up 81.38. What was the net change in the Dow Jones average for the week?

47. Sandra started one month with a balance of $875 in her savings account. During the month, she deposited $325, withdrew $430, deposited $58, and then withdrew $112. What was her balance at the end of that month?

48. Robert started one week with a balance of $1023 in his checking account. During the week he wrote a check for $85, deposited $129, wrote another check for $186, and then canceled a previously written check of $134. What was his balance at the end of that week?

Simplify each expression.

49. $|7| - |2|$

50. $|7| - |-2|$

51. $|5| - |-8|$

52. $|-5| - |-8|$

53. $|-3| + |-5| - |-2|$

54. $|-8| - |-7| + |-4|$

55. $|-3| - |-5 + 2|$

56. $|-7| - |-8 + 10|$

Simplify each expression by combining like terms.

57. $4x - 7x$

58. $7a - 12a$

59. $-9m - 10m$

60. $-12y - 15y$

61. $4x - 5x - 7x$

62. $-9b - 8b + 5b$

63. $2x - 3y + 7x + 5y$

64. $5a + 8b - 6a + 2b$

65. $-2x - (-3y) - (-4x) - 7y$

66. $-8a - 2b - (-5a) - (-6a)$

Perform the indicated multiplication.

67. $5(-2)$

68. $(-6)(-8)$

69. $(-5)(3)$

70. $(-12)(3)$

71. $(-3)(-7)$

72. $(15)(-3)$

73. $(-2)(3)(-5)$

74. $(-4)(-3)(-2)$

75. $(-5)(-2)(3)(-2)$

76. $(-6)(2)(-5)(2)$

77. $(-2)\left(-\dfrac{1}{2}\right)$

78. $(-18)\left(-\dfrac{1}{6}\right)$

79. $\left(-\dfrac{4}{5}\right)\left(-\dfrac{15}{8}\right)$

80. $\left(-\dfrac{3}{4}\right)\left(\dfrac{16}{3}\right)$

81. $(-6)^2$

82. -6^2

Perform the indicated operations.

83. $5(-4) + 6$

84. $(-30)(5) + 7$

85. $(-2)(8) - 5$

86. $(-3)(-9) - 2$

87. $-5 - (-3)(-7)$

88. $-8 - 3(-2)$

89. $(-13 + 4)(2 - 2)$

90. $(47 - 12)(13 - 13)$

91. $(-12 + 8)(-12 - 8)$

92. $(-7 + 6)(-7 - 6)$

Divide.

93. $\dfrac{12}{-4}$

94. $\dfrac{-45}{-5}$

95. $\dfrac{-49}{-7}$

96. $\dfrac{-125}{5}$

97. $\dfrac{-2 - 4 \cdot 2}{-4 - 1(-2)}$

98. $\dfrac{-3 - 2(4 - 8)}{-5 + 9(4 - 3)}$

99. $\dfrac{(-4)(5) - (-5)^2}{(-2)^2 - 3^2}$

100. $\dfrac{4^2 - (-6)^2}{(-3)^2 - (-2)^2}$

Evaluate each of the following expressions if $a = -2$, $b = 5$, $c = -4$, and $d = 6$.

101. $3a + 4c$

102. $5c - 7a$

103. $c^2 - 2d$

104. $4b^2 - 2c^2$

105. $a(b + 3c)$

106. $c(3a - d)$

107. $\dfrac{2b - 3a}{c + 2d}$

108. $\dfrac{3d - 2b}{5a + d}$

109. $d^2 - c^2$

110. $(d - c)^2$

111. $(d + c)(d - c)$

112. $d^2 - 2cd + c^2$

113. $c^3 - a^3$

114. $(c - a)^3$

115. $(c - a)(c^2 + ac + a^2)$

116. $c^3 - 3c^2a + 3ca^2 - a^3$

Transcribe Your Skills

117. Explain the order of operations that is used to evaluate an expression.

118. Why do -5^5 and $(-5)^5$ represent the same number?

119. Why do -4^2 and $(-4)^2$ represent different numbers?

120. In evaluating an expression on a calculator, the $\boxed{=}$ key is not pressed until the entire expression is evaluated. Explain why this procedure must be followed.

Think About These

Replace each blank with *always, sometimes,* or *never.*

121. The sum of two positive numbers is _____ positive.

122. The sum of two negative numbers is _____ positive.

123. The sum of two negative numbers is _____ negative.

124. The sum of a positive number and a negative number is _____ a negative number.

125. Algebra is _____ difficult.

126. Algebra students _____ study.

127. The product of two positive numbers is _____ positive.

128. The product of two negative numbers is _____ negative.

129. The product of a positive number and a negative number will _____ be negative.

130. The quotient of two numbers with opposite signs will _____ be positive.

131. $|x + 5| = |x| + |5|$ is _____ a true statement.

132. $|x - 3| = |x| - |3|$ is _____ a true statement.

133. $|x - 5| = |5 - x|$ is _____ a true statement.

134. $|x - a| = |a - x|$ is _____ a true statement.

In the following problems, planting trees represents positive numbers and harvesting trees represents negative numbers. Rewrite each problem as a sum of real numbers, and answer the question.

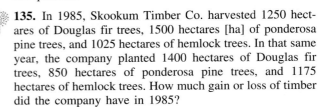

135. In 1985, Skookum Timber Co. harvested 1250 hectares of Douglas fir trees, 1500 hectares [ha] of ponderosa pine trees, and 1025 hectares of hemlock trees. In that same year, the company planted 1400 hectares of Douglas fir trees, 850 hectares of ponderosa pine trees, and 1175 hectares of hemlock trees. How much gain or loss of timber did the company have in 1985?

136. In 1988, Spar Tree Timber Co. harvested 1325 hectares of Douglas fir trees, 1175 hectares of ponderosa pine trees, and 835 hectares of hemlock trees. In that same year the company planted 1240 hectares of Douglas fir trees, 685 hectares of ponderosa pine trees, and 1275 hectares of hemlock trees. How much gain or loss of timber did the company have in 1988?

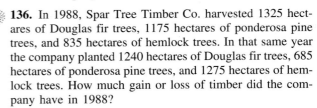

137. In 1985, McNichols Timber Co. harvested 1528 hectares of company land and replanted 1265 hectares. In 1986, the company harvested 936 hectares and replanted 1242 hectares. In 1987, the company harvested 1127 hectares and replanted 1345 hectares. Over the 3-year period, was there a net gain or loss of forested land? How much gain or loss was there?

138. In 1988, Carmichael Timber Co. harvested 1252 hectares of company land and replanted 1365 hectares. In 1986, the company harvested 1093 hectares and replanted 1254 hectares. In 1987, the company harvested 1412 hectares and replanted 1634 hectares. Over the 3-year period, was there a net gain or loss of forested land? How much gain or loss was there?

Skillscan

Evaluate each expression for the given value of x.

a. $4x - 5$; $x = 2$

b. $-2x + 7$; $x = \dfrac{1}{2}$

c. $2(x - 5) + 3$, $x = 7$

d. $-3(x - 1) - 4$; $x = -2$

e. $2x + 7 - (x - 3)$; $x = 5$

f. $-2(x + 1) - 3(-x + 2)$; $x = 3$

Sets and the Real Number System [1.1]

Definitions and Notation A *set* is a collection of objects, symbols, or numbers. We can describe a set by listing its contents in braces. Sets are often named by capital letters.

$A = \{1, 2, 3, 4\}$ is a set containing the first four counting numbers.

Any object contained in a set is called an *element,* or a *member,* of that set. We write

For the set A above

$3 \in A$

$$a \in A$$

to indicate that a is an element of set A. We can also write

$$a \notin A$$

$5 \notin A$

to indicate that a is *not* an element of set A.

Two sets are *equal* if they have exactly the same elements. We write

$$A = B$$

If $B = \{4, 3, 2, 1\}$,

$A = B$

to indicate that set A is equal to set B.

A set whose elements continue indefinitely is said to be an *infinite* set, and ellipses (. . .) are often used in describing infinite sets.

A set that has some specific number of elements is called a *finite* set.

$E = \{2, 4, 6, \ldots\}$, the set of positive even numbers, is an infinite set.

$A = \{1, 2, 3, 4\}$ is a finite set. It has four elements.

Numeric Sets The set of *natural numbers* is N, where

$5 \in N \qquad -3 \notin N \qquad \dfrac{2}{3} \notin N$

$$N = \{1, 2, 3, \ldots\}$$

The set of *integers* is denoted Z, where

$7 \in Z \qquad -8 \in Z \qquad \dfrac{3}{4} \notin Z$

$$Z = \{\ldots, -2, -1, 0, 1, 2, \ldots\}$$

The set of *rational numbers* is denoted Q, where

$\dfrac{2}{5} \in Q \qquad -\dfrac{3}{4} \in Q \qquad -5 \in Q$

$$Q = \left\{ \frac{p}{q} \,\middle|\, p, q \in Z, q \neq 0 \right\}$$

The set of *irrational numbers* is denoted Q', where Q' consists of all numbers that *cannot* be expressed as the ratio of two integers—that is, they are *not* rational.

$\sqrt{2} \in Q' \qquad \sqrt[3]{5} \in Q'$

$\pi \in Q' \qquad \dfrac{4}{5} \notin Q'$

The set of *real numbers* is denoted R, where R combines the set of rational numbers with the set of irrational numbers.

$\dfrac{2}{3} \in R \qquad -8 \in R$

$\sqrt{3} \in R \qquad \sqrt{-2} \notin R$

The set of real numbers can be pictured on the *real number line*.

Zero is neither positive nor
negative.

Every real number corresponds to exactly one point on the number line, and every point on that line corresponds to exactly one real number.

Properties of the Real Numbers [1.2]

For any real numbers a, b, and c:

Closure Properties

3 + 4 is a real number.

$a + b$ is a real number

3 · 4 is a real number.

$a \cdot b$ is a real number

Commutative Properties

7 + 5 = 5 + 7

$a + b = b + a$

7 · 5 = 5 · 7

$a \cdot b = b \cdot a$

Associative Properties

(2 + 7) + 4 = 2 + (7 + 4)

$(a + b) + c = a + (b + c)$

(2 · 7) · 4 = 2 · (7 · 4)

$(a \cdot b) \cdot c = a \cdot (b \cdot c)$

Identities There exists a unique real number 0 such that

7 + 0 = 7

$a + 0 = 0 + a = a$

−8 + 0 = −8

The number 0 is called the *additive identity*.
There exists a unique real number 1 such that

3 · 1 = 3

$a \cdot 1 = 1 \cdot a = a$

(−5)(1) = −5

The number 1 is called the *multiplicative identity*.

Inverse Properties For any real number a, there exists a unique real number $-a$ such that

3 + (−3) = 0

$a + (-a) = (-a) + a = 0$

and $-a$ is called the *additive inverse*, or the *opposite*, of a.

For any real number a ($a \neq 0$), there exists a unique number $\dfrac{1}{a}$ such that

$(3)\left(\dfrac{1}{3}\right) = 1$

$a \cdot \dfrac{1}{a} = \dfrac{1}{a} \cdot a = 1$

$\dfrac{1}{a}$ is called the *multiplicative inverse*, or the *reciprocal*, of a.

Distributive Property

$a(b + c) = ab + ac$

$(b + c)a = ba + ca$

$5(4 + 3) = 5 \cdot 4 + 5 \cdot 3$

Equality, Order, and Absolute Value [1.3]

Equality

Equality is a relation on the real numbers. An equals sign, $=$, is the symbol for equality, and we write

$a = b$

to indicate that the two expressions represent the same number.

Properties of Equality

For any real numbers a, b, and c:

Reflexive Property $a = a$

$2x^2 + 1 = 2x^2 + 1$

Symmetric Property If $a = b$, then $b = a$.

If $d = rt$, $rt = d$.

Transitive Property If $a = b$ and $b = c$, then $a = c$.

If $x = 5$ and $5 = y$, then $x = y$.

Substitution Principle If $a = b$, then a can be replaced by b, or b by a, in any statement without changing the validity of that statement.

If $x + 2y = 3$ and $y = 3x$, $x + 2(3x) = 3$.

Inequalities

The inequality relations are as follows:

1. *Is less than,* which we denote with the symbol $<$.

   We write $a < b$ to indicate that a is less than b. Graphically this means that the point corresponding to a must lie *to the left* of the point corresponding to b.

   Graph $\{x \mid x < 3\}$.

   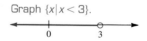

   The *open* circle means that 3 is *not included* in the graph.

2. *Is greater than,* which we denote with the symbol $>$.

   We write $a > b$ to indicate that a is greater than b. Graphically this means that the point corresponding to a must lie *to the right* of the point corresponding to b.

   Graph $\{x \mid x > -4\}$.

   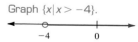

Graph $\{x \mid x \leq -1\}$.

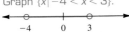

The *closed* circle means that -1 is *included* in the graph.

Graph $\{x \mid -4 < x < 3\}$.

Two other symbols, $\leq$ and $\geq$, are also used in writing inequalities.

The statement $a \leq b$ is read "a is less than or equal to b."

The statement $a \geq b$ is read "a is greater than or equal to b."

Double Inequalities

An inequality of the form

$$-4 < x < 3$$

is called a *double inequality* and combines the two inequality statements

$$x > -4 \qquad \text{and} \qquad x < 3$$

It can be read "x is between -4 and 3."

Absolute Value

The *absolute value* of a real number is the distance (on the number line) between the point named by that real number and the origin. The absolute value of a number a is denoted $|a|$. In general, we can define the absolute value of any real number a as

$|7| = 7$

$|0| = 0$

$|-7| = -(-7) = 7$

$$|a| = \begin{cases} a & \text{if } a \text{ is positive} \\ 0 & \text{if } a \text{ is zero} \\ -a & \text{if } a \text{ is negative} \end{cases}$$

Operations with Signed Numbers [1.4]

The rules for the four basic operations on the real numbers are as follows.

Addition

$5 + 3 = 8$

$(-5) + (-3) = -8$

$5 + (-3) = 2$

$(-5) + 3 = -2$

1. If the numbers have the same sign, add their absolute values. Give the sum the sign of the original numbers.

2. If the numbers have opposite signs, subtract their absolute values, the smaller from the larger. Give the sum the sign of the number with the larger absolute value.

Subtraction

$5 - 3 = 2$

$5 - (-3) = 5 + 3 = 8$

To subtract b from a, add the additive inverse (or the opposite) of b to a.

Multiplication

$(7)(3) = 21$

$(-7)(-3) = 21$

$(7)(-3) = -21$

$(-7)(3) = -21$

To multiply two real numbers, first multiply their absolute values. Then determine the proper sign for the product:

1. If the two numbers have the same sign, the product is always positive.

2. If the two numbers have opposite signs, the product is always negative.

Division

The rules of signs for division are then identical to those for multiplication.

1. The quotient of two numbers with the same sign is always positive.

2. The quotient of two numbers with opposite signs is always negative.

$$\frac{32}{8} = 4 \qquad \frac{-32}{-8} = 4$$

$$\frac{-32}{8} = -4 \qquad \frac{32}{-8} = -4$$

SUMMARY EXERCISES

This summary exercise set is provided to give you practice with each of the objectives of the chapter. Each exercise is keyed to the appropriate chapter section. The answers are provided in the instructor's manual that accompanies this text. Your instructor will provide guidelines on how to best use these exercises in your instructional program.

[1.1] Name all the numeric sets to which each of the following numbers belongs. The numeric sets consist of the natural numbers N, integers Z, rational numbers Q, irrational numbers Q', and real numbers R.

  1. 0

  2. -7

  3. $-\sqrt{2}$

  4. $\dfrac{2}{3}$

  5. 2π

  6. $-\dfrac{7}{8}$

  7. $1.\overline{63}$

  8. 2.75

[1.1] On a number line, graph the points corresponding to the elements of each of the given sets.

  9. $\{1, 5, 9\}$

  10. $\{-5, 0, 5\}$

  11. $\left\{-\dfrac{1}{3}, \dfrac{4}{3}, \dfrac{7}{3}\right\}$

  12. $\{-\sqrt{6}, -\sqrt{3}, \sqrt{2}\}$

[1.2] Apply the appropriate properties to rewrite the given expressions. Then simplify your results.

  13. $6 + (x + 3)$

  14. $4(2ab)$

  15. $-7 + (x + 7)$

  16. $4x\left(\dfrac{1}{4}\right)$

  17. $\left(\dfrac{2}{7}\right)(7x)\left(\dfrac{5}{2}\right)$

  18. $2(a + 2b + 3c)$

  19. $4a + 2a + a$

  20. $3(2x + 3y + 7x)$

  21. $3x + 8x + 4 + x$

  22. $3y + 6 + 2y + 1$

  23. $2 + 3(2z + 1) + 2z$

  24. $5 + 2x(x + 1) + 3x$

[1.2] State the property that is used to justify each of the following statements.

  25. $24 + (-24) = 0$

  26. $(24)\left(\dfrac{1}{24}\right) = 1$

  27. $x + (2x + 3) = (x + 2x) + 3$

  28. $2a + 3 + 5a = 2a + 5a + 3$

  29. $2(3a) = (2 \cdot 3)a$

  30. $3(2x + 1) = 3 \cdot 2x + 3 \cdot 1$

  31. $2x + 0 = 2x$

  32. $(1)(5y) = 5y$

[1.3] Which property of equality justifies each of the following statements?

  33. If $2x + 1 = y$, then $y = 2x + 1$.

  34. $2x^2 + 5x + 1 = 2x^2 + 5x + 1$

  35. If $2a + 1 = b$ and $b = 13$, then $2a + 1 = 13$.

  36. If $2x + 5 = y$ and $x = 5$, then $2(5) + 5 = y$.

[1.3] Graph each of the following sets.

  37. $\{x \mid x > 2\}$

  38. $\{x \mid x \leq 4\}$

  39. $\left\{x \mid x < \dfrac{5}{2}\right\}$

40. $\{x \mid x \geq -4\}$

41. $\{x \mid -2 < x < 7\}$

42. $\{x \mid -4 \leq x < -1\}$

[1.3] Write each of the following expressions without the absolute value symbol.

43. $|7|$

44. $|-3.2|$

45. $-|2|$

46. $-|-4.5|$

47. $|3| + |-3|$

48. $|-2| + |-5|$

49. $-(|2| + |-3|)$

50. $-(|-8| + |-9|)$

[1.4] Perform the indicated addition or subtraction.

51. $-2 + 4$

52. $-5 + (-7)$

53. $-2 - 5$

54. $6 - (-4)$

55. $-2.8 + 1.4$

56. $-7.9 - (-3.9)$

57. $\dfrac{7}{2} - \left(-\dfrac{5}{2}\right)$

58. $-\left(\dfrac{7}{8}\right) - \left(\dfrac{9}{8}\right)$

[1.4] Perform the indicated multiplication.

59. $(3)(-12)$

60. $(-5)(-5)(-5)$

61. $(4)(-6)\left(\dfrac{5}{8}\right)$

62. $(-1.5)(-6)$

63. $(-6)^2$

64. -6^2

65. $(-4)^3$

66. -4^3

[1.4] Simplify each expression by combining all like terms.

67. $6a - 8a$

68. $6x - (-8x)$

69. $-2y - 3y + 5y$

70. $-2b - (-5b)$

71. $-3x + 2y - 5x - 7y$

72. $-x + 4x - 5y - (-x)$

73. $2a - 5b - (-6a) - (-b)$

74. $2(3x - 7y) + 3(-2x - y)$

[1.4] Evaluate each expression if $x = -3$, $y = 6$, $z = -4$, and $w = 2$.

75. $5y - 4z$

76. $3x^2 - 2w^2$

77. $w(2x - z)$

78. $\dfrac{3x - y}{w - x}$

79. $z^3 - w^3$

80. $(z - w)^3$

SELF-TEST

The purpose of this self-test is to help you check your progress and to review for a chapter test in class. Allow yourself about an hour to take the test. When you are done, check your answers in the back of the book. If you missed any problems, be sure to go back and review the appropriate sections in the chapter and the exercises that are provided.

State the property that is used to justify each of the following statements.

1. $3(ab) = (3a)b$

2. $0.6 + (-0.6) = 0$

3. $x + (3x + 4) = (x + 3x) + 4$

4. $5(2y + 3) = 5 \cdot 2y + 5 \cdot 3$

Graph each of the following sets.

5. $\{x \mid x > -3\}$

6. $\{x \mid x \leq -2\}$

7. $\{x \mid -3 \leq x \leq 5\}$

8. $\{x \mid 0 \leq x < 6\}$

Write each expression without absolute-value symbols.

9. $|-7| - |-5|$

10. $|-3 + 1|$

Perform the indicated operations.

11. $-8 + (-7)$

12. $2.3 - (-5.3)$

13. $(-3)(-2)(-5)$

14. $\dfrac{-72}{-9}$

15. $3 + 4 \cdot 2^3$

16. $2(-3)^3$

Simplify each expression by combining like terms.

17. $-4a + 3b - 5a - 7b$

18. $7x + 3(2x + 5) - 3$

Evaluate each expression if $a = 3$, $b = -5$, and $c = 10$.

19. $\dfrac{ab + a^2b}{2ac}$

20. $(a + b)(a - b)$

CHAPTER TWO

AN INTRODUCTION TO EQUATIONS AND INEQUALITIES

LOSING OUR LANDFILLS By the year 2010, more than 75 percent of U.S. landfills are expected to close, and most will not be replaced by new ones. What will we do with all our trash if we run out of landfill space? If we are going to run out of landfill space, it might be a good idea to try to preserve the space we now have, and it might be smart to find other things to do with trash besides bury it.

Municipal trash is usually about 50 percent paper, with newspaper taking up nearly 20 percent of the space. Organic debris such as wood, yard waste, and food scraps takes up about 12.5 percent of the space, and plastics such as milk jugs, soda bottles, garbage bags, food packaging, and foam products occupy about 10 percent of U.S. landfills.

Since much of the paper that ends up in landfills comes from packaging, this is an area that might be considered. A toy or tool may be packed in a plastic bag, mounted in a bubble pack on a cardboard back, put in a box of like items for shipment, unpacked and displayed in a cardboard rack, and finally put in a paper or plastic sack at the time of sale. Much of this packaging is unnecessary. If we could eliminate the paper backing, cardboard rack, and paper bag, then we could recycle the box and the plastic is all we have left for the landfill.

Many parts of the country are experiencing a glut of recycled newsprint. Therefore it is unlikely that a large amount of newspaper can be diverted from landfills in the near future.

Plastics are another product that could be diverted from landfills. Plastics have been approximately 10 percent of U.S. municipal refuse since the early 1970s. Even though plastics aren't taking up any more space than they used to, we could save valuable space by recycling plastics. Most plastics can be easily recycled if they are separated by types.

Paper, plastics, and organics, which can be composted instead of put in a landfill, account for nearly 75 percent of the trash in landfills. If we could reduce the amount of these three groups of waste by recycling or reusing them, we could extend the life of many of our landfills.

Linear Equations in One Variable   **2.1**

OBJECTIVES: 1. To solve equations by using the addition and multiplication properties of equality
2. To recognize conditional equations, identities, and contradictions

TAPE IN2

We begin this chapter by considering one of the most important tools of mathematics, the equation. The ability to recognize and solve various types of equations and inequalities is probably the most useful algebraic skill you will learn, and we will continue to build on the methods developed here throughout the remainder of the text. To start, let's describe what we mean by an equation.

An *equation* is a mathematical statement in which two expressions represent the same quantity. An equation has three parts. Consider

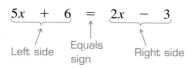

$$5x + 6 = 2x - 3$$

Left side   Equals sign   Right side

The equation simply says that the expression on the left and the expression on the right represent the same quantity.

In this chapter, we will work with a particular kind of equation.

> A *linear equation in one variable* is any equation that can be written in the form
>
> $$ax + b = 0$$
>
> where a and b are any real numbers and $a \neq 0$.

Linear equations are also called *first-degree equations* because the highest power of the variable is the first power, or first degree.

The *solution* of an equation in one variable is any number that will make the equation a true statement. The *solution set* for such an equation is simply the set consisting of all solutions.

We also say the solution *satisfies* the equation.

EXAMPLE 1 Checking Solutions

For

$$5x + 6 = 2x - 3$$

the solution is -3 because replacing x with -3 gives

$$5(-3) + 6 \stackrel{?}{=} 2(-3) - 3$$

$$-15 + 6 \stackrel{?}{=} -6 - 3$$

$$-9 = -9 \quad \text{A true statement}$$

We use the question mark over the equals sign when we are checking to see if the statement is true.

Check Yourself 1

Verify that 7 is a solution for this equation.

$$5x - 15 = 2x + 6$$

Solving linear equations in one variable will require using *equivalent equations*.

> **Two equations are *equivalent* if they have the same solution set.**

For example, the three equations

$$5x + 5 = 2x - 4 \qquad 3x = -9 \qquad \text{and} \qquad x = -3$$

You can easily verify this by replacing *x* with −3 in each equation.

are all equivalent because they all have the same solution set, $\{-3\}$. Note that replacing x with -3 will give a true statement in the third equation, but it is not as clear that -3 is a solution for the other two equations. This leads us to an equation-solving strategy of *isolating* the variable, as is the case in the equation $x = -3$.

To form equivalent equations that will lead to the solution of a linear equation, we need two properties of equality.

> **ADDITION PROPERTY OF EQUALITY**
>
> If $\qquad a = b$
>
> then $\quad a + c = b + c$

Adding the same quantity to both sides of an equation gives an equivalent equation. This is true whether *c* is positive or negative.

Since subtraction can always be defined in terms of addition,

$$a - c = a + (-c)$$

the addition property also allows us to *subtract* the same quantity from both sides of an equation. We also have this property:

> **MULTIPLICATION PROPERTY OF EQUALITY**
>
> If $\qquad a = b$
>
> then $\quad ac = bc \qquad$ where $c \neq 0$

Multiplying both sides of an equation by the same nonzero quantity gives an equivalent equation.

Since division can be defined in terms of multiplication,

$$\frac{a}{c} = a \cdot \frac{1}{c} \qquad c \neq 0$$

The multiplication property allows us to *divide* both sides of an equation by the same nonzero quantity.

EXAMPLE 2 Applying the Properties of Equality

Solve for x:

$$3x - 5 = 4 \tag{1}$$

Solution We start by using the addition property to add 5 to both sides of the equation.

$$3x - 5 + 5 = 4 + 5$$
$$3x = 9 \tag{2}$$

Why did we add 5? Because it is the *opposite* of -5 and the resulting equation will have the variable term on the left and the constant term on the right.

Now we want to get the x term alone on the left with a coefficient of 1 (we call this *isolating* the x). To do this, we use the multiplication property and multiply both sides by $\frac{1}{3}$.

$$\frac{1}{3}(3x) = \frac{1}{3}(9)$$

$$\left(\frac{1}{3} \cdot 3\right)(x) = 3$$

$$x = 3 \tag{3}$$

We choose $\frac{1}{3}$ because $\frac{1}{3}$ is the *reciprocal* of 3 and $\frac{1}{3} \cdot 3 = 1$

Since any application of the addition or multiplication properties leads to an equivalent equation, Equations (1), (2), and (3) all have the same solution, 3.

To check this result, we can replace x with 3 in the original equation:

$$3(3) - 5 \stackrel{?}{=} 4$$
$$9 - 5 \stackrel{?}{=} 4$$
$$4 = 4 \quad \text{A true statement}$$

You may prefer a slightly different approach in the last step of the solution above. From Equation (2)

$$3x = 9$$

The multiplication property can be used to *divide* both sides of the equation by 3. Then

$$\frac{3x}{3} = \frac{9}{3}$$
$$x = 3$$

Of course, the result is the same.

Check Yourself 2

Solve for x.

$$4x - 7 = 17$$

The steps involved in using the addition and multiplication properties to solve an equation are the same if more terms are involved in an equation.

EXAMPLE 3 Applying the Properties of Equality

Solve for x:

$$5x - 11 = 2x - 7$$

Solution Our objective is to use the properties of equality to isolate x on one side of an equivalent equation. We begin by adding 11 to both sides.

Again, adding 11 leaves us with the constant term on the right.

$$5x - 11 + 11 = 2x - 7 + 11$$
$$5x = 2x + 4$$

We continue by adding $-2x$ to (or subtracting $2x$ from) both sides.

If you prefer, write

$5x - 2x = 2x - 2x + 4$

Again:

$3x = 4$

$$5x + (-2x) = 2x + (-2x) + 4$$
$$3x = 4$$

To isolate x, we now multiply both sides by $\dfrac{1}{3}$.

This is the same as dividing both sides by 3. So

$$\frac{3x}{3} = \frac{4}{3}$$

$$x = \frac{4}{3}$$

$$\frac{1}{3}(3x) = \frac{1}{3}(4)$$

$$x = \frac{4}{3}$$

We leave it to you to check this result by substitution.

Check Yourself 3

Solve for x.

$$7x - 12 = 2x - 9$$

Recall the discussion of like terms in Section 1.2.

Both sides of an equation should be simplified as much as possible *before* the addition and multiplication properties are applied. If like terms are involved on one side (or on both sides) of an equation, they should be combined before an attempt is made to isolate the variable. The following example illustrates.

EXAMPLE 4 Applying the Properties of Equality with Like Terms

Solve for x:

Note the like terms on the left and right sides of the equation.

$$8x + 2 - 3x = 8 + 3x + 2$$

Solution Here we combine the like terms $8x$ and $-3x$ on the left and the like terms 8 and 2 on the right as our first step. We then have

$$5x + 2 = 3x + 10$$

We can now solve as before.

$$5x + 2 - 2 = 3x + 10 - 2 \qquad \text{Subtract 2 from both sides.}$$
$$5x = 3x + 8$$

Then

$$5x - 3x = 3x - 3x + 8 \qquad \text{Subtract } 3x \text{ from both sides.}$$
$$2x = 8$$
$$\frac{2x}{2} = \frac{8}{2} \qquad \text{Divide both sides by 2.}$$

or $\qquad x = 4$

The solution is 4, which can be checked by returning to the *original equation*.

Check Yourself 4

Solve for x.

$$7x - 3 - 5x = 10 + 4x + 3$$

If parentheses are involved on one or both sides of an equation, the parentheses should be removed by applying the distributive property as the first step. Like terms should then be combined before an attempt is made to isolate the variable. Consider the following example.

EXAMPLE 5 Applying the Properties of Equality with Parentheses

Solve for x:

$$x + 3(3x - 1) = 4(x + 2) + 4$$

Solution First, remove the parentheses on the left and right sides.

$$x + 9x - 3 = 4x + 8 + 4$$

Combine like terms on each side of the equation.

$$10x - 3 = 4x + 12$$

Now isolate variable x on the left side.

$$10x - 3 + 3 = 4x + 12 + 3 \qquad \text{Add 3 to both sides.}$$
$$10x = 4x + 15$$
$$10x - 4x = 4x - 4x + 15 \qquad \text{Subtract } 4x \text{ from both sides.}$$
$$6x = 15$$
$$\frac{6x}{6} = \frac{15}{6} \qquad \text{Divide both sides by 6.}$$
$$x = \frac{5}{2}$$

Recall that to isolate the x, we must get x alone on the left side with a coefficient of 1.

The solution is $\dfrac{5}{2}$. Again, this can be checked by returning to the original equation.

Check Yourself 5

Solve for x.

$x + 5(x + 2) = 3(3x - 2) + 18$

To solve an equation involving fractions, the first step is to multiply both sides of the equation by the *least common multiple* (LCM) of all denominators in the equation. This will clear the equation of fractions, and we can proceed as before.

EXAMPLE 6 Applying the Properties of Equality with Fractions

Solve

$$\frac{x}{2} - \frac{2}{3} = \frac{5}{6}$$

for x.

$$\frac{x}{2} - \frac{2}{3} = \frac{5}{6}$$

We multiply each side by 6, the LCM of 2, 3, and 6.

$$6\left(\frac{x}{2} - \frac{2}{3}\right) = 6\left(\frac{5}{6}\right)$$

Apply the distributive law.

$$6\left(\frac{x}{2}\right) - 6\left(\frac{2}{3}\right) = 6\left(\frac{5}{6}\right)$$

$$\overset{3}{6}\left(\frac{x}{\underset{1}{2}}\right) - \overset{2}{6}\left(\frac{2}{\underset{1}{3}}\right) = \overset{1}{6}\left(\frac{5}{\underset{1}{6}}\right)$$

The equation is now cleared of fractions.

$$3x - 4 = 5$$

$$3x = 9$$

$$x = 3$$

The solution, 3, can be checked as before by returning to the original equation.

Check Yourself 6

Solve for x.

$$\frac{x}{4} - \frac{4}{5} = \frac{19}{20}$$

Be sure that the distributive property is applied properly so that *every term* of the equation is multiplied by the LCM.

EXAMPLE 7 Applying the Properties of Equality with Fractions

Solve

$$\frac{2x - 1}{5} + 1 = \frac{x}{2}$$

for x.

$$10\left(\frac{2x - 1}{5} + 1\right) = 10\left(\frac{x}{2}\right)$$

> Multiply by 10, the LCM of 5 and 2.

$$\overset{2}{10}\left(\frac{2x - 1}{\underset{1}{5}}\right) + 10(1) = \overset{5}{10}\left(\frac{x}{\underset{1}{2}}\right)$$

> Apply the distributive law on the left.

$$2(2x - 1) + 10 = 5x$$

$$4x - 2 + 10 = 5x$$

$$4x + 8 = 5x$$

$$8 = x$$

The solution for the original equation is 8.

Check Yourself 7

Solve for x.

$$\frac{3x + 1}{4} - 2 = \frac{x + 1}{3}$$

So far, we have considered only equations of the form $ax + b = 0$, where $a \neq 0$. If we allow the possibility that $a = 0$, two additional equation forms arise. The resulting equations can be classified into three types depending on the nature of their solutions. Let's define and see how to recognize the three types.

1. An equation that is true for only particular values of the variable is called a *conditional equation*.
   This case has been illustrated in all our previous examples and exercises.

   > **Note 1** Here the equation can be written in the form $ax + b = 0$, where $a \neq 0$.

2. An equation that is true for all possible values of the variable is called an *identity*.
   This will be the case if both sides of the equation reduce to the same expression (a true statement).

   > **Note 2** In this case *both a and b* are 0, so we get the equation $0 = 0$.

3. An equation that is never true, no matter what the value of the variable, is called a *contradiction*.
   This will be the case if both sides of the equation reduce to a false statement.

   > **Note 3** Here a is 0, but b is nonzero, and we end up with something like $4 = 0$.

The following example illustrates the second and third cases.

EXAMPLE 8 Identities and Contradictions

(*a*) Solve for x:

$$2(x - 3) - 2x = -6$$
$$2x - 6 - 2x = -6$$
$$-6 = -6 \qquad \text{A } \textit{true} \text{ statement}$$

See note 2 above. By adding 6 to both sides of this equation, we have $0 = 0$.

Since the two sides reduce to the true statement $-6 = -6$, the original equation is an *identity* and the solution set is the set of all real numbers.

(*b*) Solve for x:

$$3(x + 1) - 2x = x + 4$$
$$3x + 3 - 2x = x + 4$$
$$x + 3 = x + 4$$
$$3 = 4 \qquad \text{A } \textit{false} \text{ statement}$$

See note 3 above. Subtracting 3 from both sides, we have $0 = 1$.

Since the two sides reduce to the false statement $3 = 4$, the original equation is a contradiction. There are no values of the variable that can satisfy the equation. The solution set has nothing in it. We call this the *empty set* and write $\{\,\}$ or $\varnothing$.

Check Yourself 8

Determine whether each of the following equations is a conditional equation, an identity, or a contradiction.

1. $2(x + 1) - 3 = x$

2. $2(x + 1) - 3 = 2x + 1$

3. $2(x + 1) - 3 = 2x - 1$

An *algorithm* is a step-by-step process for problem solving.

An organized step-by-step procedure is the key to an effective equation-solving strategy. The following algorithm summarizes our work in this section and should give you guidance in approaching the problems that follow.

SOLVING LINEAR EQUATIONS IN ONE VARIABLE

STEP 1 Multiply both sides of the equation by the LCM of any denominators, to clear the equation of fractions.

STEP 2 Remove any grouping symbols by applying the distributive property.

STEP 3 Combine any like terms that appear on either side of the equation.

STEP 4 Apply the addition property of equality to write an equivalent equation with the variable term on *one side* of the equation and the constant term on the *other side*.

STEP 5 Apply the multiplication property of equality to write an equivalent equation with the variable isolated on one side of the equation.

STEP 6 Check the solution in the *original* equation.

Note If the equation derived in step 5 is always true, the original equation was an identity. If the equation is always false, the original equation was a contradiction.

When you are solving an equation for which a calculator is recommended, it is often easiest to do all calculations as the last step.

EXAMPLE 9 Evaluating Expressions Using a Calculator

Solve the following equation for x.

$$\frac{185(x - 3.25) + 1650}{500} = 159.44$$

Solution Following the steps of the algorithm, we get

$185(x - 3.25) + 1650 = 159.44 \cdot 500$ Multiply by the LCM.

$185x - 185 \cdot 3.25 + 1650 = 159.44 \cdot 500$ Remove parentheses.

$185x = 159.44 \cdot 500 + 185 \cdot 3.25 - 1650$ Apply the addition property.

$x = \dfrac{159.44 \cdot 500 + 185 \cdot 3.25 - 1650}{185}$ Isolate the variable.

Now, remembering to insert parentheses around the numerator, we use a calculator to simplify the expression on the right.

$x = 425.25$

Check Yourself 9

Solve the following equation for x.

$$\frac{2200(x + 17.5) - 1550}{75} = 2326$$

CHECK YOURSELF ANSWERS

1. $5(7) - 15 \overset{?}{=} 2(7) + 6$

  $35 - 15 \overset{?}{=} 14 + 6$

   $20 = 20$

  A true statement.

2. 6 **3.** $\dfrac{3}{5}$ **4.** -8 **5.** $-\dfrac{2}{3}$ **6.** 7 **7.** 5 **8.** (1) Conditional;
(2) contradiction; (3) identity **9.** 62.5

2.1 EXERCISES

Build Your Skills

Solve each of the following equations, and check your result.

1. $5x - 8 = 17$

2. $4x + 9 = -11$

3. $8 - 7x = -41$

4. $-7 - 4x = 21$

5. $7x - 5 = 6x + 6$

6. $9x + 4 = 8x - 3$

7. $8x - 4 = 3x - 24$

8. $5x + 2 = 2x - 5$

9. $7x - 4 = 2x + 26$

10. $11x - 3 = 4x - 31$

11. $4x - 3 = 1 - 2x$

12. $8x + 5 = -19 - 4x$

13. $2x + 8 = 7x - 37$

14. $3x - 5 = 9x + 22$

Simplify and then solve each of the following equations.

15. $5x - 2 + x = 9 + 3x + 10$

16. $5x + 5 - x = -7 + x - 2$

17. $7x - 3 - 4x = 5 + 5x - 13$

18. $8x - 3 - 6x = 7 + 5x + 17$

19. $5x = 3(x - 6)$

20. $2(x - 15) = 7x$

21. $5(8 - x) = 3x$

22. $7x = 7(6 - x)$

23. $2(2x - 1) = 3(x + 1)$

24. $3(3x - 1) = 4(3x + 1)$

25. $8x - 3(2x - 4) = 17$

26. $7x - 4(3x + 4) = 9$

27. $7(3x + 4) = 8(2x + 5) + 13$

28. $-4(2x - 1) + 3(3x + 1) = 9$

29. $9 - 4(3x + 1) = 3(6 - 3x) - 9$

30. $13 - 4(5x + 1) = 3(7 - 5x) - 15$

31. $5 - 2[x - 2(x - 1)] = 55 - 4[x - 3(x + 2)]$

32. $7 - 5[x - 3(x + 2)] = 25 - 2[x - 2(x - 3)]$

Clear of fractions and then solve each of the following equations.

33. $\dfrac{2x}{3} - \dfrac{5}{3} = 3$

34. $\dfrac{3x}{4} + \dfrac{1}{4} = 4$

35. $\dfrac{x}{6} + \dfrac{x}{5} = 11$

36. $\dfrac{x}{6} - \dfrac{x}{8} = 1$

37. $\dfrac{2x}{3} - \dfrac{x}{4} = \dfrac{5}{2}$

38. $\dfrac{5x}{6} + \dfrac{2x}{3} = \dfrac{5}{6}$

39. $\dfrac{x}{5} - \dfrac{x - 7}{3} = \dfrac{1}{3}$

40. $\dfrac{x}{6} + \dfrac{3}{4} = \dfrac{x - 1}{4}$

41. $\dfrac{5x - 3}{4} - 2 = \dfrac{x}{3}$

42. $\dfrac{6x - 1}{5} - \dfrac{2x}{3} = 3$

43. $\dfrac{2x + 3}{5} - \dfrac{2x - 1}{3} = \dfrac{8}{15}$

44. $\dfrac{3x}{5} - \dfrac{3x - 1}{2} = \dfrac{11}{10}$

45. $0.5x - 6 = 0.2x$

46. $0.7x - 7 = 0.3x - 5$

Classify each of the following equations as a conditional equation, an identity, or a contradiction.

47. $3(x - 1) = 2x + 3$

48. $2(x + 3) = 2x + 6$

49. $3(x - 1) = 3x + 3$

50. $2(x + 3) = x + 5$

51. $3(x - 1) = 3x - 3$

52. $2(x + 3) = 3x + 5$

53. $3x - (x - 3) = 2(x + 1) + 2$

54. $5x - (x + 4) = 4(x - 2) + 4$

55. $\dfrac{x}{2} - \dfrac{x}{3} = \dfrac{x}{6}$

56. $\dfrac{3x}{4} - \dfrac{2x}{3} = \dfrac{x}{6}$

Transcribe Your Skills

57. What is the common characteristic of equivalent equations?

58. What is meant by a *solution* to a linear equation?

59. Define (*a*) identity and (*b*) contradiction.

60. Why does the multiplication property of equality not include multiplying both sides of the equation by 0?

Think About These

Label each of the following as true or false.

61. Adding the same value to both sides of an equation creates an equivalent equation.

62. Multiplying both sides of an equation by 0 creates an equivalent equation.

63. To clear an equation of fractions, we multiply both sides by the GCF of the denominator.

64. The multiplication property of equations allows us to divide both sides by the same nonzero quantity.

65. Some equations have more than one solution.

66. No matter what value is substituted for *x*, the expressions on either side of the equals sign have the same value.

Skillscan (Section 1.4)

The following set of exercises has been designed to help you review skills that you will need in the next section. When possible, a reference has been included to the section where similar problems were introduced. Go back to the indicated section for further review if you like.

If $a = 3$, $b = -2$, $c = 4$, and $d = -1$, evaluate each of the following expressions.

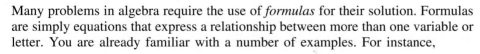

a. $\dfrac{b}{c}$ **b.** $\dfrac{ab}{c}$ **c.** $\dfrac{b}{cd}$ **d.** $\dfrac{c}{2ab}$ **e.** $\dfrac{c - b}{ad}$

f. $\dfrac{d - ac}{b}$ **g.** $\dfrac{c - 2b}{d}$ **h.** $\dfrac{2c - 3a}{b}$

Formulas and Literal Equations 2.2

OBJECTIVE: To solve literal equations for a specified variable

TAPE IN2

Many problems in algebra require the use of *formulas* for their solution. Formulas are simply equations that express a relationship between more than one variable or letter. You are already familiar with a number of examples. For instance,

$$P = R \cdot B \qquad A = \frac{1}{2}h \cdot b \qquad P = 2L + 2W$$

are formulas for percentage, the area of a triangle, and the perimeter of a rectangle, respectively.

One useful application of the equation-solving skills we considered in the previous section involves rewriting these formulas, also called *literal equations,* in more convenient equivalent forms.

Generally, that more convenient form is one in which the original formula or equation is solved for a particular variable or letter. This is called *solving the equation for a variable,* and the steps used in the process are very similar to those you saw earlier in solving linear equations.

A *literal equation* is any equation that involves more than one variable or letter.

Consider the following example.

EXAMPLE 1 Solving Literal Equations

Solve the following formula for t.

This formula gives distance d in terms of a rate r and time t.

$$d = r \cdot t$$

Solution To solve for t means to isolate t on one side of the equation. This can be done by dividing both sides by r. We use the multiplication property of equality to divide by r, the coefficient of t.

$$\frac{d}{r} = \frac{r \cdot t}{r}$$

$$\frac{d}{r} = t$$

We usually write the equation in the equivalent form with the desired variable on the left. So

This uses the symmetric property of equality: If $a = b$, then $b = a$.

$$t = \frac{d}{r}$$

We now have t in terms of d and r, as required.

Check Yourself 1

Solve the formula $C = 2\pi r$ for r.

Solving a formula for a particular variable may require the use of both properties of equality, as the following example illustrates.

EXAMPLE 2 Solving Literal Equations

Solve the following formula for L.

This formula gives the perimeter of a rectangle P in terms of its width W and its length L.

$$P = 2L + 2W$$

We want to isolate the term containing the variable we are solving for—here L.

Solution To solve for L, start by using the addition property of equality to subtract $2W$ from both sides.

$$P = 2L + 2W$$

$$P - 2W = 2L + 2W - 2W$$

$$P - 2W = 2L$$

We now use the multiplication property to divide both sides by 2:

$$\frac{P - 2W}{2} = \frac{2L}{2}$$

$$\frac{P - 2W}{2} = L \qquad \text{or} \qquad L = \frac{P - 2W}{2}$$

This gives L in terms of P and W, as desired.

Check Yourself 2

Solve the formula $ax + by = c$ for y.

You may also have to apply the distributive property in solving for a variable. Consider the following example.

EXAMPLE 3 Solving Literal Equations with Parentheses

Solve the following formula for r.

$$A = P(1 + rt)$$

Solution First, we use the distributive property to remove the parentheses on the right.

$$A = P(1 + rt) = P + Prt$$

We now subtract P from both sides.

$$A - P = P - P + Prt$$

$$A - P = Prt$$

This formula gives the amount A in an account earning simple interest, with principal P, interest rate r, and time t.

Finally, to isolate r, we divide by Pt, the coefficient of r on the right.

$$\frac{A - P}{Pt} = \frac{Prt}{Pt}$$

$$\frac{A - P}{Pt} = r \qquad \text{or} \qquad r = \frac{A - P}{Pt}$$

Check Yourself 3

Solve this equation for n.

$$S = 180(n - 2)$$

Often it is necessary to apply the multiplication property, to clear the literal equation of fractions, as the first step of the solution process. This is illustrated in Example 4.

EXAMPLE 4 Solving Literal Equations with Division Involved

Solve the formula for C.

This formula gives the yearly depreciation D for an item in terms of its cost C, its salvage value S, and the number of years n.

$$D = \frac{C - S}{n}$$

Solution As our first step, we multiply both sides of the given equation by n, to clear of fractions.

$$D = \frac{C - S}{n}$$

On the *right* notice that

$$\frac{n}{n} = 1$$

and multiplying by 1 leaves $C - S$.

$$nD = n\left(\frac{C - S}{n}\right)$$

$$nD = C - S$$

We now add S to both sides.

$$nD + S = C - S + S$$

$$nD + S = C \qquad \text{or} \qquad C = nD + S$$

and the cost C is now represented in terms of n, D, and S.

Check Yourself 4

Solve the formula $V = \frac{1}{3}\pi r^2 h$ for h.

Let's turn now to an application of our work in solving literal equations for a specified variable.

EXAMPLE 5 An Application of Example 3

Suppose a principal of $1000 is invested in a mutual fund account for 5 years. In order for the amount in the account to be $1400 at the end of that period, what must the interest rate be?

From the given problem, we know the values of A, P, and t. Two strategies are possible.

1. We can substitute the known values in $A = P(1 + rt)$ and then solve for r.

2. Or we can solve for r and then substitute. We have illustrated this approach.

Solution In Example 3, we solved the formula

$$A = P(1 + rt)$$

for r with the result

$$r = \frac{A - P}{Pt}$$

In this problem, $A = \$1400$, $P = \$1000$, and $t = 5$ years. Substituting, we have

$$r = \frac{1400 - 1000}{(1000)(5)} = \frac{400}{5000} = 0.08 = 8\%$$

The necessary interest rate is 8 percent.

Check Yourself 5

Suppose that a principal of $5000 is invested in a time deposit fund for 3 years. If the amount in the account at the end of that period is $6050, what was the annual interest rate?

Note, in the previous example and exercise, that once we solved for r in the original equation, we were able to easily use the result with different sets of data. You can compare this to first substituting the known values and then having to solve for r in each case separately.

The following exercise will provide additional experience with the two approaches.

Check Yourself 6

The formula for the area of a trapezoid is

$$A = \frac{1}{2}h(B + b)$$

where h is the height and B and b are the two bases.

1. If the area of a trapezoid is 110 square centimeters (cm^2), the height is 10 centimeters (cm), and the longer base (B) has length 14 cm, find the length of the shorter base b by direct substitution.

2. Solve the given equation for b. Then substitute to find the length of the shorter base.

Solving for a specific variable also has significance in work done with programmable calculators and computers. To find the value of a variable within a formula, the formula must be solved for that variable. That new formula is then entered in the language of the program.

In some programs, including many used with programmable calculators, the four arithmetic operations are indicated as follows:

Algebraic Expression	Program Expression
$a + b$	A + B
$a - b$	A − B
ab	A * B
$\dfrac{a}{b}$	A / B

EXAMPLE 6 Representing Literal Equations on a Computer

Solve each formula for the given variable, then express the results as a program expression. It is important to understand that, in computers, multiplication and division are done in order from left to right, then addition and subtraction are done in order from left to right. Parentheses are often needed to achieve the desired results.

(a) $d = rt$ for r

Dividing both sides by t, we find

$$r = \frac{d}{t}$$

The program expression becomes D/T.

(b) $V = \frac{1}{3}Bh$ for h

Multiplying by 3 and then dividing by B, we get

$$h = \frac{3V}{B}$$

The program expression becomes $(3 * V)/B$.

Check Yourself 7

Solve each formula for the given variable; then express the results as a program expression.

1. $V = LWH$ for H

2. $D = \frac{C - S}{n}$ for C

The following algorithm summarizes our work in solving literal equations. Following this procedure will help you organize your work in the next problem set.

SOLVING FORMULAS OR LITERAL EQUATIONS

STEP 1 Remove any fractions, using the multiplication property.

STEP 2 Remove any grouping symbols, using the distributive property.

STEP 3 Use the addition property to write all terms involving the variable you are solving for on one side of the equation and the remaining terms on the other side.

STEP 4 Use the multiplication property to divide both sides of the equation by the coefficient of the variable you are solving for. This will isolate the desired variable on one side of the equation.

CHECK YOURSELF ANSWERS

1. $r = \frac{C}{2\pi}$ **2.** $y = \frac{c - ax}{b}$ **3.** $n = \frac{S + 360}{180}$ **4.** $h = \frac{3V}{\pi r^2}$

5. 7 percent   **6.** (1) $b = 8$ cm; (2) $b = \frac{2A - Bh}{h} = 8$ cm

7. (1) V/(L * W); (2) N * D + S

Build Your Skills

Solve each of the following formulas for the indicated variable.

1. $V = Bh$ for h Volume of a prism

2. $P = RB$ for B Percentage

3. $C = 2\pi r$ for r Circumference of a circle

4. $e = mc^2$ for m Mass/energy

5. $V = LWH$ for H Volume of a rectangular solid

6. $I = Prt$ for r Simple interest

7. $V = \pi r^2 h$ for h Volume of a cylinder

8. $S = 2\pi rh$ for r Lateral surface area of a cylinder

9. $V = \dfrac{1}{3}Bh$ for B Volume of a pyramid

10. $V = \dfrac{1}{3}\pi r^2 h$ for h Volume of a cone

11. $I = \dfrac{E}{R}$ for R Electric circuits

12. $V = \dfrac{KT}{P}$ for T Volume of a gas

13. $ax + b = 0$ for x Linear equation in one variable

14. $y = mx + b$ for x Slope-intercept form for a line

15. $P = 2L + 2W$ for W Perimeter of a rectangle

16. $ax + by = c$ for y Linear equation in two variables

17. $D = \dfrac{C - S}{n}$ for S Depreciation

18. $D = \dfrac{R(100 - x)}{100}$ for R Discounted price

19. $R = C(1 + r)$ for r Retail price

20. $A = P(1 + rt)$ for t Amount at simple interest

21. $A = \dfrac{1}{2}h(B + b)$ for b Area of a trapezoid

22. $L = a + (n - 1)d$ for n General term of an arithmetic progression

23. $F = \dfrac{9}{5}C + 32$ for C Celsius-to-Fahrenheit conversion

24. $C = \dfrac{5}{9}(F - 32)$ for F Fahrenheit-to-Celsius conversion

Transcribe Your Skills

25. In the equation $V = \dfrac{1}{3}\pi R^2 h$, values are given for V and R. Describe two strategies that can be used to find the value of h.

26. Describe how to solve the equation for y:

$$Ay + Cy = D$$

Think About These

Solve each of the following problems.

27. A rectangular solid has a base with length 6 cm and width 4 cm. If the volume of the solid is 72 cubic centimeters (cm^3), find the height of the solid. See Exercise 5.

28. A cylinder has a radius of 4 inches (in). If its volume is 144π cubic inches (in^3), what is the height of the cylinder? See Exercise 7.

29. A principal of $2000 was invested in a savings account for 4 years. If the interest earned for that period was $480, what was the interest rate? See Exercise 6.

30. The retail selling price of an item was $20.70. If its cost to the store was $18, what was the markup rate? See Exercise 19.

31. The radius of the base of a cone is 3 cm. If the volume of the cone is 24π cm^3, find the height of the cone. See Exercise 10.

32. The volume of a pyramid is 30 in^3. If the height of the pyramid is 6 in, find the area of its base. See Exercise 9.

33. If the perimeter of a rectangle is 60 ft and its length is 18 ft, find its width. See Exercise 15.

34. The yearly depreciation for a piece of machinery was $1500 over 8 years. If the cost of the machinery was $15,000, what was its salvage value? See Exercise 17.

35. A principal of $5000 was invested in a time-deposit account paying 9 percent annual interest. If the amount in the account at the end of a certain period was $7250, for how long was the money invested? See Exercise 20.

36. The area of a trapezoid is 36 in^2. If its height is 4 in and the length of one of the bases is 11 in, find the length of the other base. See Exercise 21.

Write a program expression for the given formula when it is solved for the indicated variable.

37. $V = KT/P$ for T

38. $A = P + Prt$ for t

Do each of the following problems.

39. The low temperature in New York City on a given day was 59°F, while the high temperature was 86°F. That is,

$$59 \leq F \leq 86$$

expresses the temperature range for the day. Write a double-inequality statement to give that range in degrees Celsius. See Exercise 23.

40. The low temperature in London on a given day was 5°C, while the high temperature was 15°C. That is,

$$5 \leq C \leq 15$$

expresses the temperature range for the day. Write a double-inequality statement to give that range in degrees Fahrenheit. See Exercise 24.

 41. A local incinerator plant must be fenced to meet its licensing permit. The width of the plant property is 250 meters (m). If the plant has 1500 m of available fencing, how long will the enclosed property be?

42. A small community is trying to pick a new landfill site. The landfill will have the general shape of a rectangular solid. A 20-year landfill for the community requires a volume of 2 million cubic meters (m^3). The available site is a parcel of property 300 by 200 m. How deep must the landfill be to meet the community's needs?

Skillscan (Elementary Algebra)

Translate each expression to algebraic symbols.

a. The sum of x and y **b.** The difference of w and 3
c. The product of 4 and b **d.** One-half of x
e. 6 less than m **f.** 5 more than twice s **g.** 7 times p
h. 2 less than 3 times x

2.3   Applications and Problem Solving

OBJECTIVE: To solve applications involving numbers, mixtures, interest, motion, revenue, and cost

We are now ready to use the equation-solving skills acquired in previous sections to solve various applications or word problems. Being able to extend these skills to problem solving is an important goal, and the procedures developed here are used throughout the rest of the text.

Although we consider applications from a variety of areas in this section, all are approached with the same five-step strategy presented here to begin the discussion.

SOLVING APPLICATIONS

STEP 1 Read the problem carefully to determine the unknown quantities.

STEP 2 Choose a variable to represent the unknown. Express all other unknowns in terms of this variable.

STEP 3 Translate the problem to the language of algebra to form an equation.

STEP 4 Solve the equation, and answer the question of the original problem.

STEP 5 Verify your solution by returning to the original problem.

Step 3, translating the given problem to the language of algebra, generally presents the most difficulty. Some commonly used phrases occur throughout this section, and the translation of those phrases to mathematical expressions is shown below and on the next page for your review.

Verbal Expression	Algebraic Expression
Addition	
The sum of two numbers	$x + y$
3 more than a number	$x + 3$
A number increased by 8	$x + 8$
Subtraction	
The difference of two numbers	$x - y$
7 less than a number	$x - 7$
A number decreased by 5	$x - 5$
Multiplication	
The product of two numbers	$x \cdot y$
4 times a number	$4x$
Twice a number	$2x$
One-half of a number	$\dfrac{1}{2}x$
Division	
The quotient of two numbers	$\dfrac{x}{y}$
A number divided by 3	$\dfrac{x}{3}$
One-half of a number	$\dfrac{x}{2}$

(handwritten annotations: $(x + y)$; sum 2 twice the numb; $2(x+y)$)

Let's now apply our five-step algorithm for problem solving. As the following example illustrates, we label each step of the process as we work through a solution. As you build your problem-solving skills throughout this section, you may find that practice helpful.

EXAMPLE 1 An Application Involving a Sum

One number is 2 less than 3 times another. If the sum of the numbers is 38, find the two numbers.

Step 1 *What are the unknown quantities?*

We want to find the two unknown numbers.

Step 2 *Choose a variable to represent the unknown.* Express other unknowns in terms of this variable.

help ↓

You might want to review the translation chart presented earlier.

Let x be the first number. Then

3 times x
↙
$3x - 2$ ← 2 less than

represents the second number.

Step 3 *Write an equation.*

The sum of the numbers ⏞ is
$$x + 3x - 2 = 38$$
First number   Second number

The equality symbol = generally represents the word "is."

Step 4 *Solve the equation.*

$$x + 3x - 2 = 38$$

We combine like terms on the left and proceed as before.

$$4x - 2 = 38$$
$$4x = 40$$
$$x = 10$$

The first number

For the value of the second number when x is 10 we have

$$3(10) - 2 = 28$$

The desired numbers are 10 and 28.

This is an *important part* of any problem-solving strategy. Always check the conditions of the original problem, *not* in the equations written. That way you will catch possible errors in steps 2, 3, and 4 of the process.

Step 5 From step 4, the numbers should be 10 and 28. Returning to the *original* problem, we verify that 28 is 2 less than 3 times 10 and that the sum of the numbers is 38.

Check Yourself 1

One number is 5 more than twice another. If the sum of the numbers is 59, what are the two numbers?

help ↓

Consecutive integers *follow one another* in the counting order.

Applications often involve the idea of consecutive integers or consecutive even or odd integers. Let's review for a moment before looking at an example.

Two consecutive integers are 7 and 8.

Three consecutive even integers are 8, 10, and 12.

Two consecutive odd integers are 13 and 15.

Three consecutive odd integers are -5, -3, and -1.

To represent consecutive integers (or consecutive even or odd integers) algebraically, let x be the first integer and use the following patterns.

Two consecutive integers:	$x, x + 1$
Three consecutive integers:	$x, x + 1, x + 2$
Two consecutive even integers:	$x, x + 2$
Three consecutive odd integers:	$x, x + 2, x + 4$

Note that consecutive even and odd integers are represented in an identical fashion. This is because if x is even, then $x + 2$ is even or if x is odd, then $x + 2$ is odd.

Consider the following example.

Help

EXAMPLE 2 An Application with Consecutive Integers

Find three consecutive integers such that the sum of the first and second integers is 12 more than the third integer.

Step 1 We want to find the three unknown integers.

Step 2 Since the integers are consecutive, they are represented by

$x \qquad x + 1 \qquad$ and $\qquad x + 2$

Step 3 The desired equation is

Sum of first and second integers   12 more than the third integer

$$x + (x + 1) = (x + 2) + 12$$

Step 4 $\quad x + x + 1 = x + 2 + 12$
$$2x + 1 = x + 14$$
$$x = 13 \qquad \text{The first integer}$$

If $x = 13$,

$x + 1 = 14 \qquad$ and $\qquad x + 2 = 15 \qquad$ The second and third integers

The three integers are 13, 14, and 15.

Step 5 Check. The sum of the first and second integers is 27, and 12 more than the third integer, 15, is 27. The conditions of the problem are met, and the solutions are verified.

$x + x + 4 = x + 2 - 8$

Check Yourself 2

Find three consecutive even integers such that the sum of the first and third integers is 8 less than the second integer.

Applications are often drawn from the area of geometry. The following example illustrates.

EXAMPLE 3 An Application Involving Geometry

The length of a rectangle is 5 cm less than 3 times its width. If the perimeter of the rectangle is 54 cm, find the length and width of the rectangle.

Step 1 The unknowns are the dimensions (width and length) of the rectangle.

Sometimes it helps to choose letters that "suggest" the unknown quantity. Here we use w for *width*.

Step 2 Let w be the width. Then

5 less than 3 times the width

$$3w - 5$$

is the length.

Whenever geometric figures are involved, make a sketch of the information in the problem. It will help you in the next step of the process. Here we have

Sketching the information is an important part of any problem-solving strategy. Don't forget to use this tool whenever it is applicable.

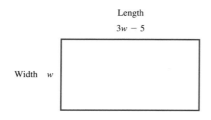

Step 3 From $P = 2W + 2L$ and our sketch above, we can write

The appropriate geometric formula is $P = 2W + 2L$, where P is the perimeter, W the width, and L the length.

$$54 = 2w + 2(3w - 5)$$

Perimeter   Width   Length

Step 4
$$54 = 2w + 2(3w - 5)$$
$$54 = 2w + 6w - 10$$
$$64 = 8w$$
$$8 = w \qquad \text{The width is 8 cm.}$$

Also $3w - 5 = 19$ The length is 19 cm.

Step 5 We leave the checking of these results to you.

Check Yourself 3

The length of a rectangular yard is 10 ft more than twice its width. If the perimeter of the yard is 110 ft, what are the dimensions of the rectangle?

We now turn to a group of applications that can be classified as *mixture problems*. The common theme is that two different quantities will be combined to form a single mixture. Our first type of mixture problem deals with combining quantities that have different monetary values. In these cases, the equation of step 3 will be formed by relating the value of each quantity to the total value of the mixture or combination. This idea is used in the following example.

EXAMPLE 4   An Application with Money

At the end of a day, a cashier had 115 $5 and $10 bills in his register. If the value of the bills was $800, how many of each denomination did he have?

Step 1 We want to find the number of $5 and $10 bills.

Step 2 Let x be the number of $5 bills. Then

Total number of bills
minus number of $5 bills

$$115 - x$$

is the number of $10 bills.

This is a common strategy and will occur frequently.

   Organizing the information is a key part of any problem-solving strategy. A table is often a useful tool for presenting the given information in a problem. A typical table is shown below:

Unit Value	Quantity	Total Value
$5	x	$5x$
$10	$115 - x$	$10(115 - x)$
Total	115	800

Note that the total value is the product of the unit value and the quantity.

Step 3 As we pointed out, we now relate the value for each of the quantities to the total value of the mixture or combination. From our table we see that

$$5x + 10(115 - x) = 800$$

Step 4 Solving as before, we find that

$$x = 70 \qquad \text{Number of \$5 bills}$$

Also $115 - x = 45$ Number of $10 bills

Step 5 You should check this solution by verifying the value of the number of bills found in step 4.

Check Yourself 4

A total of 350 tickets—$5 adult tickets and $3 student tickets—were sold for a play. If the total revenue was $1500, how many of each type of ticket were sold?

   Here is a related application. Compare the steps of Examples 4 and 5 to make sure you see the similarities in the problem-solving strategy.

EXAMPLE 5   A Mixture Problem Involving Weight

A coffee merchant has coffee beans selling for $3 per pound and for $5 per pound. The two types of beans are to be mixed to provide 100 pounds (lb) of a mixture selling for $4.50 per pound. How much of each type of coffee bean should be used to form the mixture?

We will see this problem again in Chapter 4 when we consider a different problem-solving strategy involving the use of two variables.

Step 1 The unknowns are the amounts of the two types of beans, bean 1 at $3 per pound and bean 2 at $5 per pound.

Step 2 Let x represent the amount of bean 1 at $3 per pound. Then $100 - x$ represents the amount of bean 2. Again, presenting the information in table form will help you in the next step of the process.

	Unit Value	Quantity	Total Value
Bean 1	$3	x	$3x$
Bean 2	$5	$100 - x$	$5(100 - x)$
Total	$4.50	100	4.5(100)

where did the 100 get (handwritten)

Step 3 As before, the desired equation comes from relating the value of each type of bean to the total value of the mixture.

$$3x + 5(100 - x) = 4.5(100)$$

Value of bean 1   Value of bean 2   Total value

Step 4 Solving, we find that

$$x = 25 \text{ lb} \quad \text{bean 1}$$

Also $100 - x = 75$ lb   bean 2

Step 5 We leave the checking of this problem to you.

Check Yourself 5

Peanuts which sell for $2.40 per pound and cashews which sell for $6 per pound are to be mixed to form a 60-lb mixture selling for $3 per pound. How much of each type of nut should be used?

Our next type of mixture problem involves combining two quantities with different concentrations of the same substance. In this case, the equation of step 3 relates the amount of the substance in the two quantities being combined to the total amount in the mixture. Consider the following example.

EXAMPLE 6 A Mixture Problem Involving Volume

Again, we will return to this problem in Chapter 4, approaching the solution by using two variables.

A pharmacist has a 35% alcohol solution and a 60% alcohol solution. How much of each solution should she use to form 150 milliliters (mL) of a 50% solution?

Step 1 The unknowns are the amounts of each solution to be used in forming the mixture.

Step 2 Let x be the amount of the 35% solution. Then $150 - x$ represents the amount of the 60% solution.

A table similar to that used in Example 5 will help us write an equation for the solution.

Alcohol	Quantity	Total Alcohol
35%	x	$0.35x$
60%	$150 - x$	$0.6(150 - x)$
50%	150	$0.5(150)$

A picture of the problem might look like this:

Step 3 As we mentioned above, the amount of alcohol in the two quantities being combined must equal the amount of alcohol in the mixture. From our sketch we have

$$0.35x + 0.60(150 - x) = 0.50(150)$$

Step 4 We multiply by 100 to clear the decimals in the equation of step 3.

$$35x + 60(150 - x) = 50(150)$$

Now solving as before gives

$x = 60$ mL   35% solution

Also $150 - x = 90$ mL   60% solution

Step 5 To verify this result, show that the amount of alcohol in the first solution, $(0.35)(60)$, added to the amount of alcohol in the second, $(0.60)(90)$, equals the amount in the mixture, $(0.50)(150)$.

Check Yourself 6

A chemist wishes to combine a 12% acid solution with a 60% solution to form 200 mL of a 30% acid solution. How much of each solution should be used?

Let's look now at an application involving simple interest. Typically these applications will involve two sums of money invested at different rates. Our equation will be formed by noting that the total interest from the investments must be equal to the sum of the interest from the individual amounts. Let's review for a moment.

You will see many similarities to the mixture problems we just considered.

The amount of money that is invested is called the *principal.*

The amount of money that the principal earns is called the *interest.*

The rate at which the principal earns interest is called the *interest rate.*

These three quantities are related by the formula

$$I = P \cdot R \cdot t$$

Interest   Principal   Rate   Time

For 1 year, $t = 1$ and so

$$I = P \cdot R$$

In words, interest is the product of the interest rate and the principal.
Let's consider an application of this formula in the next example.

EXAMPLE 7 An Application Involving Money

Jovita divided an inheritance of $15,000 into two investments, one a savings account paying 6 percent interest and the other a time deposit paying 8 percent. If the annual interest was $1020 from the two accounts, how much did she have invested at each rate?

Step 1 The unknowns are the amounts invested at 6 and 8 percent.

Step 2 Let x be the amount invested at 6 percent. Then $15,000 - x$ is the amount at 8 percent.

Once again, a table of the given information can be very helpful in organizing the problem.

In this table, interest is the product of the principal and the rate. Compare this to the table of Example 4.

Rate	Principal	Interest
6 percent	x	$0.06x$
8 percent	$15,000 - x$	$0.08(15,000 - x)$
Total	$15,000$	1020

Step 3 From the third column of our table we can write

$$0.06x + 0.08(15,000 - x) = 1020$$

Step 4 Multiplying by 100 to clear the decimals, we see that the equation of step 3 now has the form

$$6x + 8(15,000 - x) = 102,000$$

and solving as before, we have

$$x = 9000 \qquad \text{Amount at 6 percent}$$

$$15,000 - x = 6000 \qquad \text{Amount at 8 percent}$$

Step 5 We leave the checking again to you. Verify that the sum of the interest earned from the two accounts is $1020.

Check Yourself 7

Vanessa split an investment of $8000 into a savings account paying 5 percent interest and a mutual fund paying 9 percent. If her annual interest from the two accounts was $640, how much did she have invested at each rate?

Our next two applications fall into the category of *uniform-motion problems.* Uniform motion means that the speed of an object does not change over a certain time. To solve these problems, we will need a relationship between the distance traveled, represented by d, the rate (or speed) of the travel r, and the time of that travel t.

In general, for the distance traveled d, rate r, and time t,

$$d = r \cdot t$$

Distance   Rate   Time

The solution for uniform-motion problems always involves a relationship between the distance, rate, or time. That relationship is then used to form the necessary equation in step 3.

Consider the following examples.

EXAMPLE 8 A Distance Problem

On Friday morning Jason drove from his house to the beach in $3\frac{1}{2}$ hours (h). In coming back on Sunday afternoon, heavy traffic slowed his speed by 6 miles per hour (mi/h), and the trip took 4 h. Find his average rate in each direction.

Step 1 We want the rate in each direction.

Step 2 Let r be Jason's rate to the beach. Then $r - 6$ is his return rate.

It is always a good idea to sketch the given information in a uniform-motion problem. Here we might have

Going r (mi/h) for $3\frac{1}{2}$ h

Coming back $r - 6$ (mi/h) for 4 h

Some students also find using a table helpful in motion problems. Here we have

Rate	Time	Distance
r	$3\frac{1}{2}$	$\frac{7}{2}r$
$r - 6$	4	$4(r - 6)$

Step 3 Since we know that the distance is the same each way, our equation is formed by using the fact that the product of the rate and the time must be the same each way. We have

$$\frac{7}{2}r = 4(r - 6)$$

Distance going: $\frac{7}{2}r$

Distance coming back: $4(r - 6)$

Step 4 We now solve the equation formed in step 3.

$$\frac{7}{2}r = 4r - 24$$

$$7r = 8r - 48$$

$$-r = -48$$

$$r = 48 \text{ mi/h}$$

$$r - 6 = 42 \text{ mi/h}$$

The rate going

The rate coming back

Step 5 You should check this result for yourself.

Check Yourself 8

At 9:00 A.M., Muriel left her house to visit friends in another city. One hour later, Martin decided to join her on the visit and left along the same route, traveling 13 mi/h faster than Muriel. If Martin caught up with Muriel at 1:00 P.M., what was each person's average rate?

In the previous example and exercise, the unknown quantity was the rate of travel. Let's consider another variation of the motion problem in which the time of the travel is the unknown.

EXAMPLE 9 A Distance Problem

At noon, Linda decides to leave Las Vegas for Los Angeles, driving at 50 mi/h. At 1 P.M., Lou leaves Los Angeles for Las Vegas, driving at 55 mi/h along the same route. If the two cities are 260 mi apart, at what time will Linda and Lou meet?

Step 1 Here the time until they meet is the unknown.

Step 2 Let t be the time of Linda's travel until they meet. Then $t - 1$ is the time of Lou's travel.

Lou left 1 h later!

Again a sketch of the given information is a good idea.

You might want to use a table such as that in Example 7.

Los Angeles   55 mi/h for $(t - 1)$ h   50 mi/h for t h   Las Vegas

Lou   Linda

260 mi

Step 3 Again, since distance is the product of rate and time, from the sketch of step 2, we have

Linda's distance: $50t$

Lou's distance: $55(t - 1)$

Since the sum of those distances must be 260 mi, we can write

$$50t + 55(t - 1) = 260$$

Step 4 Solving the equation as before yields

$$t = 3 \text{ h}$$

Finally, since Linda left Las Vegas at noon, the two will meet at 3 P.M.

Step 5 Again we will leave the checking of this result to you. Verify that the sum of the distances that the two travel is 260 mi.

Check Yourself 9

At 7 A.M., a freight train leaves a city, traveling west, averaging 42 mi/h. Two hours later an express train leaves the same station, traveling east at 68 mi/h. At what time will the two trains be exactly 359 mi apart?

To complete this section, we consider an application from business. But we will need some new terminology. The total cost of manufacturing an item consists of two types of costs. First is the *fixed cost,* sometimes called the *overhead.* This includes costs such as product design, rent, and utilities. In general, this cost is constant and does not change with the number of items produced. Second is the *variable cost,* which is a cost per item, and it includes costs such as material, labor, and shipping. The variable cost depends on the number of items being produced.

A typical cost equation might be

$$C = 3.50x + 5000$$

Variable cost   Fixed cost

The total *revenue* is the income the company makes. It is calculated as the product of the selling price of the item and the number of items sold.

A typical revenue equation might be

$$R = 7.50x$$

Selling price   Number of
per item   items sold

The *break-even point* is that point at which the revenue equals the cost (the company would exactly break even without a profit or a loss).

Let's apply these concepts in the following example.

EXAMPLE 10 Finding the Break-Even Point

A firm producing videocassette tapes finds that its fixed cost is $5000 per month and that its variable cost is $3.50 per tape. The cost of producing x tapes is then given by

$$C = 3.50x + 5000$$

The firm can sell the tapes at $7.50 each, and so the revenue from selling x tapes is

$$R = 7.50x$$

Find the break-even point.

Solution Since the break-even point is that point where the revenue equals the cost, or $R = C$, from our given equations we have

$$7.50x = 3.50x + 5000$$

Revenue   Cost

Solving as before gives

$$4x = 5000$$

$$x = 1250$$

The firm will break even (no profit or loss) by producing and selling exactly 1250 tapes each month.

Check Yourself 10

A firm producing lawn chairs has fixed costs of $525 per week. The variable cost is $8.50 per chair, and the revenue per chair is $15.50. This means that the cost equation is

$$C = 8.50x + 525$$

and the revenue equation is

$$R = 15.50x$$

Find the break-even point.

CHECK YOURSELF ANSWERS

1. 18, 41   **2.** −10, −8, and −6   **3.** Width 15 ft, length 40 ft   **4.** 225 adult tickets, 125 student tickets   **5.** Peanuts 50 lb, cashews 10 lb   **6.** 12% solution: 125 mL, 60% solution: 75 mL   **7.** 5 percent: $2000, 9 percent: $6000   **8.** Muriel: 39 mi/h, Martin: 52 mi/h   **9.** 11:30 A.M.   **10.** 75 chairs

---

2.3 EXERCISES

Build Your Skills

Solve each of the following applications. Be sure to show the equation used for your solution.*

1. One number is 3 more than 4 times another. If the sum of the numbers is 38, find the two numbers.

2. One number is 5 less than twice another. The sum of the two numbers is 43. What are the two numbers?

3. One number is 3 less than another. If the sum of the smaller number and twice the larger number is 45, find the two numbers.

4. One number is 2 more than 3 times another. The sum of the smaller number and 4 times the larger number is 73. What are the two numbers?

5. Find two consecutive integers such that the sum of twice the first integer and 3 times the second integer is 28.

6. Find two consecutive odd integers such that the first integer is 5 more than twice the second.

7. Find three consecutive even integers such that the sum of the first integer and 3 times the third integer is 64 more than the second integer.

8. Find three consecutive integers such that the sum of the first integer and the second integer, increased by 5, is 11 less than 3 times the third integer.

9. The length of a playing field is 5 ft less than twice its width. If the perimeter of the field is 230 ft, find the length and width of the field.

10. The length of a rectangle is 4 cm more than 3 times its width. The perimeter of the rectangle is 56 cm. What are the dimensions of the rectangle?

11. The width of a rectangle is $\dfrac{3}{4}$ of the length. Find the dimensions of the rectangle if the perimeter is 70 cm.

12. The perimeter of a rectangle is 35 in more than the length. The width is 15 in. Find the perimeter.

*Problems marked with an asterisk will be seen again in Chapter 4, where we discuss an approach to problem solving that involves the use of two variables and a system of equations.

13. A newspaper sales stand has 200 coins at the end of a day, all dimes and quarters. If the value of the coins is $42.50, how many of each denomination coin were in the stand?

14. A grocer sold 450 lb of ground beef in one day. Two grades of beef, regular at $1.20 per pound and lean at $1.40 per pound, were sold. If the total sales were $592, how much of each type of ground beef was sold?

*15. Theater tickets were sold for $7.50 on the main floor and $5 in the balcony. The total revenue was $3250, and 100 more main-floor tickets were sold than balcony tickets. Find the number of each type of ticket sold.

16. A bank teller had 125 $10 and $20 bills to start a day. If the value of the bills was $1650, how many of each denomination bill did he have?

*17. A coffee retailer has two grades of decaffeinated beans, one selling for $4 per pound, the other for $6.50 per pound. She wishes to blend the beans to form a 150-lb mixture which will sell for $4.75 per pound. How many pounds of each grade of bean should be used in the mixture?

18. A candy merchant sells caramels at $1.30 per pound and mints at $1.90 per pound. To form a 120-lb mixture which will sell for $1.40 per pound, how many pounds of each type of candy should be used?

19. A laboratory technician wishes to combine a 15% saline solution with a 45% saline solution to form 300 mL of a 35% solution. How much of each solution should be used?

*20. A chemist mixes a 10% acid solution with a 50% acid solution to form 400 mL of a 40% solution. How much of each solution should be used in the mixture?

21. A pharmacist wants to combine a pure alcohol solution with 200 cm³ of a 25% alcohol solution to increase the concentration to 40%. How much of the pure alcohol solution should she add?

22. Jan wants to combine a 75% glycerin solution with 300 mL of a 15% solution to increase the concentration to 25%. How much of the 75% solution should be added?

*23. José decided to divide $12,000 into two investments, one a time deposit paying 8 percent annual interest and the other a bond which pays 9 percent. If his annual interest was $1010, how much did he invest at each rate?

24. Robin invested a certain amount of money in a savings account paying 6 percent interest. She then invested $1600 more than that amount in a time deposit that paid 8 percent. If her total yearly income from the investments was $828, how much did she have invested at each rate?

25. Marcus invested a portion of his $15,500 inheritance in a mutual fund paying 7 percent. The remainder was used to purchase a stock paying 10 percent. If his annual interest from the stock was $700 more than that from the mutual fund, how much did he have invested at each rate?

26. Benson sold his house for a profit of $12,000 and decided to invest that profit in a bond paying 7 percent and a money market fund paying 11 percent. If the annual interest from the money market fund was $60 more than that from the bond, how much did he have invested at each rate?

27. On her way to a business meeting, Connie drove on the freeway, and the trip took 3 h. Returning, she decided to take a side road, and her speed along that route averaged 8 mi/h slower than on the freeway. If her return trip took $3\frac{1}{2}$ h and the distance driven was the same each way, find her average speed in each direction.

28. Michele was required to make a cross-country flight in training for her pilot's license. In flying from her home airport, a steady 20 mi/h wind was behind her, and the first leg of the trip took 5 h. In returning against the same wind, the flight took 7 h. Find the plane's speed in still air and the distance traveled on each leg of the flight.

29. Mark was driving on a 245-mi trip. For the first 3 h he traveled at a steady speed. At that point, realizing that he would be late at his destination, he increased his speed by 10 mi/h for the remaining 2 h of the trip. What was his driving speed for each portion of the trip?

30. Randolph can drive to work in 45 minutes (min) whereas if he decides to take the bus, the same trip takes 1 h 15 min. If the average rate of the bus is 16 mi/h slower than his driving rate, how far does he travel to work?

31. At 10 A.M., Tony left Boston for Baltimore, traveling at 45 mi/h. One hour later, Sandra left Baltimore for Boston, traveling at 50 mi/h along the same route. If the cities are 425 mi apart, at what time did Tony and Sandra meet?

32. A passenger bus left a station at 1 P.M., traveling north at an average rate of 44 mi/h. One hour later, a second bus left the same station, traveling south at a rate of 48 mi/h. At what time will the two buses be 274 mi apart?

33. On Tuesday, Carla drove to a conference and averaged 48 mi/h for the trip. In returning on Thursday, road construction slowed her average speed by 12 mi/h. If her total driving time was 14 h, what was her driving time each way and how far was the conference from her home?

34. At 9:00 A.M., Max left on a trip, traveling at 45 mi/h. One-half hour later, Catherine discovered that Max forgot his luggage, and she left along the same route, traveling at 54 mi/h, to catch up with him. When did Catherine catch up with Max?

35. A firm producing flashlights finds that its fixed cost is $2400 per week and its variable cost is $4.50 per flashlight. The revenue is $7.50 per flashlight, so the cost and revenue equations are, respectively,

$$C = 4.50x + 2400 \quad \text{and} \quad R = 7.50x$$

Find the break-even point for the firm.

36. A company that produces portable television sets determines that its fixed cost is $8750 per month. The variable cost is $70 per set, while the revenue is $105 per set. The cost and revenue equations, respectively, are given by

$$C = 70x + 8750 \quad \text{and} \quad R = 105x$$

Find the number of sets the company must produce and sell in order to break even.

37. A firm which produces scientific calculators has a fixed cost of $1225 per week and a variable cost of $6.50 per calculator. If the company can sell the calculators for $13.50, find the break-even point.

38. A publisher finds that the fixed costs associated with a new paperback are $17,500. The book costs $2 to produce and will sell for $5.50. Find the publisher's break-even point.

For Exercises 39 and 40, you will need the following information.

An important economic application involves supply and demand. The number of units of a commodity that manufacturers are willing to *supply S* is related to the market price *p*. A typical supply equation is

$$S = 30p - 285 \quad (1)$$

(Generally the supply increases as the price increases.)

The number of units that consumers are willing to buy is called the *demand D*, and it is also related to the market price. A typical demand equation is

$$D = -55p + 1500 \quad (2)$$

(Generally the demand decreases as the price increases.)

The price where the supply and demand are equal (or *S = D*) is called the *equilibrium price* for the commodity.

39. The supply and demand equations for a certain model portable radio are given in Equations (1) and (2) above. Find the equilibrium price for the radio.

40. The supply and demand equations for a certain type of computer monitor are

$$S = 25p - 1850 \quad \text{and} \quad D = -40p + 5300$$

Find the equilibrium price for the monitor.

41. An existing landfill is built on a rectangular parcel of property. The width of the landfill is 100 m. If the landfill permit is changed to increase the length of the landfill by 50 m, its area will be increased by 5000 square meters (m^2) and its useful life span will be extended by 20 percent. What is the current length of the landfill?

42. A recycler delivers a 15-t truckload of cardboard and newspaper to his buyer and receives $350. If cardboard is worth $35 dollars per ton and newspaper is worth $10 per ton, how many tons of each commodity were in the load?

43. A recycler delivers 7 t of cardboard and 13 t of white ledger paper to her buyer and receives $830. If cardboard is worth $10 less per ton than white ledger paper, how much is each commodity worth?

44. Steel cans are worth $15 more per ton than glass jars. Plastic soda bottles are worth 3 times as much as glass jars. The combined value of 5 t of each of these materials is $1075. What is the price per ton of each material?

45. A city has 37 recycling companies, but they don't all accept the same materials. Some accept cardboard, others accept plastic, and still others accept newspaper. There are three more cardboard recyclers than plastic recyclers and twice as many newspaper recyclers as cardboard recyclers. How many of each type of company are there?

46. Steel cans are worth 5 percent as much as aluminum cans. If a 30-t load of cans is worth $15,750 and is half aluminum and half steel by weight, how much is each metal worth per ton?

Skillscan (Section 1.3)

Graph each of the following inequalities.

a. $x > 1$ **b.** $x < 3$ **c.** $x < -3$ **d.** $x > -\dfrac{7}{2}$

e. $x \geq 6$ **f.** $x \leq \dfrac{3}{2}$ **g.** $x \leq -\dfrac{5}{3}$ **h.** $x \geq -\dfrac{9}{4}$

Solving Linear Inequalities   **2.4**

TAPE IN2

OBJECTIVES: 1. To solve and graph the solution sets for linear inequalities
2. To solve and graph the solution sets for compound inequalities

In Section 2.1 we defined a linear equation in one variable as an equation that could be written in the form

$$ax + b = 0$$

where a and b are real numbers and $a \neq 0$.
   A linear inequality in one variable is defined in a similar fashion.

> A *linear inequality* can be written in the form
>
> **ax + b < 0**
>
> where *a* and *b* are real numbers and *a ≠ 0*.

The inequality symbol $<$ can be replaced with any of the other inequality symbols $>$, $\leq$, or $\geq$, so that

$$ax + b > 0 \qquad ax + b \leq 0 \qquad \text{and} \qquad ax + b \geq 0$$

are also linear inequalities.
   Fortunately your experience with linear equations in Section 2.1 provides the groundwork for solving linear inequalities. You will see many similarities.
   A *solution* of a linear inequality in one variable is any real number that will make the inequality a true statement when the variable is replaced by that number. The *solution set* of a linear inequality is the set of all solutions.

We can also say that a solution satisfies the inequality.

   Our strategy for solving linear inequalities is, for the most part, identical to that used for solving linear equations. We write a sequence of equivalent inequalities in order to isolate the variable on one side of the inequality symbol.
   Writing equivalent inequalities for this purpose requires two properties. First:

> **ADDITION PROPERTY OF INEQUALITY**
>
> If $a < b$
>
> then $a + c < b + c$

Adding the same quantity to both sides of an inequality gives an equivalent inequality.

This addition property is similar to that seen earlier in solving equations. As before, since subtraction is defined in terms of addition, the property also allows us to subtract the same quantity from both sides of an inequality without changing the solutions.
   Our second property, dealing with multiplication, has an important difference. We begin by writing the true inequality

$$2 < 3$$

Multiplying both sides of that inequality by the same *positive* number, say 3, gives

$$3(2) < 3(3)$$

$$6 < 9 \qquad \text{Another true statement!}$$

Note that the new inequality has the same sense (points in the same direction) as the original inequality.

However, if we now multiply both sides of the inequality by a *negative* number, say -3, we have

$$-3(2) < -3(3)$$

$$-6 < -9 \qquad \text{A \textit{false} statement!}$$

To make this a true statement, we must *reverse the sense* of the inequality to write

$$-6 > -9$$

This suggests that if we multiply both sides of an inequality by a negative number, we must reverse the sense of the inequality to form an equivalent inequality. From this discussion we can now state our second property.

Multiplying both sides of an inequality by a positive number gives an equivalent inequality.

Multiplying both sides of an inequality by a negative number and reversing the sense give an equivalent inequality.

MULTIPLICATION PROPERTY OF INEQUALITY

If	$a < b$	
then	$ac < bc$	where c is a *positive* number $(c > 0)$
and	$ac > bc$	where c is a *negative* number $(c < 0)$

Again, since division is defined in terms of multiplication, this property also allows us to divide both sides of an inequality by the same nonzero number. If, for example, we wish to rewrite the statement $4x < 8$, we could either multiply both sides by $\frac{1}{4}$ or divide both sides by 4. Remember to reverse the sense of the inequality if you divide by a negative number.

We will use these properties in solving inequalities in much the same way as we did in solving equations, with the one significant difference pointed out above.

The following examples illustrate the solution process for linear inequalities.

EXAMPLE 1 Graphing Linear Inequalities

Solve and graph the inequality

$$4x - 3 < 5 \tag{1}$$

Solution First, we add 3 to both sides.

As in solving equations, we apply the addition property and *then* the multiplication property to isolate the variable.

$$4x - 3 + 3 < 5 + 3$$

$$4x < 8 \tag{2}$$

We now divide both sides of the inequality by 4, to isolate the variable on the left.

$$\frac{4x}{4} < \frac{8}{4}$$

Here we use the division property.

$$x < 2 \tag{3}$$

Since inequalities (1), (2), and (3) are all equivalent, the solution set for the original inequality consists of all numbers that are less than 2. That set can be written

$$\{x \mid x < 2\}$$

The graph of the solution set is

The *open* circle means 2 is *not* included in the solution set.

Check Yourself 1

Solve and graph the inequality.

$$5x + 7 > 22$$

EXAMPLE 2 Graphing Linear Inequalities

Solve and graph the inequality

$$3x - 5 \geq 5x + 3$$

Solution Add 5 to both sides.

$$3x - 5 + 5 \geq 5x + 3 + 5$$

$$3x \geq 5x + 8$$

Subtract $5x$ from both sides.

$$3x - 5x \geq 5x - 5x + 8$$

$$-2x \geq 8$$

We must now divide both sides by -2. Since the divisor is negative, we reverse the sense of the inequality.

$$\frac{-2x}{-2} \leq \frac{8}{-2}$$

$$x \leq -4$$

The solution set consists of all numbers less than or equal to -4 and is graphed below.

$$\{x \mid x \leq -4\}$$

Here we use the *closed* circle to indicate that -4 is *included* in the solution set.

Check Yourself 2

Solve and graph.

$4x + 3 \leq 7x - 12$

In working with more complicated inequalities, as was the case with equations, any signs of grouping must be removed, and like terms combined, before the properties of inequalities are applied to isolate the variable.

EXAMPLE 3 Graphing Linear Inequalities with Parentheses

Solve and graph

$5 - 3(x - 2) \leq 1 - x$

Solution First, remove the parentheses on the left and combine like terms.

$5 - 3x + 6 \leq 1 - x$

$-3x + 11 \leq 1 - x$

We now proceed as before.

$-3x + 11 - 11 \leq 1 - 11 - x$ Subtract 11.

$-3x \leq -10 - x$

$-3x + x \leq -10 - x + x$ Add x.

$-2x \leq -10$

$\dfrac{-2x}{-2} \geq \dfrac{-10}{-2}$ Divide by -2, reversing the sense of the inequality.

$x \geq 5$

The solution set $\{x \mid x \geq 5\}$ is graphed below.

Check Yourself 3

Solve and graph.

$4 - 2(x + 5) \geq -9 - 4x$

If fractions are involved in an inequality, you should apply the multiplication property to clear the inequality of fractions as your first step.

EXAMPLE 4 Graphing Linear Inequalities with Fractions

Solve and graph the inequality.

$$\frac{3x+2}{6} - 1 < \frac{x}{3}$$

$$6\left(\frac{3x+2}{6} - 1\right) < 6\left(\frac{x}{3}\right)$$

Multiply both sides by 6, the LCM of 6 and 3.

$$6\left(\frac{3x+2}{6}\right) - 6(1) < 6\left(\frac{x}{3}\right)$$

Apply the distributive property on the left.

$$3x + 2 - 6 < 2x$$

The inequality is now cleared of fractions, and we proceed as before:

$$3x - 4 < 2x$$

$$3x < 2x + 4$$

$$x < 4$$

The solution set is graphed below.

$$\{x \mid x < 4\}$$

Check Yourself 4

Solve and graph the inequality.

$$\frac{5x-1}{4} - 2 > \frac{x}{2}$$

The following algorithm summarizes our work thus far in this section in solving linear inequalities.

SOLVING LINEAR INEQUALITIES IN ONE VARIABLE

STEP 1 Clear the inequality statement of any fractions by using the multiplication property.

STEP 2 Remove any grouping symbols, and combine like terms on each side of the equation.

STEP 3 Apply the addition property to write an equivalent inequality with the variable term on one side of the inequality and the constant term on the other side.

> **STEP 4**   Apply the multiplication property to write an equivalent inequality with the variable isolated on one side of the inequality. Be sure to reverse the sense of the inequality if you multiply or divide by a negative number.
>
> **STEP 5**   Graph the solution set of the *original* inequality.

Let's now consider two types of inequality statements which arise frequently in mathematics. Consider a statement such as

$$-2 < x < 5$$

It is called a *double inequality* because it combines the two inequalities

This is also called a *compound inequality*. In the double inequality the word "and" is understood.

$$-2 < x \qquad \text{and} \qquad x < 5$$

To solve a double inequality means to isolate the variable in the middle term, as the following example illustrates.

EXAMPLE 5 Graphing Double Inequalities

Solve and graph the double inequality

$$-3 \leq 2x + 1 \leq 7$$

Solution First, we subtract 1 from each of the three members of the double inequality.

We are really applying the additive property to each of the *two* inequalities that make up the double-inequality statement.

$$-3 - 1 \leq 2x + 1 - 1 \leq 7 - 1$$

or

$$-4 \leq 2x \leq 6$$

We now divide by 2 to isolate the variable x.

$$-\frac{4}{2} \leq \frac{2x}{2} \leq \frac{6}{2}$$

$$-2 \leq \ x \ \leq 3$$

The solution set consists of all numbers between -2 and 3, including -2 and 3, and is written

$$\{x \mid -2 \leq x \leq 3\}$$

That set is graphed below.

Because the circles at both ends of the interval are "closed," we sometimes call this a *closed interval*.

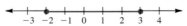

Our solution set is equivalent to

$$\{x \mid x \geq -2 \text{ and } x \leq 3\}$$

Look at the individual graphs.

$\{x \mid x \geq -2\}$

$\{x \mid x \leq 3\}$

$\{x \mid x \geq -2 \text{ and } x \leq 3\}$

Because the connecting word is "and," we want the *intersection* of the sets, that is, those numbers common to both sets.

Using set notation, we write this as

$\{x \mid x \geq -2\} \cap \{x \mid x \leq 3\}$

The $\cap$ represents the intersection of the two sets.

Check Yourself 5

Solve and graph the inequality.

$-5 \leq 2x - 3 \leq 3$

When the coefficient to the variable is a negative number, care must be taken in isolating the variable.

EXAMPLE 6 Graphing Double Inequalities

Solve and graph the double inequality

$-3 < 4 - 3x < 13$

Solution Subtract 4 from each member of the inequality.

$-7 < -3x < 9$

Now we must divide by -3. The sense of the inequality is reversed whenever we divide by a negative number.

$$\frac{-7}{-3} > \frac{-3x}{-3} > \frac{9}{-3}$$

$$\frac{7}{3} > \quad x \quad > -3$$

In the standard smallest-to-largest format, we have

$$-3 < x < \frac{7}{3}$$

The solution consists of all numbers between -3 and $\dfrac{7}{3}$ and is written

$$\left\{ x \mid -3 < x < \frac{7}{3} \right\}$$

That set is graphed below.

The circles are open at both ends; this is sometimes called an *open interval*.

Check Yourself 6

Solve and graph the double inequality.

$$-5 < 3 - 2x < 5$$

A compound inequality may also consist of two inequality statements connected by the word ''or.'' The following example illustrates the solution of that type of compound inequality.

EXAMPLE 7 Graphing Compound Inequalities

Solve and graph the inequality

$$2x - 3 < -5 \qquad \text{or} \qquad 2x - 3 > 5$$

Solution In this case we must work with each of the inequalities *separately*.

$$2x - 3 < -5 \qquad \text{or} \qquad 2x - 3 > 5$$

Add 3. $\qquad\qquad 2x < -2 \qquad\qquad\qquad 2x > 8$

Divide by 2. $\qquad\qquad x < -1 \qquad\qquad\qquad x > 4$

The graph of the solution set is shown.

$$\{x \mid x < -1 \text{ or } x > 4\}$$

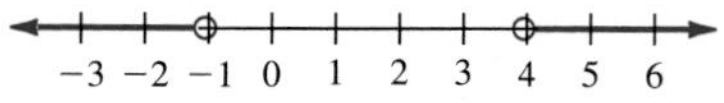

In set notation we write the union as

$$\{x \mid x < -1\} \cup \{x \mid x > 4\}$$

Note that since the connecting word is ''or'' in this case, the solution set of the original inequality is the *union* of the two sets, that is, those numbers that belong to either or both of the sets.

Check Yourself 7

Solve and graph the inequality.

$$3x - 4 \leq -7 \qquad \text{or} \qquad 3x - 4 \geq 7$$

The following chart summarizes our discussion of solving linear inequalities and the nature of the solution sets of the types of inequalities we have considered in this section.

Type of Inequality	Graph of Solution Set

$ax + b < c$

If $a > 0$:

If $a < 0$:

$-c < ax + b < c$

$ax + b < -c$ or $ax + b > c$

CHECK YOURSELF ANSWERS

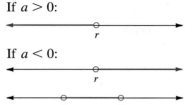

1. $x > 3$

2. $x \geq 5$

3. $x \geq -\dfrac{3}{2}$

4. $x > 3$

5. $-1 \leq x \leq 3$

6. $-1 < x < 4$

7. $x \leq -1$ or $x \geq \dfrac{11}{3}$

2.4 EXERCISES

Build Your Skills

Solve each of the following inequalities. Then graph the solution set.

1. $x - 2 < 5$

2. $x + 3 > -4$

3. $x + 5 \geq 3$

4. $x - 4 \leq -2$

5. $5x > 25$

6. $4x < -12$

7. $-3x \leq -15$

8. $-7x > 21$

9. $2x + 3 < 10$

10. $5x - 3 \leq 17$

11. $-2x - 7 \geq 5$

12. $-3x + 4 < -4$

13. $5 - 3x < 14$

14. $2 - 5x \geq 22$

15. $3x - 4 > 2x + 5$

16. $4x + 3 \leq 3x + 11$

17. $8x + 2 \leq 2x + 10$

18. $5x - 1 > x + 9$

19. $7x - 3 > 2x - 13$

20. $9x + 2 \geq 2x - 19$

21. $4x - 3 \leq 6x + 5$

22. $7x - 1 \leq 10x - 6$

do 41-65

23. $5 - 3x > 2x + 3$

24. $7 - 5x > 3x - 9$

Simplify and then solve each of the following inequalities.

25. $5(2x - 1) \leq 25$

26. $3(3x + 1) > -15$

27. $4(5x + 1) > 3(3x + 5)$

28. $3(2x + 4) \leq 5(3x - 3)$

29. $3(x - 1) - 4 < 2(3x + 1)$

30. $3(3x - 1) - 4(x + 3) \leq 15$

Clear the fractions and then solve each of the following inequalities.

31. $\dfrac{x - 4}{3} < 5$

32. $\dfrac{x + 5}{2} \geq -3$

33. $\dfrac{x + 2}{-3} \leq 3$

34. $\dfrac{x - 2}{-4} > -6$

35. $\dfrac{x}{2} - \dfrac{x}{3} \geq 2$

36. $\dfrac{x}{4} - 2 < \dfrac{x}{5}$

37. $\dfrac{x}{5} - \dfrac{x - 7}{3} < \dfrac{1}{3}$

38. $\dfrac{x}{4} - \dfrac{4x + 3}{20} < \dfrac{1}{5}$

39. $\dfrac{x - 3}{2} - \dfrac{x + 5}{5} \leq \dfrac{1}{2}$

40. $\dfrac{x + 5}{4} - \dfrac{x + 1}{3} \leq \dfrac{2}{3}$

Solve each of the following double inequalities. Then graph the solution set.

41. $3 \leq x + 1 \leq 5$

42. $-2 < x - 3 < 3$

43. $-8 < 2x < 4$

44. $-6 \leq 3x \leq 9$

45. $1 \leq 2x - 3 \leq 6$

46. $-2 < 3x - 5 < 4$

47. $-1 < 5 - 3x < 8$

48. $-7 \leq 3 - 2x \leq 8$

Solve each of the following compound inequalities. Then graph the solution set.

49. $x - 1 < -3$ or $x - 1 > 3$

50. $x + 2 < -5$ or $x + 2 > 5$

51. $2x - 1 < -7$ or $2x - 1 > 7$

52. $2x + 3 < -3$ or $2x + 3 > 3$

Transcribe Your Skills

53. Describe the similarities and differences between the process used to solve linear equations and that used to solve linear inequalities.

54. State the addition and multiplication properties of inequality in your own words.

55. Assume that $a < b < 0$. Then

$$b > a$$
$$b^2 > ab$$
$$b^2 - ab > 0$$
$$b(b - a) > 0$$

Divide both sides by $b - a$ to get $b > 0$. This implies that b is positive. We assumed initially that b was negative. Find the error in this argument.

56. Suppose the inequality $4 < \dfrac{8}{x - 1}$ is solved as follows:

$$4(x - 1) < 8 \qquad 4x < 12$$
$$4x - 4 < 8 \qquad x < 3$$

This indicates that any number less than 3 is a valid solution. Yet $x = 0$ produces $4 < -8$, which is not true. Find the error in the solution.

Think About These

Suppose that the revenue a company will receive, from producing and selling x items, is given by R and the cost of these items by C. The company will make a profit only if the revenue is greater than the cost, that is, only when $R > C$. Use this information to find the smallest number of items that must be produced and sold in order for the company to make a profit.

57. $R = 50x$, $C = 1000 + 30x$

58. $R = 800x$, $C = 24{,}000 + 500x$

Recall that the *average* of a group of test scores is the sum of those test scores divided by the number of scores. Use this information to solve the following problems.

59. Suppose that Kim has scores of 83, 94, and 91 on three 100-point tests in her chemistry class thus far. Describe the set of scores on the 100-point final test that will give her an average of 90 or above, so that she will receive an A for the course.

Hint: If x represents her final score, then

$$83 + 94 + 91 + x$$

gives her total score for the four tests.

60. Robert has scores of 78, 85, 70, and 83 on four tests. Describe the set of scores he must have on the 100-point final to average 80 or above for the course.

Solve each of the following problems.

61. A college must decide how many sections of intermediate algebra to offer during the fall quarter. Each section should contain a maximum of 35 students, and the college anticipates that a total of 400 students will enroll for the sections. How many sections should be offered?

Hint: If x represents the number of sections, then $\dfrac{400}{x}$ will give the number of students per section. Establish an inequality from the given information. Note that you can clear the inequality of fractions by multiplying by x, since x is the number of sections and must be a positive number. Also keep in mind that x must be a *whole* number.

62. A student-activities director must order buses for a football game. He anticipates that 300 students will sign up, and the capacity of each bus is 40 people. How many buses should be available?

63. The mileage markers on a freeway begin at marker 0 at the southern border of a state and continue to increase toward the northern border. The legal maximum speed on the freeway is 65 mi/h, and the legal minimum speed is 45 mi/h. If you enter the freeway at marker 100 and travel north for 2 h, what is the possible range of values for the nearest marker you could legally reach?

Hint: Since distance = rate · time, the minimum distance can be calculated as $(45)(2)$ and the maximum distance as $(65)(2)$. Let m be the marker you could legally reach, and establish a double-inequality statement for the solution.

64. You enter the freeway at marker 240 and now travel south for 3 h. What is the possible range of values for the nearest marker you could legally reach?

65. A new landfill must last at least 30 years for it to receive an operating permit from the local community. The proposed site is capable of receiving 570×10^6 metric tons of refuse over its lifespan. How much refuse can the landfill accept each year and still meet the conditions of its permit?

66. A garbage burner must receive at least 1350 t of trash per day in order to be economical enough for a community to build. Local laws restrict the truck weight to a 15-t limit. How many truck deliveries per day will be necessary to supply the burner with its daily requirement of trash?

Skillscan (Section 1.3)

Simplify each of the following expressions.

a. $|-3|$ **b.** $|7|$ **c.** $|8 - 2|$ **d.** $|3 - 7|$

e. $|-5| - |5|$ **f.** $|-5| - |-5|$ **g.** $|2 - 5| + |-3 - 1|$

h. $|-5 + 3| + |-2 + 6|$

Equations and Inequalities Involving Absolute Value   **2.5**

OBJECTIVES: 1. **To solve equations involving absolute value**
              2. **To solve and graph the solution sets for inequalities involving absolute value**

TAPE IN4

Equations and inequalities may contain absolute value notation in their statements. In this section we build on the tools developed in Sections 2.1 and 2.4 and on our earlier work with absolute value for the necessary solution techniques.

Recall from Section 1.3 that the absolute value of x, written $|x|$, is the distance between x and 0 on the number line. Consider, for example, the absolute value equation

$$|x| = 4$$

Technically we mean the distance between the point *corresponding to x and the* point corresponding *to 0, the origin.*

This means that the distance between x and 0 is 4, as pictured below.

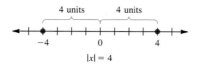

$|x| = 4$

As the sketch illustrates, $x = 4$ and $x = -4$ are the two solutions for the equation. This observation suggests the more general statement.

ABSOLUTE VALUE EQUATIONS—PROPERTY 1

For any positive number p, if

$|x| = p$

then

$x = p$ or $x = -p$

This property allows us to "translate" an equation involving absolute value to two linear equations which we can then solve separately. The following example illustrates.

EXAMPLE 1 Solving an Absolute Value Equation

Solve for x:

$|3x - 2| = 4$

Solution From Property 1 we know that $|3x - 2| = 4$ is equivalent to the equations

$$3x - 2 = 4 \quad \text{or} \quad 3x - 2 = -4$$

Add 2 to both sides of the equation.

$$3x = 6 \qquad\qquad 3x = -2$$

Divide by 3.

$$x = 2 \qquad\qquad x = -\frac{2}{3}$$

The solutions are $-\dfrac{2}{3}$ and 2. These solutions are easily checked by replacing x with $-\dfrac{2}{3}$ and 2 in the original absolute value equation.

Check Yourself 1

Solve for x.

$|4x + 1| = 9$

An equation involving absolute value may have to be rewritten before you can apply Property 1. Consider the following example.

EXAMPLE 2 Solving an Absolute Value Equation

Solve for x:

$$|2 - 3x| + 5 = 10$$

Solution To use Property 1, we must first isolate the absolute value on the left side of the equation. This is easily done by subtracting 5 from both sides for the result

$$|2 - 3x| = 5$$

We can now proceed as before by using Property 1.

$2 - 3x = 5$	or	$2 - 3x = -5$
$-3x = 3$		$-3x = -7$
$x = -1$		$x = \dfrac{7}{3}$

Subtract 2.

Divide by -3.

The solutions are -1 and $\dfrac{7}{3}$.

Check Yourself 2

Solve for x.

$$|5 - 2x| - 4 = 7$$

In some applications, there is more than one absolute value in an equation. Consider an equation of the form

$$|x| = |y|$$

Since the absolute values of x and y are equal, x and y are the same distance from 0. This means they are either *equal* or *opposite in sign.* This leads to a second general property of absolute value equations.

> **ABSOLUTE VALUE EQUATIONS—PROPERTY 2**
>
> **If** $\quad |x| = |y|$
>
> **then** $\quad x = y \quad$ **or** $\quad x = -y$

Let's look at an application of this second property in our next example.

EXAMPLE 3   Solving Equations with Two Absolute Value Expressions

Solve for x:

$|3x - 4| = |x + 2|$

Solution   By Property 2, we can write

$3x - 4 = x + 2$ or $3x - 4 = -(x + 2)$

$\qquad\qquad\qquad\qquad\qquad 3x - 4 = -x - 2$

$3x = x + 6 \qquad\qquad\quad 3x = -x + 2$

$2x = 6 \qquad\qquad\qquad 4x = 2$

$x = 3 \qquad\qquad\qquad x = \dfrac{1}{2}$

The solutions are $\dfrac{1}{2}$ and 3.

Check Yourself 3

Solve for x.

$|4x - 1| = |x + 5|$

We started this section by noting that the solution set for the equation

$|x| = 4$

consists of those numbers whose distance from the origin is equal to 4. Similarly, the solution set for the absolute value inequality

$|x| < 4$

consists of those numbers whose distance from the origin is *less than* 4, that is, all numbers between -4 and 4. The solution set is pictured below.

$|x| < 4$

The solution set would be
$\{x \mid -4 < x < 4\}$

The solution set can be described by the double inequality

$-4 < x < 4$

and this suggests the following general statement.

ABSOLUTE VALUE INEQUALITIES—PROPERTY 1

For any positive number p, if

$|x| < p$

then

$-p < x < p$

Let's look at an application of Property 1 in solving an absolute value inequality.

EXAMPLE 4 Solving Absolute Value Inequalities

Solve and graph the solution set of

$$|2x - 3| < 5$$

Solution From Property 1, we know that the given absolute value inequality is equivalent to the double inequality

$$-5 < 2x - 3 < 5$$

With Property 1 we can *translate* an absolute value inequality to an inequality *not* containing an absolute value which can be solved by our earlier methods.

Solving as before, we isolate the variable in the center term.

$-2 < 2x < 8$ Add 3 to all three parts.

$-1 < \ x \ < 4$ Divide by 2.

The solution set is

$$\{x \mid -1 < x < 4\}$$

The graph is shown below.

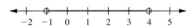

Note that the solution is an open interval on the number line.

Check Yourself 4

Solve and graph the solution set.

$$|3x - 4| \le 8$$

We know that the solution set for the absolute value inequality

$$|x| < 4$$

consists of those numbers whose distance from the origin is *less than* 4. Now what about the solution set for

$$|x| > 4$$

It must consist of those numbers whose distance from the origin is *greater than* 4. The solution set is pictured below.

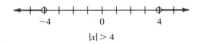

$|x| > 4$

The solution set can be described by the compound inequality

$$x < -4 \quad \text{or} \quad x > 4$$

and this suggests the following general statement.

> **ABSOLUTE VALUE INEQUALITIES—PROPERTY 2**
>
> For any positive number p, if
>
> $|x| > p$
>
> then
>
> $x < -p$ or $x > p$

Let's apply Property 2 to the solution of an absolute value inequality.

EXAMPLE 5 Solving Absolute Value Inequalities

Again we translate the absolute value inequality to the compound inequality not containing an absolute value.

Solve and graph the solution set of

$$|5x - 2| > 8$$

Solution From Property 2, we know that the given absolute inequality is equivalent to the compound inequality

$$5x - 2 < -8 \quad \text{or} \quad 5x - 2 > 8$$

Solving as before, we have

Add 2.

$$5x < -6 \quad \text{or} \quad 5x > 10$$

Divide by 5.

$$x < -\frac{6}{5} \qquad\qquad x > 2$$

You could describe the solution set as

$$\left\{x \,\middle|\, x < -\frac{6}{5}\right\} \cup \{x \mid x > 2\}$$

The solution set is $\left\{x \,\middle|\, x < -\dfrac{6}{5} \text{ or } x > 2\right\}$, and the graph is shown below.

Check Yourself 5

Solve and graph the solution set.

$$|3 - 2x| \geq 9$$

add to note

The following chart summarizes our discussion of absolute value inequalities.

As before, p must be a positive number if $p > 0$.

Type of Inequality	Equivalent Inequality	Graph of Solution Set
$|ax + b| < p$	$-p < ax + b < p$	
$|ax + b| > p$	$ax + b < -p$ or $ax + b > p$	

CHECK YOURSELF ANSWERS

1. $-\dfrac{5}{2}, 2$ **2.** $-3, 8$ **3.** $2, -\dfrac{4}{5}$

4. $-\dfrac{4}{3} \le x \le 4$

5. $x \le -3$ or $x \ge 6$

Build Your Skills

Solve each of the following absolute value equations.

(2 equations)

1. $|x| = 5$

no solution is the answer — (nes)

2. $|x| = 7$

3. $|x - 2| = 3$

4. $|x + 5| = 6$

5. $|x + 6| = 0$

6. $|x - 3| = 0$

7. $|3 - x| = 7$

8. $|5 - x| = 4$

9. $|2x - 3| = 9$

10. $|3x + 5| = 11$

11. $|5 - 4x| = 1$

12. $|3 - 6x| = 9$

13. $\left|\dfrac{1}{2}x + 5\right| = 7$

14. $\left|\dfrac{2}{3}x - 4\right| = 6$

15. $\left|4 - \dfrac{3}{4}x\right| = 8$

16. $\left|3 - \dfrac{2}{5}x\right| = 9$

17. $|3x + 1| = -2$

18. $|5x - 2| = -3$

Rewrite each of the following absolute value equations, and then solve the equations.

19. $|x| - 3 = 2$

20. $|x| + 4 = 6$

21. $|x - 2| + 3 = 5$

22. $|x + 5| - 2 = 5$

23. $|2x - 3| - 1 = 6$

24. $|3x + 5| + 2 = 4$

25. $\left|\dfrac{1}{2}x + 2\right| - 3 = 5$

26. $\left|\dfrac{1}{3}x - 4\right| + 3 = 9$

27. $8 - |x - 4| = 5$

28. $10 - |2x + 1| = 3$

29. $|3x - 2| + 4 = 3$

30. $|5x - 3| + 5 = 3$

Solve each of the following absolute value equations.

31. $|2x - 1| = |x + 3|$

32. $|3x + 1| = |2x - 3|$

33. $|5x - 2| = |2x + 4|$

34. $|7x - 3| = |2x + 7|$

35. $|x - 2| = |x + 1|$

36. $|x + 3| = |x - 2|$

37. $|2x - 5| = |2x - 3|$

38. $|3x + 1| = |3x - 1|$

39. $|x - 2| = |2 - x|$

40. $|x - 4| = |4 - x|$

Find and graph the solution set for each of the following absolute value inequalities.

41. $|x| < 5$

42. $|x| > 3$

43. $|x| \geq 7$

44. $|x| \leq 4$

45. $|x - 4| > 2$

46. $|x + 5| < 3$

47. $|x + 6| \leq 4$

48. $|x - 7| \geq 5$

49. $|3 - x| > 5$

50. $|5 - x| < 3$

51. $|x - 7| < 0$

52. $|x + 5| \geq 0$

53. $|2x - 5| < 3$

54. $|3x - 1| > 8$

55. $|3x + 4| \geq 5$

56. $|2x + 3| \leq 9$

57. $|5x - 3| > 7$

58. $|6x - 5| < 13$

59. $|2 - 3x| < 11$

60. $|3 - 2x| \geq 11$

61. $|3 - 5x| \geq 7$

62. $|7 - 3x| < 13$

63. $\left|\dfrac{3}{4}x - 5\right| < 7$

64. $\left|\dfrac{2}{3}x + 5\right| \geq 3$

Transcribe Your Skills

65. In the equation $|x| \geq p$, p must be nonnegative in order to find a solution. Why is this true?

66. Why is the equation $|x - 3| = |3 - x|$ true for all real numbers?

67. Describe graphically the set of points that satisfy the inequality $|x - 3| < 2$.

68. The inequality $|3x - 4| < -2$ has no solution. Explain why this can be determined directly from the problem statement itself.

Think About These

On some popular calculators there is a special absolute value function key. It is usually labeled $\boxed{\text{abs}}$. To register an absolute value, one presses this key and then enters the expression. For the expression $|x + 3|$, enter $\text{abs}(x + 3)$. Rewrite each expression in calculator form.

69. $|x + 2|$

70. $|x - 2|$

71. $|2x - 3|$

72. $|5x + 7|$

73. $|3x + 2| - 4$

74. $|4x - 7| + 2$

75. $2|3x - 1|$

76. $-3|2x + 8|$

Skillscan (Section 2.1)

In each of the following, determine if the two equations are equivalent or not equivalent.

a. $-3x = 18$ and $x = 6$ **b.** $-5x = -25$ and $x = 5$

c. $2(x - 1) = 6$ and $x = 4$ **d.** $-3(x + 1) = 3$ and $x = 2$ **e.** $3(2x - 7) = 12 + 3x$ and $x = 11$

f. $\dfrac{x}{3} + \dfrac{2x}{5} = 22$ and $x = 30$

2.6 Strategies in Equation Solving

TAPE IN3

OBJECTIVE: To recognize various equation types and determine the proper solution strategy

Our intent in this closing section of Chapter 2 is to reinforce and summarize our work in solving some of the basic types of equations and inequalities.

> *Recognition of equation types* is an important theme which is stressed throughout this text.

In this chapter we have dealt with linear equations and inequalities (or equations and inequalities that can be written in a linear form). Let's review for a moment.

EXAMPLE 1 Identifying Linear Equations and Inequalities

Which of the following equation types can be written in a linear form and solved, given the methods of this chapter?

(a) $7x - 3 = 2x + 5$

This equation is linear and can be solved by using the addition and multiplication properties of equality to isolate the variable.

Note that x appears only to the first power and that x is the only variable appearing in the equation.

(b) $2(x - 3) = 5(x + 2) - 7$

Again the equation is linear and can be solved by applying the distributive property as the first step.

(c) $x^2 - x = 6$

This equation is *not* linear because of the term in x^2. It cannot be solved with our present methods.

This is called a *quadratic equation* because of the term in x^2. We will see solution methods later in this text.

(d) $\dfrac{x - 1}{2} - \dfrac{x}{3} = 5$

This is a linear equation. It can be solved by using the multiplication property to clear fractions as the first step.

(e) $x^2 + 2x - 2 = x^2 + 5$

Although this equation is not linear, it can easily be written in an equivalent linear form by applying the addition property to subtract x^2 from both sides.

(f) $\dfrac{1}{x} + \dfrac{1}{x + 2} = 3$

This equation is *not* linear. We can clear the equation of fractions by using the multiplication property, but this will lead to a quadratic equation.

Note: The recognition clues discussed above are also applicable to inequalities. To determine whether an inequality is linear, use exactly the same approach.

Check Yourself 1

Which of the following equations can be written in linear form?

1. $5(x - 3) - 2 = 3(x + 5)$

2. $\dfrac{5}{x} - \dfrac{3}{x + 1} = 1$

3. $x^2 - 2x - 3 = 5x$

4. $\dfrac{x - 1}{3} + \dfrac{x}{4} = 5$

5. $x^2 - 5x + 3 = x^2 - 7$

Once you have determined that an equation (or an inequality) is linear, you can apply the equation-solving strategies developed in this chapter. Let's restate our earlier algorithm for reference and then consider illustrations for each step of the process.

SOLVING LINEAR EQUATIONS IN ONE VARIABLE

STEP 1   Multiply both sides of the equation by the LCM of any denominators to clear the equation of fractions.

STEP 2   Remove any grouping symbols by applying the distributive property.

STEP 3   Combine any like terms that appear on each side of the equation.

STEP 4   Apply the addition property of equality to write an equivalent equation with the variable term on *one side* of the equation and the constant term on the *other side*.

STEP 5   Apply the multiplication property of equality to write an equivalent equation with the variable isolated on one side of the equation.

STEP 6   Check the solution in the *original* equation.

The first step requires clearing the equation of fractions.

EXAMPLE 2 Finding Equivalent Equations

Each of our equations can be further simplified. Here we are interested in the *first step* of the process.

Find the equivalent equation that will be formed after the appropriate step of our algorithm is applied.

$$\frac{2x}{3} - \frac{x}{4} = 10$$

Solution Since fractions are involved, we start with step 1. The *least common multiple* (LCM) of the denominators is 12. Applying the multiplication property, we have

$$12\left(\frac{2x}{3}\right) - 12\left(\frac{x}{4}\right) = 12(10)$$

$$8x - 3x = 120$$

Check Yourself 2

Find the equivalent equation that will be formed after the appropriate step of our algorithm is applied.

$$\frac{5x}{6} - \frac{2x}{9} = 11$$

If there are no fractions, we proceed by removing grouping symbols.

EXAMPLE 3 Removing Grouping Symbols

(*a*) Find the equivalent equation that will be formed after the appropriate step of our algorithm is applied.

$$2(x - 5) = 3(2x + 1) - 1$$

Since grouping symbols are involved, we use step 2 and apply the distributive property.

$$2x - 10 = 6x + 3 - 1$$

(*b*) Find the equivalent equation that will be formed after the appropriate step of our algorithm is applied.

$$x(x - 2) + 5 = x^2 - 3x$$

Again we apply the distributive property.

$$x^2 - 2x + 5 = x^2 - 3x$$

Note that even though the resulting equation is not linear, it can be written in the equivalent linear form

$$-2x + 5 = -3x$$

by subtracting x^2 from both sides.

Check Yourself 3

Find the equivalent equation that will be formed after the appropriate step of our algorithm is applied.

$$2(x - 3) = 5(2x + 5) - 7$$

Combining like terms creates a simpler, yet equivalent, equation.

EXAMPLE 4 Combining Like Terms

Find the equivalent equation that will be formed after the appropriate step of our algorithm is applied.

$$x + 5 - 3x = 7 - 5x + 2$$

Solution Since like terms occur on both the left and right sides of the equation, we go to step 3 of the algorithm and combine like terms:

$$-2x + 5 = 9 - 5x$$

Check Yourself 4

Find the equivalent equation that will be formed after the appropriate step of our algorithm is applied.

$$2x - 7 + 3x = 4 - 3x - 9$$

Always isolate the variable on one side of the equation.

EXAMPLE 5   Isolating the Variable

Find the equivalent equation that will be formed after the appropriate step of our algorithm is applied.

$$7x - 3 = 2x + 12$$

Solution From step 4, we want to apply the addition property to isolate the variable term on one side of the equation. This will require two steps.
    We add 3 to both sides of the equation.

Note that we could have just as easily subtracted 2x and *then* added 3. The order of these steps does not matter.

$$7x - 3 + 3 = 2x + 12 + 3$$
$$7x = 2x + 15$$

We now subtract $2x$ from both sides.

$$7x - 2x = 2x - 2x + 15$$
$$5x = 15$$

Check Yourself 5

Find the equivalent equation that will be formed after the appropriate step of our algorithm is applied.

$$3x + 7 = 5x - 9$$

The multiplication rule is usually needed to take the last step in getting the solution.

EXAMPLE 6   Using the Multiplicative Property of Equality

Find the equivalent equation that will be formed after the appropriate step of our algorithm is applied.

$$-3x = 15$$

Solution In step 5 we use the multiplication property to divide both sides of the equation by the coefficient of the variable, here -3.

$$\frac{-3x}{-3} = \frac{15}{-3}$$

$$x = -5$$

The final step isolates the variable on one side of the equation.

This final equation has the solution -5, and since all the derived equations are equivalent, -5 is the solution for the original equation.

That solution should be checked by substitution to complete the process.

Check Yourself 6

Find the equivalent equation that will be formed after you apply the appropriate step of our algorithm.

$$-12x = -60$$

Let's conclude this section with an example that will summarize the equation-solving strategy illustrated in preceding examples.

EXAMPLE 7 Solving Equations That Contain Fractions

Solve for x:

$$\frac{x + 3}{4} - \frac{x}{3} = \frac{x + 6}{6}$$

We note that the equation is linear.

Solution First we multiply by 12, the LCM of 4, 3, and 6:

Step 1 of the algorithm

$$12\left(\frac{x + 3}{4}\right) - 12\left(\frac{x}{3}\right) = 12\left(\frac{x + 6}{6}\right)$$

$$3(x + 3) - 4x = 2(x + 6)$$

Now we use the distributive property to clear the parentheses.

Step 2

$$3x + 9 - 4x = 2x + 12$$

Combine like terms.

Step 3

$$-x + 9 = 2x + 12$$

Apply the addition property to write an equivalent equation with the variable term on the left and the constant term on the right.

Step 4

$$-x + 9 - 9 = 2x + 12 - 9$$

$$-x = 2x + 3$$

$$-x - 2x = 2x - 2x + 3$$

$$-3x = 3$$

Step 5   Use the multiplication property to divide by -3 to isolate x.

$$\frac{-3x}{-3} = \frac{3}{-3}$$

$$x = -1$$

Step 6   To check our solution, -1, we substitute -1 for x in the original equation.

$$\frac{-1+3}{4} - \frac{(-1)}{3} \overset{?}{=} \frac{-1+6}{6}$$

$$\frac{1}{2} + \frac{1}{3} \overset{?}{=} \frac{5}{6}$$

$$\frac{5}{6} = \frac{5}{6} \qquad \text{A true statement}$$

The solution is verified.

Note Again we should point out that the strategy for solving linear inequalities follows essentially the same procedures illustrated in the above example.

Check Yourself 7

Solve for x and check your result.

$$\frac{7x - 14}{4} - \frac{x}{2} = \frac{3x - 7}{8}$$

CHECK YOURSELF ANSWERS

1. (1), (4), and (5)   **2.** $15x - 4x = 198$ **3.** $2x - 6 = 10x + 25 - 7$
4. $5x - 7 = -3x - 5$ **5.** $-2x = -16$ **6.** $x = 5$ **7.** 3

2.6 EXERCISES

Build Your Skills

Which of the following equations or inequalities can be written in a linear form and solved, given the methods of this section?

1. $5x - 3 = 7x + 9$

2. $x^2 - 2x - 8 < 0$

3. $\dfrac{x-1}{2} - \dfrac{x}{3} = \dfrac{1}{6}$

4. $\dfrac{1}{x-2} - \dfrac{1}{x} = 2$

5. $x^2 - 2x + 3 = x^2 + 4$

6. $2(x - 5) + 7(x + 3) \geq 29$

7. $x - \dfrac{1}{x-2} \leq 3$

8. $x(x - 5) = x^2 + 3x + 4$

9. $5(x - 3) + 2(x + 1) = 8$

10. $\dfrac{x+2}{5} + \dfrac{x+1}{10} = 2$

11. $x^2 - 3x = x + 4$

12. $2x - 3 + 5x = 7 - 2x + 3$

Solve each of the following equations or inequalities.

13. $9x - 3 = 5x + 9$

14. $3(x + 2) = 4x$

15. $\dfrac{2x}{5} - \dfrac{x}{4} = 3$

16. $8x - 2 < 3x + 23$

17. $7(x - 3) \geq 2(x + 7)$

18. $\dfrac{x}{4} - 3 = \dfrac{x}{10}$

19. $3x - 7 = 10x + 14$

20. $2(x - 3) - (x + 2) \geq 1$

21. $3(3x + 1) - 9 = 4(2x - 1)$

22. $7x - 3 - 2x = 19 + x - 6$

Transcribe Your Skills

23. Explain why the equation $x(x - 1) = x^2 + 2x - 3$ can be written in linear form even though it contains the term x^2.

24. State the operations that you can perform on an equation to produce an equivalent equation. State one operation that does not produce an equivalent equation.

Think About These

Match each of the following equations with the equivalent equation below that will be formed after the appropriate step of the equation-solving algorithm is applied.

(a) $x = -16$ $\quad$ (b) $4x - 5 = -5x + 16$
(c) $6(x + 3) - 5(x + 8) = 30$ $\quad$ (d) $x = 6$
(e) $3(x - 1) - 4(x - 2) = 24$ $\quad$ (f) $-6x = 30$
(g) $3x + 6 - 5x + 20 = 2$ $\quad$ (h) $-3x + 12 = 5x - 6$

25. $3(x + 2) - 5(x - 4) = 2$

26. $\dfrac{x - 1}{4} - \dfrac{x - 2}{3} = 2$

27. $3x - 5 + x = 9 - 5x + 7$

28. $3x - 7 = 9x + 23$

29. $-7x = -42$

30. $\dfrac{x + 3}{5} - \dfrac{x + 8}{6} = 1$

31. $-4x = 64$

32. $5 - 3x + 7 = 2x - 6 + 3x$

Skillscan (Section 1.1)

On a number line, graph the points corresponding to each of the given numbers.

a. 4 $\quad$ **b.** -2 $\quad$ **c.** $\dfrac{1}{5}$ $\quad$ **d.** $\dfrac{-2}{3}$ $\quad$ **e.** $-\sqrt{3}$

f. $\sqrt{5}$

SUMMARY

Linear Equations in One Variable [2.1]

A *linear equation* in one variable is any equation that can be written in the form

$$ax + b = 0$$

where a and b are any real numbers and $a \neq 0$.

To solve a linear equation means to find its solution. A *solution* of an equation in one variable is any number that will make the equation a true statement. The *solution set* consists of all solutions.

Two equations are *equivalent* if they have the same solution set.

Forming a sequence of equivalent equations that will lead to the solution of a linear equation involves two properties of equality.

$5x - 6 = 3x + 2$ is a linear equation. The variable appears only to the first power.

The solution for the equation above is 4 since

$5 \cdot 4 - 6 \stackrel{?}{=} 3 \cdot 4 + 2$

$14 = 14$

is a true statement.

$5x - 6 = 3x + 2$ and $2x = 8$ are equivalent equations. Both have 4 as the solution.

Addition Property of Equality If $a = b$, then $a + c = b + c$. In words, adding the same quantity to both sides of an equation gives an equivalent equation.

If $x - 3 = 7$, then

$x - 3 + 3 = 7 + 3$.

If $2x = 8$, then $\frac{1}{2}(2x) = \frac{1}{2}(8)$.

Multiplication Property of Equality If $a = b$, then $ac = bc$, $c \neq 0$. In words, multiplying both sides of an equation by the same nonzero quantity gives an equivalent equation.

The above properties are applied in the following algorithm.

Solve

$\dfrac{x + 1}{5} - \dfrac{x}{4} = \dfrac{1}{20}$

$20\left(\dfrac{x + 1}{5}\right) - 20\left(\dfrac{x}{4}\right)$

$\qquad = 20\left(\dfrac{1}{20}\right)$

$4(x + 1) - 5x = 1$

Solving Linear Equations in One Variable

Step 1. Multiply both sides of the equation by the LCM of all denominators, to clear the equation of fractions.

Remove grouping symbols.

$4x + 4 - 5x = 1$

Step 2. Remove any grouping symbols by applying the distributive property.

Combine like terms.

$-x + 4 = 1$

Step 3. Combine like terms that appear on either side of the equation.

Subtract 4.

$-x = -3$

Step 4. Apply the addition property of equality to write an equivalent equation with the variable term on *one side* of the equation and the constant term on the *other side*.

Divide by -1.

$x = 3$

Step 5. Apply the multiplication property of equality to write an equivalent equation with the variable isolated on one side of the equation.

To check:

$\dfrac{3 + 1}{5} - \dfrac{3}{4} \stackrel{?}{=} \dfrac{1}{20}$

$\dfrac{4}{5} - \dfrac{3}{4} \stackrel{?}{=} \dfrac{1}{20}$

$\dfrac{1}{20} = \dfrac{1}{20}$

A true statement.

Step 6. Check the solution in the *original* equation.

Formulas and Literal Equations [2.2]

$P = 2L + 2W$ is a formula or a literal equation.

Formulas and *literal equations* express a relationship between more than one variable or letter.

Solving a formula or literal equation for a variable means isolating that specified variable on one side of the equation. The steps used in the process are very similar to those used in solving linear equations. The following algorithm is applied.

Solve for B:

$A = \dfrac{1}{2}h(B + b)$

Solving Formulas or Literal Equations

Step 1. Remove any fractions by using the multiplication property.

Multiply by 2.

$2A = h(B + b)$

Step 2. Remove any grouping symbols by using the distributive property.

Distribute h.

$$2A = hB + hb$$

Step 3. Use the addition property to write all terms involving the variable you are solving for on one side of the equation and the remaining terms on the other side.

Subtract hb.

$$2A - hb = hB$$

Step 4. Use the multiplication property to divide both sides of the equation by the coefficient of the variable you are solving for. This will isolate the desired variable on one side of the equation.

Divide by h.

$$\frac{2A - hb}{h} = B$$

or

$$B = \frac{2A - hb}{h}$$

Applications and Problem Solving [2.3]

The solution of applications, or word problems, can be approached with the following five-step strategy.

One number is 3 less than twice another. If the sum of the numbers is 27, find the two numbers.

Solving Applications

Step 1. Read the problem carefully to determine the unknown quantities.

1. The unknowns are the two numbers.

Step 2. Choose a variable to represent the unknown or unknowns.

2. Let x be the first number. Then $2x - 3$ is the second number.

Step 3. Translate the problem to the language of algebra, to form an equation.

3. $\underbrace{x + 2x - 3}_{\text{Sum of the numbers}} = 27$

Step 4. Solve the equation and answer the question of the original problem.

4. Solving as before gives

$$x = 10$$

and $2x - 3 = 17$

Step 5. Verify your solution by returning to the *original* problem.

5. The sum of the numbers is 27, and 17 is 3 less than twice 10.

Solving Linear Inequalities [2.4]

A *linear inequality in one variable* is any inequality that can be written in the form

$$ax + b < 0$$

$5x - 7 < 2x + 4$ is a linear inequality.

where a and b are real numbers and $a \neq 0$.

The inequalities

$$ax + b > 0 \qquad ax + b \leq 0$$

and $ax + b \geq 0$

are also linear inequalities.

)

Addition Property of Inequality If $a < b$, then $a + c < b + c$.

Multiplication Property of Inequality

If $\qquad a < b$

then $\qquad ac < bc \qquad$ where c is a *positive* number

and $\qquad ac > bc \qquad$ where c is a *negative* number.

    Solving linear inequalities involves essentially the same procedures as solving linear equations. The following algorithm is applied.

To multiply or divide both sides of an inequality by a *negative* number, *reverse the sense* of the inequality.

Solving Linear Inequalities in One Variable

Step 1. Clear the inequality statement of any fractions by using the multiplication property.

Step 2. Remove any grouping symbols and combine like terms.

$2x - 5 \geq 7x + 15$

Add 5 to both sides.

$2x \geq 7x + 20$

Subtract $7x$ from both sides.

$-5x \geq 20$

Divide both sides by -5.

$x \leq -4$

Note that we *reverse the sense* of the inequality.

The graph of the solution set $\{x \mid x \leq -4\}$ is

Step 3. Apply the addition property to write an equivalent inequality with the variable term on *one side* of the inequality and the constant term on the *other side*.

Step 4. Apply the multiplication property to write an equivalent inequality with the variable isolated on one side of the inequality. Be sure to reverse the sense of the inequality if you multiply or divide by a negative number.

Step 5. Graph the solution set of the original inequality.

$-5 < 3x - 2 < 5$

Add 2 to all terms.

$-3 < 3x < 7$

Divide by 3.

$-1 < x < \dfrac{7}{3}$

$2x + 3 < -5 \qquad$ or

$\qquad 2x < -8$

$\qquad\quad x < -4$

$2x + 3 > 5$

$\qquad 2x > 2$

$\qquad\quad x > 1$

    The *compound inequalities* considered here fall into two categories:

1. $ax + b > -c \qquad$ and $\qquad ax + b < c$

which can be written in double-inequality form as

$$-c < ax + b < c$$

This type of inequality statement is solved by isolating the variable in the middle term.

2. $ax + b < -c \qquad$ or $\qquad ax + b > c$

This type of inequality statement is solved by considering each inequality separately.

Equations and Inequalities Involving Absolute Value [2.5]

To solve absolute value equations, the following property is applied.

$|2x - 5| = 7$ is equivalent to

$2x - 5 = -7 \qquad$ or
$2x - 5 = 7$

so

$x = -1 \qquad$ or $\qquad x = 6$

Absolute Value Equations For any positive number p, if

$$|x| = p$$

then

$$x = -p \qquad \text{or} \qquad x = p$$

To solve an equation containing an absolute value, translate the equation to two equivalent linear equations. Those equations can then be solved separately.

Absolute Value Inequalities

1. For any positive number p, if

$$|x| < p$$

then

$$-p < x < p$$

To solve this form of inequality, translate to the equivalent double inequality and solve as before.

$|3x - 5| < 7$ is equivalent to

$-7 < 3x - 5 < 7$

This yields

$-2 < 3x < 12$

$-\dfrac{2}{3} < x < 4$

2. For any positive number p, if

$$|x| > p$$

then

$$x < -p \qquad \text{or} \qquad x > p$$

To solve this form of inequality, translate to the equivalent compound inequality and solve as before.

$|2 - 5x| \geq 12$ is equivalent to

$2 - 5x \leq -12 \qquad$ or
$2 - 5x \geq 12$

This yields

$x \geq \dfrac{14}{5} \qquad$ or $\qquad x \leq -2$

SUMMARY EXERCISES

This summary exercise set is provided to give you practice with each of the objectives of the chapter. Each exercise is keyed to the appropriate chapter section. The answers are provided in the instructor's manual that accompanies this text. Your instructor will provide guidelines on how best to use these exercises in your instructional program.

[2.1] Solve each of the following equations, and check your result.

1. $4x - 5 = 23$

2. $7 - 3x = -8$

3. $5x + 2 = 6 - 3x$

4. $7x - 3 = 2x + 12$

5. $2x - 7 = 9x - 35$

6. $5 - 3x = 2 - 6x$

7. $7x - 3 + 2x = 5 + 6x + 4$

8. $2x + 5 - 4x = 3 - 6x + 10$

9. $3(x - 5) = x + 1$

10. $4(2x - 1) = 6x + 5$

11. $7x - 3(x - 2) = 30$

12. $8x - 5(x + 3) = -10$

13. $7(3x + 1) - 13 = 8(2x + 3)$

14. $3(2x - 5) - 2(x - 3) = 11$

15. $\dfrac{2x}{3} - \dfrac{x}{4} = 5$

16. $\dfrac{3x}{4} - \dfrac{2x}{5} = 7$

17. $\dfrac{x}{2} - \dfrac{x + 1}{3} = \dfrac{1}{6}$

18. $\dfrac{x + 1}{5} - \dfrac{x - 6}{3} = \dfrac{1}{3}$

19. $\dfrac{5x + 7}{4} - 2 = \dfrac{x + 2}{3}$

20. $\dfrac{5x + 1}{2} - \dfrac{x + 1}{4} = 2$

[2.2] Solve each of the following literal equations for the indicated variables.

21. $P = RB$ for R

22. $I = Prt$ for t

23. $S = 2\pi rh$ for h

24. $S = \dfrac{1}{2}gt^2$ for g

25. $y = mx + b$ for m

26. $P = 2L + 2W$ for L

27. $A = P(1 + rt)$ for r

28. $A = \dfrac{1}{2}h(B + b)$ for B

[2.2] Solve each of the following problems.

29. A principal of $5000 was invested in a savings account paying 6 percent annual interest. If the interest earned over a certain period was $1200, for how long was the money invested? See Exercise 22.

30. A cylinder has lateral surface area 96π square inches (in^2). If the radius of the cylinder is 6 in, find the height of the cylinder. See Exercise 23.

31. A principal of $3000 was invested in a money market fund. If the amount in the account was $3720 at the end of 3 years, what was the annual interest rate? See Exercise 27.

32. The area of a trapezoid is 60 cm^2. If the length of one base of the trapezoid is 8 cm and the height is 3 cm, find the length of the other base. See Exercise 28.

[2.4] Solve each of the following inequalities. Then graph the solution sets.

33. $3x - 2 > 10$

34. $5x - 3 \le -18$

35. $5 - 3x \le 3$

36. $7 - 4x \ge 15$

37. $9x - 3 < 7x - 13$

38. $5 - 3x > 2 - 6x$

39. $2x - 5 \ge 7x - 10$

40. $4 - 3x < 14 + 2x$

41. $4(5x - 4) \ge 3(3x + 2)$

42. $3(2x - 1) > 2(x - 4) - 11$

43. $\dfrac{x}{2} - \dfrac{x + 8}{5} < \dfrac{1}{2}$

44. $\dfrac{x + 3}{4} - \dfrac{x - 1}{3} > \dfrac{2}{3}$

[2.4] Solve each of the following compound inequalities. Then graph the solution sets.

45. $3 < x + 5 < 7$

46. $-2 \le 3x + 4 \le 10$

47. $-5 \le 3 - 2x \le 5$

48. $-4 < 5 - 3x < 4$

49. $3x - 1 < -7$ or $3x - 1 > 7$

50. $2x + 5 \le -9$ or $2x + 5 \ge 9$

[2.5] Solve each of the following absolute value equations.

51. $|x + 3| = 5$

52. $|3x - 2| = 7$

53. $|7 - x| = 3$

54. $|5 - 3x| = 14$

55. $|2x + 1| - 3 = 6$

56. $7 - |x - 3| = 5$

57. $|3x - 1| = |x + 5|$

58. $|x - 5| = |x + 3|$

[2.5] Solve each of the following absolute value inequalities. Then graph the solution sets.

59. $|x| \le 3$

60. $|x + 3| > 5$

61. $|x - 7| > 4$

62. $|3 - x| < 6$

63. $|2x + 7| > 5$

64. $|3x - 1| < 14$

65. $|3x + 4| < 11$

66. $|5x + 2| \ge 12$

67. $|3 - 2x| \ge 15$

68. $|5 - 3x| < 11$

69. $\left| \dfrac{2x - 1}{3} \right| < 5$

70. $\left| \dfrac{2x + 1}{3} \right| \ge 5$

[2.3] Solve each of the following applications. Be sure to show the equation used for your solution.

71. One number is 4 more than 3 times another. The sum of the numbers is 52. Find the two numbers.

72. One number is 5 less than twice another. If the sum of the first number and 3 times the second number is 41, what are the two numbers?

73. Find two consecutive integers such that twice the first integer is 18 less than 3 times the second integer.

74. Find three consecutive even integers such that the sum of the first integer and twice the second integer is 5 times the third integer.

75. The length of a doubles tennis court is 6 ft more than twice its width. If the perimeter of the court is 228 ft, find the dimensions of the court.

76. The length of a rectangle is 4 ft more than its width. If the length is increased by 2 ft, the area of the rectangle is increased by 10 ft². Find the length and width of the original rectangle.

77. A movie theater sells adult tickets for $6 and student tickets for $3.50. If 320 tickets were sold for one showing, with a revenue of $1645, how many of each type of ticket were sold?

78. A candy and nut retailer wishes to mix peanuts selling for $2.25 per pound and cashews selling for $6 per pound to form a 120-lb mixture which would sell for $3 per pound. What amount of each type of nut should be used in forming the desired mixture?

79. A laboratory technician has a 20% saline solution and a 50% saline solution in her stockroom. How much of each solution should she use to form 600 mL of a 40% saline solution?

80. A chemist wants to combine a 70% acid solution with 300 mL of a 25% acid solution to increase the concentration to 40%. How much of the 70% acid solution should be added?

81. Sharon divided her $17,000 inheritance into two investments. One was a savings account paying 6 percent annual interest, the other was a bond paying 8 percent. If her annual interest from the two investments was $1200, how much did she have invested at each rate?

82. Samuel invested a portion of $15,000 in a savings account paying 5 percent annual interest and the remainder in a time deposit paying 7 percent. If his annual interest from the time deposit was $330 more than that from the savings account, how much did he have invested at each rate?

83. Lisa left Friday morning, driving on the freeway to visit friends for the weekend, and her trip took 4 h. In returning on Sunday, heavier traffic slowed her average speed by 6 mi/h, and the trip took $4\frac{1}{2}$ h. What was her average speed in each direction, and how far did she travel each way?

84. A bicyclist started on a 132-mi trip and rode at a steady rate for 3 h. He began to tire at that point and slowed his speed by 4 mi/h for the remaining 2 h of the trip. What was his average speed for each part of the journey?

85. At noon, Jan left her house, jogging at an average rate of 8 mi/h. Two hours later, Stanley left on his bicycle along the same route, averaging 20 mi/h. At what time will Stanley catch up with Jan?

86. At 9 A.M., David left New Orleans for Tallahassee, averaging 47 mi/h. Two hours later, Gloria left Tallahassee for New Orleans along the same route, driving 5 mi/h faster than David. If the two cities are 391 mi apart, at what time will David and Gloria meet?

87. A firm producing running shoes finds that its fixed costs are $3900 per week while its variable cost is $21 per pair of shoes. If the firm can sell the shoes for $47 per pair, how many pairs of shoes must be produced and sold each week for the company to break even?

88. For a certain type of scientific calculator, the supply available at price p (in dollars) is given by

$$S = 40p - 200$$

The demand for that same calculator at price p is

$$D = -80p + 1720$$

Find the equilibrium price (where the supply equals the demand).

SELF-TEST

The purpose of this self-test is to help you check your progress and to review for a chapter test in class. Allow yourself about an hour to take the test. When you are done, check your answers in the back of the book. If you missed any problems, be sure to go back and review the appropriate sections in the chapter and the exercises that are provided.

Solve each of the following equations.

1. $7 - 5x = 3$

2. $7x + 8 = 30 - 4x$

3. $5x - 3(x - 5) = 19$

4. $\dfrac{x + 3}{4} - \dfrac{x}{2} = \dfrac{3}{8}$

Solve each of the following literal equations for the indicated variables.

5. $A = P(1 + rt)$ for r

6. $A = \dfrac{1}{2}h(B + b)$ for h

Solve each of the following inequalities. Then graph the solution set.

7. $5x - 3 \le 17$

8. $3x + 7 < 5(x - 2)$

9. $\dfrac{x + 1}{2} - \dfrac{1}{3} > \dfrac{x + 3}{6}$

10. $-5 \le 3 - 2x \le 7$

Solve each of the following.

11. $|3x - 5| = 7$

12. $|2x - 3| = |x + 1|$

13. $|4x - 3| < 9$

14. $|5 - 4x| \ge 13$

Solve each of the following applications. Be sure to show the equation used for your solution.

15. Find two consecutive odd integers such that the first is 9 more than twice the second.

16. The length of a singles badminton court is 10 ft more than twice its width. If the perimeter of the court is 122 ft, find its length and width.

17. A candy merchant wants to combine chocolate kisses which sell for \$3.40 per pound with malted milk balls which sell for \$1.60 per pound to form a 225-lb mixture which will sell for \$2.40 per pound. How much of each type of candy should be used?

18. A pharmacist has a 20% and a 60% alcohol solution in stock. How much of each solution should be used to form 400 mL of a 35% alcohol solution?

19. Marsha divided her \$22,000 inheritance into two investments, a savings account paying 6 percent annual interest and a tax-free bond paying 9 percent. If her annual interest from the two accounts was \$1530, how much did she have invested at each rate?

20. At 10 A.M., Sandra left her house on a business trip and drove at an average rate of 45 mi/h. One hour later, Adam discovered that Sandra had left her briefcase behind, and he began driving at 55 mi/h along the same route. When will Adam catch up with Sandra?

GRAPHING LINEAR EQUATIONS AND INEQUALITIES

AIR POLLUTION *Air pollution is recognized as a world-wide problem, in both industrialized and developing nations. All major cities of the world must deal with the problems of smog and industrial pollutants in their air.*

Air pollution is caused by hundreds of substances, most of which can be grouped into a few major categories: smog, a combination of nitrogen oxides and volatile organic compounds which occurs in the presence of sunlight and heat; ozone, toxic to most plants and animals; sulfur oxides, the primary component, along with nitrogen oxides, of acid rain; carbon monoxide, a poisonous, odorless gas resulting from the incomplete burning of fuels; carbon dioxide, a natural by-product of the burning process and a major greenhouse gas; suspended particulate matter, solid particles such as dust, soot, asbestos, lead, or other solid products of burning, erosion, and manufacturing processes; toxic chemicals, released during manufacturing processes, which can cause cancers and other diseases; radioactive substances, any of several radioactive substances that enter the atmosphere as gases or particles; heat, produced by a variety of activities such as power generation, manufacturing, and transportation; and noise, produced by nearly all human activities, particularly urban activity.

The Clean Air Act of 1970 made the United States the first industrialized nation to pass a law regulating levels of air pollutants. Since that time, the levels of many of the major air pollutants in U.S. cities have dropped. Particulates, sulfur dioxide, and lead are three substances, in particular, which have been reduced the most as a result of the Clean Air Act of 1970. The 1990 amendments to the Clean Air Act include new regulations affecting smog, airborne toxic chemicals, and acid rain.

As of 1990, there were still nearly 100 major U.S. cities not in compliance with the clean-air regulations of the 1970 law. Whereas the original Clean Air Act regulated only seven toxic substances emitted to the atmosphere, the 1990 law lists 189 such substances to be controlled by the "maximum achievable control technology" by 2003. Many U.S. cities still have major tasks ahead if they are to meet these new, tougher standards.

3.1 The Cartesian Coordinate System

TAPE IN5

OBJECTIVES: 1. To plot points in a plane, given their coordinates
2. To give the coordinates of a point in the plane

In Chapter 1 we introduced the concept of a number line. Recall that each point on the number line corresponded to a real number. For our work in this chapter, it will be useful to establish a similar correspondence between points in the plane and ordered pairs of real numbers.

Although there are differences of opinion about the "inventor" of analytic geometry, much of the subject's development is credited to René Descartes (1596–1650), a French mathematician and philosopher. The cartesian coordinate system is named in his honor.

As we will see, this correspondence is accomplished through the use of the *cartesian coordinate system*. This coordinate system is the basis for the union of the subjects of algebra and geometry into a branch of mathematics called *analytic geometry*. Essentially, analytic geometry allows us to consider the behavior of algebraic equations through the study of their corresponding geometric figures or graphs and to study geometric figures by looking at the corresponding algebraic equations.

To summarize: Analytic geometry considers two basic types of questions:

1. If we have an algebraic equation, can we draw the corresponding geometric "picture" or graph?

2. If we have a set of geometric conditions, can we find the corresponding algebraic equation?

All our work in this chapter involves the investigation of these two questions.

We begin our work in analytic geometry by constructing the cartesian coordinate system. We start with a horizontal number line and then intersect that line with a vertical number line, so that the point of intersection of the two lines is the respective zero point of the lines.

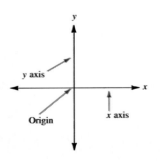

We call the horizontal line the *x axis* and the vertical line the *y axis*. The point at which the two axes intersect is called the *origin*. The normal positioning of the axes is such that the positive directions are to the *right* on the *x* axis and *upward* on the *y* axis.

That is, for each point in the plane there is a unique ordered pair, and for every ordered pair there corresponds a unique point in the plane.

With the coordinate system so constructed, we can now establish a one-to-one correspondence between points in the plane and *ordered pairs* of real numbers, written as

$$(x, y)$$

We agree that the first number is always the *x* value, or the *x coordinate*, and that the second number is always the *y* value, or the *y coordinate*. Other names are *abscissa* for the *x* coordinate and *ordinate* for the *y* coordinate.

The point *P* with coordinates (x, y)—often shortened for convenience to "the point (x, y)" and written $P(x, y)$—is located in the plane by the following rule:

LOCATING POINTS IN THE PLANE

For point *P* corresponding to the ordered pair (*x*, *y*):

1. **The first number, or *x* coordinate, gives the directed distance from the *y* axis to the point**

   To the right if the *x* coordinate is positive

   To the left if the *x* coordinate is negative

2. **The second number, or *y* coordinate, gives the directed distance from the *x* axis to the point**

   Upward if the *y* coordinate is positive

   Downward if the *y* coordinate is negative

Note from our drawing that the coordinate axes separate the plane into four regions, called *quadrants*. They are numbered in a counterclockwise direction from the upper right, as shown. Each point in the plane is either in one of the four quadrants or on one of the coordinate axes.

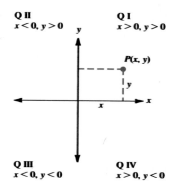

Applying the rule above to *graph* (or *plot*) the point in the plane corresponding to the ordered pair (4, 2), we can move 4 units from the origin to the *right* along the *x* axis. We then move 2 units *upward* and parallel to the *y* axis.

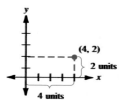

To locate the point corresponding to the ordered pair (−3, −4), we move 3 units from the origin to the *left* along the *x* axis and then move 4 units *downward* and parallel to the *y* axis.

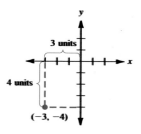

To locate the point corresponding to the ordered pair (0, 5), there is *no* movement along the *x* axis. We move 5 units *upward* along the *y* axis.

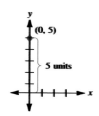

EXAMPLE 1 Plotting Points

In the plane graph (or plot) points A, B, C, D, and E having the following coordinates:

In this text, we usually use the same scale on the x and y axes, but that is *not* required. In fact, for many applications, different scales are desirable.

$A(3, 5)$ $B(-2, -3)$ $C(-4, 0)$

$D(5, -1)$ $E(0, -6)$

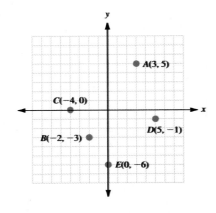

Check Yourself 1

Graph (or plot) the points with the following coordinates:

$P(2, 4)$ $Q(-5, 7)$ $R(3, 0)$

$T(0, 4)$ $S(-4, -3)$

If we are given a point in the plane, reversing the process allows us to write the coordinates of that point by observing the distance and direction of the point from the x and y axes.

EXAMPLE 2 Identifying Plotted Points

Points U, W, X, Y, and Z are located in a cartesian coordinate system, as shown. Give the coordinates of each point.

$U(4, 4)$ $W(-2, 0)$ $X(-4, -5)$

$Y(5, -5)$ $Z(-5, 6)$

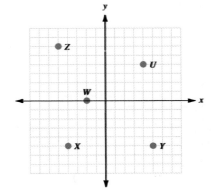

Check Yourself 2

Points *A*, *B*, *C*, *D*, and *E* are located in the cartesian coordinate system, as shown. Give the coordinates of each point.

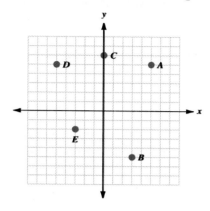

With the structure of the cartesian coordinate system established, we now have the tools to approach the first basic question of analytic geometry, presented earlier. In Section 3.2 we associate ordered pairs of real numbers with equations. Those ordered pairs will then be plotted in the plane to form the desired picture or graph.

CHECK YOURSELF ANSWERS

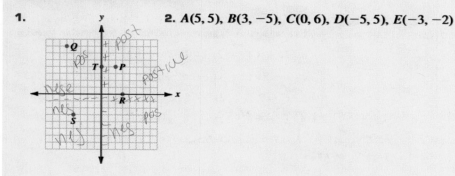

1.

2. $A(5, 5)$, $B(3, -5)$, $C(0, 6)$, $D(-5, 5)$, $E(-3, -2)$

3.1 EXERCISES

Build Your Skills

Give the quadrant in which each of the following points is located or the axis on which the point lies.

1. $(4, 5)$

2. $(-3, 2)$

3. $(-4, -3)$

4. $(2, -4)$

5. $(5, 0)$

6. $(-5, 7)$

7. $(-4, 7)$

8. $(-3, -7)$

9. $(0, -7)$

10. $(-3, 0)$

Graph each of the following points in the cartesian coordinate system.

11. $A(3, 5)$

12. $B(-4, 6)$

13. $C(-4, -6)$

14. $D(4, -6)$

15. $E(-5, 0)$

16. $F(6, 0)$

17. $G(-3, -4)$

18. $H(0, 4)$

19. $I(0, -5)$

20. $J(2, -3)$

Give the coordinates (ordered pairs) associated with each of the indicated points.

21. P

22. Q

23. R

24. S

25. T

26. U

27. V

28. W

29. X

30. Y

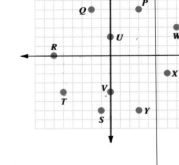

Transcribe Your Skills

All answers to Transcribe Your Skills exercises should be expressed as complete sentences.

31. The concepts of a point and a line were first described by the Greek mathematician Euclid. Use an encyclopedia to find out when Euclid lived and what his set of geometry books is called.

32. Euclid described a line as "length without width." Restate that definition in your own words.

Think About These

33. Graph the points with coordinates $(1, 2)$, $(2, 3)$, and $(3, 4)$. What do you observe? Can you give the coordinates of another point with the same property?

34. Graph points with coordinates $(1, 2)$, $(2, 4)$, and $(3, 6)$. What do you observe? Give the coordinates of another point with the same property.

35. Graph points with coordinates $(-1, 3)$, $(0, 0)$, and $(1, -3)$. What do you observe? Give the coordinates of another point with the same property.

36. Graph points with coordinates $(-1, -2)$, $(1, 2)$, and $(3, 6)$. What do you observe? Give the coordinates of another point with the same property.

Skillscan

Evaluate each expression for the given variable value.

a. $2x + 1$ $(x = 2)$ **b.** $2x + 1$ $(x = -2)$

c. $3 - 2x$ $(x = 1)$ **d.** $3 - 2x$ $(x = -1)$

e. $x^2 - 2$ $(x = 2)$ **f.** $x^2 - 2$ $(x = -2)$

g. $x^2 + 5$ $(x = 1)$ **h.** $x^2 + 5$ $(x = -1)$

3.2   An Introduction to Functions

TAPE IN6

OBJECTIVES: 1. To define a function
2. To identify the domain and range of a function

In the first two chapters, we looked at equations with one variable. One of the most important concepts in mathematics is the relation between two variables. This concept is used in virtually every field in which mathematics is applied.

Physicists are interested in the relationship between a planet's size and its gravitational pull.

Economists are interested in the relationship between taxes and unemployment.

Business people are interested in the relationship between an item's price and the number that will be purchased.

Biologists are interested in the relationship between a person's genetic makeup and her IQ.

In each of these examples, a researcher will match items from the first set (called the *domain*) with items from the second set (called the *range*). Let's look at an example in which we will match elements.

Suppose that a survey asks each person's name, age, phone number, and social security number. The following data might be gathered.

Name (Domain)	Age (Range)	Name (Domain)	Phone Number (Range)	Name (Domain)	Social Security Number (Range)
Aaron Able	19	Aaron Able	657-1234	Aaron Able	111-11-1111
Beth Bell	23	Beth Bell	252-6666	Beth Bell	222-22-2222
			254-3688		
Carl Carr	23	Carl Carr	255-0491	Carl Carr	333-33-3333
Dora Deer	57	Dora Deer	652-1453	Dora Deer	444-44-4444

Notice that, in both the first table and the third table, you could ask what this person's age is or what this person's social security number is and there would be only one right answer. On the other hand, if you ask what Beth Bell's phone number is, there could be two right answers. The first and third tables are examples of *functions*. The second table is a *relation* that is not a function.

> A *function* is a relation for which no element of the domain is matched with two or more different elements of the range.

EXAMPLE 1 Identifying Functions

Decide which of the following relations are functions.

(a) $1 \rightarrow 3$
$4 \rightarrow 3$
$7 \rightarrow 4$

(b) $2 \rightarrow -1$
$3 \rightarrow -2$
$4 \rightarrow -3$

(c) $1 \rightarrow 2$
$2 \rightarrow 3$
$2 \rightarrow 4$

Both (a) and (b) represent functions, but (c) does not. In (c), the domain element 2 is mapped to both 3 and 4.

Check Yourself 1

Decide which of the following relations is a function.

1. $1 \rightarrow 2$
$2 \rightarrow 3$
$1 \rightarrow 4$

2. $-1 \rightarrow 1$
$0 \rightarrow 1$
$1 \rightarrow 2$

A relation can also be defined as a set of points. When it is, the domain is the set of all the first components of the points, and the range is the set of all second components.

EXAMPLE 2 Identifying the Domain and Range of a Relation

The set of ordered pairs

$$\{(1, 2), (2, 3), (0, 4), (1, 3)\}$$ relatio

is a relation. The domain of the relation is

This is the set of first components.

$$\{0, 1, 2\}$$

The range of the relation is

This is the set of second components.

$$\{2, 3, 4\}$$

Check Yourself 2

Find the domain and range of the relation.

$$\{(5, 3), (4, 1), (3, -1)\}$$

We can specify the ordered pairs in a relation by three methods:

1. The ordered pairs can simply be listed, as in Example 2.

2. A rule or equation can be given to generate the ordered pairs in the relation.

3. A graph can be used to indicate the ordered pairs in the relation.

Since each ordered pair maps an element of the domain to an element of the range, we can determine from a list of points in a relation whether that relation is a function.

EXAMPLE 3 Identifying Functions

Note that in Example 3b, 1 is "paired with" both 3 and 2.

In Example 3c, both 3 and 4 are "paired with" 5. The relation is still a function since the range value can repeat.

(a) The relation $\{(1, 3), (2, 4), (3, 2)\}$ is a function.
(b) The relation $\{(1, 3), (2, 4), (1, 2)\}$ is *not* a function.
(c) The relation $\{(3, 5), (4, 5), (6, 7)\}$ is a function.

Check Yourself 3

Which of the following relations are also functions?

1. $\{(3, 5), (5, 7), (7, 9)\}$
2. $\{(3, 5), (4, 6), (3, 7)\}$
3. $\{(3, 5), (4, 5), (5, 6)\}$

Since relations are sets of ordered pairs, it seems reasonable to graph these relationships. That will lead us to a useful test for a function. Consider the graphs of two of the relations from Example 3.

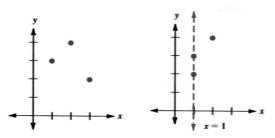

> To graph a relation (a set of ordered pairs), we must ensure that both the x and y components are real numbers.
>
> All the relations that we consider in this chapter will then consist of ordered pairs of real numbers.

Note that when a relation contains ordered pairs such as $(1, 3)$ and $(1, 2)$, with the same first components but different second components, a vertical line (here $x = 1$) will meet the graph of the relation in at least two points.

Interpreting the definition of a function geometrically then gives the following rule:

VERTICAL-LINE TEST

If a vertical line meets the graph of a relation in two or more points, the relation is *not* a function.

EXAMPLE 4 Using the Vertical-Line Test

Use the vertical-line test to determine whether the following graphs represent functions.

(a)

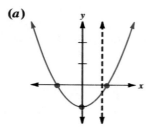

No vertical line can meet the graph in more than one point. The graph does represent a function.

(b)

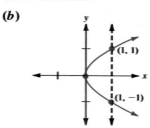

Since a vertical line can meet the graph in more than one point, this graph does not represent a function.

(c)

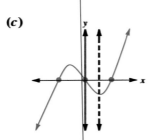

No vertical line can meet the graph in more than one point. The relation is a function.

Check Yourself 4

Use the vertical-line test to determine whether each of the following is a function.

1.

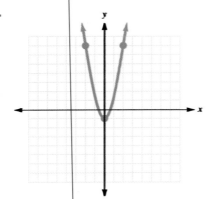

2.

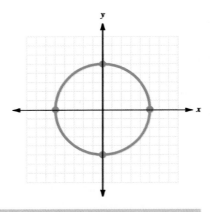

CHECK YOURSELF ANSWERS

1. (1) not a function; (2) function   **2.** Domain: $\{3, 4, 5\}$; range: $\{-1, 1, 3\}$
3. (1) Function; (2) not a function; (3) function   **4.** (1) Function; (2) not a function

3.2 EXERCISES

Build Your Skills

For each of the following relations, give the domain and range and state which are also functions.

1. $\{(1, 2), (3, 1), (5, 1)\}$

2. $\{(3, 1), (2, 5), (1, 9)\}$

3. $\{(4, 3), (2, 3), (2, 1)\}$

4. $\{(5, -2), (2, -2), (2, 4)\}$

5. $\{(-5, 4), (-7, 1), (2, 1)\}$

6. $\{(-4, 7), (-2, 5), (0, 3)\}$

7. $\{(1, 5), (2, 7), (-1, 5)\}$

8. $\{(1, 3), (5, 3), (7, 3)\}$

9. $\{(-1, 0), (-2, 0), (-1, 1)\}$

10. $\{(2, 4), (2, 5), (2, 6)\}$

Use the vertical-line test to determine whether each of the following graphs represents a function.

11. _function_

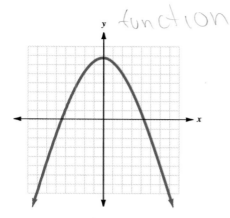

12.

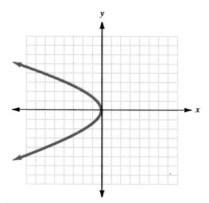

13. _function_

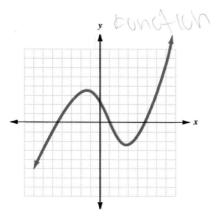

14.

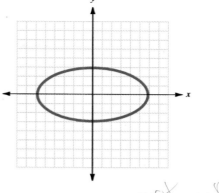

15. _not functi_

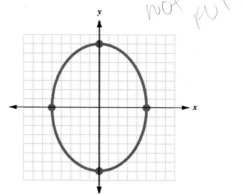

16.

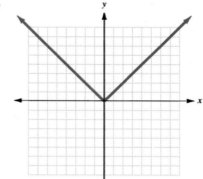

17. _fun_

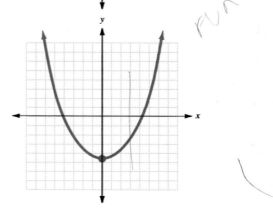

18.

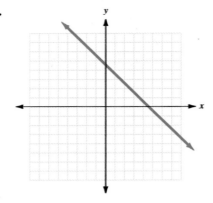

24.

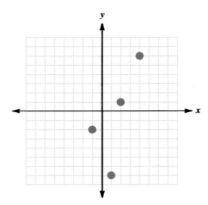

Transcribe Your Skills

19. Is the relationship between the length of a 6-in oak board and its price a function? Justify your answer.

20. Is the relationship between the length of a plane trip and its price a function? Justify your answer.

21. Assume that a graph is not a function because it does not pass the vertical-line test. What must be true of the x coordinates of the two points that the line passes through?

22. Is there a straight-line graph that is not a function? Explain.

25.

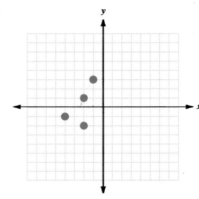

Think About These

Describe the domain and range for each of the given graphs.

23.

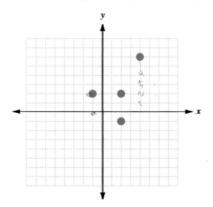

26.

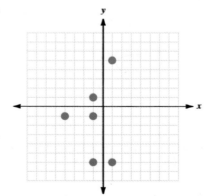

Skillscan

Evaluate the expression $2x^2 - 3x$ for each given value of x.

a. $x = -2$ **b.** $x = -1$ **c.** $x = 0$ **d.** $x = 1$

e. $x = 2$ **f.** $x = 3$

Graphing Linear Equations **3.3**

OBJECTIVES: 1. To graph linear equations
2. To graph linear equations by using intercepts
3. To graph vertical and horizontal lines

TAPE IN5

In Section 3.1, we introduced the cartesian coordinate system and located points in the plane by using that system. For the remainder of this chapter, we will show how the coordinate system can be used to provide "pictures," or graphs, of the solution set of an equation in two variables. First let's review the idea of a solution set.

We know that the solution set for an equation in one variable such as

$$2x - 3 = x + 1$$

is the set of real number values for x which will make the equation a true statement.

What about an equation in two variables such as this?

$$x + 2y = 6$$

The solution set is defined in a similar fashion.

In this case, the solution set has the single element 4. That value for x *satisfies* the equation—the equation is a true statement when 4 is substituted for x.

> ### SOLUTION SET FOR AN EQUATION IN TWO VARIABLES
>
> The *solution set* for an equation in two variables is the set containing all ordered pairs of real numbers (x, y) that will make the equation a true statement.

EXAMPLE 1 Verifying the Solution to a Linear Equation

The ordered pair $(4, 1)$ is a member of the solution set of the equation

$$x + 2y = 6$$

because substituting 4 for x and 1 for y results in a true statement.

$$4 + 2 \cdot 1 \stackrel{?}{=} 6$$
$$4 + 2 \stackrel{?}{=} 6$$
$$6 = 6$$

Check Yourself 1

Verify that $(2, -3)$ is a member of the solution set of $2x - y = 7$.

To find solutions for an equation in two variables, note that in most cases when a specific value for x is substituted in the equation, we can find the corresponding y value.

EXAMPLE 2 Finding the Missing Coordinate

We can also solve for y and write the equation as

$y = -3x + 9$

In this case, x is called the *independent variable* and y is the *dependent variable*.

As we assign values to x, we can find the corresponding values for y. The value for y *depends on* the value for x.

Suppose that $3x + y = 9$.

Let $x = 1$:

$3 \cdot 1 + y = 9$

$3 + y = 9$

$y = 6$

and the ordered pair $(1, 6)$ is a solution.

Let $x = 3$:

$3 \cdot 3 + y = 9$

$9 + y = 9$

$y = 0$

and $(3, 0)$ is a solution.

Check Yourself 2

Find three solutions of $x - 3y = 6$.

The graph of the solution set of an equation in two variables, usually called the *graph of the equation,* is the set of all points with coordinates (x, y) that will satisfy the equation.

In this section we are interested in a particular type of equation in x and y and the graph of that equation. The equations considered involve x and y to the first power, and they are called *linear equations.*

LINEAR EQUATIONS

An equation of the form

$ax + by = c$

where a and b cannot both be zero, is called the *standard form for a line*. Its graph is always a straight line.

EXAMPLE 3 Graphing by Plotting Points

Graph the equation

$x + y = 5$

Solution This is a linear equation in two variables. To draw its graph, we can begin by assigning values to x and finding the corresponding values for y. For instance, if $x = 1$, we have

$1 + y = 5$

$\quad y = 4$

Therefore $(1, 4)$ satisfies the equation and is in the graph of $x + y = 5$.

Similarly, $(2, 3)$, $(3, 2)$, and $(4, 1)$ are in the graph. Often these results are recorded in a table of values, as shown below.

$x + y = 5$

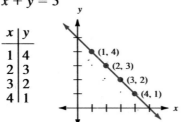

x	y
1	4
2	3
3	2
4	1

Since two points determine a straight line, technically two points are all that is needed to graph the equation. You may want to locate at least one other point as a check of your work.

If you first rewrite an equation so that y is isolated on the left side, it can be easily entered and graphed with a graphing calculator. In this case, graph the equation

$y = -x + 5$

We then plot the points determined and draw a straight line through those points. Every point on the graph of the equation $x + y = 5$ has coordinates which satisfy the equation, and every point with coordinates that satisfy the equation lies on the line.

Check Yourself 3

Graph the equation $2x - y = 6$.

The following algorithm summarizes our first approach to graphing a linear equation in two variables.

TO GRAPH A LINEAR EQUATION note book 9/17

STEP 1 Find at least three solutions for the equation, and write your results in a table of values.

STEP 2 Graph the points associated with the ordered pairs found in step 1.

STEP 3 Draw a straight line through the points plotted above to form the graph of the equation.

Two particular points are often used in graphing an equation because they are very easy to find. The *x intercept* of a line is the x coordinate of the point where the line crosses the x axis. If the x intercept exists, it can be found by setting $y = 0$ in the equation and solving for x. The *y intercept* is the y coordinate of the point where the line crosses the y axis. If the y intercept exists, it is found by letting $x = 0$ and solving for y.

EXAMPLE 4 Graphing by the Intercept Method

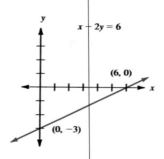 Solving for y, we get

$$y = \frac{1}{2}x - 3$$

To graph this result on your calculator, you can enter

$$\left(\frac{1}{2}\right)x - 3$$

Use the intercepts to graph the equation

$$x - 2y = 6$$

Solution To find the x intercept, let $y = 0$.

$$x - 2 \cdot 0 = 6$$
$$x = 6$$

To find the y intercept, let $x = 0$.

$$0 - 2y = 6$$
$$-2y = 6$$
$$y = -3$$

Graphing the intercepts and drawing the line through those intercepts, we have the desired graph.

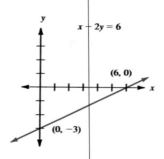

Check Yourself 4

Graph, using the intercept method.

$$4x + 3y = 12 \qquad \frac{3y}{3} = \frac{-4x}{3} + \frac{12}{3} \quad y$$

The following algorithm summarizes the steps of graphing a straight line by the intercept method.

GRAPHING BY THE INTERCEPT METHOD

STEP 1 Find the x intercept. Let $y = 0$ and solve for x.

STEP 2 Find the y intercept. Let $x = 0$ and solve for y.

STEP 3 Plot the two intercepts determined in steps 1 and 2.

STEP 4   Draw a straight line through the intercepts.

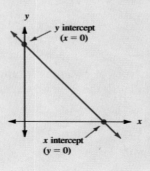

When can the intercept method not be used? Some straight lines have only one intercept. For instance, the graph of $x + 2y = 0$ passes through the origin. In this case, other points must be used to graph the equation.

EXAMPLE 5 Graphing a Line that Passes Through the Origin

Graph $x + 2y = 0$.

Solution Letting $y = 0$ gives

$$x + 2 \cdot 0 = 0$$

$$x = 0$$

Thus $(0, 0)$ is a solution, and the line has only one intercept.
    We proceed by choosing any other convenient values for x.
    If $x = 2$:

$$2 + 2y = 0$$

$$2y = -2$$

$$y = -1$$

So $(2, -1)$ is a solution. You can easily verify that $(4, -2)$ is also a solution.
    Again, plotting the points and drawing the line through those points, we have the desired graph.

Graph the equation

$$y = -\frac{1}{2}x$$

Note that the line passes through the origin.

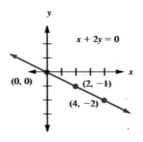

Check Yourself 5

Graph the equation $x - 3y = 0$.

There are two particular types of equations worthy of special attention. Their graphs are lines that are parallel to the x or y axis, and the equations are special cases of the general form

Again a and b cannot *both* be 0.

$$ax + by = c$$

in which either $a = 0$ or $b = 0$.

VERTICAL OR HORIZONTAL LINES

Here h and k are both constants.

1. A line with an equation of the form

$$x = h$$

is *vertical* (parallel to the y axis).

2. A line with an equation of the form

$$y = k$$

is *horizontal* (parallel to the x axis).

The following example illustrates both cases.

EXAMPLE 6 Graphing Vertical and Horizontal Lines

Since part *a* is a function, it can be graphed on your calculator. Part *b* is not a function, and can therefore not be graphed on the calculator.

(*a*) Graph the line with equation

$$y = 3$$

You can think of the equation in the equivalent form

$$0 \cdot x + y = 3$$

Note that any ordered pair of the form (___, 3) will satisfy the equation. Since x is multiplied by 0, y will always be equal to 3.

For instance, $(-2, 3)$ and $(5, 3)$ are on the graph. The graph, a horizontal line, is shown below.

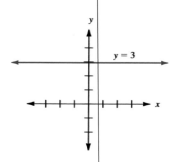

(*b*) Graph the line with equation

$x = -2$

In this case you can think of the equation in the equivalent form

$x + 0 \cdot y = -2$

Now any ordered pair of the form $(-2, \_\_)$ will satisfy the equation. Examples are $(-2, -1)$ and $(-2, 5)$. The graph, a vertical line, is shown below.

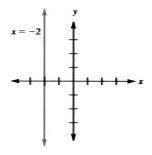

Check Yourself 6

Graph each equation.

1. $y = -3$ **2.** $x = 5$

Many applications involve linear relationships between variables, and the methods of this section can be used to picture or graph those relationships.

In Section 3.1, we mentioned that the axes may have different scales. The following example is such an application involving a linear equation.

EXAMPLE 7 Adjusting the Scale for the *x* and *y* Axes

A car rental agency advertises daily rates for a midsize automobile at $20 per day plus 10¢ per mile. The cost per day C and the distance driven in miles s are then related by the following linear equation:

$C = 0.10s + 20$ (1)

Graph the relationship between C and s.

Solution First, we proceed by finding three points on the graph.

s	C
0	20
100	30
200	40

So as the distance s varies from 0 to 200 mi, the cost C changes from $20 to $40. To draw a "reasonable" graph, it makes sense to choose a different scale for the horizontal (or s) axis than for the vertical (or C) axis.

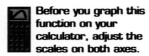 Before you graph this function on your calculator, adjust the scales on both axes.

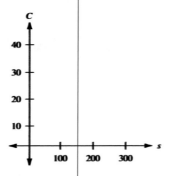

We have chosen units of 100 for the s axis and units of 10 for the C axis. The graph can then be completed as shown below.

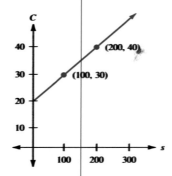

Note that the graph of Equation (1) does not extend beyond the first quadrant. This is due to the nature of our problem in which solutions are only realistic when $s \geq 0$.

Check Yourself 7

A salesperson's monthly salary S is based on a fixed salary of $1200 plus 8 percent of all monthly sales x. The linear equation relating S and x is

$$S = 0.08x + 1200$$

Graph the relationship between S and x. *Hint:* Find the monthly salary for sales of $0, $10,000, and $20,000.

CHECK YOURSELF ANSWERS

1. $2(2) - (-3) \stackrel{?}{=} 7$
$4 + 3 \stackrel{?}{=} 7$
$7 = 7$

2. $(0, -2)$, $(3, -1)$, and $(6, 0)$ are three possible solutions.

3. $2x - y = 6$

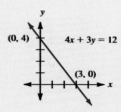

x	y
0	-6
1	-4
2	-2

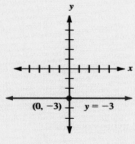

4.

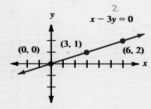

5.

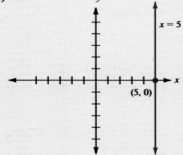

6. (1)

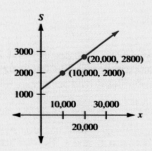

(2)

7.

home work
1-33 -57, 59

3.3 EXERCISES

Build Your Skills

Complete each table of values, and then graph the given equations.

1-33
how do you graph This

1. $y = x + 1$

x	y
-2	
-1	
0	
1	
2	

2. $x + y = 3$

x	y
-2	
-1	
0	
1	
2	

3. $y = x - 4$

x	y
-2	
-1	
0	
1	
2	

4. $2x - y = 6$

x	y
-1	
0	
1	
2	
3	

Graph each of the following equations.

5. $x + y = 6$

6. $x - y = 5$

7. $y = x - 2$

8. $y = x + 4$

9. $y = x + 1$

10. $y = 2x + 1$

11. $y = -2x + 1$

12. $y = -3x + 1$

13. $y = \frac{1}{2}x - 3$

14. $y = 2x - 3$

15. $y = -x - 3$

16. $y = -2x - 3$

17. $x + 2y = 0$

18. $x - 3y = 0$

Find the x and y intercepts, and then graph each of the following equations.

19. $x - 2y = 4$

20. $x + 3y = 6$

21. $2x - y = 6$

22. $3x + 2y = 12$

23. $2x + 5y = 10$

24. $2x - 3y = 6$

25. $x + 4y + 8 = 0$

26. $2x - y + 6 = 0$

Graph each of the following equations.

27. $x = 4$

28. $x = -3$

29. $y = 4$

30. $y = -5$

31. A car rental agency charges $12 per day and 8¢ per mile for the use of a compact automobile. The cost of the rental C and the number of miles driven per day s are related by the linear equation

$$C = 0.08s + 12$$

Graph the relationship between C and s. Be sure to select appropriate scaling for the s and C axes.

32. A bank has the following structure for charges on checking accounts. The monthly charge consists of a fixed amount of $8 and an additional charge of 5¢ per check. The monthly cost of an account C and the number of checks written per month n are related by the linear equation

$$C = 0.05n + 8$$

Graph the relationship between C and n.

33. A college has tuition charges based on the following pattern. Tuition is \$25 per credit-hour plus a fixed student fee of \$40. The total tuition charge T and the number of credit-hours taken h are related by the linear equation

$$T = 25h + 40$$

Graph the relationship between T and h.

34. A salesperson's weekly salary is based on a fixed amount of \$200 plus 10 percent of the total amount of weekly sales. The weekly salary S and the amount of weekly sales x (in dollars) are related by the linear equation

$$S = 0.1x + 200$$

Graph the relationship between S and x.

Transcribe Your Skills

35. Is the relationship described in Exercise 31 a function? Explain.

36. Is the relationship described in Exercise 32 a function? Explain.

37. Describe the characteristics of an equation that is easiest to graph with the intercept method.

38. What do we mean when we use the word "application"?

Think About These

As was pointed out earlier, linear relationships are useful in many applications. The following exercises will help build your skill in writing the corresponding equations. Express each linear relationship by using symbols.

39. y is 3 times x.

40. m is 4 more than n.

41. y is 3 more than twice x.

42. L is 5 less than 3 times W.

43. The sum of r and s is 7.

44. The sum of twice w and 3 times z is 10.

45. x is equal to the sum of 4 times y and 7.

46. The difference of twice c and d is 2.

Two distinct lines in the plane either are parallel or intersect. Graph each pair of equations on the same set of axes, and find the point of intersection, where possible.

47. $x + y = 6$
   $x - y = 4$

48. $y = x + 3$
   $y = -x + 1$

49. $y = 2x$
   $y = x + 1$

50. $2x + y = 4$
   $2x + y = 6$

51. Graph $y = x$ and $y = 2x$ on the same set of axes. What do you observe?

52. Graph $y = 2x + 1$ and $y = -2x + 1$ on the same set of axes. What do you observe?

53. Graph $y = 2x$ and $y = 2x + 1$ on the same set of axes. What do you observe?

54. Graph $y = 3x + 1$ and $y = 3x - 1$ on the same set of axes. What do you observe?

55. Graph $y = 2x$ and $y = -\dfrac{1}{2}x$ on the same set of axes. What do you observe?

56. Graph $y = \dfrac{1}{3}x + 2$ and $y = -3x + 2$ on the same set of axes. What do you observe?

57. The amount of yearly worldwide carbon dioxide emissions (C) in millions of metric tons and the number of years since 1950 (Y) are related by the linear equation

$$C = 125Y + 1600$$

Graph the relationship between C and Y.

58. Use the graph from Exercise 57 to predict the amount of carbon dioxide emissions worldwide in the year 2000.

59. The 1990 amendments to the Clean Air Act strengthened the limits on emissions of sulfur dioxide in the United States. These new regulations restrict the yearly emissions of SO_2 to 8.9 million tons. The relationship between total SO_2 emissions (S) after the year 2000 and any given year (Y) after that time will be given by the linear equation

$$S = 8.9(Y - 2000) \qquad (S \text{ is in millions of tons})$$

or $\qquad S = 8.9Y - 17{,}800$

Graph the relationship between S and Y. Notice that, in this problem, Y actually represents the year, for example, 2010.

60. Use the graph from Exercise 59 to predict the total SO_2 emissions since 2000 in the year 2005.

Graphing Utility

Use your graphing utility to graph each of the following equations.

61. $y = -3$

62. $y = 2$

63. $y = 3x - 1$

64. $y = -2x + 5$

Adjust the viewing window on your graphing utility so that x values from -50 to 50 and y values from -20 to 20 are displayed. Then graph each of the following.

65. $y = 5x - 15$

66. $y = -2x + 8$

67. $y = \dfrac{1}{5}x - 3$

68. $y = -\dfrac{1}{3}x + 10$

Skillscan (Section 1.4)

Evaluate each of the following expressions.

a. $\dfrac{6 - 3}{7 - 4}$

b. $\dfrac{11 - 2}{6 - 3}$

c. $\dfrac{-8 - 4}{-3 - 3}$

d. $\dfrac{-14 - 2}{2 - 6}$

e. $\dfrac{5 - (-2)}{4 - 2}$

f. $\dfrac{-5 - (-4)}{2 - (-2)}$

g. $\dfrac{-5 - (-5)}{7 - 2}$

h. $\dfrac{-8 + (-8)}{-4 + 4}$

3.4 The Slope of a Line

TAPE IN7

OBJECTIVES:
1. To find the slope of a line
2. To find the slope of a line parallel to a given line
3. To find the slope of a line perpendicular to a given line
4. To graph linear equations by using the slope of a line

Recall the second basic problem of analytic geometry: Given a set of geometric conditions, we must be able to find the corresponding algebraic equation that satisfies those conditions.

For instance, later in this chapter we will be concerned with finding the equation of a line, given information such as a point on the line and the inclination of that line. In this section we start by defining the *slope* of a line, which will give us a numerical measure of the steepness, or inclination, of that line.

To define a formula for slope, choose any two distinct points on the line, say, P with coordinates (x_1, y_1) and Q with coordinates (x_2, y_2). As we move along the line from P to Q, the x value, or coordinate, changes from x_1 to x_2. That change in x, also called the *horizontal change*, is $x_2 - x_1$. Similarly, as we move from P to Q, the corresponding change in y, called the *vertical change*, is $y_2 - y_1$. The *slope* is then defined as the ratio of the vertical change to the horizontal change. The letter m is used to represent the slope, which we now define as follows.

SLOPE OF A LINE

The *slope* of a line through two distinct points $P(x_1, y_1)$ and $Q(x_2, y_2)$ is given by

$$m = \frac{\text{change in } y}{\text{change in } x} = \frac{y_2 - y_1}{x_2 - x_1} \qquad \textbf{(1)}$$

where $x_1 \neq x_2$.

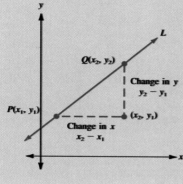

The difference $x_2 - x_1$ is often called the *run*. The difference $y_2 - y_1$ is the *rise*. So the slope can be thought of as "rise over run."

Note that $x_1 \neq x_2$ or $x_2 - x_1 \neq 0$ ensures that the denominator is nonzero, so that the slope is defined. It also means the line cannot be vertical.

Let's look at some examples using the definition.

EXAMPLE 1 Finding the Slope Through Two Points

Find the slope of the line through the points $(-3, 2)$ and $(3, 5)$.

Solution Let $(x_1, y_1) = (-3, 2)$ and $(x_2, y_2) = (3, 5)$. From the definition we have

$$m = \frac{5 - 2}{3 - (-3)} = \frac{3}{6} = \frac{1}{2}$$

Note that if the pairs are reversed, so that

$$(x_1, y_1) = (3, 5) \qquad \text{and} \qquad (x_2, y_2) = (-3, 2)$$

then we have

$$m = \frac{2 - 5}{-3 - 3} = \frac{-3}{-6} = \frac{1}{2}$$

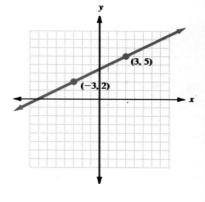

The slope in either case is the same.

The work of Example 1 suggests that no matter which point is chosen as (x_1, y_1) or (x_2, y_2), the slope formula will give the same result. You must simply stay with your choice once it is made, and use the same order of subtraction in the numerator and the denominator.

Check Yourself 1

Find the slope of the line through the points $(-2, -1)$ and $(1, 1)$.

The slope indicates both the direction of a line and its steepness. First we will compare the steepness of two examples.

EXAMPLE 2 Finding the Slope

Find the slope of the line through $(-2, -3)$ and $(2, 5)$.

Solution Again, by Equation (1),

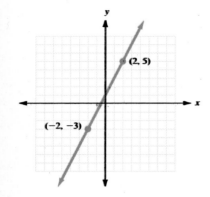

$$m = \frac{5 - (-3)}{2 - (-2)} = \frac{8}{4} = 2$$

Compare the lines of Examples 1 and 2. In Example 1 the line has slope $\frac{1}{2}$. The slope here is 2. Now look at the two lines. Do you see the idea of slope as measuring steepness? The greater the absolute value of the slope, the steeper the line.

Check Yourself 2

Find the slope of the line through the points $(-1, 2)$ and $(2, 7)$. Draw a sketch of this line and the line of the Check Yourself 1 exercise on the same coordinate axes. Compare the lines and the two slopes.

The sign of the slope indicates in which direction the line tilts. The following example illustrates.

EXAMPLE 3 Finding the Slope

Find the slope of the line through the points $(-1, 2)$ and $(4, -3)$.

Solution We see that

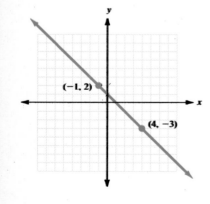

$$m = \frac{-3 - 2}{4 - (-1)} = \frac{-5}{5} = -1$$

Now the slope is negative.

Comparing this with our previous examples, we see that

1. In Examples 1 and 2, the lines were rising from left to right, and the slope was *positive*.

2. In this example, the line is falling from left to right, and the slope is *negative*.

Check Yourself 3

Find the slope of the line through the points $(-2, 5)$ and $(4, -1)$.

Let's continue by looking at the slopes of lines in two particular cases.

EXAMPLE 4 Finding the Slope of a Horizontal Line

Find the slope of the line through $(-2, 3)$ and $(5, 3)$.

$$m = \frac{3 - 3}{5 - (-2)} = \frac{0}{7} = 0$$

The slope of the line is 0. Note that the line is parallel to the x axis and $y_2 - y_1 = 0$. *The slope of any horizontal line will be 0.*

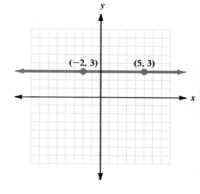

Check Yourself 4

Find the slope of the line through the points $(-2, -4)$ and $(3, -4)$.

EXAMPLE 5 Finding the Slope of a Vertical Line

Find the slope of the line through the points $(1, -3)$ and $(1, 4)$.

$$m = \frac{4 - (-3)}{1 - 1} = \frac{7}{0}$$

Here the line is parallel to the y axis, and $x_2 - x_1$ (the denominator of the slope formula) is 0. Since division by 0 is undefined, we say that the slope is *undefined,* as will be the case for *any vertical line.*

Be very careful not to confuse a slope of 0 (in the case of a horizontal line) with an undefined slope or no slope (in the case of a vertical line).

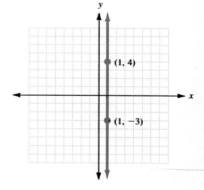

Check Yourself 5

Find the slope of the line through the points $(2, -3)$ and $(2, 7)$.

The following summarizes our work in the previous examples.

1. **If the slope of a line is positive, the line is rising from left to right.**
2. **If the slope of a line is negative, the line is falling from left to right.**
3. **If the slope of a line is 0, the line is horizontal.**
4. **If the slope of a line is undefined, the line is vertical. We can also say that the line has no slope in this case.**

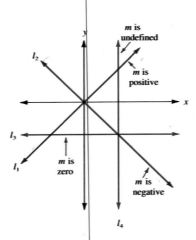

There are two more important results regarding the slope. Recall from geometry that two distinct lines in the plane either intersect at a point or never intersect. Two lines in the plane that do not intersect are called *parallel lines.* It can be shown that two distinct parallel lines will always have the same slope, and we can state the following.

SLOPE OF PARALLEL LINES

For nonvertical lines L_1 and L_2, if line L_1 has slope m_1 and line L_2 has slope m_2, then

L_1 is parallel to L_2 if and only if $m_1 = m_2$

Note All vertical lines are parallel to each other.

This means that if the lines are parallel, then their slopes are equal. Conversely, if the slopes are equal, then the lines are parallel.

Mathematicians use the symbol

$\Leftrightarrow$

to represent "if and only if."

EXAMPLE 6 Parallel Lines

Are lines L_1 through $(2, 3)$ and $(4, 6)$ and L_2 through $(-4, 2)$ and $(0, 8)$ parallel, or do they intersect?

$$m_1 = \frac{6 - 3}{4 - 2} = \frac{3}{2}$$

$$m_2 = \frac{8 - 2}{0 - (-4)} = \frac{6}{4} = \frac{3}{2}$$

Since the slopes of the lines are equal, the lines are parallel. They do *not* intersect.

Unless, of course, L_1 and L_2 are actually the *same line*. In this case a quick sketch will show that the lines are distinct.

Check Yourself 6

Are lines L_1 through $(-2, -1)$ and $(1, 4)$ and L_2 through $(-3, 4)$ and $(0, 8)$ parallel, or do they intersect?

Two lines are perpendicular if they intersect at right angles. Also if two lines (which are not vertical or horizontal) are perpendicular, their slopes are the negative reciprocals of each other. We can then state the following result for perpendicular lines.

SLOPE OF PERPENDICULAR LINES

For nonvertical lines L_1 and L_2, if line L_1 has slope m_1 and line L_2 has slope m_2, then

L_1 is perpendicular to L_2 if and only if $m_1 = -\dfrac{1}{m_2}$ *slope* (circled)

or equivalently

$m_1 \cdot m_2 = -1$

Note Horizontal lines are perpendicular to vertical lines.

EXAMPLE 7 Perpendicular Lines

Are lines L_1 through points $(-2, 3)$ and $(1, 7)$ and L_2 through points $(2, 4)$ and $(6, 1)$ perpendicular?

$$m_1 = \frac{7 - 3}{1 - (-2)} = \frac{4}{3}$$

$$m_2 = \frac{1 - 4}{6 - 2} = -\frac{3}{4}$$

Since the slopes are negative reciprocals, the lines are perpendicular.

Note

$$\left(\frac{4}{3}\right)\left(-\frac{3}{4}\right) = -1$$

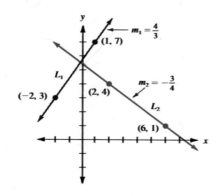

Check Yourself 7

Are lines L_1 through points $(1, 3)$ and $(4, 1)$ and L_2 through points $(-2, 4)$ and $(2, 10)$ perpendicular?

Handwritten: help 4 2

Given the equation of a line, we can also find its slope, as the following example illustrates.

EXAMPLE 8 Finding the Slope from an Equation

Note Let's try solving the original equation for y:

$3x + 2y = 6$

$2y = -3x + 6$

$y = -\dfrac{3}{2}x + 3$

Consider the coefficient of x. What do you observe?

Find the slope of the line with equation $3x + 2y = 6$.

Solution First, find any two points on the line. In this case, $(2, 0)$ and $(0, 3)$, the x and y intercepts, will work and are easy to find. From the slope formula

$$m = \frac{0 - 3}{2 - 0} = \frac{-3}{2} = -\frac{3}{2}$$

The slope of the line with equation $3x + 2y = 6$ is $-\dfrac{3}{2}$.

Check Yourself 8

Find the slope of the line with equation $3x - 4y = 12$.

Handwritten:
$-4y = 12 - 3x$
$\dfrac{-4y}{-4} = \dfrac{-3x}{-4} + \dfrac{12}{-4}$
$y = \dfrac{3}{4}x + 3$

The slope of a line can also be useful in graphing a line. In the following example, the slope of a line is used in sketching its graph.

EXAMPLE 9 Graphing a Line with a Given Slope

Suppose a line has slope $\dfrac{3}{2}$ and passes through the point $(5, 2)$. Graph the line.

Solution First locate the point $(5, 2)$ in the coordinate system. Now since the slope, $\dfrac{3}{2}$, is the ratio of the change in y to the change in x, move 2 units to the right in the x direction and then 3 units up in the y direction. This determines a second point, here $(7, 5)$, and we can draw our graph.

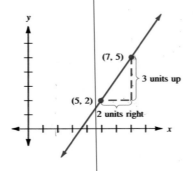

Check Yourself 9

o.k.

Graph the line with slope $-\dfrac{3}{4}$ which passes through the point $(2, 3)$.

Hint: Consider the x change as 4 units and the y change as -3 units (down).

Since, given a point on a line and its slope, we can graph the line, we also should be able to write its equation. That is, in fact, the case, as we will see in Section 3.7.

CHECK YOURSELF ANSWERS

1. $m = \dfrac{2}{3}$ **2.** $m = \dfrac{5}{3}$ **3.** $m = -1$ **4.** 0   **5.** Undefined

6. The lines intersect   **7.** The lines are perpendicular   **8.** $m = \dfrac{3}{4}$

9.

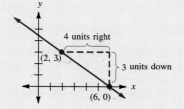

3.4 EXERCISES

$|-1|$

Build Your Skills

Find the slope (if it exists) of the line determined by the following pairs of points. Sketch each line so that you can compare the slopes.

1. $(2, 3)$ and $(4, 7)$

2. $(-1, 2)$ and $(5, 3)$

3. $(2, -3)$ and $(-2, -5)$

4. $(0, 0)$ and $(5, 7)$

5. $(2, 5)$ and $(-3, 5)$

6. $(-2, -4)$ and $(5, 3)$

7. $(-1, 4)$ and $(-1, 7)$

8. $(4, 2)$ and $(-2, 5)$

9. $(8, -3)$ and $(-2, -5)$

10. $(4, -3)$ and $(-2, 7)$

11. $(-4, -3)$ and $(2, -7)$

12. $(3, 6)$ and $(3, -4)$

Find the slope of the line determined by each equation.

13. $y = -3x - \dfrac{1}{2}$ $m - 3$ $b - \frac{1}{2}$

14. $y = \dfrac{1}{4}x + 3$

15. $y + \dfrac{1}{2}x = 2$ $y = -\frac{1}{2}x + 2$

16. $2y - 3x + 5 = 0$ $y = \frac{1}{2}x + 2$

17. $2x - 3y = 6$ $m = \frac{1}{2}$

18. $x + 4y = 4$ $b = 2$

19. $3x + 4y = 12$

20. $x - 3y = 9$

Help

Are the following pairs of lines parallel, perpendicular, or neither?

21. L_1 through $(-2, -3)$ and $(4, 3)$
L_2 through $(3, 5)$ and $(5, 7)$

22. L_1 through $(-2, 4)$ and $(1, 8)$
L_2 through $(-1, -1)$ and $(-5, 2)$

23. L_1 through $(8, 5)$ and $(3, -2)$
L_2 through $(-2, 4)$ and $(4, -1)$

24. L_1 through $(-2, -3)$ and $(3, -1)$
L_2 through $(-3, 1)$ and $(7, 5)$

25. L_1 with equation $x - 3y = 6$
L_2 with equation $3x + y = 3$

26. L_1 with equation $x + 2y = 4$
L_2 with equation $2x + 4y = 5$

27. Find the slope of any line parallel to the line through points $(-2, 3)$ and $(4, 5)$.

28. Find the slope of any line perpendicular to the line through points $(0, 5)$ and $(-3, -4)$.

? 29. A line passing through $(-1, 2)$ and $(4, y)$ is parallel to a line with slope 2. What is the value of y?
0

30. A line passing through $(2, 3)$ and $(5, y)$ is perpendicular to a line with slope $\dfrac{3}{4}$. What is the value of y?

If points P, Q, and R are collinear (lie on the same line), the slope of the line through P and Q must equal the slope of the line through Q and R. Use the slope concept to determine whether the following sets of points are collinear.

31. $P(-2, -3)$, $Q(3, 2)$, and $R(4, 3)$

32. $P(-5, 1)$, $Q(-2, 4)$, and $R(4, 9)$

33. $P(0, 0)$, $Q(2, 4)$, and $R(-3, 6)$

34. $P(-2, 5)$, $Q(-5, 2)$, and $R(1, 12)$

35. $P(2, 4)$, $Q(-3, -6)$, and $R(-4, 8)$

36. $P(-1, 5)$, $Q(2, -4)$, and $R(-2, 8)$

check the slope

Graph the lines through each of the specified points having the given slope.

37. $(0, 1)$, $m = 3$

38. $(0, -2)$, $m = -2$

39. $(3, -1)$, $m = 2$

40. $(2, -3)$, $m = -3$

41. $(2, 3)$, $m = \dfrac{2}{3}$

42. $(-2, 1)$, $m = -\dfrac{3}{4}$

43. $(4, 2)$, $m = 0$

44. $(3, 0)$, m is undefined

45. On the same graph, sketch lines with slope 2 through each of the following points: $(-1, 0)$, $(2, 0)$, and $(5, 0)$.

46. On the same graph, sketch one line with slope $\dfrac{1}{3}$ and one line with slope -3, having both pass through point $(2, 3)$.

Transcribe Your Skills

47. In Exercise 31, you used the slope concept to show that three points were collinear. Explain how this worked.

48. Given equations for two lines, how many points of intersection could they have?

Think About These

A four-sided figure (quadrilateral) is a parallelogram if the opposite sides have the same slope. If the adjacent sides are perpendicular, the figure is a rectangle.

For each of the following quadrilaterals $ABCD$, determine whether it is a parallelogram; then determine whether it is a rectangle.

49. $A(0, 0)$, $B(2, 0)$, $C(2, 3)$, $D(0, 3)$

50. $A(-3, 2)$, $B(1, -7)$, $C(3, -4)$, $D(-1, 5)$

51. $A(0, 0)$, $B(4, 0)$, $C(5, 2)$, $D(1, 2)$

52. $A(-3, -5)$, $B(2, 1)$, $C(-4, 6)$, $D(-9, 0)$

Graphing Utility

Solve each equation for y, then use your graphing utility to graph each of the following equations.

53. $2x + 5y = 10$

54. $5x - 3y = 12$

55. $x + 7y = 14$

56. $-2x - 3y = 9$

Skillscan (Section 1.2)

Simplify the following.

a. $|6 - (-1)|$ b. $|-3 - 2|$ c. $|-5 + 2|$
d. $\sqrt{3^2 + 4^2}$ e. $\sqrt{(-2)^2 + (-3)^2}$
f. $\sqrt{2^2 + (-1)^2}$

Distance and Midpoints   **3.5**

OBJECTIVES: 1. To find the distance between two points on a plane
2. To find the midpoint of two points on a plane
3. To determine if three points are collinear

TAPE IN8

Another geometric concept that can be approached through the use of the cartesian coordinate system is distance. For instance, we may want to find the length of a line segment in the plane—or the distance between two points in a plane.

Let's start with two particular cases.

In general, given (x_1, y_1) and (x_2, y_1), two points with the same y coordinate,

$$d = |x_2 - x_1|$$

The absolute value ensures a nonnegative distance.

EXAMPLE 1 Finding the Distance Between Two Points

Find the distance between points $(-3, 4)$ and $(2, 4)$.

Solution From a sketch of the problem we know that the segment between the points is horizontal, and the distance between the points is given by

$$d = |2 - (-3)| = |5| = 5$$

$$2 + 3 = |5| = 5$$

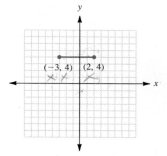

Check Yourself 1

Find the distance between points $(-3, 2)$ and $(5, 2)$.

EXAMPLE 2 Finding the Distance Between Two Points

Find the distance between points $(5, -2)$ and $(5, 7)$.

Solution Now the segment between the points is vertical, and the distance is given by

$$d = |7 - (-2)| = |9| = 9$$

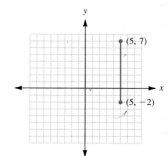

Check Yourself 2

Find the distance between points $(4, -3)$ and $(4, 7)$.

Here the distance is given by $|y_2 - y_1|$.

Now, to approach the more general problem of the distance between any two points in the plane, we need the *Pythagorean theorem.* Let's restate that theorem.

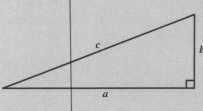

PYTHAGOREAN THEOREM

In any right triangle, the sum of the squares of the lengths of the legs is equal to the square of the length of the hypotenuse. In symbols,

$$a^2 + b^2 = c^2$$

We can write

$$c = \sqrt{a^2 + b^2}$$

Recall that the square root symbol $\sqrt{}$ represents the *positive* square root.

EXAMPLE 3 Finding the Distance Between Two Points

Find the distance between points $P(2, -3)$ and $Q(5, 4)$.

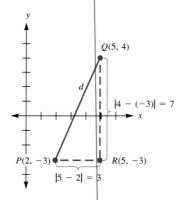

Draw a horizontal line through P and a vertical line through Q. These lines must intersect at the point $R(5, -3)$. From our sketch and the Pythagorean theorem, the distance, labeled d, is given by

$$d = \sqrt{7^2 + 3^2}$$
$$= \sqrt{49 + 9} = \sqrt{58}$$

Check Yourself 3

Use the Pythagorean theorem to find the distance between the two points $(-1, 4)$ and $(3, 1)$.

We can now apply the same strategy to derive a general formula for the distance between two points in the plane.

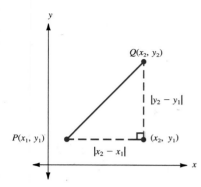

If $P(x_1, y_1)$ and $Q(x_2, y_2)$ are any two points in the plane, the distance between those points is the length of the hypotenuse of the right triangle, as shown. By the Pythagorean theorem,

$$d^2 = |x_2 - x_1|^2 + |y_2 - y_1|^2$$

or

$$d = \sqrt{(x_2 - x_1)^2 + (y_2 - y_1)^2}$$

This result is known as the *distance formula*.

Note the use of the fact that $|a|^2 = a^2$ in this derivation.

Also d is positive since the indicated root is the principal root, which is positive.

There will be further discussion of principal roots in Chapter 7.

> **DISTANCE BETWEEN TWO POINTS**
> The distance between two points $P(x_1, y_1)$ and $Q(x_2, y_2)$ is given by
> $$d = \sqrt{(x_2 - x_1)^2 + (y_2 - y_1)^2} \tag{1}$$

EXAMPLE 4 Finding the Distance Between Two Points

Find the distance between $(-2, 3)$ and $(5, 5)$.

Solution Again, in finding distance, as with slope, it makes no difference which point is chosen as (x_1, y_1) or (x_2, y_2). Applying Equation (1), we have

$$d = \sqrt{[5 - (-2)]^2 + (5 - 3)^2}$$
$$= \sqrt{7^2 + 2^2}$$
$$= \sqrt{53}$$

Check Yourself 4

Find the distance between points $(-3, -4)$ and $(2, 3)$.

Let's work through one further example which applies the distance formula.

EXAMPLE 5 Verifying an Isosceles Triangle

Show that points $P(-1, 3)$, $Q(3, 6)$, and $R(6, 2)$ form an isosceles triangle (a triangle with two sides of equal length).

Solution From the sketch of the problem, we can make the following calculations:

You may also have observed that

$$d_1{}^2 + d_2{}^2 = d_3{}^2$$

which, by the Pythagorean theorem, means that the triangle is also a right triangle. Considering the slopes of PQ and QR shows another means of verifying that triangle PQR is a right triangle. Try that for yourself.

$$d_1 = \sqrt{[3 - (-1)]^2 + (6 - 3)^2}$$
$$= \sqrt{4^2 + 3^2} = 5$$
$$d_2 = \sqrt{(2 - 6)^2 + (6 - 3)^2}$$
$$= \sqrt{(-4)^2 + 3^2} = 5$$
$$d_3 = \sqrt{[6 - (-1)]^2 + (2 - 3)^2}$$
$$= \sqrt{7^2 + (-1)^2} = \sqrt{50}$$

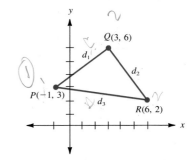

Since $d_1 = d_2 = 5$, the triangle is isosceles.

Check Yourself 5

Show that points $A(-2, 1)$, $B(3, 4)$, and $C(1, -4)$ are the vertices of an isosceles triangle.

When we connect two points, we have a line segment. The two points are called *endpoints*. The point on the line segment that is an equal distance from the two endpoints is called the *midpoint*.

EXAMPLE 6 Finding the Midpoint

Find the midpoint on the line segment connecting points $A(2, -1)$ and $B(2, 7)$.

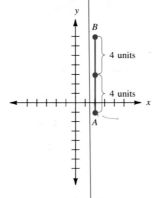

From the sketch, we can see that $(2, 3)$ is on the segment and an equal distance from the endpoints.

Check Yourself 6

Find the midpoint on the line segment connecting $A(3, 5)$ and $B(-5, 5)$.

To find the midpoint of any line segment, we find the point that has as its coordinates the average x and average y of the endpoints. The following formula is used.

> **MIDPOINT FORMULA**
>
> **The midpoint between (x_1, y_1) and (x_2, y_2) is**
>
> $$\left(\frac{x_1 + x_2}{2}, \frac{y_1 + y_2}{2} \right)$$

EXAMPLE 7 Finding the Midpoint

Find the midpoint of the segment with endpoints $(-1, 9)$ and $(5, 5)$.

Substituting into the formula, we get

$$\left(\frac{-1 + 5}{2}, \frac{9 + 5}{2} \right)$$

or $(2, 7)$.

Check Yourself 7

Find the midpoint of the segment with endpoints $(3, -6)$ and $(7, 4)$.

> **CHECK YOURSELF ANSWERS**
>
> **1.** 8 **2.** 10 **3.** 5 **4.** $\sqrt{74}$ **5.** $AB = AC = \sqrt{34}$ **6.** $(-1, 5)$
> **7.** $(5, -1)$

3.5 EXERCISES

Build Your Skills

Find the distance between each of the following pairs of points.

1. $(5, 2)$ and $(5, 4)$

2. $(1, 4)$ and $(5, 4)$

3. $(-1, 3)$ and $(9, 3)$

4. $(4, -2)$ and $(4, -3)$

5. $(2, 2)$ and $(5, -4)$

6. $(4, 7)$ and $(-2, -3)$

7. $(-2, 1)$ and $(-1, 6)$

8. $(-2, 3)$ and $(3, -4)$

9. $(-3, -4)$ and $(-2, 5)$

10. $(3, -4)$ and $(-4, 0)$

11. Using the distance formula, show that points $A(-2, 1)$, $B(2, 4)$, and $C(5, 0)$ are the vertices of a right triangle.

12. Using the distance formula, show that points $P(-1, 3)$, $Q(2, 1)$, and $R(4, 4)$ are the vertices of a right triangle.

Find the midpoint of a segment with the given endpoints.

13. $(0, 4)$ and $(-8, 4)$

14. $(2, -5)$ and $(2, 7)$

15. $(3, 6)$ and $(7, 18)$

16. $(1, 0)$ and $(13, -24)$

17. $(-2, 5)$ and $(-5, -8)$

18. $(-8, 12)$ and $(9, -3)$

19. $(2, -5)$ and $\left(\dfrac{1}{2}, -\dfrac{3}{2}\right)$

20. $\left(-\dfrac{1}{4}, -\dfrac{3}{4}\right)$ and $(3, 1)$

Think About These

21. Find the perimeter of the parallelogram with vertices at $(-2, -4)$, $(3, 8)$, $(7, 5)$, and $(2, -7)$.

22. Find the perimeter of the parallelogram with vertices at $(-3, -5)$, $(3, 3)$, $(6, -1)$, and $(0, -9)$.

23. Show that points $A(-1, 2)$, $B(2, 7)$, $C(7, 4)$, and $D(4, -1)$ are the vertices of a square. Find the slope of diagonals AC and BD. What do you observe?

Three points P, Q, and R are collinear if the sum of distances PQ and QR is equal to distance PR. Using the distance formula, determine whether the following sets of points are collinear.

24. $P(-1, -2)$, $Q(3, 1)$, and $R(7, 4)$

25. $P(-2, 3)$, $Q(1, -1)$, and $R(5, -4)$

26. $P(0, 0)$, $Q(2, 4)$, and $R(3, 6)$

27. $P(-2, 5)$, $Q(-5, 2)$, and $R(1, 12)$

28. $P(2, 4)$, $Q(-3, -6)$, and $R(-4, 8)$

29. $P(-1, 5)$, $Q(2, -4)$, and $R(-2, 8)$

Transcribe Your Skills

30. In Exercise 25, you used the distance formula to show that three points were collinear. Explain how this worked.

31. Assume that the distance from point A to point B is 5 units and the distance from point B to point C is 3 units. What is the minimum distance from A to C? What is the maximum distance?

Graphing Utility

Use the computational capability of your graphing calculator to approximate the distance between each pair of points to the nearest tenth.

32. $(\sqrt{2}, 1)$ and $(-3, 5)$

33. $\left(-1, \dfrac{3}{2}\right)$ and $(2, \sqrt{3})$

34. $(\pi, 3)$ and $(-1, -3)$

35. $(3, \sqrt{2})$ and $(\sqrt{3}, \pi)$

Skillscan (Section 1.3)

Evaluate each of the expressions for the given value of the variable.

a. $3x + 2$, $x = 1$ **b.** $-2x + 7$, $x = -1$

c. $x^2 - 2x + 1$, $x = 2$ **d.** $x^2 + 3x - 2$, $x = -3$

e. $2x^2 + x - 4$, $x = 0$ **f.** $x^3 - 3x$, $x = 1$

3.6 Function Notation

TAPE IN6

OBJECTIVE: To use the function notation

In Section 3.2, we saw that a function described a relationship between two sets of numbers, one set called the *domain* of the function and the other called the *range* of the function. Typically we used an equation in two variables to express that relationship. For example,

$y = x - 2$ defines a function where y is expressed in terms of x

$d = 50t$ defines a function where d is expressed in terms of t

Consider the function defined by the second equation above. Suppose that we want to find the distance traveled d when the time t is 1 second (s). We could write the following:

If $t = 1$ then $d = 50(1) = 50$ ft

There is a much more convenient alternative for expressing the relationship between d and t (or between x and y). That alternative is called the *function notation*.

The expression $f(x)$, which is read "f of x," can be used in place of y in defining a function. Thus

$y = 4x - 3$ and $f(x) = 4x - 3$

are equivalent statements and represent the same relationship or function. Pictorially we can show the idea as follows:

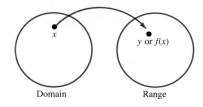

Domain   Range

This notation is more convenient when we wish to express the function value corresponding to a certain value for x. Rather than writing

If $x = 3$ then $y = 4(3) - 3$

we now have

$f(3) = 4(3) - 3 = 12 - 3 = 9$

The number in parentheses replaces x.

> **Note** Typically we use the letters f, g, and h to name functions.

> y or $f(x)$ is the *function value*. Variable x is the independent variable, and the function value is dependent on x.

> **CAUTION**
> *Be Careful! Do not confuse* the notation $f(3)$ with the multiplication "f times 3." The symbol $f(3)$ indicates the *value* of the function when x has been replaced by 3.

EXAMPLE 1 Evaluating a Function

If $f(x) = 3x - 2$:

(a) $f(2) = 3 \cdot 2 - 2 = 6 - 2 = 4$

Replace x with 2.

The function f then assigns a range value of 4 to the domain value of 2. The relationship is often pictured with a "function machine."

(b) $f(0) = 3 \cdot 0 - 2 = 0 - 2 = -2$
(c) $f(-3) = 3(-3) - 2 = -9 - 2 = -11$
(d) $f(5) - f(2) = (3 \cdot 5 - 2) - (3 \cdot 2 - 2)$
$\qquad = 13 - 4 = 9$

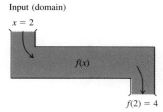

Input (domain)

$x = 2$

$f(x)$

$f(2) = 4$

Output (range)

Check Yourself 1

If $f(x) = x^2 + 3$, find each of the following.

1. $f(4)$

2. $f(0)$

3. $f(-4)$

4. $f(4) - f(2)$

There are many applications in mathematics in which we wish to replace x (or any other independent variable) with some other letter or expression. The following are some typical examples.

EXAMPLE 2 Evaluating a Function

If $f(x) = 4x + 3$:

Replace x with a.

Replace x with $x + 1$. ⟶

(a) $f(a) = 4a + 3$

(b) $f(x + 1) = 4(x + 1) + 3$
$\qquad = 4x + 4 + 3$
$\qquad = 4x + 7$

Replace x with $x + h$.

(c) $f(x + h) = 4(x + h) + 3$
$\qquad\quad = 4x + 4h + 3$

This form will have particular significance in later mathematics courses when you are finding the slope of a curve.

(d) $\dfrac{f(x + h) - f(x)}{h} = \dfrac{4(x + h) + 3 - (4x + 3)}{h}$

$\qquad\qquad = \dfrac{4x + 4h + 3 - 4x - 3}{h}$

$\qquad\qquad = \dfrac{4h}{h} = 4$

Check Yourself 2

If $f(x) = 2x - 5$, find each of the following.

1. $f(r)$

2. $f(x + 2)$

3. $\dfrac{f(x + h) - f(x)}{h}$

We mentioned earlier that the letters f, g, and h are commonly used to name functions. However, you are not limited to these letters. If an application suggests a more meaningful name (letter), by all means use it. For instance, *cost* is related to the number of items produced. We could write

We choose C to name the *cost* function.

$C(x) = 0.1x^2 + 20x + 500$

The *area* of a circle is related to its radius. We could write

We choose A to name the *area* function.

$A(r) = \pi r^2$

caltutor

EXAMPLE 3 Evaluating a Function

A company's profit, after it has produced and sold x items, is given by

$$P(x) = -0.1x^2 + 100x - 5000$$

Choose P to name the *profit* function.

(a) $P(100) = (-0.1)(100)^2 + 100(100) - 5000$
$\quad\quad\quad = -1000 + 10{,}000 - 5000$
$\quad\quad\quad = 4000$

(b) $P(500) = (-0.1)(500)^2 + 100(500) - 5000$
$\quad\quad\quad = -25{,}000 + 50{,}000 - 5000$
$\quad\quad\quad = 20{,}000$

$P(500)$ is the largest possible profit. Do you remember why?

Check Yourself 3 $16\,2^2$

The function $h(t) = -16t^2 + 96t$ gives the height of a ball thrown upward from the ground with an initial velocity of 96 ft/s. Find each of the following.

1. $h(2)$ **2.** $h(3)$ **3.** $h(6)$

All the functions that we have considered thus far have been functions of a single variable. Other functions involving more than one independent variable do exist and have particular importance in later mathematics courses, specifically calculus. Here are some examples:

The *volume* of a cylinder is a function of the height and radius of the cylinder. We can write

$$V(r, h) = \pi r^2 h$$

Volume V is now a function of the *two* variables r and h.

The *interest* paid on a loan is a function of the principal, interest rate, and time for which the money is borrowed. We can write

$$I(p, r, t) = prt$$

Interest I is a function of the *three* variables p, r, and t.

CHECK YOURSELF ANSWERS

1. (1) 19; (2) 3; (3) 19; (4) 12   **2.** (1) $2r - 5$; (2) $2x - 1$; (3) 2
3. (1) 128; (2) 144; (3) 0.

3.6 EXERCISES

Build Your Skills

If $f(x) = 4x - 3$, find each of the following.

1. $f(5)$

2. $f(4)$

3. $f(0)$

4. $f(-1)$

5. $f(-4)$

6. $f\left(\dfrac{1}{2}\right)$

If $g(x) = 3x - 4$, find each of the following.

7. $g(3)$

8. $g(5)$

9. $g(-3)$

10. $g(-5)$

11. $g(0)$

12. $g(4)$

If $h(x) = -x + 1$, find each of the following.

13. $h(2)$

14. $h(4)$

15. $h(-2)$

16. $h(-4)$

17. $h(0)$

18. $h(1)$

If $F(x) = \sqrt{x - 3}$, find (where possible) each of the following.

19. $F(4)$

20. $F(3)$

21. $F(12)$

22. $F(2)$

If $G(x) = \dfrac{1}{x - 2}$, find (where possible) each of the following.

23. $G(3)$

24. $G(-4)$

25. $G(2)$

26. $G(0)$

If $f(x) = 5x - 1$, find each of the following.

27. $f(a)$

28. $f(2r)$

29. $f(x + 1)$

30. $f(a - 2)$

31. $f(x + h)$

32. $\dfrac{f(x + h) - f(x)}{h}$

If $g(x) = x^2 + 2$, find each of the following.

33. $g(m)$

34. $g(5n)$

35. $g(x + 2)$

36. $g(s - 1)$

Transcribe Your Skills

37. Why do we use notation such as $C(r)$ for circumference and $A(r)$ for area?

38. Find a dictionary definition for the noun "function." Describe how that definition relates to the mathematical definition of the word.

Think About These

Let $f(x) = 2x + 3$ in Exercises 39 through 42.

39. Find $f(1)$.

40. Find $f(3)$.

41. Form the ordered pairs $(1, f(1))$ and $(3, f(3))$.

42. Write the equation of the line passing through the points determined by the ordered pairs of Exercise 41.

Let $f(x) = 3x$ in Exercises 43 and 44.

43. Does $f(ax) = af(x)$?

44. Does $f(a + b) = f(a) + f(b)$?

In the computer language BASIC, we can name and define functions as illustrated:

DEF FNA(X) = 2 * X + 3

DEF FNB(X) = X^2 − 3 * X + 1

[**Note** FNA(X) is equivalent to $A(x) = 2x + 3$, and FNB(X) is equivalent to $B(x) = x^2 - 3x + 1$.]

FNA(3) = 2 · 3 + 3 = 9   (Replace X with 3.)

Similarly,

FNB(−2) = $(-2)^2 - 3(-2) + 1 = 11$ (Replace X with −2.)

Using the above information, find each of the following.

45. FNA(1)

46. FNB(1)

47. FNB(−1)

48. FNA(−2)

49. FNA(5)

50. FNB(5)

51. FNB(−5)

52. FNA(−5)

53. FNA(FNB(−3))

54. FNB(FNA(−3))

The power, in joules per second, generated by a certain windmill with a 6-m-diameter rotor is given by the function

$$P(v) = 0.003(6)^2 v^3$$

where v represents wind velocity in meters per second.

Find each of the following.

 55. $P(15)$

 56. $P(20)$

 57. $P(18)$

 58. $P(22)$

Graphing Utility

Use the computational capability of your graphing calculator to evaluate $f(x)$ at the given value for x where

$$f(x) = 3x^3 - 2x^2 + 5x - 9$$

59. $f(1)$

60. $f(-1)$

61. $f(3)$

62. $f(-3)$

Skillscan (Section 2.2)

Solve each equation for y.

a. $x + y = 3$ **b.** $2x - y = 5$ **c.** $x + 3y = 9$
d. $2x + 3y = 6$ **e.** $3x - y = 0$ **f.** $x + 2y = 0$
g. $2x - 5y = 10$ **h.** $3x - 4y = 12$

Forms of Linear Equations　**3.7**

OBJECTIVES:
1. To write the equation of a line given its slope and y intercept
2. To write the equation of a line given its slope and any point on the line
3. To write the equation of a line given two points
4. To write the equation of a line given a point and a parallel or perpendicular line
5. To express the equation of a line as a linear function

TAPE IN7

Recall that the special form

$$ax + by = c$$

where a and b cannot both be zero, is called the *standard form* for a linear equation. In this section we use the slope concept developed in Section 3.4 to write the equations of lines, given various sets of geometric conditions. For this purpose we develop two further special forms for the equation of a line which will prove useful in a variety of situations.

First, suppose we know the y intercept of a line L and its slope m. Since b is the y intercept, we know that the point with coordinates $(0, b)$ is on the line. Let $P(x, y)$ be any other point on that line. Using $(0, b)$ as (x_1, y_1) and (x, y) as (x_2, y_2) in the slope formula, we have

$$m = \frac{y - b}{x - 0} \tag{1}$$

or

$$m = \frac{y - b}{x} \tag{2}$$

Just as m is used for the slope, b is used for the y intercept.

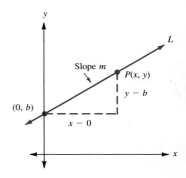

Multiplying both sides of Equation (2) by x gives

$$mx = y - b$$

or

$$y = mx + b \qquad (3)$$

Equation (3) will be satisfied by any point on line L, including $(0, b)$. It is called the *slope-intercept form* for a line, and we can state the following general result.

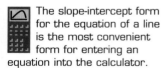
The slope-intercept form for the equation of a line is the most convenient form for entering an equation into the calculator.

SLOPE-INTERCEPT FORM FOR THE EQUATION OF A LINE

The equation of a line with y intercept b and slope m can be written as

$$y = mx + b$$

EXAMPLE 1 Finding the Equation of a Line

Write the equation of the line with slope 2 and y intercept 3.

Solution Here $m = 2$ and b (the y intercept) $= 3$. Applying the slope-intercept form, we have

The x coefficient is 2; the y intercept is 3.

$$y = 2x + 3$$

as the equation of the specified line.

It is easy to see that whenever a linear equation is written in slope-intercept form (that is, solved for y), then the slope of the line is simply the x coefficient and the y intercept is given by the constant.

Check Yourself 1

Write the equation of the line with slope $-\dfrac{2}{3}$ and y intercept -3.

Note that the slope-intercept form now gives us a second (and generally more efficient) means of finding the slope of a line whose equation is written in standard form. Recall that in Section 3.4 we determined two specific points on the line and then applied the slope formula. Now, rather than using specific points, we can simply solve the given equation for y to rewrite the equation in the slope-intercept form and identify the slope of the line as the x coefficient.

EXAMPLE 2 Finding the Equation of a Line

Find the slope and y intercept of the line with equation

$$2x + 3y = 3$$

Solution To write the equation in slope-intercept form, we solve for y:

$2x + 3y = 3$

$\qquad 3y = -2x + 3$ Subtract $2x$ from both sides.

$\qquad y = -\dfrac{2}{3}x + 1$ Divide by 3.

[handwritten: $2x + 3y = -2x + 3$ over 3; $y = \frac{-2}{3}x + \frac{3}{3} = 1$]

We now see that the slope of the line is $-\dfrac{2}{3}$ and the y intercept is 1.

Check Yourself 2

Find the slope and y intercept of the line with equation

$3x - 4y = 8$ *[handwritten: $= 4y = -3x + 8$]*

[handwritten: $3x - 4y = 8$; $-4y = -3x + 8$ over -4 and 4; $y = \frac{3}{4}x + 2$]

We can also use the slope-intercept form to determine whether the graphs of given equations will be parallel, intersecting, or perpendicular lines.

EXAMPLE 3 Verifying That Two Lines Are Parallel

Show that the graphs of $3x + 2y = 4$ and $6x + 4y = 12$ are parallel lines.

Solution First, we solve each equation for y:

$3x + 2y = 4$

$\qquad 2y = -3x + 4$

$\qquad y = -\dfrac{3}{2}x + 2$ (4)

$6x + 4y = 12$

$\qquad 4y = -6x + 12$

$\qquad y = -\dfrac{3}{2}x + 3$ (5)

Note that the slopes are the same, but the y intercepts are different. Therefore the lines are distinct.

Since the two lines have the same slope here $-\dfrac{3}{2}$, the lines are parallel.

Check Yourself 3

Show that the graphs of the equations

$-3x + 2y = 4 \qquad$ and $\qquad 2x + 3y = 9$

are perpendicular lines.

The slope-intercept form can also be used in graphing a line, as the following example illustrates.

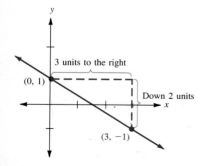

3 units to the right

(0, 1)

Down 2 units

(3, −1)

We treat $-\dfrac{2}{3}$ as $\dfrac{-2}{+3}$ to move over 3 units and down 2 units.

EXAMPLE 4 Graphing the Equation of a Line

Graph the line $2x + 3y = 3$.

Solution In Example 2, we found that the slope-intercept form for this equation was

$$y = -\frac{2}{3}x + 1$$

To graph the line, plot the y intercept at $(0, 1)$. Now since the slope m is equal to $-\dfrac{2}{3}$, from $(0, 1)$ we move *over* 3 units and then *down* 2 units, to locate a second point on the graph of the line, here $(3, -1)$.

We can now draw a line through the two points to complete the graph.

Check Yourself 4

Graph the line with equation

$$3x - 4y = 8$$

Hint: You worked with this equation in a previous Check Yourself exercise.

The following algorithm summarizes the use of graphing with the slope-intercept form.

GRAPHING BY USING THE SLOPE-INTERCEPT FORM

1. **Write the original equation of the line in slope-intercept form.**
2. **Determine the slope m and the y intercept b.**
3. **Plot the y intercept at $(0, b)$.**
4. **Use m (the change in y over the change in x) to determine a second point on the desired line.**
5. **Draw a line through the two points determined above to complete the graph.**

The desired form for the equation is

$$y = mx + b$$

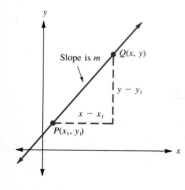

Slope is m

$Q(x, y)$

$y - y_1$

$x - x_1$

$P(x_1, y_1)$

Often in mathematics it is useful to be able to write the equation of a line, given its slope and *any* point on the line. We will now derive a third special form for a line for this purpose.

Suppose that a line has slope m and that it passes through the known point $P(x_1, y_1)$. Let $Q(x, y)$ be any other point on the line. Once again we can use the definition of slope and write

$$m = \frac{y - y_1}{x - x_1} \tag{6}$$

Multiplying both sides of Equation (6) by $x - x_1$, we have

$$m(x - x_1) = y - y_1$$

or

$$y - y_1 = m(x - x_1) \tag{7}$$

Equation (7) is called the *point-slope form* for the equation of a line, and all points lying on the line [including (x_1, y_1)] will satisfy this equation. We can state the following general result.

Know

POINT-SLOPE FORM FOR THE EQUATION OF A LINE

The equation of a line with slope m that passes through point (x_1, y_1) is given by

$$y - y_1 = m(x - x_1)$$

The equation of a line with undefined slope passing through the point (x_1, y_1) is given by $x = x_1$.

EXAMPLE 5 Finding the Equation of a Line

Write the equation for the line that passes through point $(3, -1)$ with a slope of 3.

Solution Letting $(x_1, y_1) = (3, -1)$ and $m = 3$ in point-slope form, we have

$$y - (-1) = 3(x - 3)$$

or

$$y + 1 = 3x - 9$$

We can write the final result in slope-intercept form as

$$y = 3x - 10$$

Check Yourself 5

Write the equation of the line that passes through point $(-2, -4)$ with a slope of $\frac{3}{2}$. Write your result in slope-intercept form.

Since we know that two points determine a line, it is natural that we should be able to write the equation of a line passing through two given points. Using the point-slope form together with the slope formula will allow us to write such an equation.

EXAMPLE 6 Finding the Equation of a Line

Write the equation of the line passing through $(2, 4)$ and $(4, 7)$.

Solution First, we find m, the slope of the line. Here

$$m = \frac{7 - 4}{4 - 2} = \frac{3}{2}$$

Note We could just as well have chosen to let

$(x_1, y_1) = (4, 7)$

The resulting equation will be the same in either case. Take time to verify this for yourself.

Now we apply the point-slope form with $m = \dfrac{3}{2}$ and $(x_1, y_1) = (2, 4)$:

$$y - 4 = \frac{3}{2}(x - 2)$$

$$y - 4 = \frac{3}{2}x - 3$$

$$y = \frac{3}{2}x + 1$$

Check Yourself 6

Write the equation of the line passing through $(-2, 5)$ and $(1, 3)$. Write your result in slope-intercept form.

A line with slope zero is a horizontal line. A line with an undefined slope is vertical. The next example illustrates the equations of such lines.

EXAMPLE 7 Finding the Equation of a Line

(*a*) Find the equation of a line passing through $(7, -2)$ with a slope of zero.

We could find the equation by letting $m = 0$. Substituting into the slope-intercept form, we can solve for the y intercept b.

$$y = mx + b$$
$$-2 = 0(7) + b$$
$$-2 = b$$

So,

$$y = 0 \cdot x - 2 \qquad \text{or} \qquad y = -2$$

It is far easier to remember that any line with a zero slope is a horizontal line and has the form

$$y = b$$

The value for b will always be the y coordinate for the given point.

(*b*) Find the equation of a line with undefined slope passing through $(4, -5)$.

A line with undefined slope is vertical. It will always be of the form $x = a$, where a is the x coordinate for the given point. The equation is

$$x = 4$$

Check Yourself 7

1. Find the equation of a line with zero slope that passes through point $(-3, 5)$.

2. Find the equations of a line passing through $(-3, -6)$ with undefined slope.

Alternate methods for finding the equation of a line through two points do exist and have particular significance in other fields of mathematics, such as statistics. The following example shows such an alternate approach.

EXAMPLE 8 Finding the Equation of a Line

Write the equation of the line through points $(-2, 3)$ and $(4, 5)$.

Solution First, we find m, as before:

$$m = \frac{5 - 3}{4 - (-2)} = \frac{2}{6} = \frac{1}{3}$$

We now make use of the slope-intercept equation, but in a slightly different form. Since $y = mx + b$, we can write

$$b = y - mx$$

Now letting $x = -2$, $y = 3$, and $m = \dfrac{1}{3}$, we can calculate b:

We substitute these values because the line must pass through $(-2, 3)$.

$$b = 3 - \left(\frac{1}{3}\right)(-2)$$

$$= 3 + \frac{2}{3} = \frac{11}{3}$$

With $m = \dfrac{1}{3}$ and $b = \dfrac{11}{3}$, we can apply the slope-intercept form, to write the equation of the desired line. We have

$$y = \frac{1}{3}x + \frac{11}{3}$$

Check Yourself 8

Repeat the Check Yourself 6 exercise, using the technique illustrated in Example 8.

We now know that we can write the equation of a line once we have been given appropriate geometric conditions, such as a point on the line and the slope of that line. In some applications the slope may be given not directly but through specified parallel or perpendicular lines.

EXAMPLE 9 Finding the Equation of a Line

Find the equation of the line passing through $(-4, -3)$ and parallel to the line determined by $3x + 4y = 12$.

Solution First, we find the slope of the given parallel line, as before:

$$3x + 4y = 12$$

$$4y = -3x + 12$$

The slope of the given line is $-\dfrac{3}{4}$.

$$y = -\frac{3}{4}x + 3$$

Now since the slope of the desired line must also be $-\dfrac{3}{4}$, we can use the point-slope form to write the required equation:

The line must pass through $(-4, -3)$, so let $(x_1, y_1) = (-4, -3)$.

$$y - (-3) = -\frac{3}{4}[x - (-4)]$$

This simplifies to

$$y = -\frac{3}{4}x - 6$$

and we have our equation in slope-intercept form.

Check Yourself 9

Find the equation of the line passing through $(5, 4)$ and perpendicular to the line with equation $2x - 5y = 10$.
Hint: Recall that the slopes of perpendicular lines are negative reciprocals of each other.

There are many applications of our work with linear equations in various fields. The following is just one of many typical examples.

EXAMPLE 10 An Application of a Linear Function

In producing a new product, a manufacturer predicts that the number of items produced x and the cost in dollars C of producing those items will be related by a linear equation.

Suppose that the cost of producing 100 items will be $5000 and the cost of producing 500 items will be $15,000. Find the linear equation relating x and C.

Solution To solve this problem, we must find the equation of the line passing through points $(100, 5000)$ and $(500, 15,000)$.

Even though the numbers are considerably larger than we have encountered thus far in this section, the process is exactly the same.

First, we find the slope:

$$m = \frac{15{,}000 - 5000}{500 - 100} = \frac{10{,}000}{400} = 25$$

We can now use the point-slope form as before to find the desired equation:

$$C - 5000 = 25(x - 100)$$

$$C - 5000 = 25x - 2500$$

$$C = 25x + 2500$$

To graph the equation we have just derived, we must choose the scaling on the x and C axes carefully to get a "reasonable" picture. Here we choose increments of 100 on the x axis and 2500 on the C axis, since those seem appropriate for the given information.

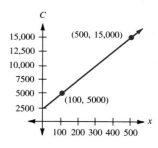

Note how the change in scaling "distorts" the slope of the line.

Check Yourself 10

A company predicts that the value in dollars V and the time that a piece of equipment has been in use t are related by a linear equation. If the equipment is valued at $1500 after 2 years and at $300 after 10 years, find the linear equation relating t and V.

A line written in slope-intercept form can be easily rewritten in function form.

EXAMPLE 11 Writing Equations in Function Form

Rewrite each line in function form.

(a) $y = 3x - 2$

Since y is already expressed in terms of x, y is a function of x. We simply write

$$f(x) = 3x - 2$$

(b) $2y + 3x = 6$

We first solve the equation for y:

$$2y = -3x + 6$$

$$y = -\frac{3}{2}x + 3$$

Now we can rewrite the equation as the function

$$f(x) = -\frac{3}{2}x + 3$$

Check Yourself 11

Rewrite each equation as a function.

1. $y = -2x + 5$ **2.** $4y - 3x = 12$

The following chart summarizes the various forms of the equation of a line that we have considered in this chapter.

Form	Equation for Line L	Conditions
Standard	$ax + by = c$	Constants a and b cannot both be zero.
Slope-intercept	$y = mx + b$	Line L has y intercept b with slope m.
Function form	$f(x) = ax + b$	Line L has y intercept b and slope a.
Point-slope	$y - y_1 = m(x - x_1)$	Line L passes through point (x_1, y_1) with slope m.
Horizontal	$y = k$	Slope is zero.
Vertical	$x = h$	Slope is undefined.

CHECK YOURSELF ANSWERS

1. $y = -\dfrac{2}{3}x - 3$ **2.** $y = \dfrac{3}{4}x - 2, m = \dfrac{3}{4}, y$ intercept is -2 **3.** $m_1 = \dfrac{3}{2},$

$m_2 = -\dfrac{2}{3}, m_1 \cdot m_2 = -1$ **4.** $y = \dfrac{3}{4}x - 2$

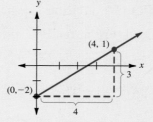

5. $y = \dfrac{3}{2}x - 1$ **6.** $y = -\dfrac{2}{3}x + \dfrac{11}{3}$ **7.** (1) $y = 5$; (2) $x = -3$

8. $y = -\dfrac{2}{3}x + \dfrac{11}{3}$ **9.** $y = -\dfrac{5}{2}x + \dfrac{33}{2}$ **10.** $V = -150t + 1800$

11. (1) $f(x) = -2x + 5$; (2) $f(x) = \dfrac{3}{4}x + 3$

Build Your Skills

In Exercises 1 to 8, match the graph with one of the given equations.

1.

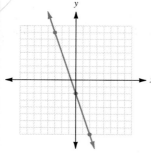

(a) $y = 2x$
(b) $y = x + 1$
(c) $y = -x + 3$
(d) $y = 2x + 1$
(e) $y = -3x - 2$
(f) $y = \dfrac{2}{3}x + 1$
(g) $y = -\dfrac{4}{3}x + 1$
(h) $y = -4x$

2.

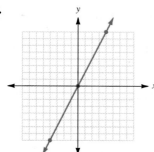

3.

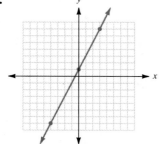

4.

5.

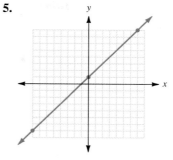

6.

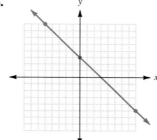

7.

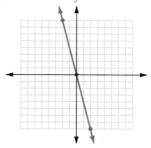

8.

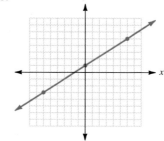

Write each equation in slope-intercept form. Give its slope and y intercept.

$y = mx + b$

9. $x + y = 5$

10. $2x + y = 3$

11. $2x - y = -2$

12. $x + 3y = 6$

13. $x + 3y = 9$

14. $4x - y = 8$

15. $2x - 3y = 6$

16. $3x - 4y = 12$

17. $2x - y = 0$

18. $3x + y = 0$

19. $y + 3 = 0$

20. $y - 2 = 0$

Write the equation of the line passing through each of the given points with the indicated slope. Give your results in slope-intercept form, where possible.

$y - y_1 = m(x - x_1)$

21. $(0, 2)$, $m = 3$

22. $(0, -4)$, $m = -2$

23. $(0, 2)$, $m = \dfrac{3}{2}$

24. $(0, -3)$, $m = -2$

25. $(0, 4)$, $m = 0$

26. $(0, 5)$, $m = -\dfrac{3}{5}$

27. $(0, -5)$, $m = \dfrac{5}{4}$

28. $(0, -4)$, $m = -\dfrac{3}{4}$

29. $(1, 2)$, $m = 3$

30. $(-1, 2)$, $m = 3$

31. $(-2, -3)$, $m = -3$

32. $(1, -4)$, $m = -4$

33. $(5, -3)$, $m = \dfrac{2}{5}$

34. $(4, 3)$, $m = 0$

35. $(2, -3)$, m is undefined

36. $(2, -5)$, $m = \dfrac{1}{4}$

37. $(5, 0)$, $m = -\dfrac{4}{5}$

38. $(-3, 0)$, m is undefined

Write the equation of the line passing through each of the given pairs of points. Write your result in slope-intercept form, where possible.

39. $(2, 3)$ and $(5, 6)$ $y = x + 1$

40. $(3, -2)$ and $(6, 4)$

41. $(-2, -3)$ and $(2, 0)$

42. $(-1, 3)$ and $(4, -2)$

43. $(-3, 2)$ and $(4, 2)$

44. $(-5, 3)$ and $(4, 1)$

45. $(2, 0)$ and $(0, -3)$

46. $(2, -3)$ and $(2, 4)$

47. $(0, 4)$ and $(-2, -1)$

48. $(-4, 1)$ and $(3, 1)$

Write the equation of the line L satisfying the given geometric conditions.

49. L has slope 4 and y intercept -2.

50. L has slope $-\dfrac{2}{3}$ and y intercept 4.

51. L has x intercept 4 and y intercept 2.

52. L has x intercept -2 and slope $\dfrac{3}{4}$.

53. L has y intercept 4 and a 0 slope.

54. L has x intercept -2 and an undefined slope.

55. L passes through point $(3, 2)$ with a slope of 5.

56. L passes through point $(-2, -4)$ with a slope of $-\dfrac{3}{2}$.

57. L has y intercept 3 and is parallel to the line with equation $y = 3x - 5$.

58. L has y intercept -3 and is parallel to the line with equation $y = \dfrac{2}{3}x + 1$.

Write the equation of each line in function form.

59. L has y intercept 4 and is perpendicular to the line with equation $y = -2x + 1$.

60. L has y intercept 2 and is parallel to the line with equation $y = -1$.

61. L has y intercept 3 and is parallel to the line with equation $y = 2$.

62. L has y intercept 2 and is perpendicular to the line with equation $2x - 3y = 6$.

63. L passes through point $(-3, 2)$ and is parallel to the line with equation $y = 2x - 3$.

64. L passes through point $(-4, 3)$ and is parallel to the line with equation $y = -2x + 1$.

65. L passes through point $(3, 2)$ and is parallel to the line with equation $y = \dfrac{4}{3}x + 4$.

66. L passes through point $(-2, -1)$ and is perpendicular to the line with equation $y = 3x + 1$.

67. L passes through point $(5, -2)$ and is perpendicular to the line with equation $y = -3x - 2$.

68. L passes through point $(3, 4)$ and is perpendicular to the line with equation $y = -\dfrac{3}{5}x + 2$.

69. L passes through $(-2, 1)$ and is parallel to the line with equation $x + 2y = 4$.

70. L passes through $(-3, 5)$ and is parallel to the x axis.

Transcribe Your Skills

71. Describe the process for finding the equation of a line if you are given two points on the line.

72. How would you find the equation of a line if you were given the slope and the x *intercept?*

Think About These

73. Find the equation of the perpendicular bisector of the segment joining $(-3, -5)$ and $(5, 9)$. *Hint:* First determine the midpoint of the segment. The perpendicular bisector passes through that point.

74. Find the equation of the perpendicular bisector of the segment joining $(-2, 3)$ and $(8, 5)$.

75. A temperature of $10°C$ corresponds to a temperature of $50°F$. Also $40°C$ corresponds to $104°F$. Find the linear equation relating F and C.

76. In planning for a new item, a manufacturer assumes that the number of items produced x and the cost in dollars C of producing these items are related by a linear equation. Projections are that 100 items will cost $10,000 to produce and that 300 items will cost $22,000 to produce. Find the equation that relates C and x.

77. A word processing station was purchased by a company for $10,000. After 4 years it is estimated that the value of the station will be $4000. If the value in dollars V and the time the station has been in use t are related by a linear equation, find the equation that relates V and t.

78. Two years after an expansion, a company had sales of $42,000. Four years later the sales were $102,000. Assuming that the sales in dollars S and the time in years t are related by a linear equation, find the equation relating S and t.

79. The concentration of carbon monoxide C in the atmosphere and the number of years t since 1978 are linearly related. If there were 9.9 parts per million (ppm) of carbon monoxide in the atmosphere in 1978 and there were 6.9 ppm in the atmosphere in 1987, find the equation that relates C and t.

80. Assuming the relationship in Exercise 79, what was the concentration of carbon monoxide in the atmosphere in 1993?

81. The index of carbon monoxide (CO) in the atmosphere is a number that relates the amount of CO in the atmosphere in a given year to an arbitrary base year. The index I and the number of years t past the base year are linearly related. If 1978 had a CO index of 100 (the base year) and 1987 had a CO index of 68, find the equation that relates I and t.

82. Assuming the relationship in Exercise 81, what will the CO index be in 1996?

Graphing Utility

Use your graphing utility to graph each of the following.

83. $3x - 5y = 30$

84. $2x + 7y = 13$

85. The line with slope $\dfrac{2}{3}$ and y intercept at 7

86. The line with slope $-\dfrac{1}{5}$ and y intercept at 3

87. The line with slope n passing through point $(1, 5)$

88. The line with slope $-\sqrt{2}$ passing through point $(2, -2)$

Skillscan (Section 2.4)

Graph each of the following inequalities.

a. $x < 3$ **b.** $x \geq -2$ **c.** $2x \leq 8$ **d.** $3x > -9$

e. $-3x < 12$ **f.** $-2x \leq 10$ **g.** $\dfrac{2}{3}x \leq 4$

h. $-\dfrac{3}{4}x \geq 6$

| 3.8 | **Graphing Linear Inequalities** |

TAPE IN7

OBJECTIVE: To graph linear inequalities in two variables

Our last three sections have dealt with linear equations in two variables. Linear inequalities in the two variables x and y are obtained from linear equations by replacing the symbol for equality ($=$) with one of the inequality symbols ($<$, $>$, $\leq$, $\geq$).

The general form for a linear inequality in two variables is

$$ax + by < c$$

where a and b cannot both be 0. The symbol $<$ can be replaced with $>$, $\leq$, or $\geq$. Some examples are

 Although your graphing calculator will not shade the entire solution set, it can help you see the solution to an inequality. If we can graph

$y = -2x + 6$

then the solution to the inequality

$y < -2x + 6$

is all the points on the plane with a y value *below that line.*

$$y < -2x + 6$$

or

$$x - 2y + 4 > 0$$

or

$$2x + 3y \leq 6$$

or

$$x \geq 3$$

As was the case with an equation, the solution set of a linear inequality is a set of ordered pairs of real numbers. However, in the case of linear inequalities, we will find that the solution sets will be all points in an entire region of the plane (called *half planes*), which have the corresponding linear equations as their border.

To determine such a solution set, let's start with the first inequality above. To graph the solution set of

$$y < -2x + 6 \tag{1}$$

we begin the graphing process by writing the corresponding linear equation

To write the equation, just replace the inequality symbol with an equals sign.

$$y = -2x + 6$$

First note that the graph of $y = -2x + 6$ is simply a straight line.

Now to graph the solution set of inequality (1), we must include all ordered pairs that satisfy that inequality. For instance, if $x = 1$, we have

$$y < -2 \cdot 1 + 6$$

$$y < 4$$

So we want to include all points of the form $(1, y)$, where $y < 4$. Of course, since $(1, 4)$ is *on* the corresponding line, this means that we want all points *below* the line along the vertical line $x = 1$. The result will be similar for any choice of x, and our solution set will then contain all points below the line $y = -2x + 6$.

We can then graph the solution set as the shaded region shown.

In general, the solution set of an inequality of the form

$$ax + by < c \qquad \text{or} \qquad ax + by > c$$

will be a half plane either above or below the corresponding line determined by

$$ax + by = c$$

How do we decide which half plane represents the desired solution set? The use of a "test point" provides an easy answer. Choose any point *not* on the line. Then substitute the coordinates of that point into the given inequality. If the coordinates satisfy the inequality (result in a true statement), then shade the region or half plane that includes the test point; if not, shade the opposite half plane. The following example illustrates the process.

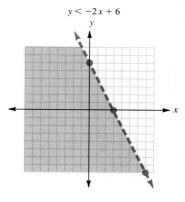

$y < -2x + 6$

The line is dashed to indicate that equality is *not* included.

We call the graph of the equation

$ax + by = c$

the *boundary line* at the half planes.

EXAMPLE 1 Graphing a Linear Inequality

Graph the linear inequality

$$x - 2y < 4$$

Solution First, we graph the corresponding equation

$$x - 2y = 4$$

to find the boundary line. Now to decide on the appropriate half plane, we need a test point *not* on the line. As long as the line *does not pass through the origin,* we can always use $(0, 0)$ as a test point. It provides the easiest computation.

Here letting $x = 0$ and $y = 0$, we have

$$0 - 2 \cdot 0 < 4$$

$$0 < 4$$

Since this is a true statement, we proceed to shade the half plane including the origin (the test point), as shown.

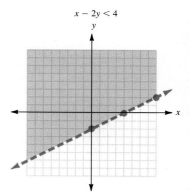

$x - 2y < 4$

Check Yourself 1

Graph the solution set of $3x + 4y > 12$.

The graphs of some linear inequalities will include the boundary line. That will be the case whenever equality is included with the inequality statement, as illustrated in the following example.

EXAMPLE 2 Graphing a Linear Inequality

Graph the inequality

$$2x + 3y \geq 6$$

Solution First, we graph the boundary line, here corresponding to $2x + 3y = 6$. Note that we use a solid line in this case since equality is included in the original statement.

Again we choose a convenient test point not on the line. As before, the origin will provide the simplest computation.

Substituting $x = 0$ and $y = 0$, we have

$$2 \cdot 0 + 3 \cdot 0 \geq 6$$
$$0 \geq 6$$

This is a *false* statement. Hence the graph will consist of all points on the *opposite* side from the origin. The graph will then be the upper half plane, as shown.

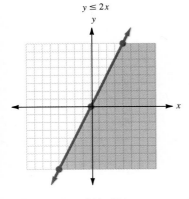

$2x + 3y \geq 6$

A solid boundary line means that equality *is included.*

Check Yourself 2

Graph the solution set of $x - 3y \leq 6$.

$y \leq 2x$

EXAMPLE 3 Graphing a Linear Inequality

Graph the solution set of

$$y \leq 2x$$

Solution We proceed as before by graphing the boundary line (it is solid since equality is included). The only difference between this and previous examples is that we *cannot use the origin* as a test point. Do you see why?

The choice of (1, 1) is arbitrary. We simply want *any* point *not* on the line.

Choosing $(1, 1)$ as our test point gives the statement

$$1 \leq 2 \cdot 1$$
$$1 \leq 2$$

Since the statement is *true,* we shade the half plane *including* the test point $(1, 1)$.

Check Yourself 3

Graph the solution set of $3x + y > 0$.

Let's consider a special case of graphing linear inequalities in the rectangular coordinate system.

EXAMPLE 4 Graphing a Linear Inequality

Graph the solution set of $x > 3$.

Solution First, we draw the boundary line (a dashed line since equality is not included) corresponding to

$x = 3$

We can choose the origin as a test point in this case, and that results in the false statement

$0 > 3$

We then shade the half plane *not* including the origin. In this case the solution set is represented by the half plane to the right of the vertical boundary line.

As you may have observed, in this special case choosing a test point is not really necessary. Since we want values of x that are *greater than* 3, we want those ordered pairs that are to the *right* of the boundary line.

Here we specify the rectangular coordinate system to indicate we want a two-dimensional graph.

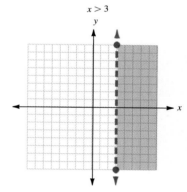

$x > 3$

Check Yourself 4

Graph the solution set of

$y \leq 2$

in the rectangular coordinate system.

Applications of linear inequalities will often involve more than one inequality condition. Consider the following example.

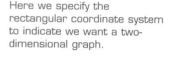

$3x + 4y \leq 12$
$x \geq 0$
$y \geq 0$

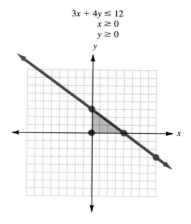

EXAMPLE 5 Graphing a Region Defined by Linear Inequalities

Graph the region satisfying the following conditions.

$3x + 4y \leq 12$

$\qquad x \geq 0$

$\qquad y \geq 0$

The solution set in this case must satisfy *all three conditions.*

Solution As before, the solution set of the first inequality is graphed as the half plane *below* the boundary line. The second and third inequalities mean that x and y must also be nonnegative. Therefore our solution set is restricted to the first quadrant (and the appropriate segments of the x and y axes), as shown.

Check Yourself 5

Graph the region satisfying the following conditions.

$$x + 4y \leq 8$$
$$x \geq 0$$
$$y \geq 0$$

The following algorithm summarizes our work in graphing linear inequalities in two variables.

TO GRAPH A LINEAR INEQUALITY

1. Replace the inequality symbol with an equality symbol to form the equation of the boundary line of the solution set.
2. Graph the boundary line. Use a dashed line if equality is not included ($<$ or $>$). Use a solid line if equality is included ($\leq$ or $\geq$).
3. Choose any convenient test point *not* on the line.
4. If the inequality is true for the test point, shade the half plane including the test point. If the inequality is false for the test point, shade the half plane *not* including the test point.

CHECK YOURSELF ANSWERS

1.

$3x + 4y > 12$

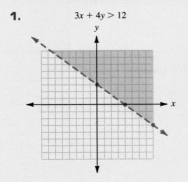

2.

$x - 3y \leq 6$

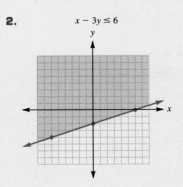

3. $3x + y > 0$

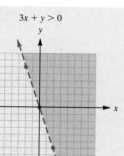

4. $y \leq 2$

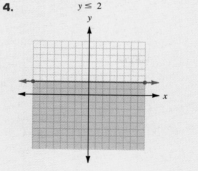

5. $x + 4y \leq 8$
$x \geq 0$
$y \geq 0$

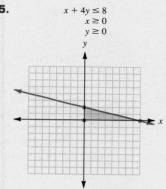

Build Your Skills

Graph the solution sets of the following linear inequalities.

1. $x + y < 4$

2. $x + y \geq 6$

3. $x - y \geq 3$

4. $x - y < 5$

5. $y \geq 2x + 1$

6. $y < 3x - 4$

7. $2x + 3y < 6$

8. $3x - 4y \geq 12$

9. $x - 4y > 8$

10. $2x + 5y \leq 10$

11. $y \geq 3x$

12. $y \leq -2x$

13. $x - 2y > 0$

14. $x + 4y \leq 0$

15. $x < 3$

16. $y < -2$

17. $y > 3$

18. $x \leq -4$

19. $3x - 6 \leq 0$

20. $-2y > 6$

21. $0 < x < 1$

22. $-2 \leq y \leq 1$

23. $1 \leq x \leq 3$

24. $1 < y < 5$

Graph the region satisfying each of the following sets of conditions.

25. $0 \le x \le 3$
$2 \le y \le 4$

26. $1 \le x \le 5$
$0 \le y \le 3$

27. $x + 2y \le 4$
$x \ge 0$
$y \ge 0$

28. $2x + 3y \le 6$
$x \ge 0$
$y \ge 0$

Transcribe Your Skills

29. Assume that you are working only with the variable x. Describe the solution to the statement $x > -1$.

30. Now, assume that you are working in two variables, x and y. Describe the solution to the statement $x > -1$.

Think About These

31. A manufacturer produces a standard model and a deluxe model of 13-in television set. The standard model requires 12 h to produce, while the deluxe model requires 18 h. The labor available is limited to 360 h per week.

If x represents the number of standard-model sets produced per week and y represents the number of deluxe models, draw a graph of the region representing the feasible values for x and y. Keep in mind that the values for x and y must be nonnegative since they represent a quantity of items. (This will be the solution set for the system of inequalities.)

32. A manufacturer produces standard record turntables and compact disc players. The turntables require 10 h of labor to produce while the disc players require 20 h. Let x represent the number of turntables produced and y the number of disc players.

If the labor hours available are limited to 300 h per week, graph the region representing the feasible values for x and y.

 33. The operating permit for a power plant restricts the combined levels of sulfur oxide (SO) and nitrogen oxide (NO) to 3000 tons. Due to the limits of the plant's pollution controls, the least possible amount of SO emissions is 1000 tons, and the least possible amount of NO emissions is 500 tons. Draw a graph of the region's possible emission levels of SO and NO.

 34. The operating permit for a steel plant restricts the combined levels of sulfur oxide (SO) and particulate matter (PM) to 4500 tons. Due to the limits of the plant's pollution controls, the least possible amount of SO emissions is 1000 tons, and the least possible amount of PM emissions is 1500 tons. Draw a graph of the region's possible emission levels of SO and PM.

Graphing Utility

Your graphing utility can be used to ''see'' the solution to a linear inequality. For example, to see where $2x > 4$, we graph the two lines $y = 2x$ and $y = 4$. The solution is the set of x values for which the first graph ($y = 2x$) is above the second graph ($y = 4$). Explain how that technique could be applied to see the solution for each of the following inequalities.

35. $2x + 1 > 5$

36. $3x - 2 < 7$

Skillscan (Section 1.4)

Evaluate each of the following expressions for the given value of x.

a. $x^2 - 2x$, $x = 1$ **b.** $x^2 + 4x$, $x = -2$
c. $-x^2 + 6x$, $x = 3$ **d.** $-x^2 - 4x$, $x = -2$

e. $x^2 + 2x - 3$, $x = -1$ **f.** $x^2 - 3x + 2$, $x = \dfrac{3}{2}$

g. $-x^2 + 5x - 4$, $x = \dfrac{5}{2}$ **h.** $-x^2 + 6x - 9$, $x = 3$

SUMMARY

The Cartesian Coordinate System [3.1]

The *cartesian coordinate system* allows us to establish a one-to-one correspondence between points in the plane and ordered pairs of real numbers:

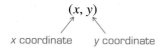

To *graph* (or *plot*) a point (x, y) in the plane:

1. Start at the origin.

2. Move to the right or left along the x axis according to the value of the x coordinate.

3. Move up or down and parallel to the y axis according to the value of the y coordinate.

An Introduction to Functions [3.2]

A *relation* is any set of ordered pairs.

> The set $\{(1, 4), (2, 5), (1, 6)\}$ is a relation.

   The set of all first components of the ordered pairs is called the *domain* of the relation. The set of all second components of the ordered pairs is called the *range* of the relation.

> The domain is $\{1, 2\}$.
>
> The range is $\{4, 5, 6\}$.

   A *function* is a relation in which no two distinct ordered pairs have the same first component.

> $\{(1, 2), (2, 3), (3, 4)\}$ is a function.
>
> $\{(1, 2), (2, 3), (2, 4)\}$ is *not* a function.

   The set of points in the plane which correspond to ordered pairs in a relation or function is called the *graph* of that relation or function.

   A useful means of determining whether a graph represents a relation that is also a function is called the *vertical-line test*.

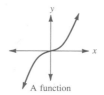

A function

If a vertical line meets the graph of a relation in two or more points, the relation is *not* a function.

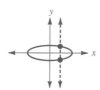

A relation—*not* a function

Graphing Linear Equations [3.3]

(1, 3) is a solution for

$3x + 2y = 9$

because

$3 \cdot 1 + 2 \cdot 3 = 9$

is a true statement.

The *solution set* of an equation in two variables is the set of ordered pairs of real numbers (x, y) which will make the equation a true statement.

An equation of the form

$$ax + by = c \qquad (1)$$

where a and b cannot both be zero, is a *linear equation in two variables*. The graph of such an equation is always a *straight line*. An equation in form (1) is called the *standard form* of the equation of a line.

To graph a linear equation:

1. Find at least three solutions for the equation, and write your results in a table of values.

2. Graph the points associated with the ordered pairs found above.

3. Draw a straight line through the points plotted above, to form the graph of the equation.

To graph

$y = 2x - 3$

(0, −3), (1, −1), and (2, 1) are solutions.

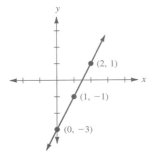

A second approach to graphing linear equations uses the x and y *intercepts* of the line. The x intercept is the x coordinate of the point where the line intersects the x axis. The y intercept is the y coordinate of the point where the line intersects the y axis.

To graph by the intercept method:

1. Find the x intercept. Let $y = 0$ and solve for x.

2. Find the y intercept. Let $x = 0$ and solve for y.

3. Plot the two intercepts determined above.

4. Draw a straight line through the points of intercept.

The Slope of a Line [3.4]

The *slope* of a line gives a numerical measure of the direction and steepness, or inclination, of the line. The slope m of a line containing the distinct points in the plane (x_1, y_1) and (x_2, y_2) is given by

$$m = \frac{y_2 - y_1}{x_2 - x_1} \qquad x_2 \neq x_1 \qquad (2)$$

The slopes of two nonvertical parallel lines are equal. The slopes of two nonvertical perpendicular lines are the negative reciprocals of each other.

Distance and Midpoints [3.5]

The distance between points (−2, 3) and (5, 1) is

$d = \sqrt{[5 - (-2)]^2 + (1 - 3)^2}$

$= \sqrt{7^2 + (-2)^2}$

$= \sqrt{53}$

The *distance* between points (x_1, y_1) and (x_2, y_2) is given by

$$d = \sqrt{(x_2 - x_1)^2 + (y_2 - y_1)^2} \qquad (3)$$

The midpoint between (x_1, y_1) and (x_2, y_2) is

$$\left(\frac{x_1 + x_2}{2}, \frac{y_1 + y_2}{2} \right)$$

(4)

The midpoint of the line segment connecting $(-2, 3)$ and $(5, 1)$ is

$$\left(\frac{-2 + 5}{2}, \frac{3 + 1}{2} \right) = \left(\frac{3}{2}, 2 \right)$$

Function Notation [3.6]

The expression $f(x)$ is read "*f of x.*" It represents the *function value,* or the value of the function, at x. Since y is another name for the function value, y and $f(x)$ are equivalent, and we write

$$y = f(x)$$

Function notation gives us a convenient means of indicating the function's value when the independent variable x is replaced by some number or expression.

$y = x^2 - 2$ and $f(x) = x^2 - 2$ are different ways of describing the same function.

If $f(x) = x^2 - 2$, then

$f(3) = 3^2 - 2 = 7$ and

$f(a) = a^2 - 2$

Forms of Linear Equations [3.7]

There are two useful special forms for the equation of a line. The *slope-intercept form* of the equation of a line is $y = mx + b$, where the line has slope m and intercept b. The *point-slope form* of the equation of a line is $y - y_1 = m(x - x_1)$, where the line has slope m and passes through the point (x_1, y_1). And $x = x_1$ is the equation of a line through (x, y) with undefined slope.

$y = \frac{2}{3}x + 4$ is in slope-intercept form. The slope m is $\frac{2}{3}$, and the y intercept is 4.

If line l has slope $m = -2$ and passes through $(-2, 3)$, its equation is

$y - 3 = -2[x - (-2)]$

$y - 3 = -2(x + 2)$

$y - 3 = -2x - 4$

$y = -2x - 1$

Graphing Linear Inequalities [3.8]

The graph of the solution set of a linear inequality in two variables is a *half plane,* either above or below the boundary line of the graph.

The boundary line is included in the graph if equality is included in the statement of the original inequality.

To graph a linear inequality:

1. Replace the inequality symbol with an equality symbol to form the equation of the boundary line of the solution set.

2. Graph the boundary line. Use a dashed line if equality is not included ($<$ or $>$). Use a solid line if equality is included ($\leq$ or $\geq$).

3. Choose any convenient test point *not* on the boundary line.

4. If the inequality is *true* for the test point, shade the half plane *including* the test point. If the inequality is *false* for the test point, shade the half plane *not including* the test point.

To graph: $x - 2y < 4$

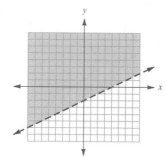

SUMMARY EXERCISES

This summary exercise set is provided to give you practice with each of the objectives of the chapter. Each exercise is keyed to the appropriate chapter section. The answers are provided in the instructor's manual that accompanies this text. Your instructor will provide guidelines on how best to use these exercises in your instructional program.

[3.1] Graph each of the following points in the cartesian coordinate system.

1. $A(2, -3)$

2. $B(4, 5)$

3. $C(0, 5)$

4. $D(-2, -6)$

5. $E(-4, 1)$

6. $F(-6, 0)$

7. $G(5, -5)$

8. $H(0, -2)$

[3.2] For each relation, give the domain and the range; then decide whether the relation is a function.

9. $\{(-1, 2), (0, 2), (1, 2)\}$

10. $\{(-3, 2), (-3, 4), (-3, 6)\}$

11. $\{(1, 2), (2, 3), (1, 4)\}$

12. $\{(0, 1), (1, 2), (2, 4), (3, 8)\}$

[3.2] Use the vertical-line test to determine whether each graph represents a function.

13.

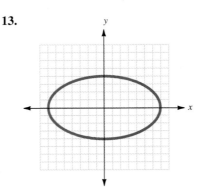

14.

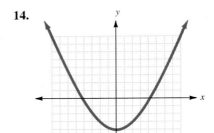

[3.3] Graph each of the following equations.

15. $x - y = 8$

16. $y = -3x - 2$

17. $x + 3y = 0$

18. $x - 4y = 0$

19. $3x - 5y = 15$

20. $4x + 3y = 12$

21. $x = 6$

22. $y = -3$

[3.4] Find the slope (if it exists) of the line determined by each of the following pairs of points.

23. $(3, 2)$ and $(5, 8)$

24. $(-2, 5)$ and $(1, -1)$

25. $(2, -4)$ and $(3, 5)$

26. $(-3, -4)$ and $(3, 0)$

27. $(4, -3)$ and $(4, 4)$

28. $(4, -2)$ and $(-2, -2)$

29. $(2, -4)$ and $(-1, -3)$

30. $(5, -2)$ and $(5, 3)$

[3.4] Find the slope of the line determined by each of the following equations.

31. $3x + 2y = 6$

32. $x - 4y = 8$

[3.4] Are the following pairs of lines parallel, perpendicular, or neither?

33. L_1 through $(-3, -2)$ and $(1, 3)$
   L_2 through $(0, 3)$ and $(4, 8)$

34. L_1 through $(-4, 1)$ and $(2, -3)$
   L_2 through $(0, -3)$ and $(2, 0)$

35. L_1 with equation $x + 2y = 6$
   L_2 with equation $x + 3y = 9$

36. L_1 with equation $4x - 6y = 18$
   L_2 with equation $2x - 3y = 6$

[3.5] Find the distance between each of the following pairs of points.

37. $(-2, 4)$ and $(3, 4)$

38. $(-2, 1)$ and $(3, 4)$

39. $(5, -2)$ and $(5, 3)$

40. $(-3, -2)$ and $(5, 1)$

41. $(2, -3)$ and $(-4, -2)$

42. $(-4, 2)$ and $(-3, 5)$

43. $(2, -4)$ and $(-4, 2)$

44. $(3, -1)$ and $(-2, -6)$

45. $\left(0, \dfrac{1}{2}\right)$ and $\left(-\dfrac{1}{2}, 0\right)$

46. $\left(4, -\dfrac{7}{2}\right)$ and $\left(\dfrac{7}{2}, -4\right)$

[3.5] Find the midpoint for the segment created by the given endpoints.

47. $(0, 0)$ and $(0, 4)$

48. $(-4, 8)$ and $(3, 8)$

49. $(-2, 4)$ and $(3, -5)$

50. $(2, 9)$ and $(-2, -9)$

51. $(3, -4)$ and $(-5, -1)$

52. $(1, -3)$ and $(12, -2)$

53. $(-12, 9)$ and $(3, -4)$

54. $(3, -3)$ and $(-5, -5)$

55. $\left(\dfrac{1}{2}, -2\right)$ and $\left(\dfrac{7}{2}, 9\right)$

56. $\left(4, -\dfrac{5}{2}\right)$ and $\left(-\dfrac{1}{2}, -2\right)$

[3.6] For each given function, find $f(-1)$, $f(0)$, and $f(2)$.

57. $f(x) = 3$

58. $f(x) = 3x - 6$

59. $f(x) = \dfrac{1}{2}x + \dfrac{7}{2}$

60. $f(x) = 2x^2 - 3x + 1$

[3.7] Write the equation of the line passing through each of the following points with the indicated slope. Give your results in slope-intercept form, where possible.

61. $(0, -5)$, $m = \dfrac{2}{3}$

62. $(0, -3)$, $m = 0$

63. $(2, 3)$, $m = 3$

64. $(4, 3)$, m undefined

65. $(3, -2)$, $m = \dfrac{5}{3}$

66. $(-2, -3)$, $m = 0$

67. $(-2, -4)$, $m = -\dfrac{5}{2}$

68. $(-3, 2)$, $m = -\dfrac{4}{3}$

69. $\left(\dfrac{2}{3}, -5\right)$, $m = 0$

70. $\left(-\dfrac{5}{2}, -1\right)$, m is undefined

[3.7] Write the equation of the line L satisfying each of the following sets of geometric conditions.

71. L passes through $(-3, -1)$ and $(3, 3)$.

72. L passes through $(0, 4)$ and $(5, 3)$.

73. L has slope $\dfrac{3}{4}$ and y intercept 3.

74. L passes through $(4, -3)$ with a slope of $-\dfrac{5}{4}$.

75. L has y intercept -4 and is parallel to the line with equation $3x - y = 6$.

76. L passes through $(3, -2)$ and is perpendicular to the line with equation $3x - 5y = 15$.

77. L passes through $(2, -1)$ and is perpendicular to the line with the equation $3x - 2y = 5$.

78. L passes through the point $(-5, -2)$ and is parallel to the line with the equation $4x - 3y = 9$.

[3.8] Graph the solution set for each of the following linear inequalities.

79. $y < 2x + 1$

80. $y \geq -2x + 3$

81. $3x + 2y \geq 6$

82. $3x - 5y < 15$

83. $y < -2x$

84. $4x - y \geq 0$

85. $y \geq -3$

86. $x < 4$

SELF-TEST

The purpose of this self-test is to help you check your progress and to review for a chapter test in class. Allow yourself about an hour to take the test. When you are done, check your answers in the back of the book. If you missed any problems, be sure to go back and review the appropriate section in the chapter and the exercises that are provided.

Graph each of the following equations.

1. $y = 3x + 2$

2. $x - 2y = 0$

3. $3x - 2y = 6$

4. $x = -4$

Find the domain and range for each function.

5. $\{(-1, 2), (0, 4), (2, 8), (4, 16)\}$

6. $\left\{ \left(-\dfrac{7}{2}, \dfrac{1}{3} \right), \left(-\dfrac{3}{2}, \dfrac{1}{3} \right), \left(-\dfrac{1}{2}, -\dfrac{1}{3} \right) \right\}$

Are the following pairs of lines parallel, perpendicular, or neither?

7. L_1 through $(-1, -4)$ and $(3, 2)$
   L_2 through $(0, 5)$ and $(3, 3)$

8. L_1 through $(-2, 1)$ and $(3, 4)$
   L_2 through $(-4, 4)$ and $(1, 7)$

Find the slope of each line.

9. $2x - 3y = 5$

10. $5x + 2y = 1$

Find the distance between each of the following pairs of points.

11. $(5, -2)$ and $(2, 3)$

12. $(-2, 0)$ and $(-4, -5)$

Find the midpoint of the segment defined by each pair of endpoints.

13. $(2, -4)$ and $(4, -12)$

14. $(-2, 7)$ and $(3, -8)$

Evaluate $f(-1)$ and $f(3)$.

15. $f(x) = -2x - 7$

16. $f(x) = \dfrac{1}{3}x - \dfrac{2}{3}$

Write the equation of the line L satisfying each of the following sets of geometric conditions.

17. L passes through the points $(-4, 3)$ and $(-1, 7)$.

18. L passes through $(-3, 7)$ and is perpendicular to $2x - 3y = 7$.

Graph the solution sets for the following linear inequalities.

19. $x - 4y < 8$

20. $3x + 5y \geq 15$

SYSTEMS OF LINEAR RELATIONS

CHAPTER FOUR

TRANSPORTATION IN THE UNITED STATES

U.S. transportation is a major contributor to many environmental problems. Motor vehicles account for 63 percent of our oil consumption and create more than 50 percent of the air pollution. Highways and parking lots cover much prime agricultural land and cause water pollution due to excessive runoff into rivers and lakes. More Americans have been killed by automobiles than in all U.S. wars combined.

Travel in the United States is overwhelmingly done by automobile: 98 percent of all urban transportation, 85 percent of travel between cities, and 84 percent of all travel to and from work are done by car or small truck. And 70 percent of the U.S. workforce drive to work alone.

Because the car is such an important part of U.S. culture, it will be very difficult to reduce its role in U.S. life. However, some changes have already been made, and more could be made in the future.

Current laws require the use of unleaded gasoline to reduce the level of lead in the atmosphere. This has resulted in a significant drop in lead in the environment because automobile exhaust was the largest source of lead pollution prior to these laws. New engine technologies resulting from California's stricter emissions laws have benefited everyone.

New engines that run on hydrogen gas will emit water vapor and greatly reduce levels of nitrogen oxides. Electric cars will decrease the air pollution from automobiles but could increase the demand for electric power plants, which have their own air pollution problems. Fuels such as natural gas and alcohol burn more cleanly than gasoline.

Besides new technology, we may need to adopt changes in our lifestyles as well. Mass transit use reduces air pollution and the need for more highways. This leaves more land available for other uses. Higher housing density could be accompanied by an increase in the use of bicycles in the city and could eliminate some of the need for automobiles. Neighborhood shops instead of shopping malls would allow people to walk to the store instead of driving.

People in the United States have adapted their lives to the automobile. In the future, the automobile must adapt to the life that we choose to lead.

TAPE IN9

4.1 Systems of Linear Equations in Two Variables

OBJECTIVES: 1. To solve systems of linear equations in two variables by graphing
2. To solve systems by the addition method
3. To solve systems by the substitution method

Our work in this chapter focuses on systems of equations and the various solution techniques available for your work with such systems. First, let's consider what we mean by a system of equations.

In many applications you will find it helpful to use two variables when labeling the quantities involved. Often this leads to a *linear equation in two variables*. A typical equation might be

It might be helpful to review Section 3.3 on graphing linear equations at this point.

$$x - 2y = 6$$

A solution for such an equation is any ordered pair of real numbers (x, y) that satisfies the equation. For example, the ordered pair $(4, -1)$ is a solution for the equation above since substituting 4 for x and -1 for y results in the true statement

Of course, there are an infinite number of solutions for an equation of this type. You might want to verify that $(2, -2)$ and $(6, 0)$ are also solutions.

$$4 - 2(-1) \stackrel{?}{=} 6$$
$$4 + 2 \stackrel{?}{=} 6$$
or $$6 = 6$$

Whenever two or more equations are considered together, they form a *system of equations*. If the equations of the system are linear, the system is called a *linear system*. Our work here and in Sections 4.2 and 4.5 involves finding solutions for such systems. Let's look at a definition.

A *solution* for a linear system of equations in two variables is an ordered pair of real numbers (x, y) that satisfies both equations in the system. For instance, given the linear system

$$x - 2y = -1$$
$$2x + y = 8$$

the pair $(3, 2)$ is a solution because after substituting 3 for x and 2 for y in the two equations of the system, we have the *two* true statements

Both equations are satisfied by $(3, 2)$.

$$3 - 2(2) \stackrel{?}{=} -1 \qquad \text{and} \qquad 2(3) + 2 \stackrel{?}{=} 8$$
$$-1 = -1 \qquad\qquad\qquad 8 = 8$$

Since a solution to a system of equations represents a point on both lines, one approach to finding the solution for a system is to graph each equation on the same set of coordinate axes and then identify the point of intersection. This is shown in the following example.

EXAMPLE 1 Solving a System by Graphing

Solve the system by graphing:

$2x + y = 4$
$x - y = 5$

We graph the lines corresponding to the two equations of the system.

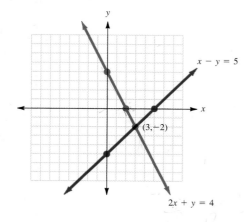

Here we used the intercept method to graph the lines.

 Solving each equation for *y*, then graphing

$y = -2x + 4$ and
$y = x - 5$

We can *approximate* the solution by tracing the curves near their intersection.

Each equation has an infinite number of solutions (ordered pairs) corresponding to points on a line. The point of intersection, here $(3, -2)$, is the *only* point lying on both lines, and so $(3, -2)$ is the only ordered pair satisfying both equations, and $(3, -2)$ is the solution for the system.

Check Yourself 1

Solve the system by graphing.

$3x - y = 2$
$x + y = 6$

In Example 1, the two lines are nonparallel and intersect at only one point. The system has a unique solution corresponding to that point. Such a system is called a *consistent system*. In our next example, we examine a system representing two lines that have no point of intersection.

EXAMPLE 2 Solving a System by Graphing

Solve the system by graphing:

$2x - y = 4$
$6x - 3y = 18$

The lines corresponding to the two equations are graphed below.

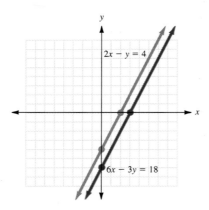

The lines are distinct and parallel. There is no point at which they intersect. So the system has no solution. We call such a system an *inconsistent system*.

Check Yourself 2

Solve the system if possible.

$$3x - y = 1$$
$$6x - 2y = 3$$

Sometimes the equations in a system have the same graph.

EXAMPLE 3 Solving a System by Graphing

Solve the system by graphing:

$$2x - y = 2$$
$$4x - 2y = 4$$

The equations are graphed.

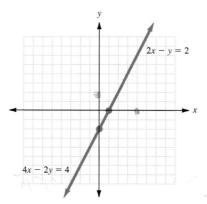

Since the lines representing the equations coincide, the system has an infinite number of solutions. Such a system is called a *dependent system*. The lines are called *coincident lines*. The solution is the set of all points on the line.

Note that every point on the line is a point of intersection.

Check Yourself 3

Solve the system, if possible.

$$x - 3y = 1$$
$$-2x + 6y = -2$$

We can summarize our consideration of the nature of the solutions of a linear system in two variables as follows.

SOLUTIONS FOR A LINEAR SYSTEM OF TWO EQUATIONS IN TWO VARIABLES

Consider the lines determined by the equations of the system. The solutions are illustrated by one of the following graphs.

1. The lines are not parallel. There is one and only one solution, and the system is *consistent*.

2. The lines are distinct and parallel. There is no solution, and the system is *inconsistent*.

3. The lines coincide. There are an infinite number of solutions, and the system is *dependent*.

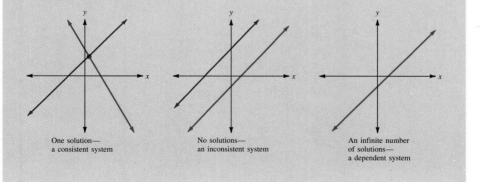

One solution— a consistent system

No solutions— an inconsistent system

An infinite number of solutions— a dependent system

The graphical method of solution is not always practical, particularly if the coordinates of the point of intersection are not integers. In that case, we can only "guess at," or estimate, the solution.

Because of this, we will turn to algebraic methods that allow us to find exact solutions. The first algebraic approach we consider is the *addition method* of solution. We will need the following definition.

There is nothing wrong with graphing, and then estimating, a solution. In fact, it is an excellent first approach to the problem. But exact solutions may not result from the graphical method.

Two systems of equations are *equivalent* if they have the same solution sets.

The addition method of solving systems is based on the following result regarding equivalent systems:

An equivalent system is formed whenever

1. One of the equations is multiplied by a nonzero number.

2. One of the equations is replaced by the sum of a constant multiple of another equation and that equation.

The following examples illustrate the addition method of solution.

EXAMPLE 4 Solving a System by the Addition Method

Solve the system by the addition method.

$$5x - 2y = 12 \tag{1}$$

$$3x + 2y = 12 \tag{2}$$

The addition method is sometimes called *solution by elimination* for this reason.

Solution In this case, adding the equations will eliminate variable y, and we have

$$8x = 24$$

$$x = 3 \tag{3}$$

Now Equation (3) can be paired with either of the original equations to form an equivalent system. We let $x = 3$ in Equation (1):

$$5(3) - 2y = 12$$

$$15 - 2y = 12$$

$$-2y = -3$$

$$y = \frac{3}{2}$$

The solution should be checked by substituting these values into Equation (2). Here

$$3(3) + 2\left(\frac{3}{2}\right) \stackrel{?}{=} 12$$

$$9 + 3 \stackrel{?}{=} 12$$

$$12 = 12$$

is a true statement.

and $\left(3, \dfrac{3}{2}\right)$ is the solution for our system.

Check Yourself 4

Solve the system by the addition method.

$$4x - 3y = 19$$

$$-4x + 5y = -25$$

Remember that multiplying one or both of the equations by a nonzero constant produces an equivalent system.

Example 4 and Check Yourself 4 were straightforward in that adding the equations of the system immediately eliminated one of the variables. The following example illustrates the fact that often we must multiply one or both of the equations by a nonzero constant before the addition method is applied.

EXAMPLE 5 Solving a System by the Addition Method

Solve the system by the addition method.

$$3x - 5y = 19 \tag{4}$$
$$5x + 2y = 11 \tag{5}$$

It is clear that adding the equations of the given system will *not* eliminate one of the variables. Therefore we must use multiplication to form an equivalent system. The choice of multipliers depends on which variable we decide to eliminate. Here we have decided to eliminate y. We multiply Equation (4) by 2 and Equation (5) by 5. We then have

$$6x - 10y = 38$$
$$25x + 10y = 55$$

Adding now eliminates y and yields

$$31x = 93$$
$$x = 3 \tag{6}$$

Pairing Equation (6) with Equation (4) gives an equivalent system, and we can substitute 3 for x in Equation (4):

$$3 \cdot 3 - 5y = 19$$
$$9 - 5y = 19$$
$$-5y = 10$$
$$y = -2$$

and $(3, -2)$ is the solution for the system.

> All these solutions can be approximated by graphing the lines and tracing near the intersection. This is particularly useful when the solutions are "ugly."

> Note that the coefficients of y are now *opposites* of each other.

> Again the solution should be checked by substitution in Equation (5).

Check Yourself 5

Solve the system by the addition method.

$$2x + 3y = -18$$
$$3x - 5y = 11$$

The following algorithm summarizes the addition method of solving linear systems of two equations in two variables.

SOLVING BY THE ADDITION METHOD

STEP 1 If necessary, multiply one or both of the equations by a constant so that one of the variables can be eliminated by addition.

STEP 2 Add the equations of the equivalent system formed in step 1.

STEP 3 Solve the equation found in step 2.

> **STEP 4** Substitute the value found in step **3** into either of the equations of the original system to find the corresponding value of the remaining variable. The ordered pair formed is the solution to the system.
>
> **STEP 5** Check the solution by substituting the pair of values found in step **4** into the other equation of the original system.

The next example illustrates two special situations.

EXAMPLE 6 Solving a System by the Addition Method

Solve each system by the addition method.

(a) $4x + 5y = 20$ (7)
     $8x + 10y = 19$ (8)

Multiply Equation (7) by -2. Then

$$-8x - 10y = -40$$
$$\underline{8x + 10y = 19}$$
$$0 = -21$$

We add the two left sides to get 0 and the two right sides to get −21.

The result $0 = -21$ is a *false* statement, a contradiction. The assumption that a point of intersection exists is then *false,* and there is no solution. This system represents two parallel lines. The system is inconsistent, and there is no solution.

(b) $5x - 7y = 9$ (9)
     $15x - 21y = 27$ (10)

Multiply Equation (9) by -3. We then have

$$-15x + 21y = -27$$
$$\underline{15x - 21y = 27}$$
$$0 = 0$$

We add the two equations.

The solution set could be written as $\{(x, y) \mid 5x - 7y = 9\}$.

Both variables have been eliminated, and the result is a *true* statement. The two lines coincide, and there are an infinite number of solutions, one for each point on that line.

Check Yourself 6

Solve each system by the addition method, if possible.

1. $3x + 2y = 8$
     $9x + 6y = 11$

2. $x - 2y = 8$
     $3x - 6y = 24$

The results of Example 6 can be summarized as follows:

> **When a system of two linear equations is solved:**
> 1. If a false statement such as 3 = 4 is obtained, then the system is inconsistent and has no solution.
> 2. If a true statement such as 8 = 8 is obtained, then the system is dependent and has an infinite number of solutions.

A third method for finding the solutions of linear systems in two variables is called the *substitution method.* You may very well find the substitution method more difficult to apply in solving certain systems than the addition method, particularly when the equations involved in the substitution lead to fractions. However, the substitution method does have important extensions to systems involving higher-degree equations, as we will see later in this text.

To outline the technique, we solve one of the equations from the original system for one of the variables. That expression is then substituted into the *other* equation of the system to provide an equation in a single variable. That equation is solved, and the corresponding value for the other variable is found as before. The following example illustrates.

EXAMPLE 7 Solving a System by the Substitution Method

(*a*) Solve the system by the substitution method.

$$2x - 3y = -3 \tag{11}$$

$$y = 2x - 1 \tag{12}$$

Since Equation (12) is already solved for y, we substitute $2x - 1$ for y in Equation (11):

$$2x - 3(2x - 1) = -3$$

We now have an equation in the single variable x.

Solving for x gives

$$2x - 6x + 3 = -3$$

$$-4x = -6$$

$$x = \frac{3}{2}$$

We now substitute $\frac{3}{2}$ for x in Equation (12):

$$y = 2\left(\frac{3}{2}\right) - 1$$

$$= 3 - 1 = 2$$

and $\left(\frac{3}{2}, 2\right)$ is the solution for our system.

To check this result, we substitute these values in Equation (11) and have

$$2\left(\frac{3}{2}\right) - 3 \cdot 2 \overset{?}{=} -3$$

$$3 - 6 \overset{?}{=} -3$$

$$-3 = -3$$

A true statement!

(*b*) Solve the system by the substitution method.

$$2x + 3y = 16 \tag{13}$$

$$3x - y = 2 \tag{14}$$

We start by solving Equation (14) for y:

Why did we choose to solve for y in Equation (14)? We could have solved for x, so that

$$x = \frac{y + 2}{3}$$

We simply chose the easier case to avoid fractions.

$$3x - y = 2$$

$$-y = -3x + 2$$

$$y = 3x - 2 \tag{15}$$

Substituting in Equation (13) yields

$$2x + 3(3x - 2) = 16$$

$$2x + 9x - 6 = 16$$

$$11x = 22$$

$$x = 2$$

We now substitute 2 for x in Equation (15):

$$y = 3 \cdot 2 - 2$$

$$= 6 - 2 = 4$$

The solution should be checked in *both* equations of the original system.

and (2, 4) is the solution for the system. We leave the check of this result to you.

Check Yourself 7

Solve each system by the substitution method.

1. $2x + 3y = 6$
$\quad\ \ x = 3y + 6$

2. $3x + 4y = -3$
$\quad\ \ x + 4y = 1$

The following algorithm summarizes the substitution method for solving linear systems of two equations in two variables.

SOLVING BY THE SUBSTITUTION METHOD

STEP 1 If necessary, solve one of the equations of the original system for one of the variables.

STEP 2 Substitute the expression obtained in step 1 into the *other* equation of the system to write an equation in a single variable.

STEP 3 Solve the equation found in step 2.

STEP 4 Substitute the value found in step 3 into the equation derived in step 1 to find the corresponding value of the remaining variable. The ordered pair formed is the solution for the system.

STEP 5 Check the solution by substituting the pair of values found in step 4 into *both* equations of the original system.

A natural question at this point is, How do you decide which solution method to use? First, the graphical method can generally provide only approximate solutions. When exact solutions are necessary, one of the algebraic methods must be applied. Which method to use depends totally on the given system.

If you can easily solve for a variable in one of the equations, the substitution method should work well. However, if solving for a variable in either equation of the system leads to fractions, you may find the addition approach more efficient.

CHECK YOURSELF ANSWERS

1. $x + y = 6$; $3x - y = 2$

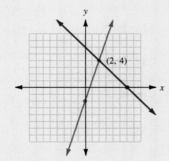

(2, 4)

2. The system is inconsistent. There is no solution.   **3.** The system is depen-
dent. The solution set is $\{(x, y) \mid x - 3y = 1\}$.   **4.** $\left(\dfrac{5}{2}, -3\right)$ **5.** $(-3, -4)$

6. (1) The system is inconsistent. There is no solution. (2) The system is depen-
dent. The solution set is $\{(x, y) \mid x - 2y = 8\}$.   **7.** (1) $\left(4, -\dfrac{2}{3}\right)$; (2) $\left(-2, \dfrac{3}{4}\right)$

4.1 EXERCISES

Build Your Skills

Solve each of the following systems by graphing. If a unique solution does not exist, state whether the system is dependent or inconsistent.

1. $x + y = 6$
$x - y = 4$

2. $x - y = 8$
$x + y = 2$

3. $x + 2y = 4$
$x - y = 1$

4. $x - 2y = 2$
$x + 2y = 6$

5. $3x - y = 3$
$3x - y = 6$

6. $3x + 2y = 12$
$y = 3$

7. $x + 3y = 12$
$2x - 3y = 6$

8. $3x - 6y = 9$
$x - 2y = 3$

Solve each of the following systems by the addition method. If a unique solution does not exist, state whether the system is dependent or inconsistent.

9. $2x - y = 1$
$-2x + 3y = 5$

10. $x + 3y = 12$
$2x - 3y = 6$

11. $x + 2y = -2$
$3x + 2y = -12$

12. $2x + 3y = 1$
$5x + 3y = 16$

13. $x + y = 3$
$3x - 2y = 4$

14. $x - y = -2$
$2x + 3y = 21$

15. $2x + y = 8$
$-4x - 2y = -16$

16. $3x - 4y = 2$
$4x - y = 20$

17. $5x - 2y = 31$
$4x + 3y = 11$

18. $2x - y = 4$
$6x - 3y = 10$

19. $3x - 2y = 7$
$-6x + 4y = -15$

20. $3x + 4y = 0$
$5x - 3y = -29$

21. $-2x + 7y = 2$
$3x - 5y = -14$

22. $5x - 2y = 3$
$10x - 4y = 6$

Solve each of the following systems by the substitution method. If a unique solution does not exist, state whether the system is inconsistent or dependent.

23. $x - y = 7$
$y = 2x - 12$

24. $x - y = 4$
$x = 2y - 2$

25. $3x + 2y = -18$
$x = 3y + 5$

26. $3x - 18y = 4$
$x = 6y + 2$

27. $10x - 2y = 4$
$y = 5x - 2$

28. $4x + 5y = 6$
$y = 2x - 10$

29. $3x + 4y = 9$
$y = 3x + 1$

30. $6x - 5y = 27$
$x = 5y + 2$

31. $x - 7y = 3$
$2x - 5y = 15$

32. $4x + 3y = -11$
$5x + y = -11$

33. $4x - 12y = 5$
$-x + 3y = -1$

34. $5x - 6y = 21$
$x - 2y = 5$

Solve each of the following systems by any method discussed in this section.

35. $2x - 3y = 4$
$x = 3y + 6$

36. $5x + y = 2$
$5x - 3y = 6$

37. $4x - 3y = 0$
$5x + 2y = 23$

38. $7x - 2y = -17$
$x + 4y = 4$

39. $3x - y = 17$
$5x + 3y = 5$

40. $7x + 3y = -51$
$y = 2x + 9$

Solve each of the following systems by any method discussed in this section. *Hint:* You should multiply to clear fractions as your first step.

41. $\frac{1}{2}x - \frac{1}{3}y = 8$
$\frac{1}{3}x + y = -2$

42. $\frac{1}{5}x - \frac{1}{2}y = 0$
$x - \frac{3}{2}y = 4$

43. $\frac{2}{3}x + \frac{3}{5}y = -3$
$\frac{1}{3}x + \frac{2}{5}y = -3$

44. $\frac{3}{8}x - \frac{1}{2}y = -5$
$\frac{1}{4}x + \frac{3}{2}y = 4$

Transcribe Your Skills

45. Graphing calculators have made it easy to display the graph of a function. Assume that a system consists of two equations expressed in function form. Would a picture that gives the approximate coordinates of the intersection of the lines be useful in solving the system? How would you use the information?

46. It is possible to have a system that consists of three equations. If each equation were linear, how many solutions would be possible? Sketch a picture to accompany your answer.

Think About These

Certain systems that are not linear can be solved with the methods of this section if we first substitute to change variables. For instance, the system

$$\frac{1}{x} + \frac{1}{y} = 4$$

$$\frac{1}{x} - \frac{3}{y} = -6$$

can be solved by the substitutions $u = \frac{1}{x}$ and $v = \frac{1}{y}$. That gives the system $u + v = 4$ and $u - 3v = -6$. The system is then solved for u and v, and the corresponding values for x and y are found. Use this method to solve the following systems.

47. $\dfrac{1}{x} + \dfrac{1}{y} = 4$

$\dfrac{1}{x} - \dfrac{3}{y} = -6$

48. $\dfrac{1}{x} + \dfrac{3}{y} = 1$

$\dfrac{4}{x} + \dfrac{3}{y} = 3$

49. $\dfrac{2}{x} + \dfrac{3}{y} = 4$

$\dfrac{2}{x} - \dfrac{6}{y} = 10$

50. $\dfrac{4}{x} - \dfrac{3}{y} = -1$

$\dfrac{12}{x} - \dfrac{1}{y} = 1$

Writing the equation of a line through two points can be done by the following method. Given the coordinates of two points, substitute each pair of values into the equation $y = mx + b$. This gives a system of two equations in variables m and b, which can be solved as before.

Write the equation of the line through each of the following pairs of points, using the method outlined above.

51. $(2, 1)$ and $(4, 4)$

52. $(-3, 7)$ and $(6, 1)$

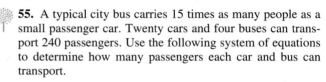

53. Twenty cars (C) and five buses (B) can transport 295 people. Thirty cars and ten buses can transport 550 people. Use the following system of equations to determine how many people each car and bus can transport.

$$20C + 5B = 295$$

$$30C + 10B = 550$$

54. Fifteen cars and four buses can transport 305 people. Five cars and twelve buses can transport 795 people. Use the following system of equations to determine how many people each car and bus can transport.

$$15C + 4B = 305$$

$$5C + 12B = 795$$

55. A typical city bus carries 15 times as many people as a small passenger car. Twenty cars and four buses can transport 240 passengers. Use the following system of equations to determine how many passengers each car and bus can transport.

$$20C + 4B = 240$$

$$B = 15C$$

56. An articulated (bendable) city bus carries 16 times as many people as a midsize passenger car. Fifteen cars and five buses can transport 380 passengers. Use the following system of equations to determine how many passengers each car and bus can transport.

$$15C + 5B = 380$$

$$B = 16C$$

▱ Graphing Utility

Use your graphing utility to approximate the solution to each system. Express your answer to the nearest tenth.

57. $y = 2x - 3$

$2x + 3y = 1$

58. $3x - 4y = -7$

$2x + 3y = -1$

Adjust the viewing window on your graphing utility so that you can see the point of intersection for the two lines representing the equations in the system. Then approximate the solution. Express your answer as the nearest integral solution.

59. $5x - 12y = 8$

$7x + 2y = 44$

60. $9x - 3y = 10$

$x + 5y = 58$

Skillscan (Section 1.4)

Simplify each of the following expressions.

a. $(3x - 2y + z) + (5x + 2y - 3z)$

b. $(7x - 6y + 3z) - (5x + 2y + 3z)$

c. $(x + 2y - 3z) + 3(2x + y + z)$

d. $2(2x + y + z) + (x - 2y + 3z)$

e. $2(x + 3y - z) + 3(x - 2y + z)$

f. $4(x - 2y - 3z) + 3(2x + y + 4z)$

g. $2(x + 3y - z) - 3(2x + 2y + z)$

h. $5(2x - y - z) - 2(5x + y + 3z)$

4.2 Systems of Linear Equations in Three Variables

TAPE IN9

OBJECTIVE: To solve systems of three linear equations in three variables by the addition method

Suppose that an application involves three quantities which we wish to label x, y, and z. A typical equation used for the solution might be

$$2x + 4y - z = 8$$

This is called a *linear equation in three variables.* The solution for such an equation is an *ordered triple* (x, y, z) of real numbers that satisfies the equation. For example, the ordered triple $(2, 1, 0)$ is a solution for the equation above since substituting 2 for x, 1 for y, and 0 for z results in the true statement

Of course, other solutions, in fact infinitely many, exist. You might want to verify that $(1, 1, -2)$ and $(3, 1, 2)$ are also solutions.

$$2 \cdot 2 + 4 \cdot 1 - 0 \overset{?}{=} 8$$
$$4 + 4 \qquad \overset{?}{=} 8$$
$$8 = 8$$

To extend the concepts of the last section, we want to consider systems of three linear equations in three variables such as

$$x + y + z = 5$$
$$2x - y + z = 9$$
$$x - 2y + 3z = 16$$

For a unique solution to exist, when three variables are involved, we must have three equations.

The solution for such a system is the set of all ordered triples that satisfy each equation of the system. In the case above, you should verify that $(2, -1, 4)$ is a solution for the system, since that ordered triple makes each equation a true statement.

The choice of which variable to eliminate is yours. Generally you should pick the variable that allows the easiest computation.

Let's turn now to the solution process itself. In this section we will consider the addition method. The central idea is to choose *two pairs* of equations from the system and, by the addition method, to eliminate the *same variable* from each of those pairs. The method is best illustrated by example. So let's proceed to see how the solution for the previous system was determined.

EXAMPLE 1 Solving a Linear System in Three Variables

Solve the system.

$$x + y + z = 5 \tag{1}$$
$$2x - y + z = 9 \tag{2}$$
$$x - 2y + 3z = 16 \tag{3}$$

Solution First we choose two pairs of equations and the variable to eliminate. Variable y seems convenient in this case. Pairing Equations (1) and (2) and then adding, we have

$$\begin{array}{r} x + y + z = 5 \\ 2x - y + z = 9 \\ \hline 3x \qquad + 2z = 14 \end{array} \tag{4}$$

Any pair of equations could have been selected.

We now want to choose a *different pair* of equations and *again eliminate y*. Pairing Equations (1) and (3) this time, we multiply Equation (1) by 2 and then add the result to Equation (3):

$$\begin{array}{r} 2x + 2y + 2z = 10 \\ x - 2y + 3z = 16 \\ \hline 3x \qquad + 5z = 26 \end{array} \tag{5}$$

We now have two equations in variables x and z:

$$3x + 2z = 14 \tag{4}$$
$$3x + 5z = 26 \tag{5}$$

Since we are now dealing with a system of two equations in two variables, any of the methods of the previous section apply. We have chosen to multiply Equation (4) by -1 and then add that result to Equation (5). This yields

$$3z = 12$$
$$z = 4$$

Substituting $z = 4$ in Equation (4) gives

$$3x + 2 \cdot 4 = 14$$
$$3x + 8 = 14$$
$$3x = 6$$
$$x = 2$$

Finally letting $x = 2$ and $z = 4$ in Equation (1) gives

$$2 + y + 4 = 5$$
$$y = -1$$

Any of the original equations could have been used.

To check, substitute these values into the other equations of the original system.

and $(2, -1, 4)$ is shown to be the solution for the system.

Check Yourself 1

Solve the system.

$$x - 2y + z = 0$$
$$2x + 3y - z = 16$$
$$3x - y - 3z = 23$$

One or more of the equations of a system may already have a missing variable. The elimination process is simplified in that case, as the following example illustrates.

EXAMPLE 2 Solving a Linear System in Three Variables

Solve the system.

$$2x + y - z = -3 \qquad (6)$$
$$y + z = 2 \qquad (7)$$
$$4x - y + z = 12 \qquad (8)$$

Solution Noting that Equation (7) involves only y and z, we must simply find another equation in those same two variables. Multiply Equation (6) by -2 and add the result to Equation (8) to eliminate x:

We now have a *second* equation in y and z.

$$\begin{array}{r} -4x - 2y + 2z = 6 \\ 4x - y + z = 12 \\ \hline -3y + 3z = 18 \\ y - z = -6 \qquad (9) \end{array}$$

We now form a system consisting of Equations (7) and (9) and solve as before:

Adding eliminates z.

$$\begin{array}{r} y + z = 2 \qquad (7) \\ y - z = -6 \qquad (9) \\ \hline 2y = -4 \\ y = -2 \end{array}$$

From Equation (7) if $y = -2$,

$$-2 + z = 2$$
$$z = 4$$

and from Equation (6) if $y = -2$ and $z = 4$,

$$2x - 2 - 4 = -3$$
$$2x = 3$$
$$x = \frac{3}{2}$$

The solution for the system is

$$\left(\frac{3}{2}, -2, 4\right)$$

Check Yourself 2

Solve the system.

$$x + 2y - z = -3$$
$$x - y + z = 2$$
$$x \quad\quad - z = 3$$

The following algorithm summarizes the procedure for finding the solutions for a linear system of three equations in three variables.

SOLVING A SYSTEM OF THREE EQUATIONS IN THREE UNKNOWNS

STEP 1 Choose a pair of equations from the system, and use the addition method to eliminate one of the variables.

STEP 2 Choose a *different* pair of equations, and eliminate the *same* variable.

STEP 3 Solve the system of two equations in two variables determined in steps 1 and 2.

STEP 4 Substitute the values found above into one of the original equations, and solve for the remaining variable.

STEP 5 The solution is the ordered triple of values found in steps 3 and 4. It can be checked by substituting into the other equations of the original system.

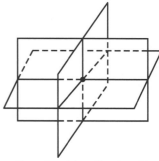

Three planes intersecting at a point

Systems of three equations in three variables may have (1) exactly one solution, (2) infinitely many solutions, or (3) no solution. Before we look at an algebraic approach in the second and third cases, let's discuss the geometry involved.

The graph of a linear equation in three variables is a plane (a flat surface) in three dimensions. Two distinct planes either will be parallel or will intersect in a line.

If three distinct planes intersect, that intersection will be either a single point (as in our first example) or a line (think of three pages in an open book—they intersect along the binding of the book).

Let's look at an example of how the solution proceeds in these cases.

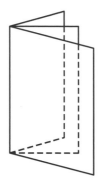

Three planes intersecting in a line

EXAMPLE 3 Solving a Dependent Linear System in Three Variables

Solve the system.

$$x + 2y - z = 5 \tag{10}$$

$$x - y + z = -2 \tag{11}$$

$$-5x - 4y + z = -11 \tag{12}$$

Solution We begin as before by choosing two pairs of equations from the system and eliminating the same variable from each of the pairs.

Adding Equations (10) and (11) gives

$$2x + y = 3 \tag{13}$$

Adding Equations (10) and (12) gives

$$-4x - 2y = -6 \tag{14}$$

Now consider the system formed by Equations (13) and (14). We multiply equation (13) by 2 and add again:

$$
\begin{array}{r}
4x + 2y = 6 \\
\underline{-4x - 2y = -6} \\
0 = 0
\end{array}
$$

There are ways of representing the solutions. You will see these in later courses.

This true statement tells us that the system has an infinite number of solutions (lying along a straight line).

Check Yourself 3

Solve the system.

$$2x - y + 3z = 3$$

$$-x + y - 2z = 1$$

$$y - z = 5$$

There is a third possibility for the solutions of systems in three variables, as Example 4 illustrates.

EXAMPLE 4 Solving an Inconsistent Linear System in Three Variables

Solve the system.

$$3x + y - 3z = 1 \tag{15}$$

$$-2x - y + 2z = 1 \tag{16}$$

$$-x - y + z = 2 \tag{17}$$

Solution This time we eliminate variable y. Adding Equations (15) and (16), we have

$$x - z = 2 \tag{18}$$

Adding Equations (15) and (17) gives

$$2x - 2z = 3 \tag{19}$$

Now multiply Equation (18) by -2 and add the result to Equation (19):

$$\begin{array}{r} -2x + 2z = -4 \\ \underline{2x - 2z = 3} \\ 0 = -1 \end{array}$$

All the variables have been eliminated, and we have arrived at a contradiction, $0 = -1$. This means that the system is *inconsistent* and has no solutions. There is *no* point common to all three planes.

Check Yourself 4

Solve the system.

$$\begin{aligned} x - y - z &= 0 \\ -3x + 2y + z &= 1 \\ 3x - y + z &= -1 \end{aligned}$$

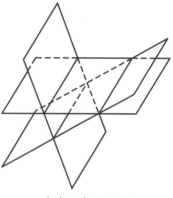

An inconsistent system

As a closing note, we have by no means illustrated all possible types of inconsistent and dependent systems. Other possibilities involve either distinct parallel planes or planes that coincide. The solution techniques in these additional cases are, however, similar to those illustrated above.

CHECK YOURSELF ANSWERS

1. $(5, 1, -3)$ **2.** $(1, -3, -2)$ **3.** An infinite number of solutions
4. An inconsistent system. There are *no* solutions.

4.2 EXERCISES

Build Your Skills

Solve each of the following systems of equations. If a unique solution does not exist, state whether the system is inconsistent or has an infinite number of solutions.

1. $\begin{aligned} x - y + z &= 3 \\ 2x + y + z &= 8 \\ 3x + y - z &= 1 \end{aligned}$

2. $\begin{aligned} x - y - z &= 2 \\ 2x + y + z &= 8 \\ x + y + z &= 6 \end{aligned}$

3. $\begin{aligned} x + y + z &= 1 \\ 2x - y + 2z &= -1 \\ -x - 3y + z &= 1 \end{aligned}$

4. $x - y - z = 6$
$-x + 3y + 2z = -11$
$3x + 2y + z = 1$

5. $x + y + z = 1$
$-2x + 2y + 3z = 20$
$2x - 2y - z = -16$

6. $x + y + z = -3$
$3x + y - z = 13$
$3x + y - 2z = 18$

7. $2x + y - z = 2$
$-x - 3y + z = -1$
$-4x + 3y + z = -4$

8. $x + 4y - 6z = 8$
$2x - y + 3z = -10$
$3x - 2y + 3z = -18$

9. $3x - y + z = 5$
$x + 3y + 3z = -6$
$x + 4y - 2z = 12$

10. $2x - y + 3z = 2$
$x - 2y + 3z = 1$
$4x - y + 5z = 5$

11. $x + 2y + z = 2$
$2x + 3y + 3z = -3$
$2x + 3y + 2z = 2$

12. $x - 4y - z = -3$
$x + 2y + z = 5$
$3x - 7y - 2z = -6$

13. $x + 3y - 2z = 8$
$3x + 2y - 3z = 15$
$4x + 2y + 3z = -1$

14. $x + y - z = 2$
$3x + 5y - 2z = -5$
$5x + 4y - 7z = -7$

15. $x + y - z = 2$
$x - 2z = 1$
$2x - 3y - z = 8$

16. $x + y + z = 6$
$x - 2y = -7$
$4x + 3y + z = 7$

17. $x - 3y + 2z = 1$
$16y - 9z = 5$
$4x + 4y - z = 8$

18. $x - 4y + 4z = -1$
$y - 3z = 5$
$3x - 4y + 6z = 1$

19. $x + 2y - 4z = 13$
$3x + 4y - 2z = 19$
$3x + 2z = 3$

20. $x + 2y - z = 6$
$-3x - 2y + 5z = -12$
$x - 2z = 3$

Transcribe Your Skills

21. Describe an example (that you can see) of three planes intersecting at a single point. Feel free to use the room you're in as a reference.

22. Describe a similar example of three parallel planes. Use the building you're in as a reference.

Think About These

The solution process illustrated in this section can be extended to solving systems of more than three variables in a natural fashion. For instance, if four variables are involved, eliminate one variable in the system and then solve the resulting system in three variables as before. Substituting those three values into one of the original equations will provide the value for the remaining variable and the solution for the system.

Solve each of the following systems, using the procedure described above.

23. $x + 2y + 3z + w = 0$
$-x - y - 3z + w = -2$
$x - 3y + 2z + 2w = -11$
$-x + y - 2z + w = 1$

24. $x + y - 2z - w = 4$
$x - y + z + 2w = 3$
$2x + y - z - w = 7$
$x - y + 2z + w = 2$

In some systems of equations there are more equations than variables. We can illustrate this situation with a system of three equations in two variables. To solve this type of system, pick any two of the equations and solve this system. Then substitute the solution obtained into the third equation. If a true statement results, the solution used is the solution to the entire system. If a false statement occurs, the system has no solution.

Use this procedure to solve each of the following systems.

25. $x - y = 5$
$2x + 3y = 20$
$4x + 5y = 38$

26. $3x + 2y = 6$
$5x + 7y = 35$
$7x + 9y = 8$

27. Experiments have shown that cars (C), trucks (T), and buses (B) emit different amounts of air pollutants. In one such experiment, a truck emitted 1.5 pounds (lb) of carbon dioxide (CO_2) per passenger-mile and 2 grams (g) of nitrogen oxide (NO) per passenger-mile. A car emitted 1.1 lb of

CO_2 per passenger-mile and 1.5 g of NO per passenger-mile. A bus emitted 0.4 lb of CO_2 per passenger-mile and 1.8 g of NO per passenger-mile. A total of 85 mi was driven by the three vehicles, and 73.5 lb of CO_2 and 149.5 g of NO were collected. Use the following system of equations to determine the miles driven by each vehicle.

$$T + \quad C + \quad B = \quad 85.0$$
$$1.5T + 1.1C + 0.4B = \quad 73.5$$
$$2T + 1.5C + 1.8B = 149.5$$

28. Experiments have shown that cars (C), trucks (T), and trains (R) emit different amounts of air pollutants. In one such experiment, a truck emitted 0.8 lb of carbon dioxide per passenger-mile and 1 g of nitrogen oxide per passenger-mile. A car emitted 0.7 lb of CO_2 per passenger-mile and 0.9 g of NO per passenger-mile. A train emitted 0.5 lb of CO_2 per passenger-mile and 4 g of NO per passenger-mile. A total of 141 mi was driven by the three vehicles, and 82.7 lb of CO_2 and 424.4 g of NO were collected. Use the following system of equations to determine the miles driven by each vehicle.

$$T + \quad C + \quad R = 141.0$$
$$0.8T + 0.7C + 0.5R = \quad 82.7$$
$$T + 0.9C + \quad 4R = 424.4$$

Skillscan (Section 2.1)

Solve each of the following equations.

a. $x - (-2) = 5$ **b.** $y - 2(3) = -14$

c. $y + 3(-4) = -8$ **d.** $x - 3\left(-\dfrac{2}{3}\right) = 5$

e. $x - 2(1) + 3(2) = 1$ **f.** $x - 3(-2) - 4(-1) = 0$

g. $x - 2\left(\dfrac{1}{2}\right) + 5\left(-\dfrac{2}{5}\right) = 3$

h. $x + 3\left(-\dfrac{2}{3}\right) - 4\left(-\dfrac{3}{2}\right) = -2$

Applications and Problem Solving   **4.3**

OBJECTIVE: To use systems of equations in the solution of applications

TAPE IN9

Thus far in this chapter we have dealt with systems of equations and the various techniques that can be used in solving such systems. These added tools provide new flexibility in approaching applications. For instance, suppose an application involves more than one unknown quantity.

In many cases, we can approach that application through our earlier single-variable methods of Chapter 2. However, we now have the option of representing the unknown quantities with different variables and establishing a system of equations for the solution.

The problem-solving algorithm, presented earlier, remains essentially the same no matter which approach to the solution we choose.

SOLVING APPLICATIONS

STEP 1   Read the problem carefully to determine the unknown quantities.

STEP 2   Choose a variable or variables to represent the unknowns.

STEP 3   Translate the problem to the language of algebra to form an equation or a system of equations.

STEP 4   Solve the equation or the system of equations.

STEP 5   Verify your solution by returning to the original problem.

So that you can compare the problem-solving approaches, we show our first example solved with the use of a single variable and then with two variables leading to a system of equations.

EXAMPLE 1 Solving a Mixture Problem

A coffee merchant has two types of coffee beans, one selling for $3 per pound and the other for $5 per pound. The beans are to be mixed to provide 100 lb of a mixture selling for $4.50 per pound.

How much of each type of coffee bean should be used to form 100 lb of the mixture?

Step 1 The unknowns are the amounts of the two types of beans.

$100 - x$ is the remainder of the desired mixture.

Step 2 Let x represent the amount of the $3 per pound coffee beans. Then $100 - x$ represents the amount of the $5 beans.

Step 3 The total value of the mixture must be $450 (100 lb at $4.50 per pound). So our equation is

$$3x + 5(100 - x) = 450$$

Value of   Value of   Total
$3 beans in $5 beans   value
mixture

Step 4 $3x + 5(100 - x) = 450$

$$3x + 500 - 5x = 450$$

$$-2x = -50$$

or $x = 25$ lb $3 beans

and $100 - x = 75$ lb $5 beans

Step 5 To check the result, show that the value of the $3 beans, added to the value of the $5 beans, equals the desired value of the mixture.

Check Yourself 1

Peanuts which sell for $2.40 per pound and cashews which sell for $6 per pound are to be mixed to form a 60-lb mixture selling for $3 per pound. How much of each type of nut should be used? Solve the problem by using a single variable.

Now let's return to the problem of Example 1, this time using two variables for the solution so that you can compare the methods.

EXAMPLE 2 Solving a Mixture Problem

Step 1 Again we want to find the amounts of $3 and $5 beans to use in the mixture.

Step 2 We use two variables to represent the two unknowns. Let x be the amount of $3 beans and y the amount of $5 beans.

Step 3 We now want to establish a system of two equations. One equation will be based on the *total amount* of the mixture, the other on the mixture's *value*:

$$x + \quad y = 100 \qquad \text{The mixture must weigh 100 lb.} \qquad (1)$$
$$3x + 5y = 450 \qquad\qquad\qquad (2)$$

Since we use *two* variables, we must form *two* equations.

Value of   Value of   Total value
$3 beans $5 beans

Step 4 An easy approach to the solution of the system is to multiply Equation (1) by -3 and add to eliminate x:

$$-3x - 3y = -300$$
$$3x + 5y = \quad 450$$
$$2y = \quad 150$$
$$y = \quad 75 \text{ lb}$$

and by substitution in Equation (1)

$$x = 25 \text{ lb}$$

Step 5 The solution is checked as before.

Check Yourself 2

Do the problem of the Check Yourself 1 exercise, this time using two variables for the solution.

A related problem is illustrated in the next example.

EXAMPLE 3 Solving a Mixture Problem

A chemist has a 25% and a 50% acid solution. How much of each solution should be used to form 200 mL of a 35% acid solution?

Step 1 The unknowns in this case are the amounts of the 25% and 50% solutions to be used in forming the mixture.

Step 2 Again we use two variables to represent the two unknowns.
   Let x be the amount of the 25% solution and y the amount of the 50% solution. Let's draw a picture before proceeding to form a system of equations.

Drawing a sketch of a problem is often a valuable part of the problem-solving strategy.

Step 3 Now to form our two equations, we want to consider two relationships—the *total amounts* combined and the *amounts of acid* combined.
From our sketch of the problem

Total amounts combined

$$x + \quad y = 200 \tag{3}$$

Amounts of acid combined

$$0.25x + 0.50y = 0.35(200) \tag{4}$$

Step 4 You encountered similar systems in the exercises of Section 4.1. Remember that the suggested strategy was to clear Equation (4) of decimals as the first step. The solution then proceeds as before, with the result

Multiply Equation (4) by 100 as your first step.

$$x = 120 \text{ mL} \quad (25\% \text{ solution})$$

$$y = \quad 80 \text{ mL} \quad (50\% \text{ solution})$$

Step 5 To check, show that the amount of acid in the 20% solution, $(0.25)(120)$, added to the amount in the 50% solution, $(0.50)(80)$, equals the correct amount in the mixture, $(0.35)(200)$.

Check Yourself 3

A pharmacist wishes to prepare 300 mL of a 20% alcohol solution. How much of a 30% solution and a 15% solution should be used to form the desired mixture?

Applications that involve a constant rate of travel, or speed, require the use of the formula seen earlier:

$$d = rt$$

Here d is the distance traveled, r is the rate, or speed, and t is the time. Example 4 illustrates.

EXAMPLE 4 Solving a Distance-Rate-Time Problem

A boat can travel 36 mi downstream in 2 h. Coming back upstream, the boat takes 3 h.
What is the rate of the boat in still water? What is the rate of the current?

Step 1 We want to find the two rates.

Step 2 Let x be the rate of the boat in still water and y the rate of the current.

Downstream the rate is then

$x + y$

Upstream, the rate is

$x - y$

Step 3 To form a system, think about the following. Downstream the rate of the boat is *increased* by the effect of the current. Upstream the rate is *decreased*.
In many applications it is helpful to lay out the information in tabular form. Let's try that strategy here.

	d	r	t
Downstream	36	$x + y$	2
Upstream	36	$x - y$	3

Now since $d = rt$, from the table we can easily form two equations:

$$36 = (x + y)(2) \qquad\qquad\qquad (5)$$

$$36 = (x - y)(3) \qquad\qquad\qquad (6)$$

Step 4 We clear Equations (5) and (6) of parentheses and simplify, to write the equivalent system

$$x + y = 18$$

$$x - y = 12$$

Solving, we have

$$x = 15 \text{ mi/h}$$

$$y = 3 \text{ mi/h}$$

Step 5 To check, verify the $d = rt$ equation in *both* the upstream and the downstream cases. We leave that to you.

Check Yourself 4

A plane flies 480 mi in an easterly direction, with the wind, in 4 h. Returning westerly along the same route, against the wind, the plane takes 6 h. What is the rate of the plane in still air? What is the rate of the wind?

The use of systems of equations in problem solving has many applications in a business setting. The following example illustrates one such application.

EXAMPLE 5 Solving a Business-Based Application

A manufacturer produces a standard model and a deluxe model of 13-in television set. The standard model requires 12 h of labor to produce, while the deluxe model requires 18 h. The company has 360 h of labor available per week. The plant's capacity is a total of 25 sets per week. If all the available time and capacity are to be used, how many of each type of set should be produced?

Step 1 The unknowns in this case are the number of standard and deluxe models that can be produced.

Step 2 Let x be the number of standard models and y the number of deluxe models.

The choices for x and y could have been reversed.

Step 3 Our system will come from the two given conditions that fix the total number of sets that can be produced and the total labor hours available:

$$x + y = 25 \longleftarrow \text{Total number of sets}$$

$$12x + 18y = 360 \longleftarrow \text{Total labor hours available}$$

Labor hours—   Labor hours—
standard sets   deluxe sets

Step 4 Solving the system of step 3, we have

$$x = 15 \qquad \text{and} \qquad y = 10$$

which tells us that to use all the available capacity, the plant should produce 15 standard sets and 10 deluxe sets per week.

Step 5 We leave the check of this result to the reader.

Check Yourself 5

A manufacturer produces standard record turntables and compact disc players. The turntables require 2 h of electronic assembly and the players 3 h. The turntables require 4 h of case assembly and the players 2 h. The company has 120 h of electronic assembly time available per week and 160 h of case assembly time. How many of each type of unit can be produced each week if all available assembly time is to be used?

Let's look at one final application that leads to a system of two equations.

EXAMPLE 6 Solving a Business-Based Application

Two car rental agencies have the following rate structures for a subcompact car. Company A charges $20 per day plus 15¢ per mile. Company B charges $18 per day plus 16¢ per mile. If you rent a car for 1 day, for what number of miles will the two companies have the same total charge?

Solution Letting c represent the total a company will charge and m the number of miles driven, we have

For company A:

$$c = 20 + 0.15m \tag{7}$$

You first saw this type of linear model in exercises in Section 3.3.

For company B:

$$c = 18 + 0.16m \tag{8}$$

The system can be solved most easily by substitution. Substituting $18 + 0.16m$ for c in Equation (7) gives

$$18 + 0.16m = 20 + 0.15m$$

$$0.01m = 2$$

$$m = 200 \text{ mi}$$

The graph of the system is shown below.

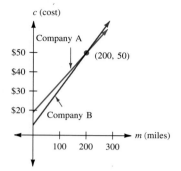

From the graph, how would you make a decision about which agency to use?

Check Yourself 6

For a compact car, the same two companies charge $27 per day plus 20¢ per mile and $24 per day plus 22¢ per mile. For a 2-day rental, when will the charges be the same?

In many instances, if an application involves three unknown quantities, you will find it useful to assign three variables to those quantities and then build a system of three equations from the given relationships in the problem. The extension of our problem-solving strategy is natural, as the following examples illustrate.

EXAMPLE 7 Solving a Number Problem

The sum of the digits of a three-digit number is 12. The tens digit is 2 less than the hundreds digit, and the units digit is 4 less than the sum of the other two digits. What is the number?

Step 1 The three unknowns are, of course, the three digits of the number.

Step 2 We now want to assign variables to each of the three digits. Let

u be the units digit

t be the tens digit

h be the hundreds digit

Step 3 There are three conditions given in the problem. That will allow us to write the necessary three equations. From those conditions

$h + t + u = 12$ (9)

$t = h - 2$ (10)

$u = h + t - 4$ (11)

Sometimes it helps to choose variable letters that relate to the words—

u for *units*

t for *tens*

h for *hundreds*

Take a moment now to go back to the original problem and pick out those conditions. That skill is a crucial part of the problem-solving strategy.

Note Although we did not discuss substitution in Section 4.2 while solving systems of three equations, it can sometimes be useful. Here, for instance, note from Equation (11) that we could substitute $h + t - 4$ for u in Equation (9). This would give an equation in h and t which could then be paired with Equation (10) and the system solved as before.

Step 4 There are various ways to approach the solution. To use addition, write the system in the equivalent form

$$h + t + u = 12$$
$$-h + t = -2$$
$$-h - t + u = -4$$

and solve by our earlier methods. The solution, which you can verify, is $h = 5$, $t = 3$, and $u = 4$, and the desired number is 534.

Step 5 To check, you should show that the digits of 534 meet each of the conditions of the original problem.

Check Yourself 7

The sum of the measures of the angles of a triangle is 180°. In a given triangle, the measure of the second angle is twice the measure of the first. The measure of the third angle is 30° less than the sum of the measures of the first two. Find the measure of each angle.

Let's continue with a slightly different application that will lead to a system of three equations.

EXAMPLE 8 Solving an Investment Application

Monica decided to divide a total of $42,000 into three investments—a savings account paying 5 percent interest, a time deposit paying 7 percent, and a bond paying 9 percent. Her total annual interest from the three investments was $2600, and the interest from the bank account was $200 less than the total interest from the other two investments. How much did she invest at each rate?

Step 1 The three amounts are the unknowns.

Again we choose letters that suggest the unknown quantities—s for savings, t for time deposit, and b for bond.

Step 2 We let s be the amount invested at 5 percent, t the amount at 7 percent, and b the amount at 9 percent. Note that the interest from the savings account is then $0.05s$, and so on.

A table will help with the next step.

For 1 year, the interest formula is

$I = Pr$

(interest equals principal times rate).

	5 percent	**7 percent**	**9 percent**
Principal	s	t	b
Interest	$0.05s$	$0.07t$	$0.09b$

Step 3 Again there are three conditions in the given problem. By using the table above, they lead to the following equations:

Total invested

$$s + t + b = 42{,}000$$

Total interest

$$0.05s + 0.07t + 0.09b = 2{,}600$$

The savings interest was $200 *less than* that from the other two investments.

$$0.05s = 0.07t + 0.09b - 200$$

Step 4 We clear of decimals and solve as before, with the result

$s = \$24,000 \qquad t = \$11,000 \qquad b = \$7000$

Step 5 We leave the check of these solutions to you.

Find the interest earned from each investment, and verify that the conditions of the problem are satisfied.

Check Yourself 8

Glenn has a total of $11,600 invested in three accounts—a savings account paying 6 percent interest, a stock paying 8 percent, and a mutual fund paying 10 percent. The annual interest from the stock and mutual fund is twice that from the bank account, and the mutual fund returned $120 more than the stock. How much did Glenn invest in each account?

For our final example, we turn to an application from economics.

EXAMPLE 9 Solving a Business-Based Application

A manufacturer can produce and sell x items per week at a cost

$C = 30x + 800$

The revenue from selling those items is given by

$R = 110x$

Find the number of units at which the revenue equals the cost. This is called the *break-even point.*

Solution We form a linear system from the given equations.

$$C = 30x + 800 \tag{12}$$
$$R = 110x \tag{13}$$

Finding the break-even point requires that the revenue equal the cost, $R = C$. From Equation (13), we can substitute $110x$ for C in Equation (12). We then have

$110x = 30x + 800$

or

$110x - 30x = 800$

$\qquad 80x = 800$

$\qquad\quad x = 10$

For $x = 10$ units, the cost (and the revenue) is $1100. This system is illustrated below.

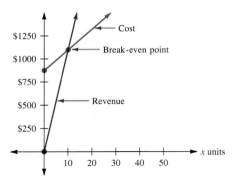

Note that if the company sells more than 10 units, it makes a profit since the revenue exceeds the cost.

Check Yourself 9

A manufacturer can produce and sell x items per week at a cost

$$C = 30x + 1800$$

The revenue from selling those items is given by

$$R = 120x$$

Find the break-even point.

CHECK YOURSELF ANSWERS

1 and **2.** Peanuts 50 lb, cashews 10 lb   **3.** 100 mL of 30% solution; 200 mL of 15% solution   **4.** Plane 100 mi/h, wind 20 mi/h   **5.** 30 turntables, 20 compact disc players   **6.** 300 mi, $114 charge   **7.** 35°, 70°, 75°   **8.** $5000 at 6 percent, $3000 at 8 percent, $3600 at 10 percent   **9.** 20 units: $R = C = $2400

4.3 EXERCISES

Each of the applications in Exercises 1 through 8 can be solved by the use of a system of linear equations. Match the application with the appropriate system below.

(a) $12x + 5y = 116$
     $8x + 12y = 112$

(b) $x + y = 8000$
     $0.06x + 0.09y = 600$

(c) $x + y = 200$
     $0.20x + 0.60y = 90$

(d) $x + y = 36$
          $y = 3x - 4$

(e) $2(x + y) = 36$
     $3(x - y) = 36$

(f) $x + y = 200$
     $5.50x + 4y = 980$

(g) $L = 2W + 3$
     $2L + 2W = 36$

(h) $x + y = 120$
     $2.20x + 5.40y = 360$

1. One number is 4 less than 3 times another. If the sum of the numbers is 36, what are the two numbers?

2. Suppose that a movie theater sold 200 adult and student tickets for a showing with a revenue of $980. If the adult tickets were $5.50 and the student tickets were $4, how many of each type of ticket were sold?

3. The length of a rectangle is 3 cm more than twice its width. If the perimeter of the rectangle is 36 cm, find the dimensions of the rectangle.

4. An order of 12 dozen roller-ball pens and 5 dozen ball-point pens cost $116. A later order for 8 dozen roller-ball pens and 12 dozen ballpoint pens cost $112. What was the cost of 1 dozen of each type of pen?

5. A candy merchant wishes to mix peanuts selling at $2.20 per pound with cashews selling at $5.40 per pound to form 120 lb of a mixed-nut blend which will sell for $3 per pound. What amount of each type of nut should be used?

6. Donald has investments totaling $8000 in two ac-counts—one a savings account paying 6 percent interest and the other a bond paying 9 percent. If the annual interest from the two investments was $600, how much did he have invested at each rate?

7. A chemist wants to combine a 20% alcohol solution with a 60% solution to form 200 mL of a 45% solution. How much of each of the solutions should be used to form the mixture?

8. Xian was able to make a downstream trip of 36 mi in 2 h. Returning upstream, he took 3 h to make the trip. How fast can his boat travel in still water? What was the rate of the river's current?

Solve each of the following problems by choosing a vari-able to represent each unknown quantity and writing a sys-tem of equations.

9. One number is 5 more than twice another. If the sum of the two numbers is 29, find the two numbers.

10. One number is 3 less than 4 times another. If the sum of the two numbers is 32, what are the two numbers?

11. Michelle has 28 coins with a value of $4.40. If the coins are all nickels and quarters, how many of each type of coin does she have?

12. A cashier has 75 $5 and $10 bills with a value of $590. How many of each denomination bill does he have?

13. Suppose that 750 tickets were sold for a concert with a total revenue of $5300. If adult tickets were $8 and student tickets were $4.50, how many of each type of ticket were sold?

14. Theater tickets sold for $7.50 on the main floor and $5 in the balcony. The total revenue was $3250, and there were 100 more main-floor tickets sold than balcony tickets. Find the number of each type of ticket sold.

15. The length of a rectangle is 3 in less than twice its width. If the perimeter of the rectangle is 84 in, find the dimensions of the rectangle.

16. The length of a rectangle is 5 cm more than 3 times its width. If the perimeter of the rectangle is 74 cm, find the dimensions of the rectangle.

17. A garden store sold 8 bags of mulch and 3 bags of fertilizer for $24. The next purchase was for 5 bags of

mulch and 5 bags of fertilizer. The cost of that purchase was $25. Find the cost of a single bag of mulch and a single bag of fertilizer.

18. The cost of an order for 10 computer disks and 3 pack-ages of paper was $22.50. The next order was for 30 disks and 5 packages of paper, and its cost was $53.50. Find the price of a single disk and a single package of paper.

19. A coffee retailer has two grades of decaffeinated beans, one selling for $4 per pound and the other for $6.50 per pound. She wishes to blend the beans to form a 150-lb mix-ture which will sell for $4.75 per pound. How many pounds of each grade of bean should be used in the mixture?

20. A candy merchant sells jelly beans at $3.50 per pound and gumdrops at $4.70 per pound. To form a 200-lb mixture which will sell for $4.40 per pound, how many pounds of each type of candy should be used?

21. Cheryl decided to divide $12,000 into two investments, one a time deposit paying 8 percent annual interest and the other a bond which pays 9 percent. If her annual interest was $1010, how much did she invest at each rate?

22. Miguel has $2000 more invested in a mutual fund pay-ing 10 percent interest than in a savings account paying 7 percent. If he received $880 in interest for 1 year, how much did he have invested in the two accounts?

23. A chemist mixes a 10% acid solution with a 50% acid solution to form 400 mL of a 40% solution. How much of each solution should be used in the mixture?

24. A laboratory technician wishes to mix a 70% saline solution and a 20% solution to prepare 500 mL of a 40% solution. What amount of each solution should be used?

25. A boat traveled 36 mi up a river in 3 h. Returning downstream, the boat took 2 h. What is the boat's rate in still water, and what is the rate of the river's current?

26. A jet flew east a distance of 1800 mi with the jetstream in 3 h. Returning west, against the jetstream, the jet took 4 h. Find the jet's speed in still air and the rate of the jetstream.

27. The sum of the digits of a two-digit number is 8. If the digits are reversed, the new number is 36 more than the original number. Find the original number. *Hint:* If u repre-sents the units digit of the number and t the tens digit, the original number can be represented by $10t + u$.

28. The sum of the digits of a two-digit number is 10. If the digits are reversed, the new number is 54 less than the origi-nal number. What was the original number?

29. A manufacturer produces a battery-powered calculator and a solar model. The battery-powered model requires 10 min of electronic assembly and the solar model 15 min. There are 450 min of assembly time available per day. Both models require 8 min for packaging, and 280 min of packaging time are available per day. If the manufacturer wants to use all the available time, how many of each unit should be produced per day?

30. A small tool manufacturer produces a standard- and a cordless-model power drill. The standard model takes 2 h of labor to assemble and the cordless model 3 h. There are 72 h of labor available per week for the drills. Material costs for the standard drill are $10, and for the cordless drill they are $20. The company wishes to limit material costs to $420 per week. How many of each model drill should be produced in order to use all the available resources?

31. In economics, a demand equation gives the quantity D that will be demanded by consumers at a given price p, in dollars. Suppose that $D = 210 - 4p$ for a particular product.

A supply equation gives the supply S that will be available from producers at price p. Suppose also that for the same product $S = 10p$.

The equilibrium point is that point where the supply equals the demand (here, where $S = D$). Use the given equations to find the equilibrium point.

32. Suppose the demand equation for a product is $D = 150 - 3p$ and the supply equation is $S = 12p$. Find the equilibrium point for the product.

33. Two car rental agencies have the following rate structure for compact cars:

Company A: $30 per day and 22¢ per mile

Company B: $28 per day and 26¢ per mile

For a 2-day rental, at what number of miles will the charges be the same?

34. Two construction companies submit the following bids:

Company A: $5000 plus $15 per square foot of building

Company B: $7000 plus $12.50 per square foot of building

For what number of square feet of building will the bids of the two companies be the same?

35. The sum of three numbers is 16. The largest number is equal to the sum of the other two, and 3 times the smallest number is 1 more than the largest. Find the three numbers.

36. The sum of three numbers is 24. Twice the smallest number is 2 less than the largest number, and the largest number is equal to the sum of the other two. What are the three numbers?

37. A cashier has 25 coins consisting of nickels, dimes, and quarters with a value of $4.90. If the number of dimes is 1 less than twice the number of nickels, how many of each type of coin does she have?

38. A theater has tickets at $6 for adults, $3.50 for students, and $2.50 for children under 12 years old. A total of 278 tickets were sold for one showing with a total revenue of $1300. If the number of adult tickets sold was 10 less than twice the number of student tickets, how many of each type of ticket were sold for the showing?

39. The perimeter of a triangle is 19 cm. If the length of the longest side is twice that of the shortest side and 3 cm less than the sum of the lengths of the other two sides, find the lengths of the three sides.

40. The measure of the largest angle of a triangle is 10° more than the sum of the measures of the other two angles and 10° less than 3 times the measure of the smallest angle. Find the measures of the three angles of the triangle.

41. Jovita divides $17,000 into three investments—a savings account paying 6 percent annual interest, a bond paying 9 percent, and a money market fund paying 11 percent. The annual interest from the three accounts is $1540, and she has 3 times as much invested in the bond as in the savings account. What amount does she have invested in each account?

42. Adrienne has $10,000 invested in a savings account paying 5 percent, a time deposit paying 7 percent, and a bond which pays 10 percent. She has $1000 less invested in the bond than in her savings account, and she earned $700 in annual interest. What has she invested in each account?

43. The sum of the digits of a three-digit number is 9, and the tens digit of the number is twice the hundreds digit. If the digits are reversed in order, the new number is 99 more than the original number. What is the original number?

44. The sum of the digits of a three-digit number is 9. The tens digit is 3 times the hundreds digit. If the digits are reversed in order, the new number is 99 less than the original number. Find the original three-digit number.

45. A manufacturer can produce and sell x items per week at a cost of $C = 20x + 3600$. The revenue from selling those items is given by $R = 140x$. Find the break-even point for this product. *Hint:* See Example 9.

46. If the cost for a second product is given by $C = 30x + 3600$ and the revenue by $R = 130x$, find the break-even point for that product.

47. To encourage carpooling, a city charges $10 per single driver or $4 per person for carpools of two or more people in its city parking lots. If one parking lot took in $2020 and 340 cars used that lot for one day, how many of each type commuter—single driver or carpool rider—used that lot that day?

48. To encourage carpooling, a city charges $12 per single driver or $5 per person for carpools of two or more people in its city parking lots. If one parking lot took in $3220 and

420 cars used that lot for one day, how many of each type commuter—single driver or carpool rider—used that lot that day?

49. Roy, Sally, and Jeff drive a total of 50 mi to work each day. Sally drives twice as far as Roy, and Jeff drives 10 mi farther than Sally. Use a system of three equations in three unknowns to find how far each person drives each day.

50. A parking lot has spaces reserved for motorcycles, cars, and vans. There are 5 more spaces reserved for vans than for

motorcycles. There are 3 times as many car spaces as van and motorcycle spaces combined. If the parking lot has 180 total reserved spaces, how many of each type are there?

Skillscan (Section 3.8)

Graph the solution sets for the following linear inequalities.

a. $x + y > 8$ **b.** $2x - y \le 6$ **c.** $3x + 4y \ge 12$
d. $y > 2x$ **e.** $y \le -3$ **f.** $x > 5$

Systems of Linear Inequalities   4.4

OBJECTIVE: To graph systems of linear inequalities

TAPE IN9

Our previous work in this chapter dealt with finding the solution set of a system of linear equations. That solution set represented the points of intersection of the graphs of the equations in the system. In this section, we extend that idea to include systems of linear inequalities.

In this case, the solution set is all ordered pairs that satisfy each inequality. *The graph of the solution set of a system of linear inequalities* is then the intersection of the graphs of the individual inequalities. Let's look at an example.

You might want to review graphing linear inequalities in Section 3.8 at this point.

EXAMPLE 1 Solving a System by Graphing

Solve the following system of linear inequalities by graphing.

$x + y > 4$

$x - y < 2$

Solution We start by graphing each inequality separately. The boundary line is drawn, and using (0, 0) as a test point, we see that we should shade the half plane above the line in both graphs.

Note that the boundary line is dashed, to indicate it is *not* included in the graph.

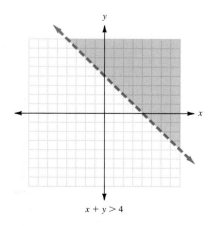
$x + y > 4$

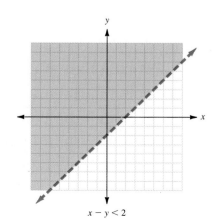
$x - y < 2$

In practice, the graphs of the two inequalities are combined on the same set of axes, as is shown below. The graph of the solution set of the original system is the intersection of the graphs drawn above.

Points on the lines are not included in the solution.

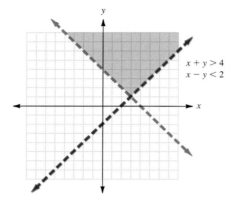

$$x + y > 4$$
$$x - y < 2$$

Check Yourself 1

Solve the following system of linear inequalities by graphing.

$$2x - y < 4$$
$$x + y < 3$$

Most applications of systems of linear inequalities lead to bounded regions. This requires a system of three or more inequalities, as shown in Example 2.

EXAMPLE 2 Solving a System by Graphing

Solve the following system of linear inequalities by graphing.

$$x + 2y \leq 6$$
$$x + y \leq 5$$
$$x \geq 2$$
$$y \geq 0$$

Solution On the same set of axes, we graph the boundary line of each of the inequalities. We then choose the appropriate half planes (indicated by the arrow that is perpendicular to the line) in each case, and we locate the intersection of those regions for our graph.

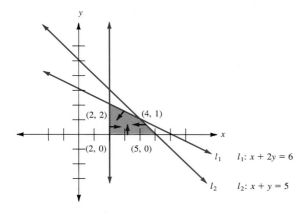

The vertices of the shaded region are given because they have particular significance in later applications of this concept. Can you see how the coordinates of the vertices were determined?

l_1: $x + 2y = 6$

l_2: $x + y = 5$

Check Yourself 2

Solve the following system of linear inequalities by graphing.

$2x - y \leq 8 \qquad x \geq 0$

$x + y \leq 7 \qquad y \geq 0$

Let's expand upon one of the exercises of Section 3.6 to see an application of our work with systems of linear inequalities. Consider the following example.

EXAMPLE 3 Solving a Business-Based Application

A manufacturer produces a standard model and a deluxe model of a 13-in television set. The standard model requires 12 h of labor to produce, while the deluxe model requires 18 h. The labor available is limited to 360 h per week. Also the plant capacity is limited to producing a total of 25 sets per week. Draw a graph of the region representing the number of sets that can be produced, given these conditions.

Solution As suggested earlier, we let x represent the number of standard-model sets produced and y the number of deluxe-model sets. Since the labor is limited to 360 h, we have

$$12x \quad + \quad 18y \quad \leq \quad 360 \tag{1}$$
$\uparrow$ $\qquad\quad$ $\uparrow$
12 h per $\qquad$ 18 h per
standard set $\quad$ deluxe set

The total labor is limited to (or less than or equal to) 360 h.

The total production, here $x + y$ sets, is limited to 25, so we can write

$$x + y \leq 25 \tag{2}$$

For convenience in graphing, we divide both members of inequality (1) by 6, to write the equivalent system:

We have $x \geq 0$ and $y \geq 0$ since the number of sets produced cannot be negative.

$$2x + 3y \leq 60$$
$$x + y \leq 25$$
$$x \geq 0$$
$$y \geq 0$$

We now graph the system of inequalities as before. The shaded area represents all possibilities in terms of the number of sets that can be produced.

The shaded area is called the *feasible region*. All points in the region meet the given conditions of the problem and represent possible production options.

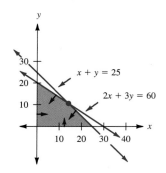

Check Yourself 3

A manufacturer produces standard record turntables and compact disc players. The turntables require 10 h of labor to produce while the disc players require 20 h. The labor hours available are limited to 300 h per week. Existing orders require that at least 10 turntables and at least 5 disc players be produced per week.

Draw a graph of the region representing the possible production options.

CHECK YOURSELF ANSWERS

1. $2x - y < 4$
$x + y < 3$

2. $2x - y \leq 8$
$x + y \leq 7$
$x \geq 0$
$y \geq 0$

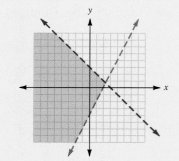

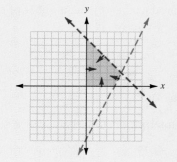

3. Let x be the number of turntables and y be the number of players. The system is

$$10x + 20y \leq 300$$
$$x \geq 10$$
$$y \geq 5$$

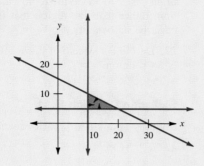

Build Your Skills

Solve each of the following systems of linear inequalities graphically.

1. $x + 2y \leq 4$
  $x - y \geq 1$

2. $3x - y > 6$
  $x + y < 6$

3. $3x + y < 6$
  $x + y > 4$

4. $2x + y \geq 8$
  $x + y \geq 4$

5. $x + 3y \leq 12$
  $2x - 3y \leq 6$

6. $x - 2y > 8$
  $3x - 2y > 12$

7. $3x + 2y \leq 12$
  $x \geq 2$

8. $2x + y \leq 6$
  $y \geq 1$

9. $2x + y < 8$
  $x > 1$
  $y > 2$

10. $3x - y \leq 6$
  $x \geq 1$
  $y \leq 3$

11. $x + 2y \leq 8$
  $2 \leq x \leq 6$
  $y \geq 0$

12. $x + y \leq 6$
  $0 \leq y \leq 3$
  $x \geq 1$

13. $3x + y \leq 6$
  $x + y \leq 4$
  $x \geq 0$
  $y \geq 0$

14. $x - 2y \geq -2$
  $x + 2y \leq 6$
  $y \geq 0$
  $y \geq 0$

15. $4x + 3y \leq 12$
  $x + 4y \leq 8$
  $x \geq 0$
  $y \geq 0$

16. $2x + y \leq 8$
  $x + y \geq 3$
  $x \geq 0$
  $y \geq 0$

17. $x - 4y \leq -4$
  $x + 2y \leq 8$
  $x \geq 2$

18. $x - 3y \geq -6$
  $x + 2y \geq 4$
  $x \leq 4$

Draw the appropriate graphs in each of the following.

19. A manufacturer produces both two-slice and four-slice toasters. The two-slice toaster takes 6 h of labor to produce and the four-slice toaster 10 h. The labor available is limited to 300 h per week, and the total production capacity is 40 toasters per week. Draw a graph of the feasible region, given these conditions, where x is the number of two-slice toasters and y is the number of four-slice toasters.

20. A small firm produces both AM and AM/FM car radios. The AM radios take 15 h to produce, and the AM/FM radios take 20 h. The number of production hours is limited to 300 h per week. The plant's capacity is limited to a total of 18 radios per week, and existing orders require that at least 4 AM radios and at least 3 AM/FM radios be produced per week. Draw a graph of the feasible region given these conditions, where x is the number of AM radios and y the number of AM/FM radios.

Transcribe Your Skills

21. When one solves a system of linear inequalities, it's often easier to shade the region that is not part of the solution, rather than the region that is. Try this method, then describe its benefits.

22. Describe a system of linear inequalities for which there is no solution.

Skillscan (Section 2.1)

Multiply both sides of each equation by the given number.

a. $2x + 3y = 4$; by -2

b. $-x + y = 3$; by 3

c. $-\dfrac{x}{2} + \dfrac{y}{5} = 5$; by 6

d. $-\dfrac{2x}{5} + \dfrac{3x}{2} = \dfrac{1}{10}$; by 10

e. $-x + y - 2z = -3$; by -4

f. $3x - y + z = 2$; by 4

Solution of Systems by Matrices

TAPE IN10

OBJECTIVE: To solve systems of linear equations by using matrices

Your work in this chapter will provide experience with the various techniques for solving systems of equations. Thus far we have considered the solution of linear systems by the algebraic methods of addition and substitution. In this section we consider yet another approach to the solution of linear systems—the use of matrices. This solution technique can be easily extended to larger systems and so has particular significance in the computer solution of such systems. First, some definitions.

The credit for inventing matrices is generally given to the English mathematician Arthur Cayley (1821–1895).

The plural of "matrix" is "matrices."

A *matrix* is a rectangular array of numbers. Each number in the matrix is called an *element* of the matrix. For example,

Brackets [ ] are used to enclose the elements of a matrix. You may also encounter the use of large parentheses to designate a matrix. In that case, the matrix would look like this:

$$\begin{pmatrix} 2 & 3 & 1 \\ -1 & 3 & 2 \end{pmatrix}$$

$$\begin{bmatrix} 2 & 3 & 1 \\ -1 & 3 & 2 \end{bmatrix}$$

is a matrix with elements 2, 3, 1, -1, 3, and 2. The matrix above has two rows and three columns. It is called a 2×3 (read "2 by 3") matrix. We always give the number of rows and *then* the number of columns.

Now let's see how a matrix of numbers is associated with a system of linear equations. For the system of equations

$$x - 2y = -1$$
$$2x + y = 8$$

the *coefficient matrix* is

$$\begin{bmatrix} 1 & -2 \\ 2 & 1 \end{bmatrix}$$

and the *constant matrix* is

$$\begin{bmatrix} -1 \\ 8 \end{bmatrix}$$

The matrix

$$\begin{bmatrix} 1 & -2 & \vdots & -1 \\ 2 & 1 & \vdots & 8 \end{bmatrix}$$

is called the *augmented matrix* of the system, and it is formed by adjoining the constant matrix to the coefficient matrix so that the matrix contains both the coefficients of the variables and the constants from the system.

> The dotted vertical line is just for convenience, to separate the coefficients and the constants.

EXAMPLE 1 Writing an Augmented Matrix

Write the augmented matrix for the given systems.

(*a*) $5x - 2y = 7$
       $3x + 4y = 8$

The augmented matrix for the system is

$$\begin{bmatrix} 5 & -2 & \vdots & 7 \\ 3 & 4 & \vdots & 8 \end{bmatrix}$$

(*b*) $2x - 3y = 9$
              $y = 6$

The coefficient matrix and constant matrix together become

$$\begin{bmatrix} 2 & -3 & \vdots & 9 \\ 0 & 1 & \vdots & 6 \end{bmatrix}$$

> Zero is used for the x coefficient in the second equation.

Check Yourself 1

Write the augmented matrices for each of the given systems.

1. $3x - 5y = -2$ **2.** $x - 3y = 3$
       $2x + 3y = 5$ $y = 2$

Your work in the Check Yourself 1 exercise led to a particular form of augmented matrix. The matrix

$$\begin{bmatrix} 1 & -3 & \vdots & 3 \\ 0 & 1 & \vdots & 2 \end{bmatrix}$$

is associated with the system

$$x - 3y = 3 \tag{1}$$

$$y = 2 \tag{2}$$

> Note the form of the coefficient matrix. We have 1s along the *main diagonal* (from the upper left to the lower right) and 0s below that diagonal.

Given a system in this form, it is particularly easy to solve for the desired variables by a process called *back substitution*.

From Equation (2) we know that $y = 2$. Substituting that value back into Equation (1), we have

$$x - 3(2) = 3$$
$$x - 6 = 3$$
$$x = 9$$

and $(9, 2)$ is the solution for our system.

Our goal in this section will be to transform the augmented matrix of a given system to the form just illustrated, so that the solutions can be easily found by back substitution.

How do we go about transforming the augmented matrix? The rules are based on our earlier work with linear systems. Recall the following:

An equivalent system is formed when.

Recall that an equivalent system has the same solutions as the original system.

1. Two equations are interchanged.

2. An equation is multiplied by a nonzero constant.

3. An equation is replaced by adding a constant multiple of another equation to that equation.

Now each property above produces a corresponding property that can be applied to the rows of the augmented matrix rather than to the actual equations of the system. These are called the *elementary row operations,* and they will always produce the augmented matrix of an equivalent system of equations.

ELEMENTARY ROW OPERATIONS

1. Two rows can be interchanged.

2. A row can be multiplied by a nonzero constant.

3. A row can be replaced by adding a nonzero multiple of another row to that row.

Let's now apply these elementary row operations to the solution of a linear system.

EXAMPLE 2 Solving a System with a Matrix

Solve the system, using elementary row operations.

$$2x - 3y = 2$$
$$x + 2y = 8$$

Solution First, we write the augmented matrix of the given system:

$$\begin{bmatrix} 2 & -3 & \vdots & 2 \\ 1 & 2 & \vdots & 8 \end{bmatrix}$$

Think about our objective. We want to use back substitution to solve an equivalent system whose augmented matrix has the form

$$\begin{bmatrix} 1 & a & \vdots & b \\ 0 & 1 & \vdots & c \end{bmatrix}$$

Compare this form to the augmented matrix in our discussion of back substitution. In the coefficient matrix we want 1s along the main diagonal and 0s below.

To arrive at this form, we start with the first column. Interchanging the rows of the augmented matrix will give a 1 in the top position, as desired:

$$\begin{bmatrix} 2 & -3 & \vdots & 2 \\ 1 & 2 & \vdots & 8 \end{bmatrix} \xrightarrow{R_1 \leftrightarrow R_2} \begin{bmatrix} 1 & 2 & \vdots & 8 \\ 2 & -3 & \vdots & 2 \end{bmatrix}$$

The notation $R_1 \leftrightarrow R_2$ is an abbreviation that we have interchanged those rows.

Now multiply row 1 by -2 and add the result to row 2. This will give a 0 in the lower position of column 1.

$$\begin{bmatrix} 1 & 2 & \vdots & 8 \\ 2 & -3 & \vdots & 2 \end{bmatrix} \xrightarrow{-2R_1 + R_2} \begin{bmatrix} 1 & 2 & \vdots & 8 \\ 0 & -7 & \vdots & -14 \end{bmatrix}$$

The notation $-2R_1 + R_2$ means R_2 has been replaced by the sum of -2 times row 1 and row 2.

We now multiply row 2 by the constant $-\dfrac{1}{7}$, to produce a 1 in the second position of the second row:

$$\begin{bmatrix} 1 & 2 & \vdots & 8 \\ 0 & -7 & \vdots & -14 \end{bmatrix} \xrightarrow{-\frac{1}{7}R_2} \begin{bmatrix} 1 & 2 & \vdots & 8 \\ 0 & 1 & \vdots & 2 \end{bmatrix}$$

The notation $-\dfrac{1}{7}R_2$ means we have multiplied row 2 by $-\dfrac{1}{7}$.

The final matrix is in the desired form with 1s along the main diagonal and 0s below. It represents the equivalent system

$$x + 2y = 8 \tag{3}$$

$$y = 2 \tag{4}$$

and we can apply back substitution for the solution. Since $y = 2$, in Equation (3) we have

$$x + 2(2) = 8$$

$$x = 4$$

The solution for the system is $(4, 2)$.

Check Yourself 2

Use elementary row operations to solve the system.

$$-3x - 2y = -1$$

$$x + 3y = -9$$

Let's extend our matrix approach to consider a system of three linear equations in three unknowns. Again our procedure is to find the augmented matrix of an equivalent system with 1s along the main diagonal of the coefficient matrix and 0s below that diagonal.

EXAMPLE 3 Solving a Three-Variable System with a Matrix

Use elementary row operations to solve the system.

$$x - 2y + z = 10$$
$$-3x + y + 2z = 5$$
$$2x + 3y - z = -9$$

Solution The augmented matrix for the system is

$$\left[\begin{array}{ccc|c} 1 & -2 & 1 & 10 \\ -3 & 1 & 2 & 5 \\ 2 & 3 & -1 & -9 \end{array}\right]$$

Again we start with the first column. The top entry is already 1, and we want zeros below that entry.

$$\xrightarrow{3R_1 + R_2} \left[\begin{array}{ccc|c} 1 & -2 & 1 & 10 \\ 0 & -5 & 5 & 35 \\ 2 & 3 & -1 & -9 \end{array}\right]$$

$$\xrightarrow{-2R_1 + R_3} \left[\begin{array}{ccc|c} 1 & -2 & 1 & 10 \\ 0 & -5 & 5 & 35 \\ 0 & 7 & -3 & -29 \end{array}\right]$$

The order of obtaining the 1s and 0s should be followed carefully.

We now want a 1 in the center position of column 2 and a 0 below that 1:

$$\xrightarrow{-\frac{1}{5}R_2} \left[\begin{array}{ccc|c} 1 & -2 & 1 & 10 \\ 0 & 1 & -1 & -7 \\ 0 & 7 & -3 & -29 \end{array}\right]$$

$$\xrightarrow{-7R_2 + R_3} \left[\begin{array}{ccc|c} 1 & -2 & 1 & 10 \\ 0 & 1 & -1 & -7 \\ 0 & 0 & 4 & 20 \end{array}\right]$$

Finally we want a 1 in the third position of the third column.

$$\xrightarrow{\frac{1}{4}R_3} \left[\begin{array}{ccc|c} 1 & -2 & 1 & 10 \\ 0 & 1 & -1 & -7 \\ 0 & 0 & 1 & 5 \end{array}\right]$$

The augmented matrix is now in the desired form and corresponds to the system

$$x - 2y + z = 10 \tag{5}$$
$$y - z = -7 \tag{6}$$
$$z = 5 \tag{7}$$

Substituting $z = 5$ in Equation (6) produces

$$y - 5 = -7$$
$$y = -2$$

Substituting $y = -2$ and $z = 5$ in Equation (5) gives

$$x - 2(-2) + 5 = 10$$
$$x + \quad 4 \quad + 5 = 10$$
$$x = \quad 1$$

and $(1, -2, 5)$ is the solution for the system.

Check Yourself 3

Use elementary row operations to solve the system.

$$x + 2y - \quad 4z = \quad -9$$
$$2x + 5y - 10z = -21$$
$$-3x - 5y + 11z = \quad 28$$

Will our work with matrices provide solutions for all linear systems? From your previous work with inconsistent and dependent systems in this chapter, you probably realize that the answer is no. The following examples illustrate.

EXAMPLE 4 Identifying an Inconsistent System

Use elementary row transformations to solve the system, if possible.

$$x + 2y = \quad 2$$
$$-3x - 6y = -5$$

Solution The augmented matrix for the system is

$$\left[\begin{array}{cc:c} 1 & 2 & 2 \\ -3 & -6 & -5 \end{array}\right]$$

Adding 3 times row 1 to row 2, we have

 $\xrightarrow{3R_1 + R_2}$ $\left[\begin{array}{cc:c} 1 & 2 & 2 \\ 0 & 0 & 1 \end{array}\right]$

Note that row 2 now gives $0x + 0y = 1$, which implies that $0 = 1$, a contradiction. This means that the given system is inconsistent. There are *no* solutions.

> Although your calculator can do row transformations, it is actually easier to do them by hand!

> In general, if we have a row with all 0s as coefficients and a nonzero constant on the right, the system is inconsistent.

Check Yourself 4

Use elementary row operations to solve the system, if possible.

$$2x - 4y = \quad 6$$
$$-x + 2y = -2$$

The next example considers one final case.

EXAMPLE 5 Identifying a Dependent System

Use elementary row operations to solve the system, if possible.

$x - 4y = 4$

$2x - 8y = 8$

Solution The augmented matrix for this system is

$$\left[\begin{array}{cc|c} 1 & -4 & 4 \\ 2 & -8 & 8 \end{array}\right]$$

In this case we multiply row 1 by -2 and add that result to row 2:

$$\xrightarrow{-2R_1 + R_2} \left[\begin{array}{cc|c} 1 & -4 & 4 \\ 0 & 0 & 0 \end{array}\right]$$

The bottom row of 0s represents the statement

$0x + 0y = 0$

which is, of course, true for any values of x and y. This means that the original system was dependent and has an infinite number of solutions, in this case all values (x, y) which satisfy the equation $x - 4y = 4$.

Check Yourself 5

Use elementary row operations to solve the system, if possible.

$3x - 6y = -9$

$-x + 2y = 3$

CHECK YOURSELF ANSWERS

1. (1) $\left[\begin{array}{cc|c} 3 & -5 & -2 \\ 2 & 3 & 5 \end{array}\right]$; (2) $\left[\begin{array}{cc|c} 1 & -3 & 3 \\ 0 & 1 & 2 \end{array}\right]$ **2.** $(3, -4)$

3. $(-3, 5, 4)$ **4.** Inconsistent system   **5.** Dependent system

4.5 EXERCISES

Build Your Skills

Write the augmented matrix for each of the following systems of equations.

1. $2x - 3y = 5$
$\ x + 4y = 2$

2. $x + 5y = 3$
$\ -2x + y = -1$

3. $x - 5y = 6$
$\ \ y = 2$

4. $x + 3y = 6$
$y = 3$

5. $x + 2y - z = 3$
$x + 3z = 1$
$y - 2z = 4$

6. $x - 2y + 5z = 3$
$3y - 2z = 1$
$z = 4$

Write the systems of equations corresponding to each of the following augmented matrices.

7. $\begin{bmatrix} 1 & 2 & \vdots & 3 \\ 1 & 5 & \vdots & -6 \end{bmatrix}$

8. $\begin{bmatrix} 1 & 2 & \vdots & 3 \\ -3 & 1 & \vdots & -3 \end{bmatrix}$

9. $\begin{bmatrix} 1 & 3 & \vdots & 5 \\ 0 & 1 & \vdots & 2 \end{bmatrix}$

10. $\begin{bmatrix} 1 & -2 & \vdots & 4 \\ 0 & 1 & \vdots & 3 \end{bmatrix}$

11. $\begin{bmatrix} 1 & 2 & 0 & \vdots & 4 \\ 0 & 1 & 5 & \vdots & 3 \\ 1 & 1 & 1 & \vdots & 1 \end{bmatrix}$

12. $\begin{bmatrix} 1 & 2 & -1 & \vdots & 3 \\ 0 & 1 & 4 & \vdots & 1 \\ 0 & 0 & 1 & \vdots & -5 \end{bmatrix}$

Solve each of the following systems, using elementary row operations if possible.

13. $x + 3y = -5$
$2x + y = 0$

14. $x - 3y = -9$
$-3x + y = 11$

15. $x - 5y = 20$
$-4x + 3y = -29$

16. $x + 2y = -10$
$5x - y = 16$

17. $3x - 2y = 14$
$x + 5y = -18$

18. $4x + 3y = 0$
$x + 5y = -17$

19. $5x + 3y = 11$
$-x - 2y = 2$

20. $6x - y = 24$
$-x + 3y = 13$

21. $x - 3y = 5$
$-4x + 12y = -18$

22. $-3x + 6y = -15$
$x - 2y = 5$

Solve each of the following systems, using elementary row operations.

23. $x + y - z = -1$
$ - 3y + 2z = 6$
$2x + 3z = 2$

24. $x + 2y = -3$
$x + 3y - 6z = -9$
$2x - 3z = 0$

25. $x - z = 2$
$2x - y + z = 3$
$ y - 2z = -2$

26. $x - 2y + 5z = 10$
$-3x + 4y = -12$
$x + z = 3$

27. $x - z = 6$
$x + 2y + z = -7$
$2x - y + 2z = 1$

28. $x + 2y + 3z = 1$
$-2x - 3y - 4z = -1$
$ y + 3z = 3$

29. $x + 2y + 3z = 1$
$x + 3y + z = 2$
$3x + 2y + z = -1$

30. $2x + 2y + z = 3$
$3x + 2y + 2z = 7$
$x + y + z = 2$

Transcribe Your Skills

31. The root word for *matrix* is the Latin "mater," meaning mother. Speculate about the connection between mother and matrix. Humorous responses are acceptable.

32. Karl Fredrich Gauss is usually named as one of the five greatest mathematicians of all time. He contributed much to the study of matrices. Find at least two other areas of mathematics to which Gauss contributed.

Skillscan (Elementary Algebra)

Evaluate each of the following expressions.

a. $4(1) - 3(-2)$ **b.** $(-2)(-3) - (4)(1)$
c. $(-3)(-4) + (-2)(1)$ **d.** $(-3)(2) - (-4)(-5)$
e. $4(-2) - (-1)(5)$ **f.** $(-5)(-3) - (-4)(-3)$

4.6 Determinants and Cramer's Rule

OBJECTIVES: 1. To evaluate second-order determinants
2. To evaluate third-order determinants
3. To solve systems of equations by using Cramer's rule

Our work in this chapter is intended to provide experience with the various solution techniques for systems of equations. We considered the algebraic methods of addition and substitution and the use of matrices. In this section we will examine determinants and see how they are used in solving systems of equations. Let's look at some definitions regarding this concept.

The idea of a determinant dates back to Seki Kowa in Japan (1683) and Gottfried Leibniz (one of the coinventors of the calculus) in Europe (1693). The name "determinant" is due to the French mathematician Augustin Cauchy and was not used until 1812.

A *determinant* is a square array of numbers written as shown below. The symbol $|\ \ |$ is used to denote the determinant.

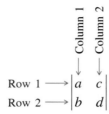

In general, a determinant of *order n* has *n* rows and *n* columns.

The *order* of a determinant is the number of rows (or columns). The determinant above is said to be of order 2. It is sometimes called a 2×2 determinant for convenience. The *value* of a second-order determinant is defined as follows:

Note that a determinant, by this definition, will be a number whenever all its entries are numbers.

$$\begin{vmatrix} a & c \\ b & d \end{vmatrix} = ad - bc$$

EXAMPLE 1 Evaluating Determinants

Find the value of each determinant.

(a) $\begin{vmatrix} 2 & 3 \\ -1 & 4 \end{vmatrix} = 2(4) - (-1)(3)$

$= 8 + 3 = 11$

(b) $\begin{vmatrix} 4 & -3 \\ 5 & 0 \end{vmatrix} = 4(0) - 5(-3)$

$= 15$

Check Yourself 1

Find the value of each determinant.

1. $\begin{vmatrix} 3 & -1 \\ 2 & 1 \end{vmatrix}$

2. $\begin{vmatrix} -2 & 2 \\ -1 & 3 \end{vmatrix}$

The use of determinants to solve linear systems is called *Cramer's rule*. To develop that rule, consider the system of two linear equations in two unknowns:

$$a_1 x + b_1 y = c_1 \qquad (1)$$

$$a_2 x + b_2 y = c_2 \qquad (2)$$

To solve for x, we multiply both sides of Equation (1) by b_2 and both sides of Equation (2) by $-b_1$. We then have

$$a_1 b_2 x + b_1 b_2 y = c_1 b_2$$

$$-a_2 b_1 x - b_1 b_2 y = -c_2 b_1$$

Adding these new equations will eliminate the variable y from the system.

$$(a_1 b_2 - a_2 b_1)x = c_1 b_2 - c_2 b_1$$

or
$$x = \frac{c_1 b_2 - c_2 b_1}{a_1 b_2 - a_2 b_1} \qquad (3)$$

We can solve for y in a similar fashion:

$$y = \frac{a_1 c_2 - a_2 c_1}{a_1 b_2 - a_2 b_1} \qquad (4)$$

To verify this, return to the original system above. Multiply Equation (1) by a_2 and Equation (2) by $-a_1$. Then add to eliminate the variable x.

To see how determinants are used, let's rewrite Equations (3) and (4) in determinant form.

$$x = \frac{\begin{vmatrix} c_1 & b_1 \\ c_2 & b_2 \end{vmatrix}}{\begin{vmatrix} a_1 & b_1 \\ a_2 & b_2 \end{vmatrix}} \qquad y = \frac{\begin{vmatrix} a_1 & c_1 \\ a_2 & c_2 \end{vmatrix}}{\begin{vmatrix} a_1 & b_1 \\ a_2 & b_2 \end{vmatrix}}$$

Using the determinant definition, you should verify that these are simply restatements of Equations (3) and (4). The equations in determinant form are the result called Cramer's rule.

We don't want you to "memorize" the above result. It is better if you observe the following patterns. Let's look at our original system again.

This rule was named for the Swiss mathematician Gabriel Cramer, who invented the idea of determinants independently of Kowa and Leibniz and published the rule for solving linear systems of equations in 1750.

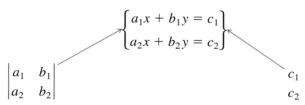

The determinant of the coefficients   The column of constants

Returning to our statement of Cramer's rule, we note that

1. The denominator for each variable is simply the determinant of the coefficients.

2. The numerator for x is formed from the determinant of the coefficients by replacing the x coefficients with the column of constants. Similarly, the numerator for y is found by replacing the y coefficients with the column of constants.

We will denote the determinant of the coefficients as D, the determinant in the numerator for x as D_x, and the determinant in the numerator for y as D_y. The following summarizes our discussion.

CRAMER'S RULE

To solve the linear system

$$a_1x + b_1y = c_1$$
$$a_2x + b_2y = c_2$$

write

$$D = \begin{vmatrix} a_1 & b_1 \\ a_2 & b_2 \end{vmatrix} \qquad \text{The determinant of the coefficients}$$

$$D_x = \begin{vmatrix} c_1 & b_1 \\ c_2 & b_2 \end{vmatrix} \qquad \text{The } x \text{ coefficients have been replaced by the constants.}$$

$$D_y = \begin{vmatrix} a_1 & c_1 \\ a_2 & c_2 \end{vmatrix} \qquad \text{The } y \text{ coefficients have been replaced by the constants.}$$

Then

$$x = \frac{D_x}{D} \qquad \text{and} \qquad y = \frac{D_y}{D} \qquad \text{when } D \neq 0$$

EXAMPLE 2 Using Cramer's Rule

Use Cramer's rule to solve the following system.

$$2x - 3y = 12$$
$$4x + 3y = 6$$

Solution

$$x = \frac{D_x}{D} = \frac{\begin{vmatrix} 12 & -3 \\ 6 & 3 \end{vmatrix}}{\begin{vmatrix} 2 & -3 \\ 4 & 3 \end{vmatrix}} = \frac{54}{18} = 3$$

$$y = \frac{D_y}{D} = \frac{\begin{vmatrix} 2 & 12 \\ 4 & 6 \end{vmatrix}}{\begin{vmatrix} 2 & -3 \\ 4 & 3 \end{vmatrix}} = \frac{-36}{18} = -2$$

The solution for the system is $(3, -2)$.

Check Yourself 2

Use Cramer's rule to solve the following system.

$$3x + 2y = 4$$
$$5x - 7y = 17$$

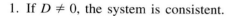

From our earlier work with systems we know that there are three possibilities for the solution of a linear system.

1. The system is *consistent* and has a unique solution.
2. The system is *inconsistent* and has no solutions.
3. The system is *dependent* and has an infinite number of solutions.

Let's see how these cases are identified in using Cramer's rule.

1. If $D \neq 0$, the system is consistent.
2. If $D = 0$ and either $D_x \neq 0$ or $D_y \neq 0$, the system is inconsistent.
3. If $D = 0$ and both $D_x = 0$ and $D_y = 0$, the system is dependent.

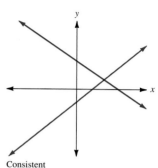

Consistent

EXAMPLE 3 Using Cramer's Rule

Use Cramer's rule to solve the system if possible.

$$2x - 3y = 4$$
$$-4x + 6y = 3$$

Solution

$$D = \begin{vmatrix} 2 & -3 \\ -4 & 6 \end{vmatrix} = 12 - 12 = 0$$

$$D_x = \begin{vmatrix} 4 & -3 \\ 3 & 6 \end{vmatrix} = 24 + 9 = 33$$

$$D_y = \begin{vmatrix} 2 & 4 \\ -4 & 3 \end{vmatrix} = 6 + 16 = 22$$

Inconsistent

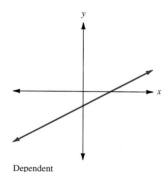

Dependent

In applying Cramer's rule, we have

$$x = \frac{33}{0} \qquad \text{and} \qquad y = \frac{22}{0}$$

Since these values are undefined, no solution for the system exists and the system is inconsistent.

Check Yourself 3

Use Cramer's rule to show that the following system is inconsistent.

$$x - 3x = 5$$
$$-2x + 6x = 8$$

EXAMPLE 4 Using Cramer's Rule

Solve, using Cramer's rule if possible.

$$x - 2y = 4$$
$$-2x + 4y = -8$$

Solution

$$D = \begin{vmatrix} 1 & -2 \\ -2 & 4 \end{vmatrix} = 4 - 4 = 0$$

$$D_x = \begin{vmatrix} 4 & -2 \\ -8 & 4 \end{vmatrix} = 16 - 16 = 0$$

$$D_y = \begin{vmatrix} 1 & 4 \\ -2 & -8 \end{vmatrix} = -8 - (-8) = 0$$

Since $D = D_x = D_y = 0$, the system is dependent and has an infinite number of solutions, in this case all ordered pairs (x, y) satisfying the equation $x - 2y = 4$.

Check Yourself 4

Use Cramer's rule to show that the following system is dependent.

$$x - 2y = -2$$
$$-2x + 4y = 4$$

The use of second-order determinants to solve systems of equations is easily extendable to systems involving three unknowns, as we will see in the remainder of this section.

To consider the solution of systems of three linear equations by determinants, we must first be able to evaluate a third-order determinant. To start, a third-order determinant is written in a similar fashion to that which we saw earlier and can be evaluated as follows:

$$\begin{vmatrix} a_1 & b_1 & c_1 \\ a_2 & b_2 & c_2 \\ a_3 & b_3 & c_3 \end{vmatrix} = a_1 b_2 c_3 + a_2 b_3 c_1 + a_3 b_1 c_2 - a_1 b_3 c_2 - a_2 b_1 c_3 - a_3 b_2 c_1 \qquad (5)$$

Even though the above formula can always be used, it is rather awkward and difficult to remember. Fortunately, a more convenient pattern is available which allows us to evaluate a 3×3 determinant in terms of 2×2 determinants. This is called *expansion of the determinant by minors*.

First a definition.

The *minor* of an element in a 3×3 determinant is the 2×2 determinant remaining when the row and column to which the element belongs have been deleted. The diagram below will help you picture the definition.

For a_1: $\begin{vmatrix} a_1 & b_1 & c_1 \\ a_2 & b_2 & c_2 \\ a_3 & b_3 & c_3 \end{vmatrix}$ The minor for a_1 is $\begin{vmatrix} b_2 & c_2 \\ b_3 & c_3 \end{vmatrix}$.   Delete the first row and the first column.

For a_2: $\begin{vmatrix} a_1 & b_1 & c_1 \\ a_2 & b_2 & c_2 \\ a_3 & b_3 & c_3 \end{vmatrix}$ The minor for a_2 is $\begin{vmatrix} b_1 & c_1 \\ b_3 & c_3 \end{vmatrix}$.   Delete the second row and the first column.

For a_3: $\begin{vmatrix} a_1 & b_1 & c_1 \\ a_2 & b_2 & c_2 \\ a_3 & b_3 & c_3 \end{vmatrix}$ The minor for a_3 is $\begin{vmatrix} b_1 & c_1 \\ b_2 & c_2 \end{vmatrix}$.   Delete the third row and the first column.

We can now evaluate a 3×3 determinant by using the three minors defined above.

To expand about the first column:

1. Write the product of each element with its minor.

2. Connect those products with a $+ - +$ sign pattern.

We will come back to this sign pattern later.

From the above we have

$$\begin{vmatrix} a_1 & b_1 & c_1 \\ a_2 & b_2 & c_2 \\ a_3 & b_3 & c_3 \end{vmatrix} = a_1 \begin{vmatrix} b_2 & c_2 \\ b_3 & c_3 \end{vmatrix} - a_2 \begin{vmatrix} b_1 & c_1 \\ b_3 & c_3 \end{vmatrix} + a_3 \begin{vmatrix} b_1 & c_1 \\ b_2 & c_2 \end{vmatrix} \qquad (6)$$

This is called the *expansion* of a 3×3 determinant about the first column. Using our earlier definition of a 2×2 determinant, you can verify that Equations (5) and (6) are the same. However, the pattern of Equation (6) is much easier to remember and is the most commonly used.

Let's look at the expansion of a 3×3 determinant, using this definition.

EXAMPLE 5 Evaluating a Determinant

Evaluate the determinant by expanding about the first column.

$\downarrow$

$\begin{vmatrix} 1 & -3 & 2 \\ -4 & 0 & 3 \\ 2 & -1 & 3 \end{vmatrix}$

If you enter the array into your calculator as a matrix, you can use the calculator to find the value of the related determinant.

Solution First we write the three minors.

Element	Minor
1	$\begin{vmatrix} 0 & 3 \\ -1 & 3 \end{vmatrix}$
-4	$\begin{vmatrix} -3 & 2 \\ -1 & 3 \end{vmatrix}$
2	$\begin{vmatrix} -3 & 2 \\ 0 & 3 \end{vmatrix}$

Now we write the product of each element and its minor, and we connect those products with the $+ \ - \ +$ sign pattern.

$$+ \ 1 \begin{vmatrix} 0 & 3 \\ -1 & 3 \end{vmatrix} - (-4) \begin{vmatrix} -3 & 2 \\ -1 & 3 \end{vmatrix} + \ 2 \begin{vmatrix} -3 & 2 \\ 0 & 3 \end{vmatrix}$$

$+ \ - \ +$ pattern

$$= 1(3) + 4(-7) + 2(-9)$$

$$= 3 - 28 - 18$$

$$= -43$$

The value of the determinant is -43.

Check Yourself 5

Evaluate the determinant by expanding about the first column.

$$\begin{vmatrix} 2 & 0 & -2 \\ 1 & -3 & -1 \\ -3 & 1 & 2 \end{vmatrix}$$

A 3×3 determinant can actually be expanded in a similar fashion about *any row or column*. First we will need the following.

The *sign array* for a 3×3 determinant is given by

$$\begin{vmatrix} + & - & + \\ - & + & - \\ + & - & + \end{vmatrix}$$

We simply place a $+$ sign in the upper left-hand corner and then alternate signs in a "checkerboard" pattern.

Now to generalize our earlier approach:

1. Choose *any* row or column for the expansion.

2. Write the product of each element, in the row or column chosen above, with its minor.

3. Connect those products with the appropriate sign pattern from the sign array shown above.

How do we pick a row or column for the expansion? The value of the determinant will be the same in any case. However, the computation will be much easier if you can choose a row or column with one or more 0s.

Let's return to our previous example, expanded in a different manner.

EXAMPLE 6 Evaluating a Determinant

Evaluate the determinant by expanding about the second row.

$$\rightarrow \begin{vmatrix} 1 & -3 & 2 \\ -4 & 0 & 3 \\ 2 & -1 & 3 \end{vmatrix}$$

The expansion is now

$$-(-4)\begin{vmatrix} -3 & 2 \\ -1 & 3 \end{vmatrix} + 0\begin{vmatrix} 1 & 2 \\ 2 & 3 \end{vmatrix} - 3\begin{vmatrix} 1 & -3 \\ 2 & -1 \end{vmatrix}$$

Note the $-\ +\ -$ from the sign array for the second row.

$$= 4(-7) + 0(-1) - 3(5)$$

$$= -28 + 0 - 15 = -43$$

Of course, the result is the same as before. Note that in practice, the 0 makes it unnecessary to evaluate (or even write) the second minor. That is why the choice of the second row for the expansion is more efficient in this case.

Check Yourself 6

Evaluate the determinant by expanding the first row.

$$\begin{vmatrix} 2 & 0 & -2 \\ 1 & -3 & -1 \\ -3 & 1 & 2 \end{vmatrix}$$

We are now ready to apply our work with 3×3 determinants to the solution of linear systems in three unknowns. Given the system

$$a_1 x + b_1 y + c_1 z = d_1$$

$$a_2 x + b_2 y + c_2 z = d_2$$

$$a_3 x + b_3 y + c_3 z = d_3$$

let D be the determinant of the coefficients.

$$D = \begin{vmatrix} a_1 & b_1 & c_1 \\ a_2 & b_2 & c_2 \\ a_3 & b_3 & c_3 \end{vmatrix}$$

We form D_x by replacing the first column of D with the constants d_1, d_2, and d_3. Similarly D_y is formed by replacing the second column with the constants and D_z by replacing the third column with the constants. Then

$$D_x = \begin{vmatrix} d_1 & b_1 & c_1 \\ d_2 & b_2 & c_2 \\ d_3 & b_3 & c_3 \end{vmatrix} \qquad D_y = \begin{vmatrix} a_1 & d_1 & c_1 \\ a_2 & d_2 & c_2 \\ a_3 & d_3 & c_3 \end{vmatrix} \quad \text{and} \quad D_z = \begin{vmatrix} a_1 & b_1 & d_1 \\ a_2 & b_2 & d_2 \\ a_3 & b_3 & d_3 \end{vmatrix}$$

With these definitions, we can now extend Cramer's rule to the case of a system of three linear equations in three unknowns.

Note If $D = 0$, the system will have either an infinite number of solutions or no solutions.

CRAMER'S RULE

Given the system

$$a_1x + b_1y + c_1z = d_1$$
$$a_2x + b_2y + c_2z = d_2$$
$$a_3x + b_3y + c_3z = d_3$$

if $D \neq 0$, the solution for the system is given by

$$x = \frac{D_x}{D} \qquad y = \frac{D_y}{D} \quad \text{and} \quad z = \frac{D_z}{D}$$

EXAMPLE 7 Using Cramer's Rule

Use Cramer's rule to solve the following system.

$$x - 2y - 3z = -8$$
$$2x + 3y + 2z = 9$$
$$-3x \qquad + 2z = 2$$

Solution We first calculate D, D_x, D_y, and D_z.

$$D = \begin{vmatrix} 1 & -2 & -3 \\ 2 & 3 & 2 \\ -3 & 0 & 2 \end{vmatrix}$$

We expand about row 3. Do you see why?

$$= -3 \begin{vmatrix} -2 & -3 \\ 3 & 2 \end{vmatrix} + 2 \begin{vmatrix} 1 & -2 \\ 2 & 3 \end{vmatrix}$$

$$= (-3)(5) + 2(7)$$

$$= -15 + 14 = -1$$

$$D_x = \begin{vmatrix} -8 & -2 & -3 \\ 9 & 3 & 2 \\ 2 & 0 & 2 \end{vmatrix}$$

Again we expand about row 3.

$$= 2 \begin{vmatrix} -2 & -3 \\ 3 & 2 \end{vmatrix} + 2 \begin{vmatrix} -8 & -2 \\ 9 & 3 \end{vmatrix}$$

$$= 2(5) + 2(-6)$$

$$= 10 - 12 = -2$$

$$D_y = \begin{vmatrix} 1 & -8 & -3 \\ 2 & 9 & 2 \\ -3 & 2 & 2 \end{vmatrix}$$

$$= 1\begin{vmatrix} 9 & 2 \\ 2 & 2 \end{vmatrix} - 2\begin{vmatrix} -8 & -3 \\ 2 & 2 \end{vmatrix} + (-3)\begin{vmatrix} -8 & -3 \\ 9 & 2 \end{vmatrix}$$

Here we expand about column 1.

$$= 1(14) - 2(-10) - 3(11)$$

$$= 14 + 20 - 33 = 1$$

$$D_z = \begin{vmatrix} 1 & -2 & -8 \\ 2 & 3 & 9 \\ -3 & 0 & 2 \end{vmatrix}$$

We expand about column 2.

$$= -(-2)\begin{vmatrix} 2 & 9 \\ -3 & 2 \end{vmatrix} + 3\begin{vmatrix} 1 & -8 \\ -3 & 2 \end{vmatrix}$$

$$= 2(31) + 3(-22)$$

$$= 62 - 66 = -4$$

We can now find x, y, and z.

$$x = \frac{D_x}{D} = \frac{-2}{-1} = 2$$

$$y = \frac{D_y}{D} = \frac{1}{-1} = -1$$

$$z = \frac{D_z}{D} = \frac{-4}{-1} = 4$$

The solution set for the system is $\{(2, -1, 4)\}$.

Remember The solution should be verified by substitution as before.

Check Yourself 7

Use Cramer's rule to solve the system.

$$x + 4y - 3z = 20$$

$$2x - 2y + z = -4$$

$$-3x + 6y + z = 10$$

The technique shown in this section for evaluating a 3 × 3 determinant can be used to evaluate a determinant of any order. Also there are rules for simplifying determinants *before* the expansion that are particularly helpful when you are working with higher-order determinants.

Some of these rules for simplification are suggested in Exercises 35 through 43.

CHECK YOURSELF ANSWERS

1. (1) 5; (2) −4   **2.** (2, −1)   **3.** 0 = 18 is a false statement, so the system is inconsistent   **4.** 0 = 0 is a true statement, so the system is dependent   **5.** 6
6. 6   **7.** (2, 3, −2)

4.6 EXERCISES

Build Your Skills

Evaluate each of the following determinants.

1. $\begin{vmatrix} 3 & 4 \\ 1 & 2 \end{vmatrix}$

2. $\begin{vmatrix} 1 & -4 \\ 5 & 2 \end{vmatrix}$

3. $\begin{vmatrix} -2 & 3 \\ 4 & -6 \end{vmatrix}$

4. $\begin{vmatrix} -5 & 2 \\ 0 & 0 \end{vmatrix}$

5. $\begin{vmatrix} 9 & 7 \\ -1 & 8 \end{vmatrix}$

6. $\begin{vmatrix} 5 & -8 \\ -7 & 12 \end{vmatrix}$

Introducing a variable in the entries of a determinant allows us to write an equation in determinant form. Solve each of the following equations for x.

7. $\begin{vmatrix} 3x & 5x \\ 2 & 4 \end{vmatrix} = 6$

8. $\begin{vmatrix} x & 2 \\ x & x \end{vmatrix} = 3$

Use Cramer's rule to find the unique solution for each of the following linear systems. Otherwise state whether the system is inconsistent or dependent.

9. $x + 5y = 15$
  $x - y = 3$

10. $2x - 3y = 9$
  $2x + 5y = -23$

11. $4x + y = -1$
  $-3x - 2y = 2$

12. $x + 2y = 7$
  $-2x - 4y = -10$

13. $2x + 3y = 8$
  $-5x + 6y = -11$

14. $x - 5y = 7$
  $-2x + 10y = -14$

15. $5x - 2y = 0$
  $4x + 3y = 0$

16. $5x - 8y = 3$
  $4x + 3y = 2$

17. $5x - 2y = 1$
  $x = 3$

18. $6x + 5y = 5$
  $3x - 4y = 2$

Given the determinant

$$\begin{vmatrix} 2 & 3 & -5 \\ 1 & 5 & 6 \\ -3 & 4 & 7 \end{vmatrix}$$

Find the value of the minor of each of the indicated elements.

19. 2

20. -5

21. 5

22. 7

23. -3

24. 4

Use expansion by minors to evaluate each of the following determinants.

25. $\begin{vmatrix} 1 & 0 & -1 \\ 2 & 1 & 3 \\ 1 & -2 & 1 \end{vmatrix}$

26. $\begin{vmatrix} 3 & 2 & 0 \\ -1 & 2 & 0 \\ 1 & 1 & 4 \end{vmatrix}$

27. $\begin{vmatrix} 4 & 2 & 1 \\ 0 & 1 & 3 \\ 0 & -2 & 5 \end{vmatrix}$

28. $\begin{vmatrix} 2 & 1 & 2 \\ 3 & 2 & 4 \\ 5 & -2 & -4 \end{vmatrix}$

29. $\begin{vmatrix} 1 & 3 & 1 \\ 5 & 2 & 7 \\ 1 & 3 & 2 \end{vmatrix}$

Solve each of the following systems by using Cramer's rule.

30. $x + 2y + z = 2$
  $x + 2z = -5$
  $x + 3y = 7$

31. $x - z = 7$
  $2x + y = 3$
  $x - 2z = 12$

32. $2x - y + 3z = -3$
  $x + 4y - 6z = 8$
  $3x - 2y + 3z = -4$

33.
$$x + 4y - 3z = 14$$
$$2x - 2y + z = -1$$
$$-3x + 6y + z = 1$$

34.
$$x + 4y + 5z = 9$$
$$-2x - 8y + 5z = 0$$
$$-x + 4y + 10z = -3$$

35. Evaluate the following determinant.
$$\begin{vmatrix} 2 & -3 & 2 \\ 0 & 0 & 0 \\ 1 & 3 & -3 \end{vmatrix}$$

36. What do you observe from the results of Exercise 35?

37. Evaluate the following determinant.
$$\begin{vmatrix} 3 & 0 & 0 \\ 0 & 2 & 0 \\ 0 & 0 & 4 \end{vmatrix}$$

38. What do you observe from the results of Exercise 37?

39. Evaluate the following determinant.
$$\begin{vmatrix} 1 & 2 & 3 \\ 0 & -1 & 2 \\ 2 & 4 & 6 \end{vmatrix}$$

40. What do you observe from the results of Exercise 39?

41. Evaluate each of the following determinants.

(a) $\begin{vmatrix} 2 & 4 & 6 \\ -1 & 0 & 3 \\ 1 & 2 & 1 \end{vmatrix}$ (b) $2\begin{vmatrix} 1 & 2 & 3 \\ -1 & 0 & 3 \\ 1 & 2 & 1 \end{vmatrix}$

42. What do you observe from the results of Exercise 41?

43. Use your observation from above to simplify your work in evaluating the following determinant.
$$\begin{vmatrix} 1 & 3 & 5 \\ 1 & 6 & -2 \\ 4 & 12 & 8 \end{vmatrix}$$

44. Solve each of the following equations for x.

(a) $\begin{vmatrix} x & 0 & -1 \\ 1 & 2 & 0 \\ -2 & 1 & 1 \end{vmatrix} = 3$ (b) $\begin{vmatrix} 1 & x & -2 \\ x & 1 & 2 \\ 1 & 0 & -3 \end{vmatrix} = 4$

45. Expand the following determinant about row 1 to write an equation in x and y.
$$\begin{vmatrix} x & y & 1 \\ 3 & 4 & 1 \\ 6 & 2 & 1 \end{vmatrix} = 0$$

46. Use the methods of Section 3.5 to write the equation of the line passing through the points with coordinates $(3, 4)$ and $(6, 2)$. Compare your results with that of Exercise 45. What do you observe?

Graphing Utility

Use your graphing utility to evaluate each of the following determinants.

47. $\begin{vmatrix} 2 & 3 & -4 & 1 \\ 3 & -1 & 2 & 5 \\ 1 & 2 & -1 & -2 \\ -2 & 0 & 5 & 1 \end{vmatrix}$

48. $\begin{vmatrix} 3 & -1 & 0 & -2 \\ -1 & -3 & 4 & 1 \\ 2 & 1 & 3 & 1 \\ 6 & -3 & 5 & 5 \end{vmatrix}$

Skillscan (Section 1.3)

Evaluate the following expressions.

a. 2^3 **b.** 3^2 **c.** $(-3)^2$ **d.** -2^4 **e.** $(-4)^3$

f. $\left(\dfrac{2}{3}\right)^2$

SUMMARY

Systems of Linear Equations in Two Variables [4.1]

A *system of linear equations* is two or more linear equations considered together. The solution for a linear system in two variables is an ordered pair of real numbers (x, y) that satisfies both equations in the system.

There are three solution techniques: the graphical method, the addition method, and the substitution method.

The solution for the system

$$2x - y = 7$$

$$x + y = 2$$

is $(3, -1)$. It is the only ordered pair that will satisfy each equation.

Solving by the Graphical Method Graph each equation of the system on the same set of coordinate axes. The solution (if one exists) will correspond to the point of intersection of the two lines. You may or may not be able to determine exact solutions for the system of equations with this method.

To solve:

$2x - y = 7$

$x + y = 2$

graphically

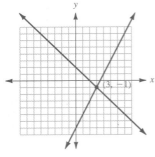

To solve:

$5x - 2y = 11$ (1)

$2x + 3y = 12$ (2)

by addition, multiply Equation (1) by 3 and Equation (2) by 2. Then add to eliminate y. This gives

$19x = 57$

$x = 3$

Substituting 3 for x in Equation (1), we have

$15 - 2y = 11$

$y = 2$

and (3, 2) is the solution for the system.

Solving by the Addition Method

Step 1. If necessary, multiply one or both of the equations by a constant so that one of the variables can be eliminated by addition.

Step 2. Add the equations of the equivalent system formed in step 1.

Step 3. Solve the equation found in step 2.

Step 4. Substitute the value found in step 3 into either of the equations of the original system, to find the corresponding value of the remaining variable. The ordered pair formed is the solution to the system.

Step 5. Check the solution by substituting the pair of values found in step 4 into the other equation of the original system.

To solve

$3x - 2y = 6$ (3)

$6x + y = 2$ (4)

by substitution, solve (4) for y.

$y = -6x + 2$ (5)

Substituting in (3) gives

$3x - 2(-6x + 2) = 6$

and

$x = \dfrac{2}{3}$

Substituting $\dfrac{2}{3}$ for x in (5) gives

$y = (-6)\left(\dfrac{2}{3}\right) + 2$

$= -4 + 2 = -2$

The solution is

$\left(\dfrac{2}{3}, -2\right)$

Solving by the Substitution Method

Step 1. If necessary, solve one of the equations of the original system for one of the variables.

Step 2. Substitute the expression obtained in step 1 into the other equation of the system, to write an equation in a single variable.

Step 3. Solve the equation found in step 2.

Step 4. Substitute the value found in step 3 into the equation derived in step 1 to find the corresponding value of the remaining variable. The ordered pair formed is the solution for the system.

Step 5. Check the solution by substituting the pair of values found in step 4 into both equations of the original system.

Consistent, Inconsistent, and Dependent Systems For a linear system of two equations in two variables, the lines corresponding to the two equations will

1. Intersect at one and only one point. The system is *consistent*.

2. Be parallel. The system is *inconsistent*.

3. Coincide. The equations have all their solutions, an infinite number, in common. The system is *dependent*.

One solution—a consistent system
No solutions—an inconsistent system
An infinite number of solutions—a dependent system

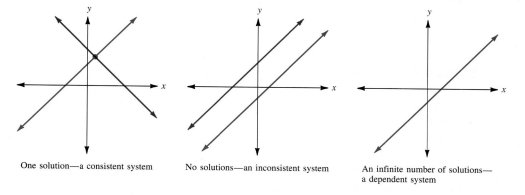

One solution—a consistent system   No solutions—an inconsistent system   An infinite number of solutions—a dependent system

Systems of Linear Equations in Three Variables [4.2]

The solution for a linear system of three equations in three variables is an ordered triple of numbers (x, y, z) which satisfies each of the equations of the system.

Solving by the Addition Method

Step 1. Choose a pair of equations from the system, and use the addition method to eliminate one of the variables.

Step 2. Choose a different pair of equations, and eliminate the same variable.

Step 3. Solve the system of two equations in two variables determined in steps 1 and 2.

Step 4. Substitute the values found above into one of the original equations, and solve for the remaining variable.

Step 5. The solution is the ordered triple of values found in steps 3 and 4. It can be checked by substituting into the other equations of the original system.

To solve:

$$x + y - z = 6 \qquad (6)$$
$$2x - 3y + z = -9 \qquad (7)$$
$$3x + y + 2z = 2 \qquad (8)$$

Adding (6) and (7) gives

$$3x - 2y = -3 \qquad (9)$$

Multiplying (6) by 2 and adding the result to (8) give

$$5x + 3y = 14 \qquad (10)$$

The system consisting of (9) and (10) is solved as before and

$$x = 1 \qquad y = 3$$

Substituting these values into (6) gives

$$z = -2$$

The solution is $(1, 3, -2)$.

Applications and Problem Solving [4.3]

To use a system of equations to solve an application, our problem-solving algorithm remains essentially the same as before.

Also determine the condition that relates the unknown quantities.
Use a different letter for each variable.
A table or a sketch often helps in writing the equations of the system.

Solving Applications

Step 1. Read the problem carefully to determine the unknown quantities.

Step 2. Choose variables to represent the unknowns.

Step 3. Translate the problem to the language of algebra to form a system of equations.

Step 4. Solve the system of equations by any of the methods discussed.

Step 5. Verify your solution by returning to the original problem.

Systems of Linear Inequalities [4.4]

A *system of linear inequalities* is two or more linear inequalities considered together. The *graph of the solution set* of a system of linear inequalities is the intersection of the graphs of the individual inequalities.

Solving Systems of Linear Inequalities Graphically

1. Graph each inequality, shading the appropriate half plane, on the same set of coordinate axes.

2. The graph of the system is the intersection of the regions shaded above.

To solve

$x + 2y \leq 8$

$x + y \leq 6$

$x \geq 0$

$y \geq 0$

graphically

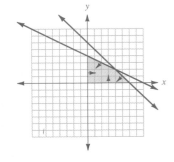

Solution of Systems by Matrices [4.5]

$\begin{bmatrix} 2 & 3 & \vdots & 1 \\ 5 & 7 & \vdots & 1 \end{bmatrix}$ is a 2 × 3 matrix

(read "2 by 3").

To solve

$3x - 2y = 5$

$x + 3y = 9$

The augmented matrix is

$\begin{bmatrix} 3 & -2 & \vdots & 5 \\ 1 & 3 & \vdots & 9 \end{bmatrix}$

A *matrix* is a rectangular array of numbers. Each number in the matrix is an *element* of the matrix. The dimensions of a matrix are stated by giving the number of rows and then the number of columns.

For the linear system

$a_1 x + b_1 y = c_1$

$a_2 x + b_2 y = c_2$

the *augmented matrix* is

$\begin{bmatrix} a_1 & b_1 & \vdots & c_1 \\ a_2 & b_2 & \vdots & c_2 \end{bmatrix}$

The solution of a system by matrices uses elementary row operations which are applied to produce the augmented matrix of an equivalent system with 1s down the main diagonal of the coefficient matrix and 0s below. That equivalent system is then solved by *back substitution*.

Determinants and Cramer's Rule [4.6]

A *determinant* is a square array of numbers. Determinants can be used to solve systems of equations by applying *Cramer's rule*.

By row operations,

$$R_1 \leftrightarrow R_2 \quad \begin{bmatrix} 1 & 3 & \vdots & 9 \\ 3 & -2 & \vdots & 5 \end{bmatrix}$$

$$-3R_1 \leftrightarrow R_2 \quad \begin{bmatrix} 1 & 3 & \vdots & 9 \\ 0 & -11 & \vdots & -22 \end{bmatrix}$$

$$-\frac{1}{11}R_2 \quad \begin{bmatrix} 1 & 3 & \vdots & 9 \\ 0 & 1 & \vdots & 2 \end{bmatrix}$$

The equivalent system is

$$x + 3y = 9$$
$$y = 2$$

Letting $y = 2$ in the first equation gives $x = 3$, so $(3, 2)$ is the solution.

CRAMER'S RULE

To solve the linear system

$$a_1 x + b_1 y = c_1$$
$$a_2 x + b_2 y = c_2$$

write

$$D = \begin{vmatrix} a_1 & b_1 \\ a_2 & b_2 \end{vmatrix} \quad \text{The determinant of the coefficients}$$

$$D_x = \begin{vmatrix} c_1 & b_1 \\ c_2 & b_2 \end{vmatrix} \quad \text{The } x \text{ coefficients have been replaced by the constants.}$$

$$D_y = \begin{vmatrix} a_1 & c_1 \\ a_2 & c_2 \end{vmatrix} \quad \text{The } y \text{ coefficients have been replaced by the constants.}$$

Then

$$x = \frac{D_x}{D} \quad \text{and} \quad y = \frac{D_y}{D} \quad \text{when } D \neq 0$$

If $D = \begin{vmatrix} 4 & 6 \\ 2 & -5 \end{vmatrix}$

then $D = 4(-5) - 2(6)$

$$= -20 - 12$$
$$= -32$$

Given the system

$$2x + 4y = 2$$
$$x - 5y = 8$$

$$D = \begin{vmatrix} 2 & 4 \\ 1 & -5 \end{vmatrix} = -14$$

$$D_x = \begin{vmatrix} 2 & 4 \\ 8 & -5 \end{vmatrix} = -42$$

$$D_y = \begin{vmatrix} 2 & 2 \\ 1 & 8 \end{vmatrix} = 14$$

$$x = 3 \quad \text{and} \quad y = 1$$

SUMMARY EXERCISES

This summary exercise set is provided to give you practice with each of the objectives of the chapter. Each exercise is keyed to the appropriate chapter section. The answers are provided in the instructor's manual that accompanies this text. Your instructor will provide guidelines on how best to use these exercises in your instructional program.

[4.1] Solve each of the following systems by graphing.

1. $x + y = 8$
$x - y = 4$

2. $x + 2y = 8$
$x - y = 5$

3. $2x + 3y = 12$
$2x + y = 8$

4. $x + 4y = 8$
$y = 1$

[4.1] Solve each of the following systems by the addition method. If a unique solution does not exist, state whether the given system is inconsistent or dependent.

5. $x + 2y = 7$
$\ x - y = 1$

6. $\ x + 3y = 14$
$4x + 3y = 29$

7. $3x - 5y = 5$
$-x + y = -1$

8. $\ x - 4y = 12$
$2x - 8y = 24$

9. $\ 6x + 5y = -9$
$-5x + 4y = 32$

10. $3x + y = -17$
$5x - 3y = -19$

11. $\ 3x + y = 8$
$-6x - 2y = -10$

12. $5x - 4y = -23$
$4x + 3y = -6$

13. $7x - 4y = 27$
$5x + 6y = 6$

14. $4x - 3y = 1$
$6x + 5y = 30$

15. $\ x - \dfrac{1}{2}y = 8$

$\dfrac{2}{3}x + \dfrac{3}{2}y = -2$

16. $\dfrac{1}{5}x - 2y = 4$

$\dfrac{3}{5}x + \dfrac{2}{3}y = -8$

17. $\ x + y = 10{,}000$
$0.06x + 0.08y = \phantom{10{,}}720$

18. $\ x + y = 300$
$0.25x + 0.50y = 120$

[4.1] Solve each of the following systems by the substitution method. If a unique solution does not exist, state whether the given system is inconsistent or dependent.

19. $2x + y = 23$
$x = y + 4$

20. $x - 5y = 26$
$y = x - 10$

21. $3x + y = 7$
$y = -3x + 5$

22. $2x - 3y = 13$
$x = 3y + 9$

23. $5x - 3y = 13$
$x - y = 3$

24. $4x - 3y = 6$
$x + y = 12$

25. $3x - 2y = -12$
$6x + y = 1$

26. $\ x - 4y = 8$
$-2x + 8y = -16$

[4.2] Solve each of the following systems by the addition method. If a unique solution does not exist, state whether the given system is inconsistent or dependent.

27. $x - y + z = 0$
$x + 4y - z = 14$
$x + y - z = 6$

28. $\ x - y + z = 3$
$3x + y + 2z = 15$
$2x - y + 2z = 7$

29. $\ x - y - z = 2$
$-2x + 2y + z = -5$
$-3x + 3y + z = -10$

30. $x - y = 3$
$2y + z = 5$
$x + 2z = 7$

31. $x + y + z = 2$
$x + 3y - 2z = 13$
$y - 2z = 7$

32. $\ x + y - z = -1$
$x - y + 2z = 2$
$-5x - y - z = -1$

33. $\ 2x + 3y + z = 7$
$-2x - 9y + 2z = 1$
$4x - 6y + 3z = 10$

34. $x - 3y + 2z = 8$
$5x + 2y - 4z = 10$
$3x - y + 2z = 6$

[4.3] Solve each of the following problems by choosing a variable to represent each unknown quantity. Then write a system of equations that will allow you to solve for each variable.

35. One number is 2 more than 3 times another. If the sum of the two numbers is 30, find the two numbers.

36. Suppose that a cashier has 78 $5 and $10 dollar bills with a value of $640. How many of each type of bill does she have?

37. Tickets for a basketball game sold at $7 for an adult ticket and $4.50 for a student ticket. If the revenue from 1200 tickets was $7400, how many of each type of ticket were sold?

38. A purchase of 8 blank cassette tapes and 4 blank videotapes costs $36. A second purchase of 4 cassette tapes and 5 videotapes costs $30. What is the price of a single cassette tape and of a single videotape?

39. The length of a rectangle is 4 cm less than twice its width. If the perimeter of the rectangle is 64 cm, find the dimensions of the rectangle.

40. A grocer in charge of bulk foods wishes to combine peanuts selling for $2.25 per pound and cashews selling for $6 per pound. What amount of each nut should be used to form a 120-lb mixture selling for $3 per pound?

41. Reggie has two investments totaling $17,000, one a savings account paying 6 percent, the other a time deposit paying 8 percent. If his annual interest is $1200, what does he have invested in each account?

42. A pharmacist mixes a 20% alcohol solution and a 50% alcohol solution to form 600 mL of a 40% solution. How much of each solution should she use in forming the mixture?

43. A jet flying east, with the wind, makes a trip of 2200 mi in 4 h. Returning, against the wind, the jet can travel only 1800 mi in 4 h. What is the plane's rate in still air? What is the rate of the wind?

44. The sum of the digits of a two-digit number is 9. If the digits are reversed, the new number is 45 more than the original number. What was the original number?

45. A manufacturer produces $5\frac{1}{4}$-in computer disk drives and $3\frac{1}{2}$-in drives. The $5\frac{1}{4}$-in drives require 20 min of component assembly time; the $3\frac{1}{2}$-in drives, 25 min. The manufacturer has 500 min of component assembly time available per day. Each of the drives requires 30 min for packaging and testing, and 690 min of that time is available per day. How many of each of the drives should be produced daily to use all the available time?

46. If the demand equation for a product is $D = 270 - 5p$ and the supply equation is $S = 13p$, find the equilibrium point. (See Exercise 23, Section 4.3.)

47. Two car rental agencies have the following rates for the rental of a compact automobile:

Company A: $18 per day plus 12¢ per mile

Company B: $20 per day plus 10¢ per mile

For a 3-day rental, at what number of miles will the charges from the two companies be the same?

48. The sum of three numbers is 15. The largest number is 4 times the smallest number, and it is also 1 more than the sum of the other two numbers. Find the three numbers.

49. The sum of the digits of a three-digit number is 16. The tens digit is 3 times the hundreds digit, and the units digit is 1 more than the hundreds digit. What is the number?

50. A theater has orchestra tickets at $10, box-seat tickets at $7, and balcony tickets at $5. For one performance, a total of 360 tickets were sold, and the total revenue was $3040. If the number of orchestra tickets sold was 40 more than that of the other two types combined, how many of each type of ticket were sold for the performance?

51. The measure of the largest angle of a triangle is 15° less than 4 times the measure of the smallest angle and 30° more than the sum of the measures of the other two angles. Find the measures of the three angles of the triangle.

52. Rachel divided $12,000 into three investments: a savings account paying 5 percent, a stock paying 7 percent, and a mutual fund paying 9 percent. Her annual interest from the investments was $800, and the amount that she had invested at 5 percent was equal to the sum of the amounts invested in the other accounts. How much did she have invested in each type of account?

53. The difference of two positive numbers is 3, and the sum of those numbers is 41. Find the two numbers.

54. The sum of two integers is 144, and the difference 42. What are the two integers?

55. The area of a rectangular building lot is 9600 ft², and the perimeter of the lot is 400 ft. Find the length and width of the lot.

56. A manufacturer's cost for producing x units of a product is given by

$C = 10x + 3600$

The revenue from selling x units of that product is given by

$R = 100x$

Find the break-even point for this product. (See Example 10, Section 4.5.)

[4.4] Solve each of the following linear inequalities graphically.

57. $x - y < 7$
   $x + y > 3$

58. $x - 2y \leq -2$
   $x + 2y \leq 6$

59. $x - 6y < 6$
   $-x + y < 4$

60. $2x + y \leq 8$
   $x \geq 1$
   $y \geq 0$

61. $2x + y \leq 6$
   $x \geq 1$
   $y \geq 0$

62. $4x + y \leq 8$
   $x \geq 0$
   $y \geq 2$

63. $4x + 2y \leq 8$
   $x + y \leq 3$
   $x \geq 0$
   $y \geq 0$

64. $3x + y \leq 6$
   $x + y \leq 4$
   $x \geq 0$
   $y \geq 0$

[4.5] Solve each of the following systems, using row operations with matrices where possible.

65. $x + 2y = -5$
   $-2x + y = -5$

66. $x + 3y = 0$
   $5x - 3y = 12$

67. $3x - y = -14$
   $x + 3y = -8$

68. $x - 4y = 1$
   $2x - 8y = 3$

69. $2x - 4y = 4$
   $-x + 2y = -2$

70. $4x + 3y = 2$
   $x + 6y = -3$

71. $x + y = 4$
   $y + z = 1$
   $x + z = -1$

72. $x + y + z = -8$
   $x + z = -4$
   $x + y = -5$

[4.6] Use Cramer's rule to find a unique solution for each linear system.

73. $5x + y = 30$
   $-x + y = 6$

74. $8x - 5y = 6$
   $3x + 4y = 4$

SELF-TEST

The purpose of this self-test is to help you check your progress and to review for a chapter test in class. Allow yourself about an hour to take the test. When you are done, check your answers in the back of the book. If you missed any problems, be sure to go back and review the appropriate sections in the chapter and the exercises that are provided.

Solve each of the following systems. If a unique solution does not exist, state whether the given system is inconsistent or dependent.

1. $3x + y = -5$
   $5x - 2y = -23$

2. $4x - 2y = -10$
   $y = 2x + 5$

3. $9x - 3y = 4$
   $-3x + y = -1$

4. $5x - 3y = 5$
   $3x + 2y = -16$

5. $x - 2y = 5$
   $2x + 5y = 10$

6. $5x - 3y = 20$
   $4x + 9y = -3$

Solve each of the following systems.

7. $x - y + z = 1$
   $-2x + y + z = 8$
   $x + 5z = 19$

8. $x + 3y - 2z = -6$
   $3x - y + 2z = 8$
   $-2x + 3y - 4z = -11$

Solve each of the following problems by choosing a variable to represent each unknown quantity. Then write a system of equations that will allow you to solve for each variable.

9. An order for 30 computer disks and 12 printer ribbons totaled $147. A second order for 12 more disks and 6 additional ribbons cost $66. What was the cost per individual disk and ribbon?

10. A candy dealer wants to combine jawbreakers selling for $2.40 per pound and licorice selling for $3.90 per pound to form a 100-lb mixture which will sell for $3 per pound. What amount of each type of candy should be used?

11. A small electronics firm assembles 5-in portable television sets and 12-in models. The 5-in set requires 9 h of assembly time; the 12-in set, 6 h. Each unit requires 5 h for packaging and testing. If 72 h of assembly time and 50 h of packaging and testing time are available per week, how many of each type of set should be finished if the firm wishes to use all its available capacity?

12. Hans decided to divide $14,000 into three investments: a savings account paying 6 percent annual interest, a bond paying 9 percent, and a mutual fund paying 13 percent. His annual interest from the three investments was $1100, and he had twice as much invested in the bond as in the mutual fund. What amount did he invest in each type?

13. The fence around a rectangular yard requires 260 ft of fencing. The length is 20 ft less than twice the width. Find the dimensions of the yard.

Solve each of the following linear inequalities graphically.

14. $x - 2y < 6$
$\quad x + y < 3$

15. $3x + 4y \geq 12$
$\qquad x \geq 1$

16. $x + 2y \leq 8$
$\quad x + y \leq 6$
$\qquad x \geq 0$
$\qquad y \geq 0$

Solve each of the following systems by using row operations with matrices.

17. $4x + 11y = 7$
$\quad x + 3y = 1$

18. $5x + 7y = 3$
$\quad 10x + 14y = 5$

19. $x - 2y - z = 5$
$x - y = 5$
$-x + 4y + 5z = -1$

20. $3x - 4y - z = 1$
$\quad 2x - 3y + z = 1$
$\quad x - 2y + 3z = 2$

CUMULATIVE REVIEW EXERCISES

This is a review of selected topics from the first four chapters.

Solve each of the following.

1. $3x - 2(x + 5) = 12 - 3x$

2. $2x - 7 < 3x - 5$

3. $|2x - 3| = 5$

4. $|3x + 5| \leq 7$

5. $|5x - 4| > 21$

6. $-8 \leq 2x - 5 \leq 7$

Graph each of the following.

7. $5x + 7y = 35$

8. $2x + 3y < 6$

9. Find the distance between the points $(-1, 2)$ and $(4, -22)$.

10. Find the slope of the line connecting $(4, 6)$ and $(3, -1)$.

Write the function form of the equation of the line satisfying each of the following given conditions.

11. L has a y intercept of 5 and passes through the point $(2, -3)$.

12. L passes through the points $(-1, 4)$ and $(5, -2)$.

13. L has a y intercept of -2 and is parallel to the line with equation $y = x + 1$.

14. L has an x intercept of 6 and is perpendicular to the line with equation $y = -3$.

Solve each of the following systems of equations.

15. $2x + 3y = 6$
$\quad 5x + 3y = -24$

16. $x + y + z = 3$
$\quad 2x - y + 2z = 0$
$-x - 3y + z = -9$

Solve each of the following, using matrices and elementary row operations.

17. $6x - y = 28$
$-x + 3y = 1$

18. $x + 2y = 0$
$x + 3y - 6z = -3$
$2x - 3z = 3$

Solve each of the following applications.

19. The length of a rectangle is 3 cm more than twice its width. If the perimeter of the rectangle is 54 cm, find the dimensions of the rectangle.

20. The sum of the digits of a two-digit number is 10. If the digits are reversed, the new number is 36 less than the original number. What was the original number?

POLYNOMIAL EXPRESSIONS CHAPTER FIVE

WATER POLLUTION *Because it is essential for life, water is one of our most important resources. Luckily, most of the planet is covered by water. While the earth's surface may be mostly water, little of this water is directly usable for human consumption.*

Over 97 percent of the earth's water is too salty for most human uses. This means that 3 percent of the water is freshwater, but even this is not all available for use. Most of our freshwater is ice, is located deep underground, is water vapor in the atmosphere, or is too polluted to use. If the world's water supply were reduced to 100 liters (L), a medium-sized aquarium, the usable freshwater would only be 0.003 L (0.5

teaspoon). Even this small fraction of our total water supply is sufficient as long as we don't pollute the water more quickly than it is replenished by natural processes.

Prior to the industrial revolution, most cities were small enough that the nearby rivers, lakes, and oceans could absorb the pollutants dumped into them without much impact. This is no longer true. Many industrial processes discharge pollutants into our water systems.

Modern agriculture causes the loss of large volumes of water. Irrigation causes water loss by evaporation, especially in hot, dry areas of the world where it is essential for growing food. Runoff from chemically treated cropland and soil erosion from poor farming practices both pollute rivers and lakes.

Water used to cool electric power plants can be polluted by too much heat. If this water is not cooled, it can cause a change in the local environment.

Household use of detergents and other chemicals can pollute water, as can improperly handled human waste.

While water pollution problems continue in many places, they are being addressed in many ways. Laws such as the Clean Water Act have caused many industries and farmers to become aware of their polluting activities and to clean the water they use before putting it back into the local water supply. Contamination of local drinking water has shocked many communities into demanding tighter regulation of polluters in communities worldwide. Water pollution prevention and control are becoming an expected part of business or industry. ■

243

5.1 | Integral Exponents and Scientific Notation

TAPE IN11

OBJECTIVES: 1. To define 0 and negative integer exponents
2. To use the properties of integer exponents
3. To use scientific notation

Exponents are a shorthand form for repeated multiplication. Instead of writing

$$a \cdot a \cdot a \cdot a \cdot a$$

we write

$$a^5$$

which we read as "a to the fifth power."

We call a the base of the expression and 5 the exponent, or power.

> **In general, for any real number a and any natural number n,**
>
> $$a^n = \underbrace{a \cdot a \cdots a}_{n \text{ factors}}$$

An expression of this type is said to be in *exponential form.* We call a the *base* of the expression and n the *exponent,* or *power.*

EXAMPLE 1 Using Exponential Notation

Write each of the following, using exponential notation.

(a) $5y \cdot 5y \cdot 5y = (5y)^3$
(b) $w \cdot w \cdot w \cdot w = w^4$

Check Yourself 1

Write each of the following, using exponential notation.

1. $3z \cdot 3z \cdot 3z$ **2.** $x \cdot x \cdot x \cdot x$

Let's consider what happens when we multiply two expressions in exponential form with the same base.

We expand the expressions and apply the associative property to regroup.

$$a^4 \cdot a^5 = (\underbrace{a \cdot a \cdot a \cdot a}_{4 \text{ factors}})(\underbrace{a \cdot a \cdot a \cdot a \cdot a}_{5 \text{ factors}})$$

$$= \underbrace{a \cdot a \cdot a \cdot a \cdot a \cdot a \cdot a \cdot a \cdot a}_{9 \text{ factors}}$$

$$= a^9$$

Notice that the product is simply the base taken to the power that is the sum of the two original exponents.

In fact, in general, the following holds:

PRODUCT RULE FOR EXPONENTS

For any nonzero real number a and positive integers m and n,

$$a^m \cdot a^n = \underbrace{(a \cdot a \cdots a)}_{m \text{ factors}}\underbrace{(a \cdot a \cdots a)}_{n \text{ factors}}$$

$$= \underbrace{a \cdot a \cdots a}_{m + n \text{ factors}}$$

$$= a^{m+n}$$

This is our *first property of exponents*

$a^m \cdot a^n = a^{m+n}$

The next example illustrates the product rule for exponents.

EXAMPLE 2 Using the Product Rule

Simplify each expression.

(a) $b^4 \cdot b^6 = b^{4+6} = b^{10}$

(b) $(2a)^3 \cdot (2a)^4 = (2a)^{3+4} = (2a)^7$

(c) $(-2)^5(-2)^4 = (-2)^{5+4} = (-2)^9$

(d) $(10^7)(10^{11}) = (10)^{7+11} = (10)^{18}$

Check Yourself 2

Simplify each product.

1. $(5b)^6(5b)^5$
2. $(-3)^4(-3)^3$
3. $10^8 \cdot 10^{12}$
4. $(xy)^2(xy)^3$

Applying the commutative and associative properties of multiplication, we know that a product such as

$2x^3 \cdot 3x^2$

can be rewritten as

$(2 \cdot 3)(x^3 \cdot x^2)$

or as

$6x^5$

We expand on the ideas illustrated above in our next example.

EXAMPLE 3 Using Properties of Exponents

Using the product rule for exponents together with the commutative and associative properties, simplify each product.

Multiply the coefficients and *add* the exponents by the product rule. With practice you will *not need to write the* regrouping step.

(a) $(5x^4)(3x^2) = (5 \cdot 3)x^4x^2 = 15x^6$

(b) $(x^2y^3)(x^2y^4) = (x^2 \cdot x^2)(y^3 \cdot y^4) = x^4y^7$

(c) $(4c^5d^3)(3c^2d^2) = (4 \cdot 3)(c^5c^2)(d^3d^2) = 12c^7d^5$

(d) $(-3p^3q^4)(2p^2q^5) = (-3)(2)(p^3p^2)(q^4q^5) = -6p^5q^9$

(e) $(-2m^2n^5)(-4m^3n^2) = (-2)(-4)(m^2m^3)(n^5n^2) = 8m^5n^7$

Check Yourself 3

Simplify each expression.

1. $(4a^2b)(2a^3b^4)$ 　　　　　　　　**2.** $(3x^4)(2x^3y)$

3. $(-2s^3t^4)(3s^2t)$ 　　　　　　　　**4.** $(-4d^5f^3)(-5d^2f^4)$

Now consider the quotient

$$\frac{a^6}{a^4}$$

If we write this in expanded form, we have

$$\overbrace{a \cdot a \cdot a \cdot a \cdot a \cdot a}^{6 \text{ factors}}$$
$$\underbrace{a \cdot a \cdot a \cdot a}_{4 \text{ factors}}$$

This can be reduced to

Divide the numerator and denominator by the four common factors of *a*.

Note that $\dfrac{a}{a} = 1$, where $a \neq 0$.

$$\frac{\cancel{a} \cdot \cancel{a} \cdot \cancel{a} \cdot \cancel{a} \cdot a \cdot a}{\cancel{a} \cdot \cancel{a} \cdot \cancel{a} \cdot \cancel{a}} \qquad \text{or} \qquad a^2$$

This means that

$$\frac{a^6}{a^4} = a^2$$

This leads to a second property of exponents.

This is our *second property of exponents.*
We write $a \neq 0$ to avoid division by 0.

QUOTIENT RULE FOR EXPONENTS

In general, for any real number a ($a \neq 0$) and positive integers m and n,

$$\frac{a^m}{a^n} = a^{m-n}$$

EXAMPLE 4 Using Properties of Exponents

Simplify each expression.

(a) $\dfrac{x^{10}}{x^4} = x^{10-4} = x^6$

Subtract the exponents, applying the quotient rule.

(b) $\dfrac{a^8}{a^7} = a^{8-7} = a \longleftarrow$ Note that $a^1 = a$; there is no need to write the exponent 1 because it is understood.

(c) $\dfrac{63w^8}{7w^5} = 9w^{8-5} = 9w^3$

We *divide* the coefficients and subtract the exponents.

(d) $\dfrac{-32a^4b^5}{8a^2b} = -4a^{4-2}b^{5-1} = -4a^2b^4$

Divide the coefficients and subtract the exponents for *each* variable.

(e) $\dfrac{10^{16}}{10^6} = 10^{16-6} = 10^{10}$

Check Yourself 4

Simplify each expression.

1. $\dfrac{y^{12}}{y^5}$ **2.** $\dfrac{x^9}{x^8}$ **3.** $\dfrac{45r^8}{-9r^6}$

4. $\dfrac{49a^6b^7}{7ab^3}$ **5.** $\dfrac{10^{13}}{10^5}$

In the quotient rule, suppose that we now allow *m to equal n*. We then have

$$\frac{a^m}{a^m} = a^{m-m} = a^0 \qquad (1)$$

$\dfrac{a^m}{a^n} = a^{m-n}$

But we know that it is also true that

$$\frac{a^m}{a^m} = 1 \qquad (2)$$

Comparing (1) and (2), we see that the following definition seems reasonable.

THE ZERO EXPONENT

For any real number a where $a \neq 0$,

$a^0 = 1$

We must have $a \neq 0$. The form 0^0 is called *indeterminate* and is considered in later mathematics classes.

EXAMPLE 5 Using Properties of Exponents

Use the above definition to simplify each expression.

(a) $17^0 = 1$ (b) $(a^3b^2)^0 = 1$

(c) $6x^0 = 6 \cdot 1 = 6$ (d) $-3y^0 = -3$

Note that in $6x^0$ the exponent 0 applies *only* to x.

Check Yourself 5

Simplify each expression.

1. 25^0 ~ I **2.** $(m^4n^2)^0$ $= 1$ **3.** $8s^0$ $= 8$ **4.** $-7t^0$ ~ 7

Now, what if we allow one of the exponents to be negative and apply the product rule? Suppose, for instance, that $m = 3$ and $n = -3$. Then

$$a^m \cdot a^n = a^3 \cdot a^{-3} = a^{3+(-3)}$$
$$= a^0 = 1$$

so

$$a^3 \cdot a^{-3} = 1$$

and dividing both sides by a^3, we get

$$a^{-3} = \frac{1}{a^3}$$

This leads us to the following general definition.

John Wallis (1616–1702), an English mathematician, was the first to fully discuss the meaning of 0, negative, and rational exponents (which we discuss in Section 7.5).

> ### NEGATIVE INTEGER EXPONENTS
>
> **For any nonzero real number a and whole number n,**
>
> $$a^{-n} = \frac{1}{a^n}$$
>
> **and a^{-n} is the *multiplicative inverse* of a^n.**

The next example illustrates this definition.

EXAMPLE 6 Using Properties of Exponents

From this point on, to *simplify* will mean to write the expression with *positive exponents only*.

Also, we will restrict all variables so that they represent nonzero real numbers.

Simplify the following expressions.

(a) $y^{-5} = \dfrac{1}{y^5}$

(b) $4^{-2} = \dfrac{1}{4^2} = \dfrac{1}{16}$

(c) $(-3)^{-3} = \dfrac{1}{(-3)^3} = \dfrac{1}{-27} = -\dfrac{1}{27}$

(d) $\left(\dfrac{2}{3}\right)^{-3} = \dfrac{1}{\left(\dfrac{2}{3}\right)^3} = \dfrac{1}{\dfrac{8}{27}} = \dfrac{27}{8}$

Check Yourself 9

Simplify each of the following expressions.

1. $x^9 \cdot x^{-5}$ x^4

2. $\dfrac{y^{-7}}{y^{-3}}$ $y^{-4} = \dfrac{1}{y^4}$

3. $\dfrac{a^{-3}a^2}{a^{-5}}$ $\dfrac{a}{a^{-5}} = a^4 = \dfrac{1}{a^4}$

Suppose that we have an expression of the form

$$(a^2)^4$$

This can be written as

$$\underbrace{(a \cdot a)(a \cdot a)(a \cdot a)(a \cdot a)}_{2 \cdot 4, \text{ or } 8, \text{ factors}} \qquad \text{or} \qquad a^8$$

This suggests, in general, the following:

POWER RULE

For any nonzero real number *a* **and integers** *m* **and** *n*,

$$(a^m)^n = a^{mn}$$

This is our *third property of exponents*. When a number with an exponent is taken to a power, we *multiply* the exponents.

Our next example illustrates the use of the third property for exponents.

EXAMPLE 10 Using the Power Rule

Using the power rule, simplify each expression.

(a) $(x^3)^5 = x^{3 \cdot 5} = x^{15}$

(b) $(a^2)^8 = a^{2 \cdot 8} = a^{16}$

(c) $(3^2)^3 = 3^{2 \cdot 3} = 3^6$

CAUTION
Be Careful! Some students confuse $(x^3)^5 = x^{15}$ with $x^3 \cdot x^5 = x^8$. In the first case we *multiply* the exponents, in the second we *add*!

Check Yourself 10

Simplify each expression.

1. $(b^4)^7$ b^{27}

2. $(10^3)^3$ 10^{30}

3. $b^4 b^7$ b^{11}

Expressions with negative exponents can also be simplified by using the power rule.

EXAMPLE 11 Using the Power Rule

Using the power rule, simplify each expression.

(a) $(x^{-3})^5 = x^{(-3)(5)} = x^{-15} = \dfrac{1}{x^{15}}$

(b) $(x^4)^{-3} = x^{(4)(-3)} = x^{-12} = \dfrac{1}{x^{12}}$

(c) $(w^{-3})^{-2} = w^{(-3)(-2)} = w^6$

Check Yourself 11

Simplify each of the following expressions.

1. $(y^3)^{-4}$ **2.** $(p^{-2})^{-5}$ **3.** $(w^{-5})^6$

Let's develop another property for exponents. An expression such as

$(2x)^5$

can be written in expanded form as

$\underbrace{(2x)(2x)(2x)(2x)(2x)}_{5\text{ factors}}$

We could use the commutative and associative properties to write this product as

$(2 \cdot 2 \cdot 2 \cdot 2 \cdot 2)(x \cdot x \cdot x \cdot x \cdot x)$

or

Note that each factor of the base has been raised to the fifth power.

$2^5 x^5 = 32x^5$

This suggests our fourth property for exponents.

RAISING PRODUCTS TO A POWER

For any nonzero real numbers a and b and any integer m,

$(ab)^m = a^m b^m$

This is our *fourth property of exponents*. Notice that the exponent is distributed over the product.

The use of the fourth property is illustrated in the next example.

EXAMPLE 12 Using Properties of Exponents

Simplify each expression.

(a) $(xy)^5 = x^5 y^5$

(b) $(3a)^4 = 3^4 \cdot a^4 = 81a^4$

We also apply the power rule in simplifying this expression.

(c) $(2p^2 q^3)^3 = 2^3 (p^2)^3 (q^3)^3$
  $= 8p^6 q^9$

Be sure to raise *each factor* to the power -2.

(d) $(x^{-2} y^3)^{-2} = (x^{-2})^{-2} (y^3)^{-2}$
  $= x^4 y^{-6} = \dfrac{x^4}{y^6}$

(e) $(2w^{-2} z^4)^3 = 2^3 (w^{-2})^3 (z^4)^3$
  $= 2^3 w^{-6} z^{12} = \dfrac{8z^{12}}{w^6}$

Check Yourself 12

Simplify each expression.

1. $(ab)^7$ $a^7 b^7$ **2.** $(4p)^3$ **3.** $(3m^4n^2)^2$

4. $(x^5y^{-2})^{-4}$ **5.** $(3a^5b^{-2})^{-2}$

Our fifth (and final) property for exponents can be established in a similar fashion to the fourth property. It deals with a power of quotients rather than the power of a product.

RAISING QUOTIENTS TO A POWER

For any nonzero real numbers a and b and integer m,

$$\left(\frac{a}{b}\right)^m = \frac{a^m}{b^m}$$

This is our fifth property of exponents.

Our next example shows the application of this property.

EXAMPLE 13 Using Properties of Exponents

Simplify each expression.

(a) $\left(\dfrac{x^2}{y}\right)^3 = \dfrac{(x^2)^3}{y^3} = \dfrac{x^6}{y^3}$ Raise the numerator and denominator to power 3.

(b) $\left(\dfrac{2a}{b^3}\right)^4 = \dfrac{(2a)^4}{(b^3)^4}$

$\qquad = \dfrac{2^4a^4}{b^{12}} = \dfrac{16a^4}{b^{12}}$

(c) $\left(\dfrac{a^3}{b^2}\right)^{-2} = \dfrac{(a^3)^{-2}}{(b^2)^{-2}}$ *Raise both the numerator and the denominator to the power -2.*

$\qquad = \dfrac{a^{-6}}{b^{-4}} = \dfrac{b^4}{a^6}$

(d) $\left(\dfrac{2w^{-4}}{z^3}\right)^{-3} = \dfrac{(2w^{-4})^{-3}}{(z^3)^{-3}}$

$\qquad = \dfrac{2^{-3}w^{12}}{z^{-9}} = \dfrac{w^{12}z^9}{2^3} = \dfrac{w^{12}z^9}{8}$

Check Yourself 13

Simplify each expression.

1. $\left(\dfrac{m^3}{n}\right)^4$

2. $\left(\dfrac{3t^2}{s^3}\right)^3$

3. $\left(\dfrac{x^5}{y^{-2}}\right)^{-3}$

4. $\left(\dfrac{3p^{-4}}{q^2}\right)^{-2}$

When a fraction is raised to a negative power (as in Example 12c and d), the work can often be simplified by the following observation.

RAISING QUOTIENTS TO A NEGATIVE POWER

$$\left(\dfrac{a}{b}\right)^{-n} = \dfrac{a^{-n}}{b^{-n}} = \dfrac{b^n}{a^n} = \left(\dfrac{b}{a}\right)^n \qquad a \neq 0,\ b \neq 0$$

This is used in the next example.

EXAMPLE 14 Using Properties of Exponents

Simplify each of the following expressions.

We will not provide a long set of rules about when to use which method. It is better for you to try both techniques and look for clues as to when one might be "better" than the other.

Try both approaches in the Check Yourself 14 that follows.

(a) $\left(\dfrac{3}{q^5}\right)^{-2} = \left(\dfrac{q^5}{3}\right)^2$

$= \dfrac{q^{10}}{9}$

(b) $\left(\dfrac{x^3}{y^4}\right)^{-3} = \left(\dfrac{y^4}{x^3}\right)^3$

$= \dfrac{(y^4)^3}{(x^3)^3} = \dfrac{y^{12}}{x^9}$

Check Yourself 14

Simplify each of the following expressions.

1. $\left(\dfrac{r^4}{5}\right)^{-2}$

2. $\left(\dfrac{a^4}{b^3}\right)^{-3}$

The chart on the next page summarizes the five properties of integer exponents introduced in this section.

General Form $a \neq 0,\ b \neq 0$	Example
1. $a^m a^n = a^{m+n}$	$x^2 \cdot x^3 = x^5$
2. $\dfrac{a^m}{a^n} = a^{m-n}$	$\dfrac{5^7}{5^3} = 5^4$
3. $(a^m)^n = a^{mn}$	$(z^5)^4 = z^{20}$
4. $(ab)^m = a^m b^m$	$(4x)^3 = 4^3 x^3 = 64x^3$
5. $\left(\dfrac{a}{b}\right)^m = \dfrac{a^m}{b^m}$	$\left(\dfrac{2}{3}\right)^6 = \dfrac{2^6}{3^6} = \dfrac{64}{729}$

As you might expect, more complicated expressions require the use of more than one of the properties, for simplification. The next examples illustrate such cases.

EXAMPLE 15 Using Properties of Exponents

Use the properties of exponents to simplify each of the following expressions.

(a) $(2x)^2(2x)^4 = (2x)^{2+4} = (2x)^6$ Product rule
$\qquad\qquad = 2^6 x^6 = 64x^6$ Product to a power

In (a) the exponents outside the parentheses apply to both 2 and x.

(b) $\dfrac{(x^4)^3}{(x^3)^2} = \dfrac{x^{12}}{x^6} = x^{12-6} = x^6$ Power rule, quotient rule

(c) $\dfrac{6a^4b^5}{3a^2b} = 2a^{4-2}b^{5-1} = 2a^2b^4$ Quotient rule, division

(d) $\dfrac{7.5 \times 10^{14}}{2.5 \times 10^3} = \dfrac{7.5}{2.5} \times 10^{14-3}$ Quotient rule

$\qquad\qquad = 3 \times 10^{11}$ Divide.

Check Yourself 15

Simplify each expression.

1. $(3y)^2(3y)^3$

2. $\dfrac{(3a^2)^3}{9a^3}$

3. $\dfrac{25x^3y^4}{5x^2y}$

EXAMPLE 16 Using Properties of Exponents

Simplify each of the following expressions.

Apply the power rule to each factor.

$$(a) \quad \frac{(a^2)^{-3}(a^3)^4}{(a^{-3})^3} = \frac{a^{-6} \cdot a^{12}}{a^{-9}}$$

Apply the product rule.

$$= \frac{a^{-6+12}}{a^{-9}} = \frac{a^6}{a^{-9}}$$

Apply the quotient rule.

$$= a^{6-(-9)} = a^{6+9} = a^{15}$$

It may help to separate the problem into three fractions, one for the coefficients and one for each of the variables.

$$(b) \quad \frac{8x^{-2}y^{-5}}{12x^{-4}y^3} = \frac{8}{12} \cdot \frac{x^{-2}}{x^{-4}} \cdot \frac{y^{-5}}{y^3}$$

$$= \frac{2}{3}x^{-2-(-4)} \cdot y^{-5-3}$$

$$= \frac{2}{3}x^2 \cdot y^{-8} = \frac{2x^2}{3y^8}$$

$$(c) \quad \left(\frac{pr^3s^{-5}}{p^3r^{-3}s^{-2}} \right)^{-2} = (p^{1-3}r^{3-(-3)}s^{-5-(-2)})^{-2}$$

Apply the power rule inside the parentheses.

$$= (p^{-2}r^6s^{-3})^{-2}$$

Apply the rule for a product to a power.

$$= (p^{-2})^{-2}(r^6)^{-2}(s^{-3})^{-2}$$

Apply the power rule.

$$= p^4r^{-12}s^6 = \frac{p^4s^6}{r^{12}}$$

Be Careful! Another possible first step (and generally an efficient one) is to rewrite an expression by using our earlier definitions.

$$a^{-n} = \frac{1}{a^n} \qquad \text{and} \qquad \frac{1}{a^{-n}} = a^n$$

For instance, in Example 16*b*, we would *correctly* write

$$\frac{8x^{-2}y^{-5}}{12x^{-4}y^3} = \frac{8x^4}{12x^2y^3y^5}$$

CAUTION

A *common error* is to write

$$\frac{8x^{-2}y^{-5}}{12x^{-4}y^3} = \frac{12x^4}{8x^2y^3y^5} \qquad \text{This is } not \text{ correct.}$$

The coefficients should not have been moved along with the factors in *x*. Keep in mind that the negative exponents apply *only* to the variables. The coefficients remain *where they were* in the original expression when the expression is rewritten by using this approach.

Check Yourself 16

Simplify each of the following expressions.

1. $\dfrac{(x^5)^{-2}(x^2)^3}{(x^{-4})^3}$

2. $\dfrac{12a^{-3}b^{-2}}{16a^{-2}b^3}$

3. $\left(\dfrac{xy^{-3}z^{-5}}{x^{-4}y^{-2}z^3} \right)^{-3}$

You may have noticed that throughout this section we have frequently used 10 as a base in the examples. You will find that experience useful as we discuss scientific notation.

We begin the discussion with a calculator exercise. On most (scientific) calculators, if you find 2.3 times 1000, the display will read

2300.

Multiply by 1000 a second time. Now you will see

2300000.

Multiplying by 1000 a third time will result in the display

2.3   09

> This must equal 2,300,000,000.

And multiplying by 1000 again yields

2.3   12

Can you see what is happening? This is the way calculators display very large numbers: The number on the left is always between 1 and 10, and the number on the right indicates the number of places the decimal point must be moved to the right to put the answer in standard (or decimal) form.

> Consider the following table:
>
> $$2.3 = 2.3 \times 10^0$$
> $$23 = 2.3 \times 10^1$$
> $$230 = 2.3 \times 10^2$$
> $$2300 = 2.3 \times 10^3$$
> $$23{,}000 = 2.3 \times 10^4$$
> $$230{,}000 = 2.3 \times 10^5$$

This notation is used frequently in science. It is not uncommon, in scientific applications of algebra, to find yourself working with very large or very small numbers. Even in the time of Archimedes (287–212 B.C.), the study of such numbers was not unusual. Archimedes estimated that the universe was 23,000,000,000,000,000 m in diameter.

In scientific notation, his estimate for the diameter of the universe would be

$$2.3 \times 10^{16} \text{ m}$$

In general, we can define scientific notation as follows:

SCIENTIFIC NOTATION

Any number written in the form

$$a \times 10^n$$

where $1 \le a < 10$ and n is an integer, is written in scientific notation.

EXAMPLE 17 Using Scientific Notation

Write each of the following numbers in scientific notation.

(a) 120,000. $= 1.2 \times 10^5$
  5 places   The power is 5.

(b) 88,000,000. $= 8.8 \times 10^7$
  7 places   The power is 7.

(c) 520,000,000. $= 5.2 \times 10^8$
  8 places

> Note the pattern for writing a number in scientific notation.
>
> The exponent on 10 shows the *number of places* we must move the decimal point so that the multiplier will be a number between 1 and 10. A positive exponent tells us to move right, while a negative exponent indicates to move left.
>
> **Note** To convert back to standard or decimal form, the process is simply reversed.

(d) $\underset{\text{9 places}}{\underline{4,000,000,000.}} = 4 \times 10^9$

(e) $\underset{\text{4 places}}{\underline{0.0005}} = 5 \times 10^{-4}$

If the decimal point is to be moved to the left, the exponent will be negative.

(f) $\underset{\text{9 places}}{\underline{0.0000000081}} = 8.1 \times 10^{-9}$

Check Yourself 17

Write in scientific notation. 2.12×10^{17}

1. 212,000,000,000,000,000

2. 5,600,000

3. 0.00079 $7.9 \times 10^{-}$

4. 0.0000007

EXAMPLE 18 An Application of Scientific Notation

(a) Light travels at a speed of 3.05×10^8 meters per second (m/s). There are approximately 3.15×10^7 s in a year. How far does light travel in a year?

We multiply the distance traveled in 1 s by the number of seconds in a year. This yields

$$(3.05 \times 10^8)(3.15 \times 10^7) = (3.05 \cdot 3.15)(10^8 \cdot 10^7)$$

Multiply the coefficients, add the exponents.

$$= 9.6075 \times 10^{15}$$

Note that $9.6075 \times 10^{15} \approx 10 \times 10^{15} = 10^{16}$.

For our purposes we round the distance light travels in 1 year to 10^{16} m. This unit is called a *light-year,* and it is used to measure astronomical distances.

(b) The distance from earth to the star Spica (in Virgo) is 2.2×10^{18} m. How many light-years is Spica from earth?

We divide the distance (in meters) by the number of meters in 1 light-year.

$$\frac{2.2 \times 10^{18}}{10^{16}} = 2.2 \times 10^{18-16}$$

$$= 2.2 \times 10^2 = 220 \text{ light-years}$$

Check Yourself 18

The farthest object that can be seen with the unaided eye is the Andromeda galaxy. This galaxy is 2.3×10^{22} m from earth. What is this distance in light-years?

CHECK YOURSELF ANSWERS

1. (1) $(3z)^3$; (2) x^4 **2.** (1) $(5b)^{11}$; (2) $(-3)^7$; (3) 10^{20}; (4) $(xy)^5$
3. (1) $8a^5b^5$; (2) $6x^7y$; (3) $-6s^5t^5$; (4) $20d^7f^7$ **4.** (1) y^7; (2) x; (3) $-5r^2$;
(4) $7a^5b^4$; (5) 10^8 **5.** (1) 1; (2) 1; (3) 8; (4) -7 **6.** (1) $\dfrac{1}{a^{10}}$; (2) $\dfrac{1}{2^4} = \dfrac{1}{16}$;

(3) $\dfrac{1}{(-4)^2} = \dfrac{1}{16}$; (4) $\dfrac{4}{25}$ **7.** (1) $\dfrac{3}{w^4}$; (2) $\dfrac{10}{x^5}$; (3) $\dfrac{1}{(2y)^4} = \dfrac{1}{16y^4}$; (4) $\dfrac{-5}{t^2}$

8. (1) x^4; (2) $3^3 = 27$; (3) $\dfrac{2a^2}{3}$; (4) $\dfrac{d^7}{c^5}$ **9.** (1) x^4; (2) $y^{-4} = \dfrac{1}{y^4}$; (3) a^4

10. (1) b^{28}; (2) 10^9; (3) b^{11} **11.** (1) $y^{-12} = \dfrac{1}{y^{12}}$; (2) p^{10}; (3) $w^{-30} = \dfrac{1}{w^{30}}$

12. (1) a^7b^7; (2) $64p^3$; (3) $9m^8n^4$; (4) $\dfrac{y^8}{x^{20}}$; (5) $\dfrac{b^4}{9a^{10}}$ **13.** (1) $\dfrac{m^{12}}{n^4}$; (2) $\dfrac{27t^6}{s^9}$;

(3) $\dfrac{1}{x^{15}y^6}$; (4) $\dfrac{q^4p^8}{9}$ **14.** (1) $\dfrac{25}{r^8}$; (2) $\dfrac{b^9}{a^{12}}$ **15.** (1) $243y^5$; (2) $3a^3$; (3) $5xy^3$

16. (1) x^8; (2) $\dfrac{3}{4ab^5}$; (3) $\dfrac{z^{24}y^3}{x^{15}}$ **17.** (1) 2.12×10^{17}; (2) 5.6×10^6;

(3) 7.9×10^{-4}; (4) 7×10^{-7} **18.** 2,300,000 light-years

5.1 EXERCISES

Build Your Skills

Simplify each of the following expressions.

1. $x^4 \cdot x^5$

2. $p^4 \cdot p^7$

3. $y^5 \cdot y^4 \cdot y^3$

4. $z^2 \cdot z^3 \cdot z^4 \cdot z^5$

5. $3^5 \cdot 3^0$

6. $(-2)^7(-2)^4$

7. $(-5)^2(-5)^3(-5)^5$

8. $6^5 \cdot 6^4 \cdot 6^3 \cdot 6^0$

9. $3 \cdot s^4 \cdot s^3 \cdot s^4$

10. $5 \cdot a^6 \cdot a^3 \cdot a^4$

11. x^{-5}

12. 3^{-3}

13. 5^{-2}

14. y^{-8}

15. $(-5)^{-2}$

16. $(-3)^{-3}$

17. $(-2)^{-3}$

18. $(-2)^{-4}$

19. $\left(\dfrac{2}{3}\right)^{-3}$

20. $\left(\dfrac{3}{4}\right)^{-2}$

21. $3x^{-2}$

22. $4w^{-3}$

23. $-5a^{-4}$

24. $(-2b)^{-4}$

25. $(-3x)^{-2}$

26. $-5x^{-2}$

27. $\dfrac{1}{x^{-3}}$

28. $\dfrac{1}{m^{-5}}$

29. $\dfrac{2}{5x^{-3}}$

30. $\dfrac{3}{4w^{-4}}$

31. $\dfrac{x^{-3}}{y^{-4}}$

32. $\dfrac{m^{-5}}{n^{-3}}$

Use the product rule of exponents together with the commutative and associative properties to simplify the following products.

33. $(a^2b^3)(a^4b^2)$

34. $(p^4q)(p^2q^3)$

35. $(x^3y^2)(x^4y^2)(x^2y^3)$

36. $(r^2s^3)(r^3s)(r^4s^0)$

37. $(2x^4)(3x^3)(-4x^3)$

38. $(2m^3)(-3m)(-4m^4)$

39. $(5a^2)(3a^3)(a^0)(-2a^3)$ —

40. $(4s^2)(2s)(s^2)(2s^3)$

$\left(ab\right)^m = a^m b^m$

41. $(5ab^3)(2a^2b)(3ab)$

42. $(-3xy)(5x^2y)(-2x^3y^0)$

43. $(ab^2c)(a^3b^5c)(a^4bc)$ $a^8 b^6 c^3$

44. $(rst)(r^8s^3t^6)(r^2st)(rst^4)$

45. $x^5 \cdot x^{-3}$

46. $y^{-4} \cdot y^5$

47. $a^{-9} \cdot a^6$

48. $w^{-5} \cdot w^3$

49. $z^{-2} \cdot z^{-8}$

50. $b^{-7} \cdot b^{-1}$

51. $a^{-5} \cdot a^5$

52. $x^4 \cdot x^{-4}$

Use the quotient rule of exponents to simplify each expression.

$\left(\dfrac{a}{b}\right)^{-m} = \dfrac{a^m}{b^m}$

53. $\dfrac{x^{10}}{x^7}$

54. $\dfrac{b^{23}}{b^{18}}$

55. $\dfrac{x^7y^{11}}{x^4y^3}$

56. $\dfrac{a^5b^9}{ab^4}$

57. $\dfrac{a^5b^4c^2}{ab^2c^0}$

58. $\dfrac{x^8y^6z^4}{x^3yz^0}$

59. $\dfrac{21s^4t^5}{7st^2}$ $3s^3t^3$

60. $\dfrac{48w^6z^6}{12w^3z}$

61. $\dfrac{x^{-5}}{x^{-2}}$ $^{-3}$ $\dfrac{x^2}{x^5}$

62. $\dfrac{m^{-3}}{m^{-6}}$

Use the first three properties of exponents to simplify each of the following.

63. $(x^5)^3$

64. $(w^4)^6$

65. $(2x^3)(x^2)^4$

66. $(p^4)(3p^3)^2$

67. $(3a^4)(a^3)(a^2)$

68. $(5y^2)(2y)(y^5)$

69. $(x^4y)(x^2)^3(y^3)^0$

70. $(r^4)^2(r^2s)(s^3)^2$

71. $(ab^2c)(a^4)^4(b^2)^3(c^3)^4$

72. $(p^2qr^2)(p^2)(q^3)^2(r^2)^0$

73. $(w^5)^{-3}$

74. $(x^{-2})^{-3}$

75. $(b^{-4})^{-2}$

76. $(a^0b^{-4})^3$

77. $(x^5y^{-3})^2$

78. $(p^{-3}q^2)^{-2}$

79. $(m^{-4}n^{-2})^{-3}$

80. $(3x^{-2}y^{-2})^3$

81. $(2r^{-3}s^0)^{-5}$

82. $\left(\dfrac{a^{-3}}{b^{-2}}\right)^2$

83. $\left(\dfrac{w^{-2}}{z^{-4}}\right)^3$

84. $\left(\dfrac{x^{-3}}{y^2}\right)^{-3}$

85. $\left(\dfrac{m^{-4}}{n^{-2}}\right)^{-2}$

86. $(3x^{-4})^2(2x^2)$

87. $(4y^{-2})^2(3y^{-4})$

Simplify each expression.

88. $(2a^5)^4(a^3)^2$

89. $(3b^2)^3(b^2)^4(b^2)$

90. $(2x^3)^3(3x^3)^2$

91. $(2 \times 10^4)^2 \cdot (1 \times 10^3)^4$

92. $(a^2b^3)^4(ab^3)^0$

93. $(xy^5z)^4(xyz^2)^8(x^6yz)^5$

94. $(a^2b^2c^2)^0(ab^2c)^2(a^3bc^2)$

95. $(3a^2)(5a^2)^2$

96. $(2a^3)^2(a^0)^5$

97. $(2a^3)^4(3a^5)^2$

98. $(3x^3)^2(2x^4)^5$

99. $\left(\dfrac{3a^6}{2b^9}\right)^3$

100. $\left(\dfrac{x^8}{y^6}\right)\left(\dfrac{2y^9}{x^3}\right)$

101. $(-7a^2b)(-3a^5b^6)^4$

102. $\left(\dfrac{2w^5z^3}{3x^3y^9}\right)\left(\dfrac{x^5y^4}{w^4z^0}\right)^2$

103. $(2a^2b^{-3})(3a^{-4}b^{-2})$

104. $(-5m^{-2}n^{-4})(2m^5n^0)$

105. $\dfrac{(y^{-3})^{-2}(y^2)^{-4}}{y^{-3}}$

106. $\dfrac{6p^3q^{-4}}{24p^{-2}q^{-2}}$

107. $\dfrac{15x^{-3}y^2z^{-4}}{20x^{-4}y^{-3}z^2}$

108. $\dfrac{24p^{-5}q^{-3}r^2}{36p^{-2}q^3r^{-2}}$

109. $\left(\dfrac{x^{-5}y^{-7}}{x^0y^{-4}}\right)^3$

110. $\left(\dfrac{xy^3w^{-4}}{x^{-3}y^{-2}w^2}\right)^{-2}$

111. $\left(\dfrac{a^{-2}b^2}{a^3b^{-2}}\right)^{-2}\left(\dfrac{a^{-4}b^2}{a^{-2}b^{-2}}\right)^2$

112. $\left(\dfrac{p^{-3}q^3}{p^{-4}q^2}\right)^3\left(\dfrac{p^{-2}q^{-2}}{pq^4}\right)^{-1}$

Express each number in scientific notation.

113. The distance from the earth to the sun: 93,000,000 mi

114. The diameter of a grain of sand: 0.000021 m

115. The diameter of the sun: 130,000,000,000 cm

116. The number of molecules in 22.4 L of a gas: 602,000,000,000,000,000,000,000 (Avogadro's number)

117. The mass of the sun is approximately 1.98×10^{30} kg. If this were written in standard or decimal form, how many 0s would follow the digit 8?

118. Scientists estimate the mass of our galaxy to be 2.95×10^{41} kg. If this number were written in standard, or decimal, form, how many 0s would follow the digit 5?

119. Archimedes estimated the universe to be 2.3×10^{19} millimeters (mm) in diameter. If this number were written in standard, or decimal, form, how many 0s would follow the digit 3?

Write each of the following numbers in standard notation.

120. 8×10^{-3}

121. 7.5×10^{-6}

122. 2.8×10^{-5}

123. 5.21×10^{-4}

Write each of the following numbers in scientific notation.

124. 0.0005

125. 0.000003

126. 0.00037

127. 0.000051

Compute each of the following, using scientific notation, and write your answer in that form.

128. $(4 \times 10^{-3})(2 \times 10^{-5})$

129. $(1.5 \times 10^{-6})(4 \times 10^2)$

130. $\dfrac{9 \times 10^3}{3 \times 10^{-2}}$

131. $\dfrac{7.5 \times 10^{-4}}{1.5 \times 10^2}$ 7.5×

132. Scientists now estimate the diameter of the observable universe to be 1.3×10^{29} mm. If this number were written in standard, or decimal, form, how many 0s would follow the digit 3?

In the expressions below, perform the indicated calculations. Write your result in scientific notation.

133. $(2 \times 10^5)(4 \times 10^4)$

134. $(2.5 \times 10^7)(3 \times 10^5)$

135. $\dfrac{6 \times 10^9}{3 \times 10^7}$

136. $\dfrac{4.5 \times 10^{12}}{1.5 \times 10^7}$

137. $\dfrac{(3.3 \times 10^{15})(6 \times 10^{15})}{(1.1 \times 10^8)(3 \times 10^6)}$

138. $\dfrac{(6 \times 10^{12})(3.2 \times 10^8)}{(1.6 \times 10^7)(3 \times 10^2)}$

139. Megrez, the nearest of the Big Dipper stars, is 6.6×10^{17} m from earth. Approximately how long does it take light, traveling at 10^{16} m/year, to travel from Megrez to earth?

140. Alkaid, the most distant star in the Big Dipper, is 2.1×10^{18} m from earth. Approximately how long does it take light to travel from Alkaid to earth?

141. The number of liters of water on earth is 15,500 followed by 19 zeros. Write this number in scientific notation.

142. Use the number of liters of water on earth from Exercise 141 to find out how much water is available for each person on earth. The population of the earth is 5.3 billion.

143. If there are 5.3×10^9 people on earth and there is enough freshwater to provide each person with 8.79×10^5 L, how much freshwater is there on earth?

144. The United States uses an average of 2.6×10^6 L of water per person each year. The United States has 2.5×10^8 people. How many liters of water does the United States use each year?

Transcribe Your Skills

145. State the five properties of exponents.

146. What is the purpose of scientific notation?

147. Explain why -5^0 does not equal 1 even though the exponent is 0.

148. Can $(a + b)^{-1}$ be written as $\dfrac{1}{a} + \dfrac{1}{b}$ by using the properties of exponents? If not, why not?

Think About These

Simplify each of the following expressions.

149. $x^{2n} \cdot x^{3n}$

150. $a^{n+1} \cdot a^{3n}$

151. $\dfrac{w^{n+3}}{w^{n+1}}$

152. $\dfrac{r^{n-4}}{r^{n-1}}$

153. $(y^n)^{3n}$

154. $(p^{n+1})^n$

155. $\dfrac{(a^{2n})(a^{n+2})}{a^{3n}}$

156. $\dfrac{(x^n)(x^{3n+5})}{x^{4n}}$

Skillscan (Section 1.1)

Use the distributive property to simplify each expression.

a. $2(3x + 5)$ **b.** $3(4x - 1)$ **c.** $2(2x - 3y)$
d. $8x + 3x$ **e.** $4y^2 + 2y^2$ **f.** $4x + 8 + 3x + 2$

5.2   Polynomials—Addition and Subtraction

TAPE IN12

OBJECTIVES: 1. To use the language of polynomials
                   2. To add and subtract polynomials

In Chapter 1, we studied the properties of the real numbers. Much of that work can be extended to an important class of algebraic expressions called *polynomials*. First, we offer some definitions.

> A *term* is an indicated product or quotient of numbers and variable factors.

EXAMPLE 1 Identifying Terms

Now, $5x^2$, 7, $4xy$, and $\dfrac{9}{x}$ are all terms. But $3x^2 + 2y$ is *not* a term; it is a sum of two terms, $3x^2$ and $2y$.

Check Yourself 1

Which of the following are terms?

1. $4x^3$ **2.** $\dfrac{8}{x}$ **3.** $2 - 3x$ **4.** $5x^2y^2$

Some terms are given special names. For instance,

> A *monomial* is a term in which only *whole numbers* can appear as exponents.

The set of *whole numbers* consists of the natural numbers and 0; so in a monomial, the allowable exponents are 0, 1, 2, 3, and so on.

Now, $4x$, -12, $3ab$, and $5m^3n$ are all monomials. But $\dfrac{9}{x}$ and $5\sqrt{m}$ are *not* monomials because they do not involve whole-number exponents $\left(\text{here the exponents are } -1 \text{ and } \dfrac{1}{2}\right)$.

The numerical factor in each monomial is called the *numerical coefficient.* For the monomials in the preceding paragraph, 4, -12, 3, and 5 are the numerical coefficients.

We often refer to the numerical coefficients as the *coefficients* for convenience.

> A *polynomial* is a monomial or any finite sum (or difference) of monomials.

A polynomial is then a sum of *several terms* which are monomials. That's what the prefix "poly" means.

The expressions

$$5x^2 + 2x \qquad 3m^3 \qquad \text{and} \qquad 5a^2 + 3ab - 2b^2 + 4$$

are all polynomials.

Certain polynomials occur often enough that they are given special names according to the number of terms that they have.

A polynomial with one term is a *monomial.* A polynomial with two terms is a *binomial.* A polynomial with three terms is called a *trinomial.*

The prefix "mono" means 1, "bi" means 2, and "tri" means 3.

EXAMPLE 2 Classifying Polynomials

Which of the following are polynomials? Classify the polynomials as monomial, binomial, or trinomial.

(*a*) $5x^2y$

(*b*) $3m + 5n$

(*c*) $4a^3 + 3a - 2$

(*d*) $5y^2 - \dfrac{2}{x}$

Solution

(*a*) Monomial.
(*b*) Binomial.
(*c*) Trinomial.
(*d*) $5y^2 - \dfrac{2}{x}$ is not a polynomial since the exponent on x is -1.

Remember that

$\dfrac{1}{x} = x^{-1}$

Check Yourself 2

Which of the following are polynomials? Classify the polynomials as monomial, binomial, or trinomial.

1. $5x^2 - 6x$ *Poly*

2. $8x^5$ *mon*

3. $5x^3 - 3xy + 7y^2$ *tri*

4. $9x - \dfrac{3}{x}$ *not at all*

It is also useful to classify polynomials by their *degree*.

> The *degree* of a monomial is the sum of the exponents of the variable factors.

EXAMPLE 3 Determining the Degree of a Monomial

(*a*) $5x^2$ has degree 2.
(*b*) $7n^5$ has degree 5.
(*c*) $4a^2b^4$ has degree 6. (The sum of the powers, 2 and 4, is 6.)
(*d*) 9 has degree 0 (because $9 = 9 \cdot 1 = 9x^0$).

Check Yourself 3

Give the degree of each monomial.

1. $4x^2$ **2.** $7x^3y^2$ **3.** $8p^2s$ **4.** 5

> The *degree* of a polynomial is that of the term with the highest degree.

EXAMPLE 4 Determining the Degree of a Polynomial

(*a*) $7x^3 - 5x^2 + 5$ has degree 3.
(*b*) $5y^7 - 3y^2 + 5y - 7$ has degree 7.
(*c*) $4a^2b^3 - 5ab^2$ has degree 5.

Polynomials such as those in Example 4*a* and *b* are called *polynomials in one variable,* and they are usually written in descending form so that the power of the variable decreases from left to right. In that case, the coefficient of the first term is called the *leading coefficient*.

Check Yourself 4

Give the degree of each polynomial. For those polynomials in one variable, write in descending form and give the leading coefficient.

1. $7x^4 - 5xy + 2$

2. 5

3. $4x^2 - 7x^3 - 8x + 5$

Before we consider the operations of addition and subtraction of polynomials, let's first review two concepts. Recall that two terms that have exactly the same variable or literal factors are called *like terms*.

EXAMPLE 5 Identifying Like Terms

(a) $5x$ and $-3x$ are like terms.
(b) $4x^2y$ and $-x^2y$ are like terms.
(c) $-2ab^2$ and $5a^2b$ are not like terms.

Check Yourself 5

Identify whether each pair of terms is composed of like terms.

1. $2ab$ and $-7ab$ **2.** $3xy$ and $-2xy^2$ **3.** $-x^2y^3$ and $3x^2y^3$

Recall the distributive property from Chapter 1.

DISTRIBUTIVE PROPERTY

For any real numbers *a*, *b*, and *c*,

$$a(b + c) = ab + ac$$

The distributive property also applies to polynomials. Like terms can always be combined by applying the distributive property. That property can also be extended to combine more than two like terms.

EXAMPLE 6 Combining Like Terms

Combine like terms.

(a) $7x + 5x = (7 + 5)x = 12x$
(b) $4a^2b - 2a^2b = (4 - 2)a^2b = 2a^2b$
(c) $9m^3 - 5m^3 + 7m^3 = (9 - 5 + 7)m^3 = 11m^3$

Check Yourself 6

Combine like terms.

1. $7p^2q - 8p^2q$

2. $3xy^3 + 5xy^3 - 7xy^3$

EXAMPLE 7 Adding Polynomials

(*a*) Add $5x^2 + 3x + 7$ and $4x^2 - 5x - 2$.

Using a horizontal format, write

$$(5x^2 + 3x + 7) + (4x^2 - 5x - 2)$$

The second step of regrouping like terms will become unnecessary with practice. You can learn to mentally combine like terms from the original problem.

$= (5x^2 + 4x^2) + (3x - 5x) + (7 - 2)$ Commutative and associative properties

$= 9x^2 - 2x + 5$

(*b*) Add $4x^3 - 2x$ and $3x^2 + 5x$.

Write

$(4x^3 - 2x) + (3x^2 + 5x) = 4x^3 + 3x^2 + (-2x + 5x)$

$= 4x^3 + 3x^2 + 3x$

(*c*) Add $x^3 + 2x^2y - 3xy^2 + y^3$ and $4x^3 - 2x^2y - y^3$.

Write

$$(x^3 + 2x^2y - 3xy^2 + y^3) + (4x^3 - 2x^2y - y^3)$$

$= (x^3 + 4x^3) + (2x^2y - 2x^2y) + (-3xy^2) + (y^3 - y^3)$

$= 5x^3 - 3xy^2$

Check Yourself 7

Add each of the following polynomials.

1. $5x^2 - 3x + 4$ and $-7x^2 + 3x - 4$ **2.** $5x^3 - 4x$ and $8x^2 + 6x$

3. $5x^3 - 3xy^2 + 2x^2y - y^3$ and $-3x^3 - xy^2 - 2x^2y + y^3$

Subtraction of polynomials proceeds in a similar fashion. We view the subtraction of a quantity as adding the opposite of that quantity.

For polynomials, we use the facts that:

Both statements are just applications of the distributive property. This shows us that the *opposite* of $a + b$ is $-a - b$ and that the *opposite* of $a - b$ is $-a + b$.

DISTRIBUTION OF THE NEGATIVE

$-(a + b) = -a - b$

and

$-(a - b) = -a + b$

We can now go on to subtracting polynomials.

EXAMPLE 8 Subtracting Polynomials

(a) Subtract $4x^2 + 5x$ from $7x^2 - 2x$.

Using the horizontal format, write

$(7x^2 - 2x) - (4x^2 + 5x)$

$= 7x^2 - 2x - 4x^2 - 5x$

Note the sign changes.

$= 3x^2 - 7x$

(b) Find the difference of $4y^2 - 2y + 3$ and $5y^2 - 2y + 7$.

The second polynomial is subtracted from the first.

Write

$(4y^2 - 2y + 3) - (5y^2 - 2y + 7)$

$= 4y^2 - 2y + 3 - 5y^2 + 2y - 7$

Note the sign changes.

$= -y^2 - 4$

(c) Find the difference of $6x^3y - 3xy^2 + 5x^2y^2$ and $2x^3y - 2xy^2 - 3x^2y^2$.

Write

$(6x^3y - 3xy^2 + 5x^2y^2) - (2x^3y - 2xy^2 - 3x^2y^2)$

$= 6x^3y - 3xy^2 + 5x^2y^2 - 2x^3y + 2xy^2 + 3x^2y^2$

$= 4x^3y - xy^2 + 8xy^2$

Check Yourself 8

1. Subtract $3a^2 - 5a$ from $2a^2 + 7a$.

2. Find the difference of $8y^2 - 6y - 10$ and $-7y^2 - 6y + 8$.

We have used parentheses as signs of grouping in subtracting polynomials. Other signs of grouping—brackets [ ] and braces { }—are also used in algebra. If one set of grouping symbols is contained within another, we use our earlier rules to remove the *innermost grouping symbol first*. Then we work outward.

EXAMPLE 9 Using the Distributive Property

(a) Simplify $3x - 3[5 - (2x - 4)]$.

Remove the parentheses first.

$3x - 3[5 - (2x - 4)]$

$= 3x - 3[5 - 2x + 4]$ Combine the like terms inside the brackets.

$= 3x - 3[-2x + 9]$ Distribute -3.

$= 3x + 6x - 27$

$= 9x - 27$

(b) Simplify $(4x + 3) - [(2x + 3) - (4x - 7)]$.

Both signs inside the brackets change.

$(4x + 3) - [(2x + 3) - (4x - 7)]$ Remove the parentheses *inside* the brackets.

$= (4x + 3) - [2x + 3 - 4x + 7]$

$= (4x + 3) - [-2x + 10]$

$= 4x + 3 + 2x - 10$

$= 6x - 7$

Check Yourself 9

Simplify $(7a - b) - [-(3a + 2b) - (5a - 3b)]$.

FUNCTION NOTATION

All polynomials are functions, so the notation for functions developed in Section 3.3 can be used to represent polynomials in the following fashion:

$$P(x) = x^3 - 4x^2 + 5x - 6 \quad \text{or} \quad Q(x) = 4x^4 - 7x^3 + 9$$

This notation allows us a very useful shorthand to indicate the numerical value of the polynomial for a given value of the variable. For instance, suppose we want the value of the polynomial $P(x)$ when x has the value of 2. We can simply write

Simply replace x with 2 in $P(x)$ to find $P(2)$—this is read "P of 2."

$$P(2) = (2)^3 - 4(2)^2 + 5(2) - 6$$

$$= 8 - 4(4) + 5(2) - 6$$

The statement

$$= 8 - 16 + 10 - 6$$

$P(2) = -4$

$$= -4$$

says that the value of the polynomial P, when $x = 2$, is -4.

To find the value of $Q(x)$ when $x = -2$, we write

$$Q(-2) = 4(-2)^4 - 7(-2)^3 + 9$$

$$= 4(16) - 7(-8) + 9$$

$$= 64 + 56 + 9$$

$$= 129$$

EXAMPLE 10 Evaluating a Function

If $T(x) = -3x^3 + 2x^2 - 3x + 8$, find $T(-1)$.

$$T(-1) = -3(-1)^3 + 2(-1)^2 - 3(-1) + 8$$

$$= -3(-1) + 2(1) - 3(-1) + 8$$

$$= 3 + 2 + 3 + 8$$

$$= 16$$

Check Yourself 10

If $R(x) = 2x^3 - 4x^2 + 7x - 7$, find each of the following.

1. $R(3)$ **2.** $R(-3)$

CHECK YOURSELF ANSWERS

1. (1) Term; (2) term; (3) not a term; (4) term   **2.** Binomial; (2) monomial; (3) trinomial; (4) not a polynomial   **3.** (1) 2; (2) 5; (3) 3; (4) 0   **4.** (1) 4; (2) 0; (3) 3   **5.** (1) Like terms; (2) not like terms; (3) like terms   **6.** (1) $-p^2q$; (2) xy^3 **7.** (1) $-2x^2$; (2) $5x^3 + 8x^2 + 2x$; (3) $2x^3 - 4xy^2$ **8.** (1) $-a^2 + 12a$; (2) $15y^2 - 18$ **9.** $15a - 2b$ **10.** (1) 32; (2) -118

Build Your Skills

Identify each of the following as a monomial, binomial, or trinomial. Give the degree of each polynomial.

1. $4x - 3$

2. -5

3. $4y^3$

4. $5x^7 - 3$

5. $4x^6 + 2x - 5$

6. $w^2 - 2w^4 + 3$

7. $5 - 2a$

8. x^2y^3z

9. $2m^2n^3 + 3m^2n$

10. $5a^3b - 3a^2b + 7ab$

Add each of the following polynomials.

11. $5a - 3$ and $4a + 7$

12. $7x^2 + 3x$ and $5x^2 - 7x$

13. $5m^2 + 3mn$ and $4mn + 2n^2$

14. $9x^2y - 2xy$ and $3xy + 5xy^2$

15. $8x^2 - 3xy + 5y^2$ and $9x^2 - 5xy - 8y^2$

16. $6r^2 - 5rs - 2s^2$ and $-8r^2 + 6rs + 2s^2$

17. $8x - 3$, $4x - 3$, and $-2x + 5$

18. $-3a^2 + 4a + 2$, $5a - 2$, and $-7a^2 - 5$

Subtract each of the following polynomials.

19. $5x - 3$ from $7x + 10$

20. $8a - 7$ from $5a - 12$

21. $-5y^2 - 2y$ from $7y^2 + 3y$

22. $-3r^2 - 3r$ from $-10r^2 + 3r$

23. $5x^2 - 3x - 2$ from $8x^2 - 5x - 7$

24. $7y^3 - 3y^2 + 7y$ from $5y^3 - 2y^2 - 8y$

25. $8a^2 - 7a$ from $5a^2 + 5$

26. $9z^2 - 3$ from $5z^2 - 7z$

Perform the indicated operations.

27. $(2x - 3) + (4x - 7) - (2x - 3)$

28. $(5a - 2) - (2a + 3) + (7a - 7)$

29. $(5x^2 - 2x + 7) - (2x^2 + 3x + 1) - (4x^2 + 2x + 3)$

30. $(8m^2 - 8m + 5) + (3m^2 + 2m - 7) - (m^2 - 6m + 4)$

31. $(8x^2 - 3) + (5x^2 + 7x) - (7x - 8)$

32. $(9r^2 + 7) - (5r^2 - 8r) - (4r^2 + 3)$

Simplify each of the following expressions by removing the innermost grouping symbols and working outward.

33. $x - [5x - (x - 3)]$

34. $m^2 - [7m^2 - (m^2 + 7)]$

35. $(2x - 3) - [x - (2x + 7)]$

36. $(3y^2 + 5y) - [y^2 - (2y^2 - 3y)]$

37. $2x - \{3x + 2[x - 2(x - 3)]\}$

38. $3r - \{5r - 2[r - 3(r + 4)]\}$

If $P(x) = 3x^3 - 2x^2 + 5x - 3$ and $Q(x) = -x^2 + x - 2$, find the following.

39. $P(2)$

40. $P(-3)$

41. $P(0)$

42. $Q(1)$

43. $Q(3)$

44. $Q(-2)$

45. $P(-2) - Q(-1)$

46. $Q(-2) - P(-2)$

Transcribe Your Skills

47. Explain why the expression $2x^2 + 3x - \dfrac{1}{x}$ is not a polynomial.

48. What is the difference between the degree of a polynomial and the degree of a term?

Think About These

Suppose that revenue is given by the polynomial $R(x)$ and cost is given by the polynomial $C(x)$. Profit $T(x)$ can then be found with the formula

$$T(x) = R(x) - C(x)$$

Find the polynomial representing profit in each of the following.

49. $R(x) = 100x$
$C(x) = 2000 + 50x$

50. $R(x) = 250x$
$C(x) = 5000 + 175x$

51. $R(x) = 100x + 2x^2$
$C(x) = 2000 + 50x + 5x^2$

52. $R(x) = 250x + 5x$
$C(x) = 5000 + 175x + 10x^2$

53. If $P(x) = 3x^2 - 2x + 1$, find $P(a + h) - P(a)$.

54. If $P(x) = 4x^2 + 5x$, find $\dfrac{P(a + h) - P(a)}{h}$.

55. Find the difference when $4x^2 + 2x + 1$ is subtracted from the sum of $x^2 - 2x - 3$ and $3x^2 + 5x - 7$.

56. Subtract $7x^2 + 5x - 3$ from the sum of $2x^2 - 5x + 7$ and $-9x^2 - 2x + 5$.

57. Subtract $8x^2 - 2x$ from the sum of $x^2 - 5x$ and $7x^2 + 5$.

58. Find the difference when $9a^2 - 7$ is subtracted from the sum of $5a^2 - 5$ and $-2a^2 - 2$.

59. The length of a rectangle is 1 cm more than twice its width. Represent the width of the rectangle by w, and write a polynomial to express the perimeter of the rectangle in terms of w. Be sure to simplify your result.

60. One integer is 2 more than twice the first. Another is 3 less than 3 times the first. Represent the first integer by x, and then write a polynomial to express the sum of the three integers in terms of x. Be sure to simplify your result.

Let $P(x) = 2x^3 - 3x^2 + 5x - 5$ and $Q(x) = -x^2 + 2x - 2$. Find each of the following.

61. $P[Q(1)]$
 Hint: First find $Q(1)$. Then evaluate $P(x)$ for that value.

62. $Q[P(1)]$

63. While studying the effects of water pollution on living organisms in the water, a chemist obtains the following expression:

$$(kC - x) - (3kC + 2x)$$

Simplify the expression.

64. While investigating the effect of temperature changes caused by the discharge of warm water from a power plant, a biologist obtains the following expression:

$$(3T^2 - 4T + 12) + (5T^2 + 3T - 15)$$

Simplify the expression.

Graphing Utility

Use your graphing calculator to approximate the value of $f(x)$ for each given x. Express your answer to the nearest integer.

$$f(x) = 12x^5 - 16x^3 + 3x^2 + 5x - 9$$

65. $f(-4)$

66. $f\left(\dfrac{7}{3}\right)$

67. $f(\sqrt{2})$

68. $f(\pi)$

Skillscan (Section 5.1)

Perform the indicated operations.

a. $(2x^3)(3x^2)$ **b.** $(-3y^3)(2y^5)$ **c.** $(4m^2n^3)(8mn^2)$

d. $(-3a^3b^3)(-7a^2b^5)$ **e.** $\dfrac{15w^3}{5w}$ **f.** $\dfrac{-24x^5}{3x^4}$ **g.** $\dfrac{30x^3y^5}{-6x^2y}$

h. $\dfrac{49p^5q^2}{7p^4q}$

Multiplication of Polynomials and Special Products 5.3

OBJECTIVES: 1. To multiply polynomials
2. To square a binomial
3. To find a product of binomials as a difference of squares

TAPE IN12

You have already had experience in multiplying polynomials. In Section 5.1, we stated the first exponent law and used that law to multiply monomials. Let's review.

EXAMPLE 1 Multiplying Monomials

Multiply.

Remember: $a^m a^n = a^{m+n}$

Add exponents.

$(8x^2y)(4x^3y^4) = (8 \cdot 4)(x^{2+3})(y^{1+4})$

Multiply.

Note the use of the commutative and associative properties to "regroup" and "reorder" the factors.

$= 32x^5y^5$

Check Yourself 1

Multiply.

1. $(4a^3b)(9a^3b^2)$ $36\,a^9\,b^3$

2. $(-5m^3n)(7mn^5)$ $-35\,m^4\,n^6$

We want now to extend the process to multiplying other polynomials. The distributive property is the key. The simplest case is the product of a monomial and a polynomial.

EXAMPLE 2 Multiplying a Monomial and a Binomial

Multiply $3x^2$ and $5x^3 - 3x^2 + 4$.

$(3x^2)(5x^3 - 3x^2 + 4)$

Now distribute $3x^2$ over the terms of the trinomial:

$(3x^2)(5x^3 - 3x^2 + 4)$

$= (3x^2)(5x^3) + (3x^2)(-3x^2) + (3x^2)(4)$

$= 15x^5 - 9x^4 + 12x^2$

Check Yourself 2

Multiply.

1. $4a^3(2a^2 - 3a)$ $8a^5 - 12a^4$

2. $-5x^2(x^4 + 2x + 3)$ $-5x^6 - 10x^3 - 15x^2$

The distributive property is also used to multiply a polynomial by a polynomial. To consider the patterns, let's start with the product of two binomials.

EXAMPLE 3 Multiplying Binomials

Multiply $x + 3$ and $2x + 5$.

$(x + 3)(2x + 5)$

We distribute $x + 3$ over the sum $2x + 5$. So

$(x + 3)(2x + 5)$ $\quad 2x^2 + 5x + 6x + 15$

Apply the distributive property again.

$= (x + 3)(2x) + (x + 3)(5)$

$= (x)(2x) + (3)(2x) + (x)(5) + (3)(5)$

$= 2x^2 + 6x + 5x + 15$

$= 2x^2 + 11x + 15$

Foil

Notice that this ensures that each term in the first polynomial is multiplied by each term in the second polynomial.

Check Yourself 3

$6x^2 \quad 14x \ -12x + 28$

Use the distributive property to multiply. $\quad 6x^2 + 2x - 28$

$(2x - 4)(3x + 7)$

Finding the product of two binomials is one of the most common forms of multiplication in algebra. Fortunately a convenient pattern emerges that will allow us to write these products directly. Let's consider another product.

We are given the product

$(3x + 2)(2x - 7)$

This time we distribute $2x - 7$ over the sum $3x + 2$:

The second factor is being distributed over the first.

$(3x + 2)(2x - 7)$

$= 3x(2x - 7) + 2(2x - 7)$

$= (3x)(2x) + (3x)(-7) + (2)(2x) + (2)(-7)$

$= 6x^2 - 21x + 4x - 14$

Before we combine the second and third terms, let's look at the process in more detail.

1. $6x^2$ is the product of the *first* terms of the binomials:

$(3x)(2x) \qquad = 6x^2$

$(3x + 2)(2x - 7)$

2. $-21x$ is the product of the *outer* terms of the binomials:

$$(3x)(-7) \quad = -21x$$

$$(3x + 2)(2x - 7)$$

3. $4x$ is the product of the *inner* terms of the binomials:

$$(2)(2x) = 4x$$

$$(3x + 2)(2x - 7)$$

4. -14 is the product of the *last* terms of the binomials:

$$(2)(-7) = -14$$

$$(3x + 2)(2x - 7)$$

So $(3x + 2)(2x - 7)$ is

$$6x^2 - 21x + 4x - 14 \qquad \text{or} \qquad 6x^2 - 17x - 14$$

This shortened process is called the *FOIL method* for multiplying binomials, and it should allow you to find such products quickly and easily.

Of course, it is called FOIL to help you remember the pattern: First, Outer, Inner, Last.

EXAMPLE 4 Multiplying Binomials

Use the FOIL method to multiply the binomials.

(*a*) $(2x - 3)(5x + 2)$

$$= 10x^2 + 4x - 15x - 6$$
$$\quad\;\; \text{F} \quad\; \text{O} \quad\; \text{I} \quad\; \text{L}$$
$$= 10x^2 - 11x - 6$$

(*b*) $(3m - 7n)(5m - 9n)$

$$= 15m^2 - 27mn - 35mn + 63n^2$$
$$\quad\;\; \text{F} \qquad \text{O} \qquad \text{I} \qquad \text{L}$$
$$= 15m^2 - 62mn + 63n^2$$

(*c*) $(8a^3 - b)(3a^3 + 2b)$

$$= 24a^6 + 16a^3b - 3a^3b - 2b^2$$
$$= 24a^6 + 13a^3b - 2b^2$$

When possible, you can combine the *outer* and *inner* terms mentally and write the product directly.

(*d*) $(5x^2 - y)(3x + 2y^2)$

$$= 15x^3 + 10x^2y^2 - 3xy - 2y^3$$

Here the outer and inner terms *cannot* be combined.

Check Yourself 4

Use the FOIL method to multiply.

1. $(5x - 3y)(3x + 7y)$

2. $(5r^2 + 2s)(3r + s^2)$

Certain products occur frequently enough in algebra that it is worth learning special formulas for dealing with them. Consider these products of two equal binomial factors.

$a^2 + 2ab + b^2$ and
$a^2 - 2ab + b^2$

are called *perfect-square trinomials*.

$$(a + b)^2 = (a + b)(a + b)$$
$$= a^2 + 2ab + b^2 \tag{1}$$
$$(a - b)^2 = (a - b)(a - b)$$
$$= a^2 - 2ab + b^2 \tag{2}$$

We can summarize these statements as follows:

SQUARING A BINOMIAL

The square of a binomial is the sum of (1) the square of the first term, (2) twice the product of the two terms, and (3) the square of the last term.

$(a + b)^2 = a^2 + 2ab + b^2$

and

$(a - b)^2 = a^2 - 2ab + b^2$

EXAMPLE 5 Squaring a Binomial

Find each of the following binomial squares.

Be sure to write out the expansion in detail.

(*a*) $(x + 5)^2 = x^2 + 2(x)(5) + 5^2$

Square of first term   Twice the product of the two terms   Square of last term

$$= x^2 + 10x + 25$$

(*b*) $(2a - 7)^2 = (2a)^2 - 2(2a)(7) + (-7)^2$
$$= 4a^2 - 28a + 49$$

CAUTION

Be Careful! A very common mistake in squaring binomials is to forget the middle term!

$(y + 7)^2$ is not equal to $y^2 + (7)^2$

The correct square is $y^2 + 14y + 49$.

$(b - 6)^2$ is not equal to $b^2 - (6)^2$

The correct square is $b^2 - 12b + 36$. The square of a binomial is *always* a trinomial.

Check Yourself 5

Find each of the following binomial squares.

1. $(x + 8)^2$ **2.** $(3x - 5)^2$

Another special product involves binomials that differ only in sign. It will be extremely important in your work later in this chapter on factoring. Consider the following:

$$(a + b)(a - b) = a^2 - ab + ab - b^2$$
$$= a^2 - b^2$$

> **PRODUCT OF BINOMIALS DIFFERING IN SIGN**
>
> $(a + b)(a - b) = a^2 - b^2$
>
> In words, the product of two binomials that differ only in the signs of their second terms is the difference of the squares of the two terms of the binomials.

EXAMPLE 6 Finding a Special Product

Multiply.

(a) $(x - 3)(x + 3) = x^2 - (3)^2$
$\qquad\qquad\qquad = x^2 - 9$

(b) $(2x - 3y)(2x + 3y) = (2x)^2 - (3y)^2$
$\qquad\qquad\qquad\qquad = 4x^2 - 9y^2$

(c) $(5a + 4b^2)(5a - 4b^2) = (5a)^2 - (4b)^2$
$\qquad\qquad\qquad\qquad\quad = 25a^2 - 16b^2$

CAUTION
The entire term 2x is squared, not just the x.

Check Yourself 6

Find each of the following products.

1. $(y + 5)(y - 5)$ 　　　　　　**2.** $(2x - 3)(2x + 3)$

3. $(4r + 5s^2)(4r - 5s^2)$

Multiplication of binomials is almost always done by the FOIL method or by one of our special product formulas. However, if one (or both) of the factors involves three or more terms, it is often easier to use a vertical format similar to that used in the multiplication of whole numbers. The following example illustrates.

This format ensures that each term of one polynomial multiplies each term of the other.

EXAMPLE 7 Multiplying Polynomials

Multiply $3x^3 - 2x^2 + 5$ and $3x + 2$.

Step 1 $\begin{array}{r} 3x^3 - 2x^2 + 5 \\ 3x + 2 \\ \hline 6x^3 - 4x^2 + 10 \end{array}$ Multiply by 2.

Step 2 $\begin{array}{r} 3x^3 - 2x^2 + 5 \\ 3x + 2 \\ \hline 6x^3 - 4x^2 + 10 \\ 9x^4 - 6x^3 + 15x \\ \hline \end{array}$ Multiply by 3x.

Note that we align like terms in the partial products.

Step 3

$$3x^3 - 2x^2 + 5$$
$$3x + 2$$
$$\overline{6x^3 - 4x^2 + 10}$$
$$9x^4 - 6x^3 + 15x$$
$$\overline{9x^4 - 4x^2 + 15x + 10}$$

Add the partial products.

Check Yourself 7

Find the following product, using the vertical method.

$(4x^3 - 6x + 7)(3x - 2)$

A horizontal approach to the multiplication of Example 7 is also possible by the distributive property. As we see below, we distribute $3x$ over the trinomial and then over 2.

EXAMPLE 8 Multiplying Polynomials

Multiply $(3x + 2)(3x^3 - 2x^2 + 5)$, using a horizontal format.

Again this ensures that each term of one polynomial multiplies each term of the other.

Step 1

$(3x + 2)(3x^3 - 2x^2 + 5)$

Step 2

Combine like terms, and write the product in descending form.

$$= \underbrace{9x^4 - 6x^3 + 15x}_{\text{Step 1}} + \underbrace{6x^3 - 4x^2 + 10}_{\text{Step 2}}$$

$$= 9x^4 - 4x^2 + 15x + 10$$

Check Yourself 8

Find the product of Check Yourself 7, using a horizontal format.

Multiplication sometimes involves the product of more than two polynomials. In such cases, the associative property of multiplication allows us to choose the order of multiplication that we find easiest. Generally we choose to start with the product of binomials. The following example illustrates.

EXAMPLE 9 Multiplying Polynomials

Find the product $(x + 3)(x - 3)$. Then distribute x as the last step.

Find the products.

(a) $x(x + 3)(x - 3) = x(x^2 - 9)$
$$= x^3 - 9x$$

Find the product of the binomials.

(b) $2x(x + 3)(2x - 1) = 2x(2x^2 + 5x - 3)$
$$= 4x^3 + 10x^2 - 6x$$

Distribute $2x$.

Check Yourself 9

Find each of the following products.

1. $m(2m + 3)(2m - 3)$ **2.** $3a(2a + 5)(a - 3)$

CHECK YOURSELF ANSWERS

1. (1) $36a^6b^3$; (2) $-35m^4n^6$ **2.** (1) $8a^5 - 12a^4$; (2) $-5x^6 - 10x^3 - 15x^2$
3. $6x^2 + 2x - 28$ **4.** (1) $15x^2 + 26xy - 21y^2$; (2) $15r^3 + 5r^2s^2 + 6rs + 2s^3$
5. (1) $x^2 + 16x + 64$; (2) $9x^2 - 30x + 25$ **6.** (1) $y^2 - 25$; (2) $4x^2 - 9$;
(3) $16r^2 - 25s^4$ **7.** $12x^4 - 8x^3 - 18x^2 + 33x - 14$ **8.** $12x^4 - 8x^3 -$
$18x^2 + 33x - 14$ **9.** (1) $4m^3 - 9m$; (2) $6a^3 - 3a^2 - 45a$

5.3 EXERCISES

Build Your Skills

Multiply each of the following polynomials.

1. $(4x)(5y)$

2. $(-3m)(5n)$

3. $(6x^2)(-3x^3)$

4. $(5y^4)(3y^2)$

5. $(5r^2s)(6r^3s^4)$

6. $(-8a^2b^5)(-3a^3b^2)$

7. $3x(2x^2 - 3x)$

8. $-5c(2c^2 - d^2)$

9. $5a^2(a^2 - 3a + 1)$

10. $3p^2(2p^3 + 5p^2 - p)$

11. $5x^2y^3(x^2 - 3xy + y^2)$

12. $4m^3n(8m^2n - 2mn + 5mn^2)$

13. $-2ab(a^2 - ab + 3ab^2 + b^3)$

14. $4cd(2c^3 - 3c^2d + 5cd^2 - d^3)$

Multiply, using the FOIL method.

15. $(x + y)(x + 3y)$

16. $(a - 3b)(a - 5b)$

17. $(m - 2n)(m + 7n)$

18. $(p + 7q)(p - 3q)$

19. $(5a - 7b)(5a - 9b)$

20. $(3r - 5s)(7r + 2s)$

21. $(7c - 5d)(3c - 4d)$

22. $(9w + 7v)(3w - 2v)$

23. $(5x^2 - 2y)(3x + 2y^2)$

24. $(6s^2 - 5t^2)(3s^2 - 2t)$

Multiply, using the special product formulas.

25. $(x + 5)^2$

26. $(y - 7)^2$

27. $(2a - 3)^2$

28. $(5s + 3)^2$

29. $(4c - 3d)^2$

30. $(7r - 5s)^2$

31. $(4x + 3y^2)^2$

32. $(3p^3 - 7q)^2$

33. $(x - 3y)(x + 3y)$

34. $(x + 5y)(x - 5y)$

35. $(2p - 3q)(2p + 3q)$

36. $(5c + 3d)(5c - 3d)$

37. $(4a^2 + 3b)(4a^2 - 3b)$

38. $(7u - 6w^2)(7u + 6w^2)$

Multiply, using the vertical format.

39. $(3x - y)(x^2 + 3xy - y^2)$

40. $(5a + b)(a^2 - 3ab + b^2)$

41. $(x - 2y)(x^2 + 2xy + 4y^2)$

42. $(m + 3n)(m^2 - 3mn + 9n^2)$

Multiply each of the following polynomials.

43. $x(x - 3)(x + 1)$

44. $a(a - 4)(a - 2)$

45. $2x(x - 3y)(x + 4y)$

46. $y^2(2y + 1)(2y + 3)$

Transcribe Your Skills

47. Does $(2x + 3)^2 = 4x^2 + 9$? If not, explain why and give the correct answer.

48. Explain the FOIL method.

49. What are the benefits of using the horizontal format in multiplying polynomials?

50. Explain why the products $(x - y)^2$ and $(y - x)^2$ are the same.

Think About These

51. You are given three integers such that the second integer is 3 more than the first and the third integer is 1 less than twice the first. Represent the first integer by x. Then write and simplify the polynomial that represents the product of the three integers.

52. The length of a box is 2 cm more than its width. The height is 3 cm less than twice its width. Represent the width of the box by w. Then write and simplify the polynomial that represents the volume of the box.

If the polynomial $P(x)$ represents the selling price of an object, then the polynomial $R(x)$, where $R(x) = x \cdot P(x)$, is the revenue produced by selling x objects.

53. If $P(x) = 100 - 0.2x$, find $R(x)$. Find $R(50)$.

54. If $P(x) = 250 - 0.5x$, find $R(x)$. Find $R(20)$.

Note that $(28)(32) = (30 - 2)(30 + 2) = 900 - 4 = 896$. Use the difference-of-squares formula to find the following products.

55. $(49)(51)$

56. $(27)(33)$

57. $(34)(26)$

58. $(98)(102)$

59. $(55)(65)$

60. $(56)(64)$

61. A water treatment plant needs to increase the size of its settling tank because of a new housing development. The expression relating the old tank size to the new size is

$$\pi(R + r)(R - r)$$

where r is the original radius of the tank and R is the radius for the new tank. Simplify the expression.

62. A farmer is irrigating a field with a length 10 m more than 3 times its width. If the width is W, the area of the field is $W(3W + 10)$. The farmer wants to increase the dimensions of the field by 20 m each.
(*a*) Write the polynomial describing the area of the larger field.
(*b*) Write and simplify the polynomial describing the difference in areas of the two fields.

Skillscan

List all the factors for each of the following integers.

a. 10   **b.** 18   **c.** 25   **d.** 23   **e.** 28   **f.** 52
g. 100   **h.** 72

5.4   Common Factors and Factoring by Grouping

TAPE IN13

OBJECTIVES: 1. To remove the greatest common factor (GCF) of a polynomial
2. To factor by grouping

Recall that a prime number is any integer greater than 1 that has only itself and 1 as factors. Writing

$15 = 3 \cdot 5$

as a product of prime factors is called the *completely factored form* for 15.

When the integers 3 and 5 are multiplied, the product is 15. We call 3 and 5 the *factors* of 15.

Writing $3 \cdot 5 = 15$ indicates multiplication, but when we write $15 = 3 \cdot 5$, we say we have *factored* 15. In general, factoring is the reverse of multiplication. We can extend this idea to algebra.

From the last section on multiplying polynomials and special products, we know that

$$(2x + 3)(x - 2) = 2x^2 - x - 6$$

Our work here will involve starting with $2x^2 - x - 6$ and finding the factors $2x +$ 3 and $x - 2$. As in arithmetic, this is called *factoring,* and here it means writing a polynomial as a product of other polynomials.

Let's start with the simplest method, *factoring out* (or removing) *the greatest common factor.*

In fact, we will see that factoring out the GCF is the *first method* to try in any of the factoring problems we will discuss.

> The *greatest common factor* (GCF) of a polynomial is the monomial with the highest degree and the largest numerical coefficient that is a factor of each term of the polynomial.

Once the GCF is found, we apply the distributive property to write the original polynomial as a product of the GCF and the polynomial formed by dividing each term by that GCF. The following examples illustrate.

EXAMPLE 1 Factoring Out a Monomial

(*a*) Factor $4x^3 - 12x$.

Note that the numerical coefficient of the GCF is 4 and the variable factor is x (the highest power common to each term). So

$$4x^3 - 12x = 4x \cdot x^2 - 4x \cdot 3$$
$$= 4x(x^2 - 3)$$

Here 4 is the GCF of the numerical coefficients, and the highest common power of x is 1, so the GCF of $4x^3 - 12x$ is $4x$.

(*b*) Factor $6a^3b^2 - 12a^2b^3 + 24a^4b^4$.

Here the GCF is $6a^2b^2$, and we can write

$$6a^3b^2 - 12a^2b^3 + 24a^4b^4$$
$$= 6a^2b^2 \cdot a - 6a^2b^2 \cdot 2b + 6a^2b^2 \cdot 4a^2b^2$$
$$= 6a^2b^2(a - 2b + 4a^2b^2)$$

Here 6 is the GCF of the numerical coefficients, the highest common power of a is 2, and the highest common power of b is 2.

(*c*) Factor $8m^4n^2 - 16m^2n^2 + 24mn^3 - 32mn^4$.

Here the GCF is $8mn^2$, and we have

$$8m^4n^2 - 16m^2n^2 + 24mn^3 - 32mn^4$$
$$= 8mn^2 \cdot m^3 - 8mn^2 \cdot 2m + 8mn^2 \cdot 3n - 8mn^2 \cdot 4n^2$$
$$= 8mn^2(m^3 - 2m + 3n - 4n^2)$$

Notice that in Example 1*b* it is also true that

$$6a^3b^2 - 12a^2b^3 + 24a^4b^4 = 3ab(2a^2b - 4ab^2 + 8a^3b^3)$$

However, this is not in *completely factored form* since we agree that this means factoring out the GCF (that monomial with the largest possible coefficient and degree). In this case we must remove $6a^2b^2$.

Check Yourself 1

Write each of the following in completely factored form.

1. $7x^3y - 21x^2y^2 + 28xy^3$

2. $15m^4n^4 - 5mn^3 + 20mn^2 - 25m^2n^2$

The following example shows a related factoring method that is used in the remainder of this section as we discuss *factoring by grouping*. It is sometimes possible to factor a binomial as the GCF.

EXAMPLE 2 Finding a Common Factor

(*a*) Factor $3x(x + y) + 2(x + y)$.

We see that *the binomial $x + y$ is a common factor* and can be removed:

> Because of the commutative property, the factors can be written in either order.

$3x(x + y) + 2(x + y)$

$= (x + y) \cdot 3x + (x + y) \cdot 2$

$= (x + y)(3x + 2)$

(*b*) Factor $3x^2(x - y) + 6x(x - y) + 9(x - y)$.

We note that here the GCF is $3(x - y)$. Factoring as before, we have

$3(x - y)(x^2 + 2x + 3)$

Check Yourself 2

Completely factor each of the polynomials.

1. $7a(a - 2b) + 3(a - 2b)$

2. $4x^2(x + y) - 8x(x + y) - 16(x + y)$

If the terms of a polynomial have no common factor (other than 1), we can sometimes use a technique called *factoring by grouping*. The following example illustrates the technique.

EXAMPLE 3 Factoring by Grouping Terms

Suppose we want to factor the polynomial

$ax - ay + bx - by$

> You will note that our example has *four terms*. That is the clue for trying the grouping method.

As you can see, the polynomial has no common factors. However, look at what happens if we separate the polynomial into *two groups of two terms*.

$ax - ay + bx - by$

$= \underbrace{ax - ay}_{(1)} + \underbrace{bx - by}_{(2)}$

Now *each* group has a common factor, and we can write the polynomial as

$a(x - y) + b(x - y)$

In this form we can see that $x - y$ is the GCF, and factoring out $x - y$, we get

$a(x - y) + b(x - y) = (x - y)(a + b)$

Check Yourself 3

Use the grouping method to factor.

$x^2 - 2xy + 3x - 6y$

Be particularly careful of your treatment of algebraic signs in applying the method of grouping. Look at our next example.

EXAMPLE 4 Factoring by Grouping Terms

Factor $2x^3 - 3x^2 - 6x + 9$.

Solution We group the polynomial as follows.

$$\underbrace{2x^3 - 3x^2}_{(1)} \underbrace{- 6x + 9}_{(2)}$$
Remove the common factor of -3 from the second two terms.   Note that $9 = (-3)(-3)$.

$= x^2(2x - 3) - 3(2x - 3)$

$= (2x - 3)(x^2 - 3)$

Check Yourself 4

Factor by grouping.

$3y^3 + 2y^2 - 6y - 4$

It may also be necessary to change the order of the terms as they are grouped. Look at the following example.

EXAMPLE 5 Factoring by Grouping Terms

Factor $x^2 - 6yz + 2xy - 3xz$.

Solution Grouping the terms as before, we have

$$\underbrace{x^2 - 6yz}_{(1)} \underbrace{+ 2xy - 3xz}_{(2)}$$

Do you see that we have accomplished nothing because there are no common factors in the first group?

We can, however, rearrange the terms to write the original polynomial as

$$\underbrace{x^2 + 2xy}_{(1)} \underbrace{- 3xz - 6yz}_{(2)}$$

$$= x(x + 2y) - 3z(x + 2y)$$ We can now remove the common factor
$$= (x + 2y)(x - 3z)$$ of $x + 2y$ in group (1) and group (2).

Note It is often true that the grouping can be done in more than one way. The factored form will be the same.

Check Yourself 5

We can write the polynomial of Example 5 as

$$x^2 - 3xz + 2xy - 6yz$$

Factor, and verify that the factored form is the same in either case.

CHECK YOURSELF ANSWERS

1. (1) $7xy(x^2 - 3xy + 4y^2)$; (2) $5mn^2(3m^3n^2 - n + 4 - 5m)$
2. (1) $(a - 2b)(7a + 3)$; (2) $4(x + y)(x^2 - 2x - 4)$ **3.** $(x - 2y)(x + 3)$
4. $(3y + 2)(y^2 - 2)$ **5.** $(x - 3z)(x + 2y)$

5.4 EXERCISES

Build Your Skills

11-33, 41, 43

Completely factor each of the following polynomials.

1. $6x + 9y$

2. $7a - 21b$

3. $4x^2 - 12x$

4. $5a^2 + 25a$

5. $18m^2n + 27mn^2$

6. $24c^2d^3 - 30c^3d^2$

7. $9x^2 - 6x + 3$

8. $8y^2 + 12y - 4$

9. $5x^3 - 15x^2 + 25x$

10. $28r^3 - 21r^2 + 7r$ need 1 for 7.

11. $12m^3n - 6mn + 18mn^2$

12. $18w^2z + 27wz - 36wz^2$

13. $4a^3b^2 - 8a^2b + 12ab^2 - 4ab$

14. $9r^2s^3 + 27r^3s^2 - 6r^2s^2 + 3rs$

15. $x(y - z) + 3(y - z)$

16. $2a(c - d) - b(c - d)$

17. $3(m - n) + 5(m - n)^2$

18. $4(r + 2s) - 3(r + 2s)^2$

19. $5x^2(x - y) - 10x(x - y) + 15(x - y)$

20. $7a^2(a + 2b) + 21a(a + 2b) - 14(a + 2b)$

Factor each of the following polynomials by grouping.

21. $ab - ac + b^2 - bc$

22. $ax + 2a + bx + 2b$

23. $6r^2 + 12rs - r - 2s$

24. $2mn - 4m^2 + 3n - 6m$

25. $ab^2 - 2b^2 + 3a - 6$

26. $r^2s^2 - 3s^2 - 2r^2 + 6$

27. $16a^3 - 4a^2b^2 - 4ab + b^3$

28. $2x^3 + 10x^2y - 7x - 35y$

29. $3x^2 - 2xy + 3x - 2y$

30. $xy - 5y^2 - x + 5y$

Factor each of the following polynomials by grouping. *Hint:* Consider a rearrangement of terms.

31. $x^2 - 10y - 5xy + 2x$

32. $a^2 - 12b + 3ab - 4a$

33. $m^2 - 6n^3 + 2mn^2 - 3mn$

34. $r^2 - 3rs^2 - 12s^3 + 4rs$

35. The Clean Water Act states that storage tanks which might cause groundwater pollution must be sealed against leakage or removed. A company needs to know what the cost will be to seal its cylindrical storage tanks. The formula for the surface area of a cylinder is $A = 2\pi rh + 2\pi r^2$. This formula will be easier to use if the right side is factored. Factor the right-hand side.

36. In studying water pollution applications, the following expression is obtained:

$$ac + ax - bc - bx$$

Factor this expression.

Transcribe Your Skills

37. What is meant by the "completely factored form" of a polynomial?

38. Define the term "greatest common factor" of a polynomial.

39. How can we check to see if the factored form of a polynomial is correct?

40. What is a prime polynomial?

Think About These

Factor each of the following.

Hint: Consider *two* groups of *three* terms.

41. $x^3 - x^2 + 3x + x^2y - xy + 3y$

42. $m^3 - m^2 - 4m + 2m^2n - 2mn - 8n$

Hint: Consider *three* groups of *two* terms.

43. $a^3 - a^2b - 3a^2 + 3ab + 3a - 3b$

44. $r^3 + 2r^2s + r^2 + 2rs - 3r - 6s$

Skillscan (Section 5.3)

Multiply.

a. $(x - 2)(x + 2)$ **b.** $(2a + 3)(2a - 3)$

c. $(5m + n)(5m - n)$ **d.** $(3c - 5d^2)(3c + 5d^2)$

e. $(x + 1)(x^2 - x + 1)$ **f.** $(w - 2)(w^2 + 2w + 4)$

g. $(a - b)(a^2 + ab + b^2)$

h. $(2p + q)(4p^2 - 2pq + q^2)$

Factoring Special Polynomials   5.5

OBJECTIVES: 1. To factor the difference of two squares
2. To factor the sum and difference of two cubes

TAPE IN14

We continue our work on factoring with a discussion of factoring polynomials with special forms. Certain patterns occur frequently enough that it is worth developing special techniques to work with them.

In Section 5.3, we looked at some special products in multiplication. Let's go back now and look at those patterns again, this time from a factoring standpoint. Recall that

$$(a + b)(a - b) = a^2 - b^2$$

Now to use the same formula in factoring, we can write the following:

The product of the sum and difference of two terms gives the *difference of two squares*.

THE DIFFERENCE OF TWO SQUARES

$$a^2 - b^2 = (a + b)(a - b) \tag{1}$$

Formula (1) is easy to apply in factoring. It is just a matter of recognizing a binomial as the difference of two squares.

EXAMPLE 1 Factoring the Difference of Two Squares

Note that our example has two terms—a clue to try factoring as the difference of two squares.

We are looking for perfect squares—the exponents must be multiples of 2 and the coefficients perfect squares—1, 4, 9, 16, and so on.

(a) Factor $x^2 - 25$.

$$x^2 - 25 = (x)^2 - (5)^2$$
$$= (x + 5)(x - 5)$$

(b) Factor $9a^2 - 16$.

$$9a^2 - 16 = (3a)^2 - (4)^2$$
$$= (3a + 4)(3a - 4)$$

(c) Factor $25m^4 - 49n^2$.

$$25m^4 - 49n^2 = (5m^2)^2 - (7n)^2$$
$$= (5m^2 + 7n)(5m^2 - 7n)$$

In Example 1 we factored the difference of two squares. What about the sum of two squares, such as

$$x^2 + 25$$

CAUTION

In general, it is *not possible* to factor (using real numbers) a sum of two squares. So

$$(x^2 + 25) \neq (x + 5)(x + 5)$$

Check Yourself 1

Factor each of the following binomials.

1. $y^2 - 36$
$(y + 6)(y - 6)$

2. $25m^2 - n^2$
$(5m + n)(5m - n)$

3. $16a^4 - 9b^2$
$(4a^2 + 3b)(4a^2 - 3b)$

We mentioned earlier that factoring out a common factor should always be considered as your first step. Then other techniques may be obvious. Consider the following example.

EXAMPLE 2 Factoring the Difference of Two Squares

Factor $a^3 - 16ab^2$.

Solution First note the common factor of a. Removing that factor, we have

$$a^3 - 16ab^2 = a(a^2 - 16b^2)$$

We now see that the binomial factor is a difference of squares, and we can continue to factor as before. So

$$a^2 - 16ab^2 = a(a + 4b)(a - 4b)$$

Check Yourself 2

Factor $2x^3 - 18xy^2$.

You may also have to apply the difference-of-squares formula *more than once* to completely factor a polynomial.

EXAMPLE 3 Factoring the Difference of Two Squares

Factor $m^4 - 81n^4$.

$$m^4 - 81n^4 = (m^2 + 9n^2)(m^2 - 9n^2)$$

Do you see that we are not done in this case? Since $m^2 - 9n^2$ is still factorable, we can continue to write

$$m^4 - 81n^4 = (m^2 + 9n^2)(m + 3n)(m - 3n)$$

in completely factored form.

Note The other binomial factor is $m^2 + 9n^2$, a *sum of two squares*. It cannot be factored further.

Check Yourself 3

Factor $x^4 - 16y^4$.

Two additional formulas are available for factoring certain binomials.

> **THE SUM AND DIFFERENCE OF TWO CUBES**
>
> $a^3 + b^3 = (a + b)(a^2 - ab + b^2)$ **(2)**
>
> $a^3 - b^3 = (a - b)(a^2 + ab + b^2)$ **(3)**

Of course, you can use multiplication to verify both formulas. The use of the formulas is quite similar to the use of the difference-of-squares formula we saw earlier.

See Exercises 49 and 50 at the end of this section.

EXAMPLE 4 Factoring the Sum or Difference of Two Cubes

(*a*) Factor $x^3 + 27$.

$$x^3 + 27 = (x)^3 + (3)^3$$

The first term is the cube of x, and the second is the cube of 3, so we can apply Equation (2). Letting $a = x$ and $b = 3$, we have

$$x^3 + 27 = (x + 3)(x^2 - 3x + 9)$$

We are now looking for perfect cubes—the exponents must be multiples of 3 and the coefficients perfect cubes—1, 8, 27, 64, and so on.

Note again that our example has *two terms*—this leads us to consider the sum or difference of two cubes as a factoring method.

(*b*) Factor $8w^3 - 27z^3$.

$8w^3 - 27z^3 = (2w)^3 - (3z)^3$ This is a difference of cubes, so use formula (3).

$$= (2w - 3z)[(2w)^2 + (2w)(3z) + (3z)^2]$$
$$= (2w - 3z)(4w^2 + 6wz + 9z^2)$$

(*c*) Factor $5a^3b - 40b^4$.

Again, looking for a *common factor* should be your first step.

First note the common factor of $5b$. So

$5a^3b - 40b^4 = 5b(a^3 - 8b^3)$ The binomial is the difference of cubes.

$$= 5b(a - 2b)(a^2 + 2ab + 4b^2)$$

Remember to write the GCF as a part of the final factored form.

Check Yourself 4

Factor completely.

1. $27x^3 + 8y^3$ **2.** $3a^4 - 24ab^3$

CHECK YOURSELF ANSWERS

1. (1) $(y + 6)(y - 6)$;   (2) $(5m + n)(5m - n)$;   (3) $(4a^2 + 3b)(4a^2 - 3b)$
2. $2x(x + 3y)(x - 3y)$ **3.** $(x^2 + 4y^2)(x + 2y)(x - 2y)$
4. (1) $(3x + 2y)(9x^2 - 6xy + 4y^2)$; (2) $3a(a - 2b)(a^2 + 2ab + 4b^2)$

5.5 EXERCISES

Build Your Skills

Factor each binomial completely.

1. $x^2 - 49$

2. $m^2 - 64$

3. $a^2 - 81$

4. $b^2 - 36$

5. $9p^2 - 1$

6. $4x^2 - 9$

7. $25a^2 - 16$

8. $16m^2 - 49$

9. $x^2y^2 - 25$

10. $m^2n^2 - 9$

11. $4c^2 - 25d^2$

12. $9a^2 - 49b^2$

13. $49p^2 - 64q^2$

14. $25x^2 - 36y^2$

15. $x^4 - 16y^2$

16. $a^2 - 25b^4$

17. $a^3 - 4ab^2$

18. $9p^2q - q^3$

19. $2b^2c - 18c^3$

20. $3r^3 - 27rs^2$

21. $a^4 - 16b^4$

22. $81x^4 - y^4$

23. $x^3 + 64$ sum and different

24. $y^3 - 8$ of two cubes

25. $m^3 - 125$

26. $b^3 + 27$

27. $a^3b^3 - 27$

28. $p^3q^3 - 64$

29. $8w^3 + z^3$

30. $c^3 - 27d^3$

31. $r^3 - 64s^3$

32. $125x^3 + y^3$

33. $8x^3 - 27y^3$

34. $64m^3 - 27n^3$

35. $8x^3 + y^6$

36. $m^6 - 27n^3$

37. $4x^3 - 32y^3$

38. $3a^3 + 81b^3$

39. $5m^4 + 40mn^3$

40. $2p^3q - 54q^4$

Go over.

Factor each polynomial completely. *Hint:* Try factoring by grouping as your first step.

41. $x^3 + 3x^2 - 4x - 12$

42. $a^3 - 5a^2 - 9a + 45$

43. $8x^3 + 12x^2 - 2x - 3$

44. $18b^3 - 9b^2 - 2b + 1$

45. $9a^3 + 27a^2 - 4a - 12$

46. $4m^3 + 12m^2 - 25m - 75$

47. $x^4 - x^3y + xy^3 - y^4$

48. $a^4 + a^3b - 8ab^3 - 8b^4$

49. Verify the formula for factoring the sum of two cubes by finding the product $(a + b)(a^2 - ab + b^2)$.

50. Verify the formula for factoring the difference of two cubes by finding the product $(a - b)(a^2 + ab + b^2)$.

Transcribe Your Skills

51. What are the characteristics of a monomial that is a perfect cube?

52. Suppose you factored the polynomial $4x^2 - 16$ as follows:

$4x^2 - 16 = (2x + 4)(2x - 4)$

Would this be in completely factored form? If not, what would be the final form?

Think About These

53. Completely factor $x^6 - y^6$ by considering the binomial as the difference of squares

$(x^3)^2 - (y^3)^2$

54. Completely factor $x^6 - y^6$ by considering the binomial as the difference of cubes

$(x^2)^3 - (y^2)^3$

55. Compare your results in Exercises 53 and 54. What can you conclude about the factors of

$x^4 + x^2y^2 + y^4$

56. In the drawing below, squares of length $\dfrac{y}{2}$ are cut from the corners of a sheet of cardboard with sides of length x.

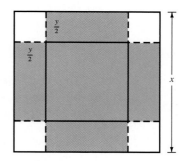

Show that the shaded area can be represented by $x^2 - y^2$.

57. Show that the shaded area in Exercise 56 can also be represented by

$(x - y)^2 + 2y(x - y)$

58. Factor and simplify your result in Exercise 57. Compare your result with that in Exercise 56. What factoring technique does this verify?

Skillscan (Section 5.3)

Multiply.

a. $(x - 3)(x + 5)$ **b.** $(m + 5)(m + 8)$

c. $(2a - 3)(a - 5)$ **d.** $(3d + 2)(2d - 5)$

e. $(5w + 3z)(w - z)$ **f.** $(7c - d)(3c + 2d)$

g. $m(3m - 2)(m + 1)$ **h.** $2y(y - 3)(2y - 7)$

| 5.6 | **Factoring Trinomials** |

TAPE IN14

OBJECTIVE: To factor trinomials

Recall that the product of two binomials may be a trinomial of the form

$$ax^2 + bx + c$$

This suggests that some trinomials may be factored as the product of two binomials. And, in fact, factoring trinomials in this way is probably the most common type of factoring that you will encounter in algebra. One process for factoring a trinomial into a product of two binomials is called *trial and error*.

As before, let's introduce the factoring technique with an example from multiplication. Consider

$$(x + 3)(x + 4) = x^2 + 4x + 3x + 12$$
$$= x^2 + 7x + 12$$

Product of first terms, x and x

Sum of inner and outer products, $3x$ and $4x$

Product of last terms, 3 and 4

To reverse the multiplication process to one of factoring, we see that the product of the *first* terms of the binomial factors is the *first* term of the given trinomial, the product of the *last* terms of the binomial factors is the *last* term of the trinomial, and the *middle* term of the trinomial must equal the sum of the *outer* and *inner* terms of those factors. That leads us to the following sign patterns in factoring a trinomial.

FACTORING TRINOMIALS

		Factoring Sign Pattern
$x^2 + bx + c$	Both signs are positive.	$(x + \ )(x + \ )$
$x^2 - bx + c$	The constant is positive, and the x coefficient is negative.	$(x - \ )(x - \ )$
$x^2 + bx - c$ or $x^2 - bx - c$	The constant is negative.	$(x + \ )(x - \ )$

Given the above information let's work through an example.

EXAMPLE 1 Factoring a Trinomial

Factor $x^2 + 7x + 10$.

Solution The desired sign pattern is

$$(x + \quad)(x + \quad)$$

From the constant, 10, and the x coefficient of our original trinomial, 7, for the second terms of the binomial factors, we want two numbers whose product is 10 and whose sum is 7.

Consider the following:

Factors of 10	Sum
1, 10	11
2, 5	7

We can see that the correct factorization is

$x^2 + 7x + 10 = (x + 2)(x + 5)$

With practice, you will do much of this work mentally. We show the factors and their sums here, and in later examples, to emphasize the process.

Note To check, multiply the factors by using the method of Section 5.3.

Check Yourself 1

Factor $x^2 + 8x + 15$.

$x + 3x + 5)$

$(x + 3)(x + 5)$

The next example shows how to factor a polynomial with a positive constant term and a negative middle term.

EXAMPLE 2 Factoring a Trinomial

Factor $x^2 - 9x + 14$.

Solution Do you see that the sign pattern must be as follows?

$(x - \quad)(x - \quad)$

We then want two factors of 14 whose sum is -9.

Factors of 14	Sum
$-1, -14$	-15
$-2, -7$	-9

Here we use two negative factors of 14 since the coefficient of the x term is negative while the constant is positive.

Since the desired middle term is $-9x$, the correct factors are

$x^2 - 9x + 14 = (x - 2)(x - 7)$

Check Yourself 2

Factor $x^2 - 12x + 32$.

$(x - 8)(x + 4)$

Let's turn now to applying our factoring technique to a trinomial whose constant term is negative. Consider the following example.

EXAMPLE 3 Factoring a Trinomial

Factor $x^2 + 4x - 12$.

Solution In this case, the sign pattern is

Since the constant is now negative, the signs in the binomial factors must be *opposite*.

$(x - \quad)(x + \quad)$

Here we want two numbers whose product is -12 and whose sum is 4. Again, let's look at the possible factors:

Factors of -12	Sum
1, -12	-11
$-1, 12$	11
3, -4	-1
$-3, 4$	1
2, -6	-4
$-2, 6$	4

From the information above, we see that the correct factors are

$$x^2 + 4x - 12 = (x - 2)(x + 6)$$

Check Yourself 3

Factor $x^2 - 7x - 18$.

Thus far we have considered only trinomials of the form $x^2 + bx + c$. Suppose that the leading coefficient is *not* 1. In general, to factor the trinomial $ax^2 + bx + c$ (with $a \neq 1$), we must consider binomial factors of the form

$(x + \quad)(x + \quad)$

where one or both of the coefficients of x in the binomial factors are greater than 1. Again, let's look at a multiplication example for some clues to the technique. Consider

$$(2x + 3)(3x + 5) = 6x^2 + 19x + 15$$

Product of $2x$ and $3x$ Sum of outer and inner products, $10x$ and $9x$ Product of 3 and 5

Now, to reverse the process to factoring, we can proceed as in the following example.

EXAMPLE 4 Factoring a Trinomial

To factor $5x^2 + 9x + 4$, we must have the pattern

Both binomials must have positive signs. Why?

$(x + \quad)(x + \quad)$ This product must be 4.

This product must be 5.

Factors of 5	Factors of 4
1, 5	1, 4
	2, 2

Therefore the possible binomial factors are

$(x + 1)(5x + 4)$

$(x + 4)(5x + 1)$

$(x + 2)(5x + 2)$

The sum of the outer and inner products must be $9x$.

Checking the middle terms of each product, we see that the proper factorization is

$5x^2 + 9x + 4 = (x + 1)(5x + 4)$

Check Yourself 4

Factor $6x^2 - 17x + 7$.

$\left(2x - 10\right)\left(3x + 2\right)$

The sign patterns discussed earlier remain the same when the leading coefficient is not 1. Look at the following example involving a trinomial with a negative constant.

EXAMPLE 5 Factoring a Trinomial

Factor $6x^2 + 7x - 3$.

Solution The sign patterns are

$(\ \ x + \ \ )(\ \ x - \ \ )$

Factors of 6	Factors of −3
1, 6	1, −3
2, 3	−1, 3

There are eight possible binomial factors:

Factors	Middle Term (I + O)
$(x + 1)(6x - 3)$	$6x - 3x = 3x$
$(x - 1)(6x + 3)$	$-6x + 3x = -3x$
$(x + 3)(6x - 1)$	$18x - x = 17x$
$(x - 3)(6x + 1)$	$-18x + x = -17x$
$(2x + 1)(3x - 3)$	$3x - 6x = -3x$
$(2x - 1)(3x + 3)$	$-3x + 6x = 3x$
$(3x + 1)(2x - 3)$	$2x - 9x = -7x$
$(3x - 1)(2x + 3)$	$-2x + 9x = 7x$

Again, as the number of factors for the first coefficient and the constant increase, the number of possible factors becomes larger. Can we reduce the search? One clue: If the trinomial has no common factors (other than 1), then a binomial factor can have no common factor. This means that $6x - 3$, $6x + 3$, $3x - 3$, and $3x + 3$ need not be considered. They are shown here to completely illustrate the possibilities.

Again, checking the middle terms, we find the correct factors:

$$6x^2 + 7x - 3 = (3x - 1)(2x + 3)$$

Factoring certain trinomials in more than one variable involves similar techniques, as is illustrated below.

Check Yourself 5

Factor $10x^2 - 7x - 12$.

EXAMPLE 6 Factoring a Trinomial

Factor $4x^2 - 16xy + 7y^2$.

Solution From the first term of the trinomial we see that possible first terms for our binomial factors are $4x$ and x or $2x$ and $2x$. The last term of the trinomial tells us that the only choices for the last terms of the binomial factors are y and $7y$. So given the sign of the middle and last terms, the only possible factors are

Find the middle term of each product.

Factors	Middle Term
$(4x - 7y)(x - y)$	$-7xy - 4xy = -11xy$
$(4x - y)(x - 7y)$	$-xy - 28xy = -29xy$
$(2x - 7y)(2x - y)$	$-14xy - 2xy = -16xy$

From the middle term of the original trinomial we see that $2x - 7y$ and $2x - y$ are the correct factors.

Check Yourself 6

Factor $6a^2 + 11ab - 10b^2$.

Recall our earlier comment that the *first step* in any factoring problem is to remove any existing common factors. As before, it may be necessary to combine common-term factoring with other methods (such as factoring a trinomial into a product of binomials) to completely factor a polynomial. Look at the following example.

EXAMPLE 7 Factoring a Trinomial

(*a*) Factor $2x^2 - 16x + 30$.

First note the common factor of 2. So we can write

"Remove" the common factor of 2.

$$2x^2 - 16x + 30 = 2(x^2 - 8x + 15)$$

Now, as the second step, examine the trinomial factor. By our earlier methods we know that

$$x^2 - 8x + 15 = (x - 3)(x - 5)$$

and we have

$$2x^2 - 16x + 30 = 2(x - 3)(x - 5)$$

in completely factored form.

(b) Factor $6x^3 + 15x^2y - 9xy^2$.

As before, note the common factor of $3x$ in each term of the trinomial. Removing that common factor, we have

$$6x^3 + 15x^2y - 9xy^2 = 3x(2x^2 + 5xy - 3y^2)$$

Again, considering the trinomial factor, we see that $2x^2 + 5xy - 3y^2$ has factors of $2x - y$ and $x + 3y$. And our original trinomial becomes

$$3x(2x - y)(x + 3y)$$

in completely factored form.

Check Yourself 7

Factor each of the following.

1. $9x^2 - 39x + 36$ **2.** $24a^3 + 4a^2b - 8ab^2$

Consider the trinomial $x^2 + 8x + 16$. Using the methods developed in this section, we can factor this trinomial as follows:

$$x^2 + 8x + 16 = (x + 4)(x + 4) = (x + 4)^2$$

Note that the final result is the square of a binomial. We say that $x^2 + 8x + 16$ is a *perfect-square trinomial.*

The *trial-and-error* process can be used to factor a perfect-square trinomial, but a faster technique is possible if we can recognize a trinomial as a perfect square.

A trinomial is a perfect square if

1. The first and third terms are squares such as a^2 and b^2.

2. The middle term is $2ab$.

3. The signs of the first and third terms are positive.

The patterns for factoring a perfect-square trinomial are as follows:

PERFECT-SQUARE TRINOMIALS

$$a^2 + 2ab + b^2 = (a + b)^2$$
$$a^2 - 2ab + b^2 = (a - b)^2$$

EXAMPLE 8 Factoring a Trinomial

(*a*) Factor $x^2 - 10x + 25$.

The sign of the middle term in the trinomial matches the sign in the binomial.

$$x^2 - 10x + 25 = (x)^2 - 2(5)(x) + (5)^2$$
$$= (x - 5)(x - 5)$$
$$= (x - 5)^2$$

(*b*) Factor $9x^2 + 24x + 16$.

$$9x^2 + 24x + 16 = (3x)^2 + 2(3x)(4) + (4)^2$$
$$= (3x + 4)^2$$

Check Yourself 8

Factor.

1. $4x^2 - 20x + 25$ **2.** $16x^2 + 24x + 9$

One Final Note When factoring, we require that all coefficients be integers. Given this restriction, not all polynomials are factorable over the integers. The following illustrates.

To factor $x^2 - 9x + 12$, we know that the only possible binomial factors (using integers as coefficients) are

$(x - 1)(x - 12)$

$(x - 2)(x - 6)$

$(x - 3)(x - 4)$

You can easily verify that *none* of these pairs gives the correct middle term of $-9x$. We then say that the original trinomial is not factorable by using integers as coefficients.

CHECK YOURSELF ANSWERS

1. $(x + 3)(x + 5)$ **2.** $(x - 4)(x - 8)$ **3.** $(x - 9)(x + 2)$
4. $(2x - 1)(3x - 7)$ **5.** $(2x - 3)(5x + 4)$ **6.** $(3a - 2b)(2a + 5b)$
7. (1) $3(x - 3)(3x - 4)$; (2) $4a(3a + 2b)(2a - b)$ **8.** (1) $(2x - 5)^2$;
(2) $(4x + 3)^2$

5.6 EXERCISES

7-45

Build Your Skills

Completely factor each of the following trinomials.

1. $x^2 + 7x + 12$

2. $a^2 + 9a + 20$

3. $b^2 - 9b + 8$

4. $m^2 - 11m + 10$

5. $y^2 - 15y + 50$

6. $x^2 - 13x + 40$

7. $m^2 + 7m - 30$

8. $a^2 - 7a - 18$

9. $x^2 - 10x + 24$

10. $r^2 + 13r - 30$

11. $a^2 - 7a - 44$

12. $y^2 - 15y - 54$

13. $x^2 + 8xy + 15y^2$

14. $a^2 - 9ab + 20b^2$

15. $m^2 - 16mn + 55n^2$

16. $p^2 - 9pq - 22q^2$

17. $3x^2 + 11x - 20$

18. $2b^2 + 9b - 18$

19. $5x^2 + 18x - 8$

20. $3m^2 - 20m - 7$

21. $12y^2 + 23y + 5$ help

22. $8x^2 + 30x + 7$

23. $4x^2 + 20x + 25$

24. $9x^2 - 24x + 16$

25. $5a^2 + 19a - 30$

26. $3p^2 + 17p - 28$

27. $5b^2 + 24b - 36$

28. $3a^2 - 14a - 24$

29. $10x^2 - 7x - 12$ check

30. $6y^2 + 5y - 21$

31. $16y^2 + 40y + 25$

32. $18x^2 + 45x + 7$

33. $7r^2 - 17rs + 6s^2$ check

34. $5a^2 + 17ab - 12b^2$

35. $8x^2 - 30xy + 7y^2$

36. $8c^2 - 14cd - 15d^2$

37. $3x^2 - 24x + 45$

38. $2a^2 + 10a - 28$

39. $2n^2 - 26n + 72$

40. $3y^2 + 39y + 120$

41. $6x^3 - 31x^2 + 5x$

42. $8a^3 + 25a^2 + 3a$

43. $5b^3 + 14b^2 - 24b$

44. $3x^4 + 17x^3 - 28x^2$

45. $3m^3 - 15m^2n - 18mn^2$

46. $2a^3 - 10a^2b - 72ab^2$

Transcribe Your Skills

47. Outline all the steps required to completely factor a trinomial.

48. Explain why a positive constant term and a negative linear term of a trinomial such as $x^2 - 3x + 2$ indicate that the signs of the binomial factors are both negative.

49. Describe the characteristics of a perfect-square trinomial.

50. Describe the trial-and-error process of factoring a trinomial.

Think About These

Certain trinomials in *quadratic form* can be factored with similar techniques. For instance, we can factor

$$x^4 - 5x^2 - 6 \qquad \text{as} \qquad (x^2 - 6)(x^2 + 1)$$

Apply a similar method to completely factor each of the following polynomials.

51. $x^4 + 3x^2 + 2$

52. $a^4 - 7a^2 + 10$

53. $m^4 - 8m^2 - 33$

54. $p^4 + 5p^2 - 14$

55. $y^6 - 2y^3 - 15$

56. $m^6 + 10m^3 + 21$

57. $x^5 - 6x^3 - 16x$

58. $b^6 - 8b^4 + 15b^2$

59. $a^4 - 5a^2 - 36$

60. $y^4 - 5y^2 + 4$

61. $y^6 - 6y^3 - 16$

62. $x^6 - 2x^3 - 3$

Skillscan (Section 5.3)

Multiply.

a. $2x^2(x^2 + 3x - 5)$ **b.** $(5a - 3)(2a + 4)$

c. $(5m - 3n)(5m + 3n)$ **d.** $(x - 2y)(x^2 + 2xy + 4y^2)$

e. $(2w - 3z)(5w - z)$ **f.** $x(x + 5y)(x - 5y)$

g. $(a + 3b)(a^2 - 3ab + 9b^2)$ **h.** $2s(3s - r)(2s + r)$

5.7 Strategies in Factoring

OBJECTIVE: To recognize types of polynomials and apply the proper factoring strategy

In the past three sections you have seen a variety of techniques for factoring polynomials. This section reviews those techniques and presents some guidelines for choosing an appropriate strategy or a combination of strategies.

1. Always look for a greatest common factor. If you find a GCF (other than 1), factor out the GCF as your first step.

    To factor $5x^2y - 10xy + 25xy^2$, the greatest common factor is $5xy$, so

    $$5x^2y - 10xy + 25xy^2 = 5xy(x - 2 + 5y)$$

2. Now look at the number of terms in the polynomial you are trying to factor.

    a. If the polynomial is a *binomial,* consider the difference-of-squares and the sum- or difference-of-cubes formulas. Recall that a sum of squares will not, in general, factor over the real numbers.

    (i) To factor $x^2 - 49y^2$, recognize the difference of squares, so

    $$x^2 - 49y^2 = (x + 7y)(x - 7y)$$

    (ii) To factor $a^3 - 64m^3$, recognize the difference of cubes, so

    $$a^3 - 64m^3 = (a - 4m)(a^2 + 4am + 16m^2)$$

    (iii) To factor $8p^3 + 27q^3$, recognize the sum of cubes, so

    $$8p^3 + 27q^3 = (2p + 3q)(4p^2 - 6pq + 9q^2)$$

    b. If the polynomial is a *trinomial,* try to factor the trinomial as a product of two binomials, using the trial-and-error method.

    To factor $2x^2 - x - 6$, a consideration of possible factors of the first and last terms of the trinomial will lead to

    $$2x^2 - x - 6 = (2x + 3)(x - 2)$$

    c. If the polynomial has *more than three terms,* try factoring by grouping.

    To factor $2x^2 - 3xy + 10x - 15y$, group the first two terms, then the last two, and remove common factors.

    $$2x^2 - 3xy + 10x - 15y = x(2x - 3y) + 5(2x - 3y)$$

    Now remove the common factor $2x - 3y$.

    $$2x^2 - 3xy + 10x - 15y = (2x - 3y)(x + 5)$$

3. You should always factor the given polynomial completely. So after you apply one of the above techniques, another may be necessary.

    a. To factor

    $$6x^3 + 22x^2 - 40x$$

    first remove the common factor of $2x$. So

    $$6x^3 + 22x^2 - 40x = 2x(3x^2 + 11x - 20)$$

    Now continue to factor the trinomial as before, and

    $$6x^3 + 22x^2 - 40x = 2x(3x - 4)(x + 5)$$

b. To factor

$$x^3 - x^2y - 4x + 4y$$

first we proceed by grouping:

$$x^3 - x^2y - 4x + 4y = x^2(x - y) - 4(x - y)$$
$$= (x - y)(x^2 - 4)$$

Now since $x^2 - 4$ is a difference of squares, we continue to factor and

$$x^3 - x^2y - 4x + 4y = (x - y)(x + 2)(x - 2)$$

5.7 EXERCISES

Build Your Skills

Factor each polynomial completely. To begin, state which method should be applied as the first step, given the guidelines of this section. Then continue the exercise and factor each polynomial completely.

1. $x^2 - 3x$ Factor

2. $4y^2 - 9$

3. $x^2 - 5x - 24$ Foil

4. $8x^3 - 1$

5. $x(x - y) + 2(x - y)$ –

6. $5a^2 - 10a + 25$ C.F.C = Foil

7. $2x^2y - 6xy + 8y^2$

8. $2p - 6q + pq - 3q^2$ grouping.

9. $y^2 - 13y + 40$

10. $m^3 + 27n^3$ Sub of a cube

11. $3b^2 + 17b - 28$

12. $3x^2 + 6x - 5xy - 10y$ grouping

13. $3x^2 - 14xy - 24y^2$ grouping

14. $16c^2 - 49d^2$

15. $2a^2 + 11a + 12$ Foil

16. $m^3n^3 - 64$

17. $125r^3 + s^3$

18. $(x - y)^2 - 16$

19. $3x^2 - 30x + 63$

20. $3a^2 - 108$

21. $40a^3 + 5$

22. $4p^2 - 8p - 60$

23. $2w^2 - 14w - 36$

24. $xy^3 - 27x$

25. $3a^2b - 48b^3$

26. $12b^3 - 86b^2 + 14b$

Transcribe Your Skills

27. Outline a strategy for completely factoring a polynomial.

28. Describe two different ways of completely factoring the expression $(x + 3)^2 - 16$.

Think About These

Factor completely.

29. $(x - 5)^2 - 169$

30. $(x - 7)^2 - 81$

31. $x^2 + 4xy + 4y^2 - 16$

32. $9x^2 + 12xy + 4y^2 - 25$

33. $6(x - 2)^2 + 7(x - 2) - 5$

34. $12(x + 1)^2 - 17(x + 1) + 6$

35. $\dfrac{x^3}{8} - \dfrac{1}{27}$

36. $\dfrac{x^3}{27} + \dfrac{1}{64}$

Skillscan (Section 5.1)

Divide.

a. $\dfrac{3x^2}{3x}$

b. $\dfrac{5a^3}{5a}$

c. $\dfrac{10m^4}{5m^2}$

d. $\dfrac{24y^3}{6y^2}$

e. $\dfrac{20x^2y}{10x}$

f. $\dfrac{36m^3n}{12m^2}$

g. $\dfrac{28a^2b^3}{7ab^2}$

h. $\dfrac{32x^3y^4}{8x^2y^2}$

5.8 | Division of Polynomials

TAPE IN12

OBJECTIVES: 1. To divide a polynomial by a monomial
2. To use long division

Our objective in this section is to examine the operation of division with various types of polynomials along with some applications of the methods used in the division process.

Recall that a monomial consists of a single term.

The easiest case for division occurs when the divisor is a monomial.

TO DIVIDE A POLYNOMIAL BY A MONOMIAL

$$\frac{P + Q}{R} = \frac{P}{R} + \frac{Q}{R}$$

In words, to divide a polynomial by a monomial, we divide *each term* of the polynomial by the monomial.

The following example illustrates.

EXAMPLE 1 Dividing a Polynomial by a Monomial

(*a*) Divide $5x^2 - 10x + 25$ by 5.

$$\frac{5x^2 - 10x + 25}{5}$$ Write the division problem as a rational expression.

$$= \frac{5x^2}{5} - \frac{10x}{5} + \frac{25}{5}$$ Divide each term by 5 and simplify.

$$= x^2 - 2x + 5$$

(*b*) Divide $24a^2b^3 + 16ab^2 - 8ab$ by $8ab$.

$$\frac{24a^2b^3 + 16ab^2 - 8ab}{8ab}$$

With practice you may very well be able to write the quotient directly.

$$= \frac{24a^2b^3}{8ab} + \frac{16ab^2}{8ab} - \frac{8ab}{8ab}$$

$$= 3ab^2 + 2b - 1$$

Check Yourself 1

Divide.

1. $24y^2 - 36y + 18$ by 6

2. $21m^2n^2 - 49m^2n + 7mn$ by $7mn$

The division problems of Example 1 can be approached in a different fashion by first factoring the numerator (the dividend). As long as the denominator (the divisor) appears as a factor, we can divide by that common factor to write the desired quotient. Let's consider the problem of Example 1b again.

EXAMPLE 2 Dividing a Polynomial by a Monomial

$$\frac{24a^2b^3 + 16ab^2 - 8ab}{8ab}$$

$$= \frac{8ab(3ab^2 + 2b - 1)}{8ab} \qquad \text{Factor and divide by the common factor } 8ab.$$

$$= 3ab^2 + 2b - 1 \qquad \text{The quotient is the same as before.}$$

Check Yourself 2

Do the division problems of Check Yourself 1 by first factoring the dividend.

A similar approach to division also allows us (in certain cases) to divide a polynomial by another polynomial that is *not* a monomial. First, we factor the numerator, if possible. If the denominator appears as a factor, we can again divide by that common factor to form the desired quotient. The following example illustrates the process.

EXAMPLE 3 Dividing a Polynomial by a Binomial

(a) Divide $6x^2 + 5x - 6$ by $3x - 2$.

$$\frac{6x^2 + 5x - 6}{3x - 2}$$

$$= \frac{(3x - 2)(2x + 3)}{3x - 2} = 2x + 3 \qquad \text{Factor and divide by } 3x - 2.$$

(b) Divide $49x^2 - 64y^2$ by $7x - 8y$.

$$\frac{49x^2 - 64y^2}{7x - 8y}$$

$$= \frac{(7x + 8y)(7x - 8y)}{7x - 8y} \qquad \text{Divide by the common factor.}$$

$$= 7x + 8y$$

Check Yourself 3

Perform the following divisions by first factoring the dividend.

1. $(4x^2 - 5x - 6) \div (4x + 3)$

2. $(27x^3 + y^3) \div (3x + y)$

Finding the quotients in Example 3 was relatively easy, and it illustrated a special case of dividing by a polynomial, because the dividend was factorable and the divisor was one of the factors of the dividend. If that is not the case, we turn to a method called *algebraic long division* which can be used to divide any polynomial by any other polynomial as long as the degree of the divisor is less than or equal to the degree of the dividend. As you will see, the process is very much like that used for long division with whole numbers.

Let's return to the division problem of Example 3a for our first look at algebraic long division.

EXAMPLE 4 Dividing a Polynomial by a Binomial

Divide $6x^2 + 5x - 6$ by $3x - 2$.

Solution First, we write

$$3x - 2 \overline{)6x^2 + 5x - 6}$$

Now we begin the actual long-division process:

Divide $6x^2$ by $3x$ to get $2x$.

$$\textbf{Step 1}\quad 3x - 2 \overline{)\overset{2x}{6x^2 + 5x - 6}}$$

$$\textbf{Step 2}\quad 3x - 2 \overline{)\overset{2x}{6x^2 + 5x - 6}}$$
$$6x^2 - 4x$$

Multiply the divisor, $3x - 2$, by $2x$. So
$$2x(3x - 2) = 6x^2 - 4x$$

Remember To subtract $6x^2 - 4x$, change *each sign* to $-6x^2 + 4x$ and add. Take your time and be careful here. It's where most errors are made.

$$\textbf{Step 3}\quad 3x - 2 \overline{)\overset{2x}{6x^2 + 5x - 6}}$$
$$\underline{6x^2 - 4x}$$
$$9x - 6 \leftarrow \text{Subtract and bring down } -6.$$

Step 4 We repeat steps 1, 2, and 3 to complete the process.

$$3x - 2 \overline{)\overset{2x + 3 \; \leftarrow \text{Quotient}}{6x^2 + 5x - 6}}$$
$$\underline{6x^2 - 4x}$$
$$9x - 6$$
$$\underline{9x - 6}$$
$$0 \leftarrow \text{Remainder}$$

The quotient is $2x + 3$, as before. Note that to check the division, we multiply.

$(2x + 3)(3x - 2) = 6x^2 + 5x - 6$

Check Yourself 4

Divide, using long division.

$(4x^2 - 5x - 6) \div (4x + 3)$

In Examples 2 and 3, we approached the same problem through factoring and then with the long-division method. However, when the remainder is not 0 in the division, long division is your only option. That is the case in the next example.

EXAMPLE 5 Dividing a Polynomial by a Binomial

Divide $2x^3 - x^2 + 8x + 25$ by $2x + 3$.

Solution To start:

Step 1: Divide $2x^3$ by $2x$.

$$
\begin{array}{r}
x^2 \\
2x + 3\overline{)2x^3 - x^2 + 8x + 25} \\
2x^3 + 3x^2 \\
\hline
- 4x^2 + 8x
\end{array}
$$

Step 3: Subtract and bring down $8x$.

Step 2: Multiply $x^2(2x + 3)$.

Now repeat the three steps above to complete the process.

$$
\begin{array}{r}
x^2 - 2x + 7 \\
2x + 3\overline{)2x^3 - x^2 + 8x + 25} \\
2x^3 + 3x^2 \\
\hline
- 4x^2 + 8x \\
- 4x^2 - 6x \\
\hline
14x + 25 \\
14x + 21 \\
\hline
4
\end{array}
$$

We can write this result as

Quotient   Remainder

$$\frac{2x^3 - x^2 + 8x + 25}{2x + 3} = x^2 - 2x + 7 + \frac{4}{2x + 3}$$

Divisor

To check, you can multiply the divisor, $2x + 3$, by the quotient, $x^2 - 2x + 7$, and add the remainder, 4. The result must be the original dividend $(2x^3 - x^2 + 8x + 25)$.

Note that this form is exactly like long division with whole numbers. For example,

$$
\begin{array}{r}
5 \\
12\overline{)67} \\
60 \\
\hline
7
\end{array}
$$

so

Quotient

$$\frac{67}{12} = 5 + \frac{7}{12}$$

Remainder

Divisor

Check Yourself 5

Divide, using long division.

$(6x^3 - 7x^2 + 4x - 20) \div (3x - 5)$

When the long-division process is used, there are several important points to remember. The dividend and the divisor must be written in descending form (so that the powers of the variable decrease). Also, the division process should continue until the result of the subtraction is a polynomial with degree less than the degree of the original divisor. Both ideas are illustrated in the next example.

EXAMPLE 6 Dividing a Polynomial by a Binomial

Divide $5x^2 - 4x + x^3 - 3$ by $-1 + x^2$.

Solution Write the divisor as $x^2 - 1$ and the dividend as $x^3 + 5x^2 - 4x - 3$.

As you do the division, be sure the like terms are aligned.

$$
\begin{array}{r}
x + 5 \\
x^2 - 1\overline{)x^3 + 5x^2 - 4x - 3} \\
\underline{x^3 \qquad\; - x} \\
5x^2 - 3x - 3 \\
\underline{5x^2 \qquad - 5} \\
- 3x + 2
\end{array}
$$

The division is complete since the degree of the remainder is less than that of the divisor.

So

$$\frac{x^3 + 5x^2 - 4x - 3}{x^2 - 1} = x + 5 + \frac{-3x + 2}{x^2 - 1}$$

Check Yourself 6

Divide, using long division.

$(5x^2 + 7 + 2x^3 + 3x) \div (2 + x^2)$

Example 7 illustrates a special situation in which the dividend has a "missing" term. One approach in that case is to fill in the missing term by using a coefficient of 0.

EXAMPLE 7 Dividing a Polynomial by a Binomial

Divide $4a^4 + 2a^3 + 3a - 7$ by $2a^2 - 3a$.

Solution Since the dividend has a missing a^2 term, we use a placeholder $0a^2$, as shown below.

Some students prefer to simply leave a blank space to provide for the a^2 column.

$$
\begin{array}{r}
2a^2 + 4a + 6 \\
2a^2 - 3a\overline{)4a^4 + 2a^3 + 0a^2 + 3a - 7} \\
\underline{4a^4 - 6a^3} \\
8a^3 + 0a^2 \\
\underline{8a^3 - 12a^2} \\
12a^2 + 3a \\
\underline{12a^2 - 18a} \\
21a - 7
\end{array}
$$

So

$$\frac{4a^4 + 2a^3 + 3a - 7}{2a^2 - 3a} = 2a^2 + 4a + 6 + \frac{21a - 7}{2a^2 - 3a}$$

The division is complete. Do you see why?

Check Yourself 7

Divide, using long division.

$(6y^4 + 3y^2 - 37y + 5) \div (3y^2 - 6y)$

One important application of the long-division method is in factoring higher-degree polynomials. The following example illustrates the use of long division in factoring a third-degree polynomial.

EXAMPLE 8 Factoring a Polynomial

Given that $x - 2$ is one factor, completely factor $2x^3 + x^2 - 13x + 6$.

Solution Since $x - 2$ is a factor, it will divide into the original polynomial *exactly*, that is, with a 0 remainder. We therefore start with long division.

$$
\begin{array}{r}
2x^2 + 5x - 3 \\
x - 2\overline{)2x^3 + x^2 - 13x + 6} \\
\underline{2x^3 - 4x^2} \\
5x^2 - 13x \\
\underline{5x^2 - 10x} \\
-3x + 6 \\
\underline{-3x + 6} \\
0 \leftarrow \text{Remainder is 0.}
\end{array}
$$

From the division above, we now know that our original polynomial can be written as

$2x^3 + x^2 - 13x + 6 = (x - 2)(2x^2 + 5x - 3)$

To complete the problem, we factor the trinomial so that

$2x^3 + x^2 - 13x + 6 = (x - 2)(2x - 1)(x + 3)$

and the original polynomial is completely factored.

Note that

$2x^2 + 5x - 3 = (2x - 1)(x + 3)$

by our earlier factoring methods.

Check Yourself 8

Use long division to completely factor $3x^3 + 4x^2 - 35x - 12$ given that $x + 4$ is one factor.

CHECK YOURSELF ANSWERS

1. (1) $4y^2 - 6y + 3$; (2) $3mn - 7m + 1$ **2.** Answers are the same as in **1.**

3. (1) $x - 2$; (2) $9x^2 - 3xy + y^2$ **4.** $x - 2$ **5.** $2x^2 + x + 3 + \dfrac{-5}{3x - 5}$

6. $2x + 5 + \dfrac{-x - 3}{x^2 + 2}$ **7.** $2y^2 + 4y + 9 + \dfrac{17y + 5}{3y^2 - 6y}$

8. $(x + 4)(x - 3)(3x + 1)$

5.8 EXERCISES

Build Your Skills

Perform the indicated division.

1. $\dfrac{9b^2 - 12b}{3b}$

2. $\dfrac{10x^2 + 15x}{5x}$

3. $\dfrac{16a^3 + 24a^2}{4a^2}$

4. $\dfrac{9x^4 - 12x^2}{3x^2}$

5. $\dfrac{18y^4 + 12y^3 - 6y^2}{6y^2}$

6. $\dfrac{21x^5 - 28x^4 + 14x^3}{7x}$

7. $\dfrac{8a^7 - 16a^5 + 4a^3}{-4a^3}$

8. $\dfrac{27m^6 + 18m^4 - 9m^2}{-9m^2}$

9. $\dfrac{20x^4y^2 - 15x^2y^3 + 10x^3y}{5x^2y}$

10. $\dfrac{16w^3z^3 + 24w^2z^2 - 40wz^3}{8wz^2}$

Perform the indicated division by factoring the numerators and then dividing by the common factors.

11. $\dfrac{x^2 - x - 6}{x + 2}$

12. $\dfrac{a^2 - 2a - 8}{a - 4}$

13. $\dfrac{2y^2 - 3y - 5}{2y - 5}$

14. $\dfrac{3m^2 - 2m - 8}{3m + 4}$

15. $\dfrac{3a^2 + 5ab - 2b^2}{3a - b}$

16. $\dfrac{4x^2 - 8xy - 5y^2}{2x - 5y}$

17. $\dfrac{x^3 + 27}{x + 3}$

18. $\dfrac{y^3 - 8}{y - 2}$

19. $\dfrac{m^4 - 16}{m - 2}$

20. $\dfrac{a^4 - 81}{a + 3}$

21. $\dfrac{x^3 - x^2 + 3x - 3}{x - 1}$

22. $\dfrac{x^3 + 3x^2 - 4x - 12}{x + 3}$

23. $\dfrac{3x^3 - 4x^2 + 15x - 20}{x^2 + 5}$

24. $\dfrac{4y^3 + 3y^2 - 28y - 21}{y^2 - 7}$

Perform the indicated division, using long division.

25. $\dfrac{x^2 - x - 6}{x + 2}$

26. $\dfrac{a^2 - 2a - 8}{a - 4}$

27. $\dfrac{y^2 + 8y - 15}{y - 5}$

28. $\dfrac{x^2 - 2x - 30}{x + 6}$

29. $\dfrac{4x^2 + 6x - 25}{2x + 7}$

30. $\dfrac{6y^2 - y - 10}{3y - 5}$

31. $\dfrac{2m^3 + 6m^2 - 15m + 15}{m + 5}$

32. $\dfrac{3x^3 - 14x^2 - 18x - 20}{x - 6}$

33. $\dfrac{2x^3 - 3x^2 + 4x + 4}{2x + 1}$

34. $\dfrac{6w^3 + w^2 + 3w - 10}{3w - 4}$

35. $\dfrac{x^3 - x^2 + 5}{x - 2}$

36. $\dfrac{x^3 + 4x - 3}{x - 3}$

37. $\dfrac{25x^3 + x}{5x - 2}$

38. $\dfrac{16y^3 + 7y}{4y + 3}$

39. $\dfrac{2x^2 - 8 - 3x + x^3}{x - 2}$

40. $\dfrac{x^2 - 18x + 2x^3 + 32}{x + 4}$

41. $\dfrac{y^4 + y^2 - 16}{y + 2}$

42. $\dfrac{w^4 - 5w^2 - 24}{w - 3}$

43. $\dfrac{x^4 - 16}{x - 2}$

44. $\dfrac{a^4 + 81}{a + 3}$

45. $\dfrac{w^4 + w^2 + 2w + 4}{w^2 + 3w + 1}$

46. $\dfrac{x^4 - 2x^3 + x^2 - 3x + 5}{x^2 - 2x + 2}$

Transcribe Your Skills

47. If the dividend has a "missing term," what is one approach that is used to carry out the long-division process?

48. Outline a procedure to check the results of the long-division process.

Think About These

49. Use long division to completely factor $2x^3 + 9x^2 - 2x - 24$, given that $x + 2$ is one factor.

50. Use long division to completely factor $3x^3 - 5x^2 - 16x + 12$, given that $x - 3$ is one factor.

51. Use long division to completely factor $4x^3 - 24x^2 + 17x + 15$, given that $x - 5$ is one factor.

52. Use long division to completely factor $6x^3 + 31x^2 + 4x - 5$, given that $x + 5$ is one factor.

53. Substitute 5 for y in the polynomial $y^2 + 8y - 15$ and evaluate. Compare your result with the remainder from the division of Exercise 27.

54. Substitute -6 for x in the polynomial $x^2 - 2x - 30$ and evaluate. Compare your result with the remainder from the division of Exercise 28.

55. Substitute -5 for x in the polynomial $2x^3 + 6x^2 - 15x + 15$ and evaluate. Compare your result with the remainder from the division of Exercise 31.

56. Substitute 6 for x in the polynomial $3x^3 - 14x^2 - 18x - 20$ and evaluate. Compare your result with the remainder from the division of Exercise 32.

Skillscan (Section 5.2)

Evaluate each polynomial as indicated.

a. $P(x) = x^2 - x + 1$; $P(1)$ **b.** $P(x) = 2x^2 + x - 2$; $P(-1)$ **c.** $P(x) = x^3 + 2x^2 - 3x + 1$; $P(2)$
d. $P(x) = -x^3 + x^2 - x + 1$; $P(3)$
e. $P(x) = x^2 + 4x$; $P(0)$ **f.** $P(x) = x^3 + 3x - 5$; $P(-2)$

5.9   Synthetic Division

OBJECTIVES: 1. To use the notation of synthetic substitution
2. To find remainders of polynomial division with synthetic division
3. To completely factor polynomials by using synthetic division

In Section 5.8, we learned to divide polynomials by using long division. Now we will examine an abbreviated method that will allow us to find divisors, and thus factors, for polynomials of any degree.

First, recall that in Section 5.2 we used the notation $P(x)$ to name a polynomial. As an example, we might say

$$P(x) = 3x^2 + 5x - 4$$

This is equivalent to $y = 3x^2 + 5x - 4$.

If you were to try to find $P(2)$ by the methods of Section 5.2, you would replace x with 2 and simplify the expression, thus

$$P(2) = 3(2)^2 + 5(2) - 4$$
$$= 12 + 10 - 4$$
$$= 18$$

Let's factor an x out of the first two terms of the polynomial and look at a slightly different approach.

$$P(x) = (3x + 5)x - 4$$

Now, we substitute 2 for x and simplify.

$$P(2) = [3(2) + 5](2) - 4$$
$$= (6 + 5)(2) - 4$$
$$= (11)(2) - 4$$
$$= 22 - 4$$
$$= 18$$

This process can be duplicated in the following way. We begin by writing the number to be substituted (separated as shown) and the coefficients of the polynomial. The leading coefficient is rewritten below the line.

$$\underline{2|} \quad 3 \quad\quad 5 \quad\quad -4$$
$$\overline{}$$
$$\ 3$$

We multiply the number to be substituted and the leading coefficient. The result is added to the second coefficient.

$2 \times 3 = 6$

$5 + 6 = 11$

$$\underline{2|} \quad 3 \quad\quad 5 \quad\quad -4$$
$$ 6$$
$$\overline{3 \quad\quad 11}$$

The process is repeated until we have used all the coefficients.

$2 \times 11 = 22$

$-4 + 22 = 18$

$$\underline{2|} \quad 3 \quad\quad 5 \quad\quad -4$$
$$ 6 \quad\quad 22$$
$$\overline{3 \quad\quad 11 \quad\quad 18}$$

The last number, 18, is $P(2)$.

When we use this notation to do the substitution, we refer to it as *synthetic substitution.*

EXAMPLE 1 Evaluating a Function by Synthetic Substitution

If $P(x) = 3x^3 - 4x^2 + 2x - 1$, find $P(-1)$.

$$
\begin{array}{r|rrrr}
-1 & 3 & -4 & 2 & -1 \\
   &   & -3 & 7 & -9 \\
\hline
   & 3 & -7 & 9 & -10
\end{array}
$$

Therefore $P(-1) = -10$.

Check Yourself 1

If $P(x) = 2x^3 + 3x^2 - 2x - 5$, find $P(2)$.

What does it mean when $P(k) = 0$? We know then that k is a solution of the equation $P(x) = 0$.

EXAMPLE 2 Evaluating a Function by Synthetic Substitution

If $P(x) = x^2 + 6x + 8$, find $P(-4)$.

$$
\begin{array}{r|rrr}
-4 & 1 & 6 & 8 \\
   &   & -4 & -8 \\
\hline
   & 1 & 2 & 0
\end{array}
$$

Since $P(-4) = 0$, we know that -4 is a solution of the equation

$x^2 + 6x + 8 = 0$

Check Yourself 2

If $P(x) = 2x^3 + 7x^2 + 12x - 8$, find $P(3)$.

We can also use synthetic substitution when we wish to divide a polynomial and the divisor is of the form $x - k$. First let us recall that when we divide a polynomial by a lesser-degree polynomial, we get a quotient, which we call $Q(x)$, and a remainder, which we will call r, so

$P(x) = (x - k) \cdot Q(x) + r$

This must be true for every value of x, so let $x = k$ and

$P(k) = (k - k) \cdot Q(k) + r$

$\quad = 0 \cdot Q(k) + r$

$\quad = r$

Think of dividing 29 by 4.

$29 = 4 \cdot 7 + 1$

We now know that the value we get by synthetic substitution is the remainder when we divide $P(x)$ by $x - k$.

EXAMPLE 3 Finding the Remainder

Find the remainder when $x^3 + 3x^2 - 8x + 1$ is divided by $x - 3$.

You could check your result by using long division. And $x - 3$ is of the form $x - k$, where $k = 3$.

$$
\begin{array}{r|rrrr}
3 & 1 & 3 & -8 & 1 \\
 & & 3 & 18 & 30 \\
\hline
 & 1 & 6 & 10 & 31
\end{array}
$$

So the remainder is 31.

Check Yourself 3

Find the remainder when $2x^3 + 7x^2 + 12x - 8$ is divided by $x - 1$.

To complete the division, we must find the quotient. Let's look at long division as discussed in Section 5.8.

Let us divide $2x^2 + 5x - 3$ by $x + 2$.

$$
\require{enclose}
\begin{array}{r}
2x + 1 \\
x + 2 \enclose{longdiv}{2x^2 + 5x - 3} \\
\underline{2x^2 + 4x} \\
x - 3 \\
\underline{x + 2} \\
-5
\end{array}
$$

So,

$$\frac{2x^2 + 5x - 3}{x + 2} = 2x + 1 + \frac{-5}{x + 2}$$

Let's compare this to synthetic substitution.

$x + 2$ is of the form $x - k$, since

$x + 2 = x - (-2)$

$k = -2$

$$
\begin{array}{r|rrr}
-2 & 2 & 5 & -3 \\
 & & -4 & -2 \\
\hline
 & 2 & 1 & -5
\end{array}
$$

You will notice several similarities between the long division and the synthetic substitution. The most interesting is along the bottom row of the substitution. Note that the first two numbers are the coefficients of the quotient and that the last number is the remainder. This will always be the case when we use synthetic substitution. When we are using this process to replace long division, we call it *synthetic division*.

EXAMPLE 4 Using Synthetic Division

Use synthetic division to divide $2x^3 - 4x^2 - 7x + 5$ by $x - 3$.

$$
\begin{array}{r|rrrr}
3 & 2 & -4 & -7 & 5 \\
 & & 6 & 6 & -3 \\
\hline
 & 2 & 2 & -1 & 2 \leftarrow \text{Remainder}
\end{array}
$$

$\underbrace{}_{\text{Coefficients of the quotient}}$

So $2x^3 - 4x^2 - 7x + 5 = (x - 3)(2x^2 + 2x - 1) + 2$.

Check Yourself 4

Divide $2x^3 + 7x^2 + 12x - 8$ by $x - 1$.

As was true with long division, we must be careful when one of the coefficients is 0. Use a placeholder of 0 when this occurs.

EXAMPLE 5 Using Synthetic Division

Divide $4x^4 - 3x^2 + 2x - 7$ by $x + 1$.

$$
\begin{array}{r|rrrrr}
\underline{1} & 4 & 0 & -3 & 2 & -7 \\
  &   & -4 & 4 & -1 & -1 \\
\hline
  & 4 & -4 & 1 & 1 & -8 \\
\end{array}
$$

Coefficients of the quotient   Remainder

Zero represents the coefficient of the x^3 term.

So $4x^4 - 3x^2 + 2x - 7 = (x + 1)(4x^3 - 4x^2 + x + 1) + (-8)$.

Check Yourself 5

Divide $3x^4 - 3x^3 + 2$ by $x + 1$.

place holder

$\underline{1\rfloor}\ 3\ -3\ +0\ +0\ 2$

Synthetic division is commonly used to determine whether $x - k$ is a factor of a given polynomial.

EXAMPLE 6 Using Synthetic Division

Show that $x - 2$ is a factor of $2x^3 - 7x^2 + 10x - 8$.

$$
\begin{array}{r|rrrr}
\underline{2} & 2 & -7 & 10 & -8 \\
  &   & 4 & -6 & 8 \\
\hline
  & 2 & -3 & 4 & 0 \\
\end{array}
$$

The remainder is 0.

This tells us that $2x^3 - 7x^2 + 10x - 8 = (x - 2)(2x^2 - 3x + 4)$.

Check Yourself 6

synthic division

Show that $x + 5$ is a factor of $x^3 + 5x^2 - 3x - 15$.

By using synthetic division we will completely factor third-degree polynomials in the next example.

EXAMPLE 7 Using Synthetic Division

(a) Complete factor $x^3 + 2x^2 - 9x - 18$, given that $x + 2$ is a factor.

$$
\begin{array}{r|rrrr}
-2 & 1 & 2 & -9 & -18 \\
   &   & -2 & 0 & 18 \\
\hline
   & 1 & 0 & -9 & 0
\end{array}
$$

We now know that

$$
\begin{aligned}
x^3 + 2x^2 - 9x - 18 &= (x + 2)(x^2 - 9) \\
&= (x + 2)(x + 3)(x - 3)
\end{aligned}
$$

(b) Completely factor $2x^3 - 3x^2 - 8x - 3$, given that $x + 1$ is a factor.

$$
\begin{array}{r|rrrr}
-1 & 2 & -3 & -8 & -3 \\
   &   & -2 & 5 & 3 \\
\hline
   & 2 & -5 & -3 & 0
\end{array}
$$

So we have

$$2x^3 - 3x^2 - 8x - 3 = (x + 1)(2x^2 - 5x - 3)$$

Using the techniques of Section 5.5, we find

$$2x^2 - 5x - 3 = (2x + 1)(x - 3)$$

So

$$2x^3 - 3x^2 - 8x - 3 = (x + 1)(2x + 1)(x - 3)$$

Check Yourself 7

Completely factor $2x^3 - 3x^2 - 8x + 12$ given $x - 2$ is a factor.

CHECK YOURSELF ANSWERS

1. 19   **2.** 145   **3.** 13   **4.** $(x - 1)(2x^2 + 9x + 21) + 13$
5. $(x + 1)(3x^3 - 6x^2 + 6x - 6) + 8$ **6.** Remainder = 0
7. $(x - 2)(x + 2)(2x - 3)$

5.9 EXERCISES

Build Your Skills

For $P(x) = 3x^3 - 2x^2 + x - 4$, find each of the following by using synthetic substitution.

1. $P(2)$

2. $P(3)$

3. $P(-3)$

4. $P(-5)$

Using synthetic division, find the remainder for each division.

5. $(x^2 - 3x + 5) \div (x - 3)$

6. $(x^2 + 2x + 4) \div (x - 2)$

7. $(2x^3 + x^2 - 5x - 1) \div (x + 1)$

8. $(3x^3 - 2x^2 + x - 2) \div (x + 3)$

9. $(3x^3 - 4x + 1) \div (x - 3)$

10. $(5x^3 - 6x + 2) \div (x - 2)$

For each of the following, use synthetic division to find the quotient and remainder.

11. $(x^2 + 4x + 5) \div (x - 1)$

12. $(x^2 - 3x + 4) \div (x - 3)$

13. $(3x^3 + 2x^2 - 5x + 2) \div (x + 1)$

14. $(2x^3 - 5x^2 + 2x - 2) \div (x + 2)$

15. $(4x^3 - 3x + 2) \div (x - 3)$

16. $(3x^3 - 6x + 2) \div (x - 4)$

Transcribe Your Skills

17. Can the quotient and remainder for $\dfrac{x^3 - 4x^2 + 5x - 6}{x^2 - 4}$ be found by using synthetic division? If not, why not? If so, what are they?

18. What are the advantages of using synthetic division to find the value of a polynomial for a specific number?

Think About These

Use synthetic division to confirm each of the following.

19. $x + 2$ is a factor of $x^2 - x - 6$.

20. $x - 4$ is a factor of $x^2 - 2x - 8$.

21. $x + 3$ is a factor of $x^3 + 27$.

22. $x - 2$ is a factor of $x^4 - 16$.

Skillscan (Section 5.5)

Factor each of the following completely

a. $x^2 - 9$ **b.** $x^2 - x - 2$ **c.** $2x^2 - x - 3$

d. $4x^2 - 16$ **e.** $x^3 + 2x^2 - 3x - 6$

f. $6x^3 - 7x^2 - 7x$

SUMMARY

Integral Exponents and Scientific Notation [5.1]

Properties of Exponents

For any nonzero real numbers a and b and integers m and n:

Product Rule

$$a^m \cdot a^n = a^{m+n}$$

$$x^5 \cdot x^7 = x^{5+7} = x^{12}$$

Quotient Rule

$$\frac{a^m}{a^n} = a^{m-n}$$

$$\frac{x^7}{x^5} = x^{7-5} = x^2$$

Power Rule

$$(a^m)^n = a^{m \cdot n}$$

$$(x^5)^3 = x^{5 \cdot 3} = x^{15}$$

Raising Products to a Power

$$(ab)^m = a^m b^m$$

$$(2xy)^3 = 2^3 x^3 y^3$$
$$= 8x^3 y^3$$

Raising Quotients to a Power

$$\left(\frac{a}{b}\right)^m = \frac{a^m}{b^m}$$

$$\left(\frac{x^2}{3}\right)^2 = \frac{(x^2)^2}{3^2}$$
$$= \frac{x^4}{9}$$

Definition

$5^0 = 1$

$x^{-3} = \dfrac{1}{x^3}$

$2^{-4} = \dfrac{1}{2^4} = \dfrac{1}{16}$

$2y^{-5} = \dfrac{2}{y^5}$

$(2y)^{-5} = \dfrac{1}{(2y)^5} = \dfrac{1}{32y^5}$

$\dfrac{1}{x^{-4}} = x^4$

$\dfrac{1}{3^{-3}} = 3^3 = 27$

$$a^0 = 1 \qquad a \neq 0$$

Expressions with negative exponents can be simplified by the following definition

$$a^{-n} = \dfrac{1}{a^n}$$

where a is any nonzero real number and n is any integer. We also have

$$\dfrac{1}{a^{-n}} = a^n$$

Scientific Notation

Scientific notation is a useful way of expressing very large or very small numbers through the use of powers of 10. Any number written in the form

$$a \times 10^n$$

where $1 \leq a < 10$ and n is an integer, is said to be written in scientific notation.

$38{,}000{,}000. = 3.8 \times 10^7$

7 places

$0.0025 = 2.5 \times 10^{-3}$

3 places

Polynomials—Addition and Subtraction [5.2]

A *monomial* is a term in which only whole numbers can appear as exponents.

A *polynomial* may be a monomial or any finite sum (or difference) of monomials. We call a polynomial with two terms a *binomial*, one with three terms a *trinomial*.

To add

$4x^2 + 3x + 2$ and
$3x^2 - 5x - 7$

write

$(4x^2 + 3x + 2) +$
$(3x^2 - 5x - 7)$

$= 7x^2 - 2x - 5$

Adding Polynomials To add polynomials, simply combine any like terms by using the distributive property.

Signs of Grouping If parentheses are preceded by a negative sign, you may remove the parentheses and the preceding sign by changing the sign of each term inside the parentheses.

If one set of grouping symbols is contained within another, remove the innermost grouping symbols first.

To simplify:

$3x - 2[4 - (3x + 5)]$
$= 3x - 2[4 - 3x - 5]$
$= 3x - 2[-3x - 1]$
$= 3x + 6x + 2 = 9x + 2$

Subtracting Polynomials To subtract polynomials, enclose the polynomial being subtracted in parentheses, preceded by a negative sign. Then remove the parentheses according to the previous rule and combine like terms.

To subtract

$3x^2 - 2x - 5$ from
$4x^2 - 6x + 2$

write

$(4x^2 - 6x + 2) -$
$(3x^2 - 2x - 5)$
$= 4x^2 - 6x + 2 -$
  $3x^2 + 2x + 5$
$= x^2 - 4x + 7$

Multiplication of Polynomials and Special Products [5.3]

Multiplying Polynomials To multiply two polynomials, multiply each term of the first polynomial by each term of the second polynomial.

To multiply:

$$
\begin{array}{r}
3x^2 - 2x + 5 \\
5x - 2 \\
\hline
-6x^2 + 4x - 10 \\
15x^3 - 10x^2 + 25x \\
\hline
15x^3 - 16x^2 + 29x - 10
\end{array}
$$

Multiplying Binomials To multiply two binomials, use the FOIL (First, Outer, Inner, Last) pattern.

$(2x + 3)(x - 4) = 2x^2 - 5x - 12$

Special Products Certain special products can be found by applying the following formulas.

$$(a + b)(a - b) = a^2 - b^2$$

$$(a + b)^2 = a^2 + 2ab + b^2$$

$$(a - b)^2 = a^2 - 2ab + b^2$$

$(2x - 3)(2x + 3) = 4x^2 - 9$

$(2x + 5)^2 = 4x^2 + 20x + 25$

$(3a - b)^2 = 9a^2 - 6ab + b^2$

Factoring Polynomials [5.4 to 5.7]

The *greatest common factor* (GCF) of a polynomial is the monomial with highest degree and largest numerical coefficient that is a factor of each term of the polynomial.

$5x^2y^2 - 10xy^2 + 35x^2y$ has GCF $5xy$.

To Factor a Polynomial In general, you can apply the following steps.

1. If a polynomial has a GCF other than 1, factor out that *greatest common factor.*

Factor:

$5x^2y^2 - 10xy^2 + 35x^2y$
$= 5xy(xy - 2y + 7x)$

2. If a polynomial is a binomial, try factoring it as a *difference of two squares* or as a *sum or difference of two cubes.*

Factor:

$16x^2 - 9y^2$
$= (4x + 3y)(4x - 3y)$

Factor:

$x^3 + 8y^3$
$= (x + 2y)(x^2 - 2xy + 4y^2)$

Factor:

$x^2 - 8x + 12$

The possible factors are

$(x - 12)(x - 1)$

$(x - 4)(x - 3)$

$(x - 6)(x - 2)$

The correct factors are
$(x - 6)(x - 2)$.

Factor:

$2x^2 + 8x - 3x - 12$
$= 2x(x + 4) - 3(x + 4)$
$= (x + 4)(2x - 3)$

3. If a polynomial is a quadratic trinomial, try factoring it by using the *trial-and-error* process developed in Section 5.6. You should also consider the possibility of a *perfect-square trinomial,* in which case the factored form will be the square of a binomial.

4. If a polynomial has more than three terms, try *factoring by grouping.*

5. Finally, check each factor to be sure that your original polynomial has been completely factored.

Division of Polynomials [5.8]

To divide:

$$\frac{14x^2y - 21xy^2 + 7xy}{7xy}$$

$$= \frac{14x^2y}{7xy} - \frac{21xy^2}{7xy} + \frac{7xy}{7xy}$$

$$= 2x - 3y + 1$$

To divide a polynomial by a monomial, apply the following rule:

$$\frac{P + Q}{R} = \frac{P}{R} + \frac{Q}{R}$$

where $R \neq 0$.

In words, to divide a polynomial by a monomial, we divide each term of the polynomial by the monomial.

To divide:

$$\frac{2x^3 + 3x^2 - 2x - 41}{2x - 5}$$

$$\require{enclose}
\begin{array}{r}
x^2 + 4x + 9 \\
2x - 5 \overline{\smash{)}2x^3 + 3x^2 - 2x - 41} \\
\underline{2x^3 - 5x^2} \\
8x^2 - 2x \\
\underline{8x^2 - 20x} \\
18x - 41 \\
\underline{18x - 45} \\
4
\end{array}$$

$$\frac{2x^3 + 3x^2 - 2x - 41}{2x - 5}$$

$$= x^2 + 4x + 9 + \frac{4}{2x - 5}$$

Long Division

When we wish to divide one polynomial by another polynomial, one approach is to factor the numerator (the dividend) to determine whether the denominator (the divisor) appears as a factor of the numerator. If that is the case, we can divide by the common factor, and the resulting polynomial will be the desired quotient.

If that is not the case, the process of algebraic long division must be used. The process is similar to that used in the division of whole numbers, and it follows an almost identical algorithm.

Synthetic Division [5.9]

To find $P(k)$, we can use *synthetic substitution*. If $P(k) = 0$, then k is a root of the polynomial $P(x)$.

The value we get by synthetic substitution with k is the remainder when $P(x)$ is divided by $x - k$.

When we use the same process to do *synthetic division*, we find the coefficients to the quotient along the bottom row.

If $P(x) = x^2 + x - 1$, find $P(1)$.

$$
\begin{array}{r|rrr}
1 & 1 & 1 & -1 \\
  &   & 1 & 2 \\
\hline
  & 1 & 2 & 1 = P(1)
\end{array}
$$

Divide $2x^3 + 3x - 4$ by $x - 2$.

$$
\begin{array}{r|rrr}
2 & 2 & 3 & -4 \\
  &   & 4 & 14 \\
\hline
  & 2 & 7 & 10
\end{array}
$$

$Q(x) = 2x + 7$; $R = 10$

SUMMARY EXERCISES

This summary exercise set is provided to give you practice with each of the objectives of the chapter. Each exercise is keyed to the appropriate chapter section. The answers are provided in the instructor's manual that accompanies this text. Your instructor will provide guidelines on how best to use these exercises in your instructional program.

[5.1] Simplify each expression, using the properties of exponents.

1. $r^4 \cdot r^9$

2. $(-3)^2(-3)^3$

3. $(a^3b^2)(a^8b^3)$

4. $(6c^0d^4)(-3c^2d^2)$

5. $\dfrac{x^{12}}{x^5}$

6. $\dfrac{(2x-1)^8}{(2x-1)^5}$

7. $(x^2y)^3$

8. $(2c^3d^4)^3$

9. $(2a^3)^0(-3a^4)^2$

10. $\left(\dfrac{x}{y^2}\right)^2$

11. $\left(\dfrac{3m^2n^3}{p^4}\right)^3$

12. $\left(\dfrac{a^3}{b^4}\right)\left(\dfrac{b^2}{2a^2}\right)^3$

13. y^{-6}

14. $4x^{-5}$

15. $(2w)^{-3}$

16. $\dfrac{3}{m^{-4}}$

17. $\dfrac{a^{-5}}{b^{-4}}$

18. $y^{-5} \cdot y^2$

19. $\dfrac{w^{-7}}{w^{-3}}$

20. $(m^{-6})^{-2}$

21. $(m^3n^{-5})^{-2}$

22. $\left(\dfrac{a^{-4}}{b^{-2}}\right)^3$

23. $\left(\dfrac{r^{-5}}{s^4}\right)^{-2}$

24. $(5w^{-2})^2(2w^{-2})$

25. $(5a^2b^{-3})(2a^{-2}b^{-6})$

26. $\dfrac{7a^{-4}b^4}{28a^{-3}b^{-3}}$

27. $\left(\dfrac{m^{-3}n^{-3}}{m^{-4}n^4}\right)^3$

28. $\left(\dfrac{x^{-4}y^{-3}z^2}{x^{-3}y^2z^{-4}}\right)^{-2}$

[5.1] Write each of the following numbers in scientific notation.

29. 4,500,000

30. 230,000,000,000

31. 0.00036

32. 0.0000002

[5.1] Write each of the following numbers in decimal form.

33. 2×10^5

34. 7.3×10^7

35. 3×10^{-4}

36. 8.2×10^{-5}

[5.1] Perform the indicated operations, and express the result in scientific form.

37. $(2.2 \times 10^6)(4 \times 10^8)$

38. $\dfrac{7.5 \times 10^8}{2.5 \times 10^2}$

39. $(3.6 \times 10^{-2})(2 \times 10^8)$

40. $\dfrac{8.8 \times 10^7}{2.2 \times 10^{-3}}$

[5.2] Identify each of the following polynomials as a monomial, binomial, or trinomial. Give the degree of each.

41. $7x^4 - 8x^6$

42. $5x^4$

43. 5

44. $4x^2 + 5x^4 + 6x^6$

[5.2] Add the following polynomials.

45. $4x^2 - 2xy + 3y^2$ and $-3x^2 + 5xy - 4y^2$

46. $8x^2 - 5$, $4x + 3$, and $5x^2 - 4x$

[5.2] Subtract each of the following polynomials.

47. $-5x^2 - 3x$ from $6x^2 - 4x$

48. $4x^2 - 3x$ from $9x^2 + 5$

49. $4x^2 - 2x + 1$ from the sum of $x^2 + 3x + 2$ and $3x^2 + 5x - 7$

[5.2] Simplify each of the following expressions.

50. $x^2 - [5x^2 - (x^2 - 3)]$

51. $4y - \{3y + 2[y - 5(y - 3)]\}$

[5.2] Evaluate each polynomial for the indicated value of the variable.

52. $5x^2 - 3x - 4$ for $x = 3$

53. $2x^2 + 4x + 6$ for $x = -2$

54. $x^3 - 3x^2 + 7$ for $x = -3$

[5.2] If $P(x) = -x^2 + 3x + 7$, find each value.

55. $P(2)$

56. $P(0)$

57. $P(-2)$

[5.3] Multiply each of the following polynomials.

58. $5x(3x^2 - 4x)$

59. $5y^2(2y^3 - 3y^2 + 5y)$

60. $(x - 2y)(x + 3y)$

61. $(a - 5b)(a - 6b)$

62. $(3c - 5d)(5c + 2d)$

63. $(4x^2 - y)(2x + 3y^2)$

64. $x(x - 3)(x + 2)$

65. $2y(2y + 3)(3y + 2)$

[5.3] Multiply each of the following polynomials, using the special product formulas.

66. $(x + 8)^2$

67. $(y - 5)^2$

68. $(2a - 3b)^2$

69. $(5x + 2y)^2$

70. $(x - 4y)(x + 4y)$

71. $(2c - 3d)(2c + 3d)$

[5.3] Multiply each of the following polynomials, using the vertical format.

72. $(2x - 3)(x^2 - 5x + 2)$

73. $(5a - b)(2a^2 - 3ab - 2b^2)$

[5.4] Factor each of the following polynomials completely.

74. $18x^2y + 24xy^2$

75. $35a^3 - 28a^2 + 7a$

76. $18m^2n^2 - 27m^2n + 45m^2n^3$

77. $x(2x - y) + y(2x - y)$

78. $5(w - 3z) - 10(w - 3z)^2$

[5.4] Factor each of the polynomials completely, using grouping.

79. $2x^2 + 2xy + 3x + 3y$

80. $2a^2 - 2ab - 5a + 5b$

81. $c^3 - 6d^2 + 3c^2d - 2cd$

82. $x^2 + xy + 2xz - 3x - 3y - 6z$

[5.5] Factor each of the following binomials completely.

83. $x^2 - 64$

84. $25a^2 - 16$

85. $16m^2 - 49n^2$

86. $3w^3 - 12wz^2$

87. $a^4 - 16b^4$

88. $m^3 - 64$

89. $8x^3 + 1$

90. $8c^3 - 27d^3$

91. $125m^3 + 64n^3$

92. $2x^4 + 54x$

[5.6] Factor each of the following trinomials completely.

93. $x^2 + 12x + 20$

94. $a^2 - a - 12$

95. $w^2 - 13w + 40$

96. $r^2 - 9r - 36$

97. $x^2 - 8xy - 48y^2$

98. $a^2 + 17ab + 30b^2$

99. $5x^2 + 13x - 6$

100. $2a^2 + 3a - 35$

101. $4r^2 + 20r + 21$

102. $6c^2 - 19c + 10$

103. $6m^2 - 19mn + 10n^2$

104. $8x^2 + 14xy - 15y^2$

105. $9x^2 - 15x - 6$

106. $5w^2 - 25wz + 30z^2$

107. $3c^3 + 18c^2 + 15c$

108. $2a^3 + 4a^2b - 6ab^2$

109. $x^4 + 6x^2 + 5$

110. $a^4 - 3a^2b^2 - 4b^4$

[5.8] Perform the indicated division.

111. $\dfrac{32x^3 + 24x}{8x}$

112. $\dfrac{35a^3b^2 - 21a^2b^3 + 14a^3b}{7a^2b}$

113. $\dfrac{x^2 + 3x - 10}{x - 2}$

114. $\dfrac{2m^2 + 3mn - 9n^2}{2m - 3n}$

115. $\dfrac{y^3 - 64}{y - 4}$

116. $\dfrac{2x^3 + x^2 - 6x - 3}{x^2 - 3}$

[5.8] Perform the indicated division, using long division.

117. $\dfrac{x^2 + 3x - 10}{x - 2}$

118. $\dfrac{z^2 - 6z + 7}{z + 3}$

119. $\dfrac{2y^2 + 9y - 25}{2y - 5}$

120. $\dfrac{m^3 - 3m^2 - m - 8}{m - 4}$

121. $\dfrac{4w^3 + w + 3}{2w - 1}$

122. $\dfrac{x^4 + 5x^2 - 6}{x^2 + 6}$

[5.8] Let $P(x) = 4x^3 - 2x^2 + 3x - 7$. Find each value.

123. $P(2)$

124. $P(1)$

125. $P(-3)$

126. $P(-4)$

[5.9] Use synthetic substitution to find the remainder for each division.

127. $(x^2 - 2x + 7) \div (x - 2)$

128. $(x^2 + 3x - 5) \div (x - 1)$

129. $(2x^2 - x + 9) \div (x + 3)$

130. $(3x^2 + 2x - 8) \div (x + 2)$

131. $(3x^3 + x - 2) \div (x + 1)$

132. $(2x^3 + 3x - 1) \div (x - 3)$

[5.9] For each of the following, use synthetic division to find the quotient and remainder.

133. $(x^2 + 3x + 4) \div (x - 1)$

134. $(2x^2 - x + 7) \div (x - 3)$

135. $(2x^2 + x - 2) \div (x + 2)$

136. $(x^2 + 3x - 5) \div (x + 4)$

137. $(4x^3 + 3x - 5) \div (x - 2)$

138. $(2x^3 + x - 3) \div (x - 3)$

SELF-TEST

The purpose of this self-test is to help you check your progress and to review for a chapter test in class. Allow yourself about an hour to take the test. When you are done, check your answers in the back of the book. If you missed any problems, be sure to go back and review the appropriate sections in the chapter and the exercises that are provided.

Simplify each expression, using the properties of exponents.

1. $(3x^2y)(-2xy^3)$

2. $\left(\dfrac{8m^2n^5}{2p^3}\right)^2$

3. $(x^4y^{-5})^2$

4. $\dfrac{9c^{-5}d^3}{18c^{-7}d^{-4}}$

Write the following numbers in scientific notation.

5. 4,230,000,000

6. 0.000025

Add the following polynomials.

7. $4x^2 - 3xy + 5y^2$ and $5x^2 + 3xy - 7y^2$

Subtract the following polynomials.

8. $5a^2 - 3ab$ from $4a^2 + 7b^2$

Simplify the following expression.

9. $5x - \{4x + 2[x - 3(x + 2)]\}$

Evaluate the following polynomial if $x = -3$.

10. $x^3 + 2x^2 - 5x - 2$

Multiply each of the following polynomials.

11. $(2a - 5b)(3a + 7b)$

12. $(5m - 3n)(5m + 3n)$

13. $(2a + 3b)^2$

14. $(2x - 5)(x^2 - 4x + 3)$

Factor each of the following polynomials completely.

15. $14a^2b^2 - 21a^2b + 35ab^2$

16. $x^2 - 3xy + 5x - 15y$

17. $25c^2 - 64d^2$

18. $27x^3 - 1$

19. $16a^4 + 2ab^3$

20. $x^2 - 2x - 48$

21. $10x^2 - 39x + 14$

22. $6x^3 + 3x^2 - 45x$

Divide each of the following polynomials, using long division.

23. $\dfrac{2x^2 + 3x - 5}{x - 1}$

24. $\dfrac{3x^2 - 4x - 14}{x - 2}$

Perform the indicated division.

25. $\dfrac{36x^3 + 26x}{2x}$

26. $\dfrac{36x^3 + 15x^2}{3x^2}$

27. Let $P(x) = 3x^3 - 2x^2 + 7x - 5$. Find $P(-2)$.

28. Use synthetic substitution to find the remainder if $x^2 - 3x + 8$ is divided by $x + 2$.

For each of the following, use synthetic division to find the quotient and remainder.

29. $(3x^2 + 5x - 1) \div (x + 1)$

30. $(2x^3 - x + 2) \div (x - 2)$

This is a review of selected topics from the first five chapters.

Solve each of the following.

1. $4x - 5(3x + 7) = 3 + 8x$

2. $|9 - 2x| = 7$

3. $|2x + 7| \leq 9$

4. $3x - 7 \geq 9 + 5x$

5. Find the slope of the line connecting $(-3, 4)$ and $(2, 8)$.

Find the equation of the line satisfying each of the following given conditions.

6. L has a slope of -2 and passes through the point $(-1, 3)$.

7. L passes through the point $(2, -1)$ and is parallel to the line with equation $3x + 4y = 12$.

8. Graph the equation $8x + 3y = 24$.

Solve each of the following systems of equations.

9. $5x + y = 5$
$3x - 2y = 29$

10. $x - y + z = 8$
$5x + y - z = 10$
$2x + y - 3z = -5$

Solve the following system of inequalities.

11. $x - 2y < 3$
$2x + y > 8$

Solve the following system, using matrices and elementary row operations.

12. $2x + 3y = 1$
$x - 2y = 4$

Simplify each of the following expressions.

13. $\dfrac{(x^5)^{-2}(x^2)^3}{(x^{-4})^3}$

14. $(2x^2y^{-3})^{-3}(-2x^{-3}y^2)^2$

Simplify each of the following.

15. $(x^2 + 3x - 2) + (-5x^2 + 2x + 9) - (3x - 2x + 5)$

16. $(3x - 5y)(2x + 3y)$

Completely factor each polynomial.

17. $6x^2 + 7xy - 3y^2$

18. $y^3 - 3y^2 - 5y + 15$

19. Use synthetic division to find the quotient and remainder if $x^3 + 2x^2 - 3x - 8$ is divided by $x - 2$.

20. At a local movie, admission is $6 for adults and $4 for children. If 155 people attended and paid a total of $808, how many of each type of ticket were sold?

CHAPTER SIX RATIONAL EXPRESSIONS

HUNGER AND WORLD FOOD SUPPLIES

Experts disagree about the causes of hunger in the world. Some say the population is increasing faster than the food supply. Others claim that poor agricultural practices keep many developing nations from producing enough food to feed their peoples or that the type of government in a region affects the availability and distribution of food. All these reasons and more can affect the food supply.

During the 1970s, research indicated there was not a food shortage in regions where people were suffering from hunger but rather problems with land ownership, concentration of power in the hands of an elite few, and government decisions that increased wealth rather than relieved hunger.

This research claims world hunger has not been caused by natural limits on the food supply. Hunger is a result of political and social institutions that have failed to meet the needs of all the people.

Current research appears to show that food production is being adversely affected by nearly all the major forms of environmental degradation. Soil erosion, deforestation, air pollution, and acid rain have all affected worldwide food production. Increased use of pesticides and fertilizers appears to create changes in the soil and insect populations that decrease the productivity of some crops.

Cropland is also being lost each year to the expansion of cities, roads, and industries. The land replacing the areas lost to expansion is seldom as fertile. This results in a net loss of crop productivity and less food for more people.

Although the rate of expansion of the world's food supply is beginning to decrease, we still have adequate supplies to feed our current population. The causes of hunger are still primarily political and social. Transport and distribution problems can be corrected if there is a desire by those in power to do so.

Continued global environmental degradation will certainly decrease the world food supply over time. If we can reduce the amount of degradation, we will increase the available food by a corresponding amount. It is important that we attempt to reduce the degradation as much as possible. Less environmental damage will mean fewer hungry people. ∎

Simplification of Rational Expressions   6.1

OBJECTIVE: To simplify a rational expression

Our work in this chapter will expand your experience with algebraic expressions to include algebraic fractions or *rational expressions*. We consider the four basic operations of addition, subtraction, multiplication, and division in the next sections.

Again, the word "rational" comes from "ratio."

Fortunately, you will observe many parallels to your previous work with arithmetic fractions.

First, let's define what we mean by a rational expression. In Chapter 1, we defined a rational number as the ratio of two integers. Similarly, a rational expression can be written as the ratio of two polynomials, where the denominator cannot have the value 0.

> A *rational expression* is the ratio of two polynomials. It can be written as
>
> $\dfrac{P}{Q}$ where P and Q are polynomials and Q cannot have the value 0

The expressions

$$\frac{x-3}{x+1} \qquad \frac{x^2+5}{x-3} \qquad \text{and} \qquad \frac{x^2-2x}{x^2+3x+1}$$

are all rational expressions. The restriction that the denominator of the expressions not be 0 means that certain values for the variable may have to be excluded. This is because division by 0 is undefined.

EXAMPLE 1 Precluding Division by Zero

(*a*) For what values of x is the following expression undefined?

$$\frac{x}{x-5}$$

To answer this question, we must find where the denominator is 0. Set

$$x - 5 = 0$$

or

$$x = 5$$

The expression $\dfrac{x}{x-5}$ is undefined for $x = 5$.

A fraction is undefined when its denominator is equal to 0.

Note that when $x = 5$, $\dfrac{x}{x-5}$

becomes $\dfrac{5}{5-5}$, or $\dfrac{5}{0}$.

(*b*) For what values of y is the following expression undefined?

$$\frac{3}{x+5}$$

Again, set the denominator equal to 0:

$x + 5 = 0$

or

$x = -5$

The expression $\dfrac{3}{x + 5}$ is undefined for $x = -5$.

Check Yourself 1

For what values of the variable are the following expressions undefined?

1. $\dfrac{1}{r + 7}$ **2.** $\dfrac{5}{2x - 9}$

Scientific calculators are often used to evaluate rational expressions for values of the variable. The *parentheses keys* help in this process.

EXAMPLE 2 Evaluating Algebraic Expressions

Using a scientific calculator, evaluate the following expressions for the given value of the variable.

(a) $\dfrac{3x}{2x - 5}$ for $x = 4$

CAUTION
Be sure to use the parenthesis keys before the 2 and after the 5.

Enter the expression in your calculator as follows:

$3 \boxed{\times} 4 \boxed{\div} \boxed{(} 2 \boxed{\times} 4 \boxed{-} 5 \boxed{)} \boxed{=}$

The display will read the value 4.

(b) $\dfrac{2x + 7}{4x - 11}$ for $x = 4$

Enter the expression as follows:

$\boxed{(} 2 \boxed{\times} 4 \boxed{+} 7 \boxed{)} \boxed{\div} \boxed{(} 4 \boxed{\times} 4 \boxed{-} 11 \boxed{)} \boxed{=}$

The display will read the value 3.

Check Yourself 2

Using a scientific calculator, evaluate each of the following.

1. $\dfrac{5x}{3x - 2}$ for $x = 4$ **2.** $\dfrac{2x + 9}{3x - 4}$ for $x = 3$

Generally, we want to write rational expressions in the simplest possible form. To begin our discussion of simplifying rational expressions, let's review for a moment. As we pointed out above, there are many parallels to your work with arithmetic fractions. Recall that

$$\frac{3}{5} = \frac{3 \cdot 2}{5 \cdot 2} = \frac{6}{10}$$

We multiply by $\frac{2}{2}$, or 1.

so

$$\frac{3}{5} \quad \text{and} \quad \frac{6}{10}$$

name equivalent fractions. In a similar fashion,

$$\frac{10}{15} = \frac{5 \cdot 2}{5 \cdot 3} = \frac{2}{3}$$

Divide numerator and denominator by the common factor 5.

so

$$\frac{10}{15} \quad \text{and} \quad \frac{2}{3}$$

We use the fact that $\frac{5}{5} = 1$.

name equivalent fractions.

We can always multiply or divide the numerator and denominator of a fraction by the same nonzero number. The same pattern is true in algebra.

FUNDAMENTAL PRINCIPLE OF RATIONAL EXPRESSIONS

For polynomials **P**, **Q**, and **R**,

$$\frac{P}{Q} = \frac{PR}{QR} \qquad Q \neq 0, \ R \neq 0$$

This principle can be used in two ways. We can multiply or divide the numerator and denominator of a rational expression by the same nonzero polynomial. The result will always be an expression that is equivalent to the original one.

In simplifying arithmetic fractions, we used this principle to divide the numerator and denominator by all common factors. With arithmetic fractions, those common factors are generally easy to recognize. Given rational expressions where the numerator and denominator are polynomials, we must determine those factors as our first step. The most important tools for simplifying expressions are the factoring techniques of Chapter 5.

In fact, you will see that most of the methods of this chapter depend on factoring polynomials.

EXAMPLE 3 Simplifying Rational Expressions

Simplify each rational expression. Assume denominators are not 0.

$$(a) \quad \frac{4x^2y}{12xy^2} = \frac{4xy \cdot x}{4xy \cdot 3y}$$

$$= \frac{x}{3y}$$

We find the common factors 4, x, and y in the numerator and denominator.

We divide the numerator and denominator by the common factor $4xy$. Note that

$$\frac{4xy}{4xy} = 1$$

Factor the numerator and the denominator.

(b) $\dfrac{3x - 6}{x^2 - 4} = \dfrac{3(x - 2)}{(x + 2)(x - 2)}$

We can now divide the numerator and denominator by the common factor $x - 2$:

We have *divided* the numerator and denominator by the common factor $x - 2$. Again note that

$\dfrac{x - 2}{x - 2} = 1$

$\dfrac{3(x - 2)}{(x + 2)(x - 2)} = \dfrac{3}{x + 2}$

and the rational expression is in simplest form.

Be Careful! Given the expression

$\dfrac{x + 2}{x + 3}$

CAUTION
Pick any value other than 0 for the variable x, and substitute. You will quickly see that

$\dfrac{x + 2}{x + 3} \neq \dfrac{2}{3}$

students are often tempted to divide by variable x, as in

$\dfrac{x + 2}{x + 3} = \dfrac{2}{3}$

This is not a valid operation. We can only divide by common *factors,* and in the expression above the variable x is a *term* in both the numerator and the denominator. The numerator and denominator of a rational expression must be factored *before* common factors are divided out.

$\dfrac{x + 2}{x + 3}$

is in its simplest possible form.

Check Yourself 3

Simplify each expression.

1. $\dfrac{36a^3b}{9ab^2}$ 　　　　　　　 **2.** $\dfrac{x^2 - 25}{4x + 20}$

The same techniques are used when trinomials need to be factored. Here are some further examples that illustrate the simplification of rational expressions.

EXAMPLE 4 Simplifying Rational Expressions

Simplify each rational expression.

Divide by the common factor $x + 1$, using the fact that

$\dfrac{x + 1}{x + 1} = 1$

where $x \neq -1$.

(a) $\dfrac{5x^2 - 5}{x^2 - 4x - 5} = \dfrac{5(x - 1)(x + 1)}{(x - 5)(x + 1)}$

$= \dfrac{5(x - 1)}{x - 5}$

(b) $\dfrac{2x^2 + x - 6}{2x^2 - x - 3} = \dfrac{(x + 2)(2x - 3)}{(x + 1)(2x - 3)}$

$\qquad\qquad = \dfrac{x + 2}{x + 1}$

(c) $\dfrac{x^3 + 2x^2 - 3x - 6}{x^3 + 8} = \dfrac{(x + 2)(x^2 - 3)}{(x + 2)(x^2 - 2x + 4)} = \dfrac{x^2 - 3}{x^2 - 2x + 4}$

Here we factor by grouping in the numerator and use the sum of cubes in the denominator. Note that

$x^3 + 2x^2 - 3x - 6$
$= x^2(x + 2) - 3(x + 2)$
$= (x + 2)(x^2 - 3)$

Check Yourself 4

Simplify each rational expression.

1. $\dfrac{x^2 - 5x + 6}{3x^2 - 6x}$

2. $\dfrac{3x^2 + 14x - 5}{3x^2 + 2x - 1}$

Simplifying certain algebraic expressions involves recognizing a particular pattern. Verify for yourself that

$$3 - 9 = -(9 - 3)$$

In general, it is true that

$$a - b = -(-a + b) = -(b - a)$$

or, by dividing both sides of the equation by $b - a$,

$$\frac{a - b}{b - a} = \frac{-(b - a)}{b - a} = -1$$

Note that

$\dfrac{a - b}{a - b} = 1$

but

$\dfrac{a - b}{b - a} = -1$

where $a \neq b$.

The following examples make use of this result.

EXAMPLE 5 Simplifying Rational Expressions

Simplify each rational expression.

Note

$\dfrac{x - 2}{2 - x} = -1$

(a) $\dfrac{2x - 4}{4 - x^2} = \dfrac{2 \overset{-1}{\cancel{(x - 2)}}}{(2 + x)\cancel{(2 - x)}}$

$\qquad = \dfrac{2(-1)}{2 + x} = \dfrac{-2}{2 + x}$

(b) $\dfrac{9 - x^2}{x^2 + 2x - 15} = \dfrac{(3 + x)\overset{-1}{\cancel{(3 - x)}}}{(x + 5)\cancel{(x - 3)}}$

$\qquad = \dfrac{(3 + x)(-1)}{x + 5}$

$\qquad = \dfrac{-x - 3}{x + 5}$

Check Yourself 5

Simplify each rational expression.

1. $\dfrac{5x - 20}{16 - x^2}$

2. $\dfrac{x^2 - 6x - 27}{81 - x^2}$

The following algorithm will summarize our work with simplifying rational expressions.

SIMPLIFYING RATIONAL EXPRESSIONS

1. Completely factor both the numerator and the denominator of the expression.
2. Divide the numerator and denominator by *all* common factors.
3. The resulting expression will be in simplest form.

CHECK YOURSELF ANSWERS

1. (1) $r = -7$; (2) $x = \dfrac{9}{2}$ **2.** (1) 2; (2) 3   **3.** (1) $\dfrac{4a^2}{b}$; (2) $\dfrac{x - 5}{4}$

4. (1) $\dfrac{x - 3}{3x}$; (2) $\dfrac{x + 5}{x + 1}$ **5.** (1) $\dfrac{-5}{x + 4}$; (2) $\dfrac{-x - 3}{x + 9}$

6.1 EXERCISES

Build Your Skills

For what values of the variable are each of the following rational expressions undefined?

1. $\dfrac{x}{x - 5}$

2. $\dfrac{y}{y + 7}$

3. $\dfrac{x + 5}{3}$

4. $\dfrac{a - 6}{4}$

5. $\dfrac{2m - 3}{2m - 1}$

6. $\dfrac{3x - 2}{3x + 1}$

7. $\dfrac{2x + 5}{x}$

8. $\dfrac{3c - 7}{c}$

9. $\dfrac{x(x + 1)}{x + 2}$

10. $\dfrac{x + 2}{3x - 7}$

11. $\dfrac{4 - x}{x}$

12. $\dfrac{2a + 7}{3x + \dfrac{1}{3}}$

Evaluate each of the following expressions, using a scientific calculator.

13. $\dfrac{3x}{2x-1}$ for $x=2$

14. $\dfrac{5x}{4x-3}$ for $x=2$

15. $\dfrac{2x+3}{x+3}$ for $x=-6$

16. $\dfrac{4x-7}{2x-1}$ for $x=-2$

Simplify each of the following fractions. Assume the denominators are not 0.

17. $\dfrac{14}{21}$

18. $\dfrac{45}{75}$

19. $\dfrac{4x^5}{6x^2}$

20. $\dfrac{25x^6}{20x^2}$

21. $\dfrac{10x^2y^5}{25xy^2}$

22. $\dfrac{18a^2b^3}{24a^4b^3}$

23. $\dfrac{-42x^3y}{14xy^3}$

24. $\dfrac{-15x^3y^3}{-20xy^2}$

25. $\dfrac{28a^5b^3c^2}{84a^2bc^4}$

26. $\dfrac{-52p^5q^3r^2}{39p^3q^5r^2}$

Simplify each of the following rational expressions.

27. $\dfrac{6x-24}{x^2-16}$

28. $\dfrac{x^2-25}{3x-15}$

29. $\dfrac{x^2+2x+1}{6x+6}$

30. $\dfrac{5y^2-10y}{y^2+y-6}$

31. $\dfrac{x^2-5x-14}{x^2-49}$

32. $\dfrac{2m^2+11m-21}{4m^2-9}$

33. $\dfrac{3b^2-14b-5}{b-5}$

34. $\dfrac{a^2-9b^2}{a^2+8ab+15b^2}$

35. $\dfrac{2y^2+3yz-5z^2}{2y^2+11yz+15z^2}$

36. $\dfrac{6x^2-x-2}{3x^2-5x+2}$

37. $\dfrac{x^3-64}{x^2-16}$

38. $\dfrac{r^2-rs-6s^2}{r^3+8s^3}$

39. $\dfrac{a^4-81}{a^2+5a+6}$

40. $\dfrac{c^4-16}{c^2-3c-10}$

41. $\dfrac{xy-2x+3y-6}{x^2+8x+15}$

42. $\dfrac{cd-3c+5d-15}{d^2-7d+12}$

43. $\dfrac{x^2+3x-18}{x^3-3x^2-2x+6}$

44. $\dfrac{y^2+2y-35}{y^2-5y-3y+15}$

45. $\dfrac{2m-10}{25-m^2}$

46. $\dfrac{5x-20}{16-x^2}$

47. $\dfrac{49-x^2}{2x^2-13x-7}$

48. $\dfrac{2x^2-7x+3}{9-x^2}$

Transcribe Your Skills

49. Explain why the following statement is false.

$$\dfrac{6m^2+2m}{2m}=6m^2+1$$

50. State and explain the fundamental principle of fractions.

51. The rational expression $\dfrac{x^2 - 4}{x + 2}$ can be simplified to $x - 2$. Is this reduction true for all values of x? Explain.

52. What is meant by a rational expression in lowest terms?

Think About These

Simplify each of the following.

53. $\dfrac{2(x + h) - 2x}{(x + h) - x}$

54. $\dfrac{-3(x + h) - (-3x)}{(x + h) - x}$

55. $\dfrac{3(x + h) - 3 - (3x - 3)}{(x + h) - x}$

56. $\dfrac{2(x + h) + 5 - (2x + 5)}{(x + h) - x}$

57. $\dfrac{(x + h)^2 - x^2}{(x + h) - x}$

58. $\dfrac{(x + h)^3 - x^3}{(x + h) - x}$

Skillscan

Perform the indicated operations.

a. $\dfrac{2}{3} \cdot \dfrac{4}{5}$ **b.** $\dfrac{5}{6} \cdot \dfrac{4}{11}$ **c.** $\dfrac{4}{7} \div \dfrac{8}{5}$ **d.** $\dfrac{1}{6} \div \dfrac{7}{9}$

e. $\dfrac{5}{8} \cdot \dfrac{16}{15}$ **f.** $\dfrac{15}{21} \div \dfrac{10}{7}$ **g.** $\dfrac{15}{8} \cdot \dfrac{24}{25}$

h. $\dfrac{28}{16} \div \dfrac{21}{20}$

6.2   Multiplication and Division of Rational Expressions

TAPE IN15

OBJECTIVES: 1. To multiply rational expressions
2. To divide rational expressions

Once again, let's turn to an example from arithmetic to begin our discussion of multiplying rational expressions. Recall that to multiply two fractions, we multiply the numerators and multiply the denominators. For instance,

$$\frac{2}{5} \cdot \frac{3}{7} = \frac{2 \cdot 3}{5 \cdot 7} = \frac{6}{35}$$

In algebra, the pattern is exactly the same.

For all problems with rational expressions, assume denominators are not 0.

MULTIPLYING RATIONAL EXPRESSIONS

For polynomials P, Q, R, and S,
$$\frac{P}{Q} \cdot \frac{R}{S} = \frac{PR}{QS} \quad \text{where } Q \neq 0 \text{ and } S \neq 0$$

EXAMPLE 1 Multiplying Rational Expressions

Multiply.

$$\frac{2x^3}{5y^2} \cdot \frac{10y}{3x^2} = \frac{20x^3y}{15x^2y^2} = \frac{5x^2y \cdot 4x}{5x^2y \cdot 3y} = \frac{4x}{3y}$$

We divide by the common factor $5x^2y$.

Check Yourself 1

Multiply.

$$\frac{9a^2b^3}{5ab^4} \cdot \frac{20ab^2}{27ab^3}$$

Generally, you will find it best to divide by any common factors before you multiply. That is why the factoring methods of Chapter 5 play such an important role here. The following example illustrates.

EXAMPLE 2 Multiplying Rational Expressions

Multiply as indicated.

(a) $\dfrac{x}{x^2 - 3x} \cdot \dfrac{6x - 18}{9x}$ Factor.

$= \dfrac{x}{x(x - 3)} \cdot \dfrac{6(x - 3)}{9x}$ Divide by the common factors of 3, x, and $x - 3$

$= \dfrac{2}{3x}$

(b) $\dfrac{x^2 - y^2}{5x^2 - 5xy} \cdot \dfrac{10xy}{x^2 + 2xy + y^2}$ Factor and divide by the common factors of 5, x, $x + y$, and $x - y$.

$= \dfrac{(x + y)(x - y)}{5x(x - y)} \cdot \dfrac{10xy}{(x + y)(x + y)}$

$= \dfrac{2y}{x + y}$

(c) $\dfrac{4}{x^2 - 2x} \cdot \dfrac{10x - 5x^2}{8x + 24}$

$= \dfrac{4}{x(x - 2)} \cdot \dfrac{\overset{-1}{5x(2 - x)}}{8(x + 3)}$ Note that

$= \dfrac{-5}{2(x + 3)}$ $\dfrac{2 - x}{x - 2} = -1$

Check Yourself 2

Multiply as indicated.

1. $\dfrac{x^2 - 5x - 14}{4x^2} \cdot \dfrac{8x + 56}{x^2 - 49}$ **2.** $\dfrac{x}{2x - 6} \cdot \dfrac{3x - x^2}{2}$

The following algorithm summarizes our work in multiplying rational expressions.

MULTIPLYING RATIONAL EXPRESSIONS

STEP 1   Write each numerator and denominator in completely factored form.

STEP 2   Divide by any common factors appearing in both a numerator and a denominator.

STEP 3   Multiply as needed to form the desired product.

In dividing rational expressions, you can again use your experience from arithmetic. Recall that

We invert the *divisor* (the second fraction) and multiply.

$$\frac{3}{5} \div \frac{2}{3} = \frac{3}{5} \cdot \frac{3}{2} = \frac{9}{10}$$

Once more, the pattern in algebra is identical.

DIVIDING RATIONAL EXPRESSIONS

For polynomials *P*, *Q*, *R*, and *S*,

$$\frac{P}{Q} \div \frac{R}{S} = \frac{P}{Q} \cdot \frac{S}{R} = \frac{PS}{QR}$$

where $Q \neq 0$, $R \neq 0$, and $S \neq 0$.

To divide rational expressions, invert the divisor and multiply as before. The following example illustrates.

EXAMPLE 3 Dividing Rational Expressions

Divide as indicated.

Invert the divisor and multiply.

(a) $\dfrac{3x^2}{8x^3y} \div \dfrac{9x^2y^2}{4y^4} = \dfrac{3x^2}{8x^3y} \cdot \dfrac{4y^4}{9x^2y^2} = \dfrac{y}{6x^3}$

CAUTION
Be Careful Invert the divisor, then factor.

(b) $\dfrac{2x^2 + 4xy}{9x - 18y} \div \dfrac{4x + 8y}{3x - 6y} = \dfrac{2x^2 + 4xy}{9x - 18y} \cdot \dfrac{3x - 6y}{4x + 8y}$

$$= \frac{\overset{}{2}x(x + 2y)}{\underset{3}{9}(x - 2y)} \cdot \frac{\overset{}{3}(x - 2y)}{\underset{2}{4}(x + 2y)} = \frac{x}{6}$$

(c) $\dfrac{2x^2 - x - 6}{4x^2 + 6x} \div \dfrac{x^2 - 4}{4x} = \dfrac{2x^2 - x - 6}{4x^2 + 6x} \cdot \dfrac{4x}{x^2 - 4}$

$$= \frac{(2x + 3)(x - 2)}{2x(2x + 3)} \cdot \frac{4x^2}{(x + 2)(x - 2)} = \frac{2x}{x + 2}$$

Check Yourself 3

Divide and simplify.

1. $\dfrac{5xy}{7x^3} \div \dfrac{10y^2}{14x^3}$

2. $\dfrac{3x - 9y}{2x + 10y} \div \dfrac{x^2 - 3xy}{4x^2 + 20xy}$

3. $\dfrac{x^2 - 9}{x^3 - 27} \div \dfrac{x^2 - 2x - 15}{2x^2 - 10x}$

Let's summarize our work in dividing fractions with an algorithm.

DIVIDING RATIONAL EXPRESSIONS

STEP 1   Invert the divisor (the *second* rational expression) to write the problem as one of multiplication.

STEP 2   Proceed as in the algorithm for multiplication of rational expressions.

CHECK YOURSELF ANSWERS

1. $\dfrac{4a}{3b^2}$ **2.** (1) $\dfrac{2(x + 2)}{x^2}$; (2) $\dfrac{-x^2}{4}$ **3.** (1) $\dfrac{x}{y}$; (2) 6; (3) $\dfrac{2x}{x^2 + 3x + 9}$

6.2 EXERCISES

Build Your Skills

Multiply or divide as indicated. Express your result in simplest form.

1. $\dfrac{x^2}{3} \cdot \dfrac{6x}{x^4}$

2. $\dfrac{-y^3}{10} \cdot \dfrac{15y}{y^6}$

3. $\dfrac{a}{7a^3} \div \dfrac{a^2}{21}$

4. $\dfrac{p^5}{8} \div \dfrac{-p^2}{12p}$

5. $\dfrac{4xy^2}{15x^3} \cdot \dfrac{25xy}{16y^3}$

6. $\dfrac{3x^3y}{10xy^3} \cdot \dfrac{5xy^2}{-9xy^3}$

7. $\dfrac{8b^3}{15ab} \div \dfrac{2ab^2}{20ab^3}$

8. $\dfrac{4x^2y^2}{9x^3} \div \dfrac{-8y^2}{27xy}$

9. $\dfrac{m^3n}{2mn} \cdot \dfrac{6mn^2}{m^3n} \div \dfrac{3mn}{5m^2n}$

10. $\dfrac{4cd^2}{5cd} \cdot \dfrac{3c^3d}{2c^2d} \div \dfrac{9cd}{20cd^3}$

11. $\dfrac{5x + 15}{3x} \cdot \dfrac{9x^2}{2x + 6}$

12. $\dfrac{a^2 - 3a}{5a} \cdot \dfrac{20a^2}{3a - 9}$

13. $\dfrac{3b - 15}{6b} \div \dfrac{4b - 20}{9b^2}$

14. $\dfrac{7m^2 + 28m}{4m} \div \dfrac{5m + 20}{12m^2}$

15. $\dfrac{x^2 - 3x - 10}{5x} \cdot \dfrac{15x^2}{3x - 15}$

16. $\dfrac{y^2 - 8y}{4y} \cdot \dfrac{12y^2}{y^2 - 64}$

17. $\dfrac{c^2 + 2c - 8}{6c} \div \dfrac{5c + 20}{18c}$

18. $\dfrac{m^2 - 49}{5m} \div \dfrac{3m + 21}{20m^2}$

19. $\dfrac{x^2 - 2x - 8}{4x - 16} \cdot \dfrac{10x}{x^2 - 4}$

20. $\dfrac{y^2 + 7y + 10}{y^2 + 5y} \cdot \dfrac{2y}{y^2 - 4}$

21. $\dfrac{d^2 - 3d - 18}{16d - 96} \div \dfrac{d^2 - 9}{20d}$

22. $\dfrac{b^2 + 6b + 8}{b^2 + 4b} \div \dfrac{b^2 - 4}{2b}$

23. $\dfrac{2x^2 - x - 3}{3x^2 + 7x + 4} \cdot \dfrac{3x^2 - 11x - 20}{4x^2 - 9}$

24. $\dfrac{4p^2 - 1}{2p^2 - 9p - 5} \cdot \dfrac{3p^2 - 13p - 10}{9p^2 - 4}$

25. $\dfrac{a^2 - 9}{2a^2 - 6a} \div \dfrac{2a^2 + 5a - 3}{4a^2 - 1}$

26. $\dfrac{2x^2 - 5x - 7}{4x^2 - 9} \div \dfrac{5x^2 + 5x}{2x^2 + 3x}$

27. $\dfrac{2w - 6}{w^2 + 2w} \cdot \dfrac{3w}{3 - w}$

28. $\dfrac{3y - 15}{y^2 + 3y} \cdot \dfrac{4y}{5 - y}$

29. $\dfrac{a - 7}{2a + 6} \div \dfrac{21 - 3a}{a^2 + 3a}$

30. $\dfrac{x - 4}{x^2 + 2x} \div \dfrac{16 - 4x}{3x + 6}$

31. $\dfrac{x^2 - 9y^2}{2x^2 - xy - 15y^2} \cdot \dfrac{4x + 10y}{x^2 + 3xy}$

32. $\dfrac{2a^2 - 7ab - 15b^2}{2ab - 10b^2} \cdot \dfrac{2a^2 - 3ab}{4a^2 - 9b^2}$

33. $\dfrac{3m^2 - 5mn + 2n^2}{9m^2 - 4n^2} \div \dfrac{m^3 - m^2n}{9m^2 + 6mn}$

34. $\dfrac{2x^2y - 5xy^2}{4x^2 - 25y^2} \div \dfrac{4x^2 + 20xy}{2x^2 + 15xy + 25y^2}$

35. $\dfrac{x^3 + 8}{x^2 - 4} \cdot \dfrac{5x - 10}{x^3 - 2x^2 + 4x}$

36. $\dfrac{a^3 - 27}{a^2 - 9} \div \dfrac{a^3 + 3a^2 + 9a}{3a^3 + 9a^2}$

Transcribe Your Skills

37. Explain the process used to divide two rational expressions.

38. Outline two different procedures that can be used to multiply rational expressions.

Think About These

Evaluate each of the following for the given variable value.

39. $\dfrac{x^3 - 3x^2 + 2x - 6}{x^2 - 9} \cdot \dfrac{5x^2 + 15x}{20x}$, $x = 4$

40. $\dfrac{3a^3 + a^2 - 9a - 3}{15a^2 + 5a} \cdot \dfrac{3a^2 + 9}{a^4 - 9}$, $a = 2$

41. $\dfrac{x^4 - 16}{x^2 + x - 6} \div (x^3 + 4x)$, $x = -5$

42. $\dfrac{w^3 + 27}{w^2 + 2w - 3} \div (w^3 - 3w^2 + 9w)$, $w = 3$

43. $\dfrac{x^2 - 4}{5x + 10} \cdot \dfrac{x^2 + 2x - 8}{2x^2 + 8x} \cdot \dfrac{10x^2}{x^2 - 4x + 4}$, $x = \pi$

44. $\dfrac{x^3 + 27}{x^2 - 9} \cdot \dfrac{x^2 + 3x - 18}{x^3 - 3x^2 + 9x} \cdot \dfrac{7x}{5x^2 + 30x}$, $x = -3$

45. $\dfrac{w^2 - 2w - 8}{2w - 8} \cdot \dfrac{w^2 + 5w}{w^2 + 5w + 6} \div \dfrac{w^2 + 2w - 15}{w^2 - 9}$,
$w = 6$

46. $\dfrac{14m - 7}{m^2 + 3m - 4} \cdot \dfrac{m^2 + 6m + 8}{2m^2 + 5m - 3} \div \dfrac{m^2 + 2m}{m^2 + 2m - 3}$,
$m = 7$

47. If the price of x units of food is $\dfrac{5x + 20}{4}$ and the demand for x units of food is $\dfrac{8000}{5x^2 + 20x}$, write the expression for the product of the price and demand in simplest terms.

48. In determining the minimum cost per hectare to produce a certain export crop, a farmer finds that the product of the factors 5.4, x, and $x - 3$ must be divided by $x^2 - 9$. Write the expression for minimum cost in simplest form.

Skillscan

Perform the indicated operations.

a. $\dfrac{7}{16} + \dfrac{3}{16}$ **b.** $\dfrac{5}{12} - \dfrac{1}{12}$ **c.** $\dfrac{3}{4} + \dfrac{1}{2}$

d. $\dfrac{7}{10} - \dfrac{3}{5}$ **e.** $\dfrac{5}{6} + \dfrac{3}{8}$ **f.** $\dfrac{7}{8} - \dfrac{3}{5}$

g. $\dfrac{5}{12} + \dfrac{7}{16}$ **h.** $\dfrac{9}{10} - \dfrac{2}{15}$

Addition and Subtraction of Rational Expressions 6.3

OBJECTIVES: 1. To add rational expressions
2. To subtract rational expressions

TAPE IN15

Recall that adding or subtracting two arithmetic fractions with the same denominator is straightforward. The same is true in algebra. To add or subtract two rational expressions with the same denominator, we add or subtract their numerators and then write that sum or difference over the common denominator. In symbols:

ADDING OR SUBTRACTING RATIONAL EXPRESSIONS

$$\frac{P}{R} + \frac{Q}{R} = \frac{P + Q}{R}$$

and

$$\frac{P}{R} - \frac{Q}{R} = \frac{P - Q}{R}$$

where $R \neq 0$.

EXAMPLE 1 Adding and Subtracting Rational Expressions

Perform the indicated operations.

$$\frac{3}{2a^2} - \frac{1}{2a^2} + \frac{5}{2a^2} = \frac{3 - 1 + 5}{2a^2}$$

$$= \frac{7}{2a^2}$$

Since we have common denominators, we simply perform the indicated operations on the numerators.

Check Yourself 1

Perform the indicated operations.

$$\frac{5}{3y^2} + \frac{4}{3y^2} - \frac{7}{3y^2}$$

The sum or difference of rational expressions should always be expressed in simplest form. Consider the following example.

EXAMPLE 2 Adding and Subtracting Rational Expressions

Add or subtract as indicated.

(a) $\dfrac{5x}{x^2 - 9} + \dfrac{15}{x^2 - 9}$ Add the numerators.

$$= \frac{5x + 15}{x^2 - 9}$$

$$= \frac{5(x + 3)}{(x - 3)(x + 3)} = \frac{5}{x - 3}$$ Factor and divide by the common factor.

(b) $\dfrac{3x + y}{2x} - \dfrac{x - 3y}{2x} = \dfrac{(3x + y) - (x - 3y)}{2x}$ Be sure to *enclose the second numerator* in parentheses.

$$= \frac{3x + y - x + 3y}{2x}$$ Remove the parentheses by *changing each sign.*

$$= \frac{2x + 4y}{2x} = \frac{2(x + 2y)}{2x}$$ Factor and divide by the common factor of 2.

$$= \frac{x + 2y}{x}$$

Check Yourself 2

Perform the indicated operations.

1. $\dfrac{6a}{a^2 - 2a - 8} + \dfrac{12}{a^2 - 2a - 8}$ **2.** $\dfrac{5x - y}{3y} - \dfrac{2x - 4y}{3y}$

Now, what if our rational expressions *do not* have common denominators? In that case, we must use the least common denominator (LCD). The *least common denominator* is the simplest polynomial that is divisible by each of the individual denominators. Each expression in the desired sum or difference is then "built up" to an equivalent expression having that LCD as a denominator. We can then add or subtract as before.

While in many cases we can find the LCD by inspection, we can state an algorithm for finding the LCD that is similar to the one used in arithmetic.

FINDING THE LEAST COMMON DENOMINATOR

STEP 1 Write each of the denominators in completely factored form.

STEP 2 Write the LCD as the product of each prime factor to the highest power to which it appears in the factored form of any individual denominators.

Again, we see the key role that factoring plays in the process of working with rational expressions.

The following example illustrates the procedure.

EXAMPLE 3 Finding the LCD for Two Rational Expressions

Find the LCD for each of the following pairs of rational expressions.

(a) $\dfrac{3}{4x^2}$ and $\dfrac{5}{6xy}$

Factor the denominators.

$4x^2 = 2^2 \cdot x^2$

$6xy = 2 \cdot 3 \cdot x \cdot y$

You may very well be able to find this LCD by inspecting the numerical coefficients and the variable factors.

The LCD must have the factors

$2^2 \cdot 3 \cdot x^2 \cdot y$

and so $12x^2y$ is the desired LCD.

(b) $\dfrac{7}{x-3}$ and $\dfrac{2}{x+5}$

Here neither denominator can be factored. The LCD must have the factors $x - 3$ and $x + 5$. So the LCD is

$(x - 3)(x + 5)$

It is generally best to leave the LCD in this factored form.

Check Yourself 3

Find the LCD for the following pairs of rational expressions.

1. $\dfrac{3}{8a^3}$ and $\dfrac{5}{6a^2}$

2. $\dfrac{4}{x+7}$ and $\dfrac{3}{x-5}$

Let's see how factoring techniques are applied in the following example.

EXAMPLE 4 Finding the LCD for Two Rational Expressions

Find the LCD for the following pairs of rational expressions.

(a) $\dfrac{2}{x^2 - x - 6}$ and $\dfrac{1}{x^2 - 9}$

Factoring, we have

$$x^2 - x - 6 = (x + 2)(x - 3)$$

and

$$x^2 - 9 = (x + 3)(x - 3)$$

The LCD of the given expressions is then

The LCD must contain *each* of the factors appearing in the original denominators.

$$(x + 2)(x - 3)(x + 3)$$

(b) $\dfrac{5}{x^2 - 4x + 4}$ and $\dfrac{3}{x^2 + 2x - 8}$

Again we factor:

$$x^2 - 4x + 4 = (x - 2)^2$$
$$x^2 + 2x - 8 = (x - 2)(x + 4)$$

The LCD must contain $(x - 2)^2$ as a factor since $x - 2$ appears *twice* as a factor in the first denominator.

The LCD is then

$$(x - 2)^2(x + 4)$$

Check Yourself 4

Find the LCD for the following pairs of rational expressions.

1. $\dfrac{3}{x^2 - 2x - 15}$ and $\dfrac{5}{x^2 - 25}$

2. $\dfrac{5}{y^2 + 6y + 9}$ and $\dfrac{3}{y^2 - y - 12}$

Let's look at some examples where the concept of the LCD is applied in adding or subtracting rational expressions.

EXAMPLE 5 Adding and Subtracting Rational Expressions

Add or subtract as indicated.

(a) $\dfrac{5}{4xy} + \dfrac{3}{2x^2}$

The LCD for $2x^2$ and $4xy$ is $4x^2y$. We rewrite each of the rational expressions with the LCD as a denominator.

$$\frac{5}{4xy} + \frac{3}{2x^2} = \frac{5 \cdot x}{4xy \cdot x} + \frac{3 \cdot 2y}{2x^2 \cdot 2y}$$

$$= \frac{5x}{4x^2y} + \frac{6y}{4x^2y} = \frac{5x + 6y}{4x^2y}$$

Note that in each case we are multiplying by 1, $\frac{x}{x}$ in the first fraction and $\frac{2y}{2y}$ in the second fraction. That's why the resulting fractions are equivalent to the original ones.

(b) $\dfrac{3}{a - 3} - \dfrac{2}{a}$

The LCD for a and $a - 3$ is $a(a - 3)$. We rewrite each of the rational expressions with that LCD as a denominator.

$$\frac{3}{a - 3} - \frac{2}{a}$$

$$= \frac{3a}{a(a - 3)} - \frac{2(a - 3)}{a(a - 3)} \qquad \text{Subtract the numerators.}$$

$$= \frac{3a - 2(a - 3)}{a(a - 3)} \qquad \text{Remove the parentheses, and combine like terms.}$$

$$= \frac{3a - 2a + 6}{a(a - 3)} = \frac{a + 6}{a(a - 3)}$$

Check Yourself 5

Perform the indicated operations.

1. $\dfrac{3}{2ab} + \dfrac{4}{5b^2}$ **2.** $\dfrac{5}{y + 2} - \dfrac{3}{y}$

Let's proceed to an example in which factoring will be required in forming the LCD.

EXAMPLE 6 Adding and Subtracting Rational Expressions

Add or subtract as indicated.

(a) $\dfrac{-5}{x^2 - 3x - 4} + \dfrac{8}{x^2 - 16}$

We first factor the two denominators.

$$x^2 - 3x - 4 = (x + 1)(x - 4)$$

$$x^2 - 16 = (x + 4)(x - 4)$$

We see that the LCD must be

$$(x + 1)(x + 4)(x - 4)$$

Again, rewriting the original expressions gives

We use the facts that

$$\frac{x+4}{x+4} = 1 \quad \text{and}$$

$$\frac{x+1}{x+1} = 1$$

Now add the numerators.

$$\frac{-5}{(x+1)(x-4)} + \frac{8}{(x-4)(x+4)}$$

$$= \frac{-5(x+4)}{(x+1)(x-4)(x+4)} + \frac{8(x+1)}{(x-4)(x+4)(x+1)}$$

$$= \frac{-5(x+4) + 8(x+1)}{(x+1)(x-4)(x+4)}$$

$$= \frac{-5x - 20 + 8x + 8}{(x+1)(x-4)(x+4)}$$

Combine like terms in the numerator.

$$= \frac{3x - 12}{(x+1)(x-4)(x+4)}$$

Factor and divide by the common factor $x - 4$.

$$= \frac{3(x-4)}{(x+1)(x-4)(x+4)}$$

$$= \frac{3}{(x+1)(x+4)}$$

(b) $\dfrac{5}{x^2 - 5x + 6} - \dfrac{3}{4x - 12}$

Again, factor the denominators.

$$x^2 - 5x + 6 = (x - 2)(x - 3)$$

$$4x - 12 = 4(x - 3)$$

The LCD is $4(x - 2)(x - 3)$, and proceeding as before, we have

$$\frac{5}{(x-2)(x-3)} - \frac{3}{4(x-3)}$$

$$= \frac{5 \cdot 4}{4(x-2)(x-3)} - \frac{3(x-2)}{4(x-2)(x-3)}$$

Simplify the numerator, and combine like terms.

$$= \frac{20 - 3(x-2)}{4(x-2)(x-3)} = \frac{20 - 3x + 6}{4(x-2)(x-3)} = \frac{-3x + 26}{4(x-2)(x-3)}$$

Check Yourself 6

Add or subtract as indicated.

1. $\dfrac{-4}{x^2 - 4} + \dfrac{7}{x^2 - 3x - 10}$

2. $\dfrac{5}{3x - 9} - \dfrac{2}{x^2 - 9}$

Our next example will look slightly different from those you have seen thus far, but the reasoning involved in performing the subtraction is exactly the same.

EXAMPLE 7 Adding and Subtracting Rational Expressions

Subtract

$$3 - \frac{5}{2x - 1}$$

Solution To perform the subtraction, remember that 3 is equivalent to the fraction $\frac{3}{1}$, so

$$3 - \frac{5}{2x - 1} = \frac{3}{1} - \frac{5}{2x - 1}$$

The LCD for 1 and $2x - 1$ is just $2x - 1$. We now rewrite the first expression with that denominator:

$$3 - \frac{5}{2x - 1} = \frac{3(2x - 1)}{2x - 1} - \frac{5}{2x - 1}$$

$$= \frac{3(2x - 1) - 5}{2x - 1} = \frac{6x - 8}{2x - 1}$$

Check Yourself 7

Subtract

$$\frac{4}{3x + 1} - 3$$

Our final example uses an observation from Section 6.1. Recall that

$$a - b = -(b - a)$$
$$= -1(b - a)$$

Let's see how this is used in adding rational expressions.

EXAMPLE 8 Adding and Subtracting Rational Expressions

Add

$$\frac{x^2}{x - 5} + \frac{3x + 10}{5 - x}$$

Solution Your first thought might be to use a denominator of $(x - 5)(5 - x)$. However, we can simplify our work considerably by using the observation above. Multiply the numerator and denominator of the second fraction by -1:

Use

$$\frac{-1}{-1} = 1$$

Note that

$$(-1)(5 - x) = x - 5$$

The fractions now have a common denominator, and we can add as before.

$$\frac{x^2}{x - 5} + \frac{3x + 10}{5 - x}$$

$$= \frac{x^2}{x - 5} + \frac{(-1)(3x + 10)}{(-1)(5 - x)}$$

$$= \frac{x^2}{x - 5} + \frac{-3x - 10}{x - 5}$$

$$= \frac{x^2 - 3x - 10}{x - 5}$$

$$= \frac{(x + 2)(x - 5)}{x - 5}$$

$$= x + 2$$

Check Yourself 8

Add

$$\frac{x^2}{x - 7} + \frac{10x - 21}{7 - x}$$

CHECK YOURSELF ANSWERS

1. $\dfrac{2}{3y^2}$ **2.** (1) $\dfrac{6}{a - 4}$; (2) $\dfrac{x + y}{y}$ **3.** (1) $24a^3$; (2) $(x + 7)(x - 5)$

4. (1) $(x - 5)(x + 5)(x + 3)$; (2) $(y + 3)^2(y - 4)$ **5.** (1) $\dfrac{8a + 15b}{10ab^2}$;

(2) $\dfrac{2y - 6}{y(y + 2)}$ **6.** (1) $\dfrac{3}{(x - 2)(x - 5)}$; (2) $\dfrac{5x + 9}{3(x + 3)(x - 3)}$

7. $\dfrac{-9x + 1}{3x + 1}$ **8.** $x - 3$

6.3 EXERCISES

Build Your Skills

Perform the indicated operations. Express your results in simplest form.

1. $\dfrac{7}{2x^2} + \dfrac{5}{2x^2}$

2. $\dfrac{11}{3b^3} - \dfrac{2}{3b^3}$

3. $\dfrac{5}{3a + 7} + \dfrac{2}{3a + 7}$

4. $\dfrac{6}{5x + 3} - \dfrac{3}{5x + 3}$

5. $\dfrac{2x}{x - 3} - \dfrac{6}{x - 3}$

6. $\dfrac{7w}{w + 3} + \dfrac{21}{w + 3}$

7. $\dfrac{y^2}{2y + 8} + \dfrac{3y - 4}{2y + 8}$

8. $\dfrac{x^2}{4x - 12} - \dfrac{9}{4x - 12}$

9. $\dfrac{4m - 7}{m - 5} - \dfrac{2m + 3}{m - 5}$

10. $\dfrac{3b - 8}{b - 6} + \dfrac{b - 16}{b - 6}$

11. $\dfrac{x - 7}{x^2 - x - 6} + \dfrac{2x - 2}{x^2 - x - 6}$

12. $\dfrac{5x - 12}{x^2 - 8x + 15} - \dfrac{3x - 2}{x^2 - 8x + 15}$

13. $\dfrac{5}{3x} + \dfrac{3}{2x}$

14. $\dfrac{4}{5w} - \dfrac{3}{4w}$

15. $\dfrac{6}{a} + \dfrac{3}{a^2}$

16. $\dfrac{3}{p} - \dfrac{7}{p^2}$

17. $\dfrac{2}{m} - \dfrac{2}{n}$

18. $\dfrac{3}{x} + \dfrac{3}{y}$

19. $\dfrac{3}{4b^2} - \dfrac{5}{3b^3}$

20. $\dfrac{4}{5x^3} - \dfrac{3}{2x^2}$

21. $\dfrac{2}{a} - \dfrac{1}{a - 2}$

22. $\dfrac{4}{c} + \dfrac{3}{c + 1}$

23. $\dfrac{2}{x + 1} + \dfrac{3}{x + 2}$

24. $\dfrac{4}{y - 1} + \dfrac{2}{y + 3}$

25. $\dfrac{5}{y - 3} - \dfrac{1}{y + 1}$

26. $\dfrac{4}{x + 5} - \dfrac{3}{x - 1}$

27. $\dfrac{2w}{w - 7} + \dfrac{w}{w - 2}$

28. $\dfrac{3n}{n + 5} + \dfrac{n}{n - 4}$

29. $\dfrac{3x}{3x - 2} - \dfrac{2x}{2x + 1}$

30. $\dfrac{5c}{5c - 1} + \dfrac{2c}{2c - 3}$

31. $\dfrac{6}{m - 7} + \dfrac{2}{7 - m}$

32. $\dfrac{5}{a - 5} - \dfrac{3}{5 - a}$

33. $\dfrac{3}{x^2 - 16} + \dfrac{2}{x - 4}$

34. $\dfrac{5}{y^2 + 5y + 6} + \dfrac{2}{y + 2}$

35. $\dfrac{4m}{m^2 - 3m + 2} - \dfrac{1}{m - 2}$

36. $\dfrac{x}{x^2 - 1} - \dfrac{2}{x + 1}$

37. $\dfrac{6y}{y^2 - 8y + 15} + \dfrac{9}{y - 3}$

38. $\dfrac{8a}{a^2 - 8a + 12} + \dfrac{4}{a - 2}$

39. $\dfrac{6x}{x^2 - 10x + 24} - \dfrac{18}{x - 6}$

40. $\dfrac{21p}{p^2 - 3p - 10} - \dfrac{15}{p - 5}$

41. $\dfrac{2}{z^2 - 4} + \dfrac{3}{z^2 + 2z - 8}$

42. $\dfrac{5}{x^2 - 3x - 10} + \dfrac{2}{x^2 - 25}$

$9 = 41$

Transcribe Your Skills

43. Explain how to find the LCD for a group of denominators.

44. Could a rational expression be added or subtracted by using a common denominator instead of a least common denominator?

Think About These

Evaluate each expression at the given variable value(s).

45. $\dfrac{5x + 5}{x^2 + 3x + 2} - \dfrac{x - 3}{x^2 + 5x + 6}$, $x = -4$

46. $\dfrac{y - 3}{y^2 - 6y + 8} + \dfrac{2y - 6}{y^2 - 4}$, $y = 3$

47. $\dfrac{2m + 2n}{m^2 - n^2} + \dfrac{m - 2n}{m^2 + 2mn + n^2}$, $m = 3$, $n = 2$

48. $\dfrac{w - 3z}{w^2 - 2wz + z^2} - \dfrac{w + 2z}{w^2 - z^2}$, $w = 2$, $z = 1$

49. $\dfrac{1}{a - 3} - \dfrac{1}{a + 3} + \dfrac{2a}{a^2 - 9}$, $a = 4$

50. $\dfrac{1}{m + 1} + \dfrac{1}{m - 3} - \dfrac{4}{m^2 - 2m - 3}$, $m = -2$

51. $\dfrac{3w^2 + 16w - 8}{w^2 + 2w - 8} + \dfrac{w}{w + 4} - \dfrac{w - 1}{w - 2}$, $w = 3$

52. $\dfrac{4x^2 - 7x - 45}{x^2 - 6x + 5} - \dfrac{x + 2}{x - 1} - \dfrac{x}{x - 5}$, $x = -3$

53. $\dfrac{a^2 - 9}{2a^2 - 5a - 3} \cdot \left(\dfrac{1}{a - 2} + \dfrac{1}{a + 3} \right)$, $a = -3$

54. $\dfrac{m^2 - 2mn + n^2}{m^2 + 2mn - 3n^2} \cdot \left(\dfrac{2}{m - n} - \dfrac{1}{m + n} \right)$, $m = 4$, $n = -3$

55. One farmer can plow a field in t hours (h). A second farmer can plow the same field in $(t + 1)$ hours and a third farmer can plow the same field in $(t + 2)$ hours. If the three farmers work together to plow the same field, their plowing time is described by the expression.

$$\frac{1}{t} + \frac{1}{t + 1} + \frac{1}{t + 2}$$

Add the three fractions.

56. One crew can load a grain ship in h hours. A second crew can load the same ship in $(h - 1)$ hours, and a third crew can load the same ship in $(h + 2)$ hours. If the three crews work together to load the ship, their time to load the ship is described by the expression.

$$\frac{1}{h} + \frac{1}{h - 1} + \frac{1}{h + 2}$$

Add the three fractions.

Skillscan (Section 6.2)

Multiply.

a. $\dfrac{3}{10} \cdot 20$ **b.** $\dfrac{7}{12} \cdot 24$ **c.** $\dfrac{6}{a^2} \cdot a^2$ **d.** $\dfrac{9}{w^2} \cdot w^3$

e. $\dfrac{2}{mn} \cdot mn^2$ **f.** $\dfrac{3}{x^2} \cdot x^2 y^2$ **g.** $\dfrac{5}{rs} \cdot r^2 s^2$

h. $\dfrac{3}{a^2 b} \cdot a^2 b^2$

6.4   **Complex Fractions**

TAPE IN15

OBJECTIVE: To simplify complex fractions

Our work in this section deals with simplifying complex fractions. A complex fraction is a fraction that has a fraction in its numerator or denominator (or both). Some examples are

$$\frac{\dfrac{5}{6}}{\dfrac{3}{4}} \qquad \frac{\dfrac{4}{x}}{\dfrac{3}{x + 1}} \qquad \text{and} \qquad \frac{1 + \dfrac{1}{x}}{1 - \dfrac{1}{x}}$$

Fundamental principle:

$$\frac{P}{Q} = \frac{PR}{QR}$$

where $Q \neq 0$ and $R \neq 0$

Method 1 Recall that by the fundamental principle we can always multiply the numerator and denominator of a fraction by the same nonzero quantity. In simplifying a complex fraction, we multiply the numerator and denominator by the LCD of all fractions that appear within the complex fraction.

Here the denominators are 5 and 10, so we can write

$$\frac{\dfrac{3}{5}}{\dfrac{7}{10}} = \frac{\dfrac{3}{5} \cdot 10}{\dfrac{7}{10} \cdot 10} = \frac{6}{7}$$

Again we are multiplying by $\dfrac{10}{10}$, or 1.

Method 2 Our second approach interprets the complex fraction as indicating division and applies our earlier work in dividing fractions. The numerator and denominator of the complex fraction are written as single fractions. The division step follows.

$$\frac{\dfrac{3}{5}}{\dfrac{7}{10}} = \frac{3}{5} \div \frac{7}{10} = \frac{3}{5} \cdot \frac{10}{7} = \frac{6}{7}$$

Invert and multiply.

Which method is better? The answer depends on the expression you are trying to simplify. Both approaches are effective, and you should be familiar with both. With practice you will be able to tell which method may be easier to use in a particular situation.

Let's look at the same two methods applied to the simplification of an algebraic complex fraction.

EXAMPLE 1 Simplifying Complex Fractions

Simplify

$$\frac{1 + \dfrac{2x}{y}}{2 - \dfrac{x}{y}}$$

Use $\dfrac{y}{y} = 1$.

Method 1 The LCD of 1, $\dfrac{2x}{y}$, 2, and $\dfrac{x}{y}$ is y. So we multiply the numerator and denominator by y:

$$\frac{1 + \dfrac{2x}{y}}{2 - \dfrac{x}{y}} = \frac{\left(1 + \dfrac{2x}{y}\right) \cdot y}{\left(2 - \dfrac{x}{y}\right) \cdot y}$$

Distribute y over the numerator and denominator.

$$= \frac{1 \cdot y + \dfrac{2x}{y} \cdot y}{2 \cdot y - \dfrac{x}{y} \cdot y}$$

Simplify.

$$= \frac{y + 2x}{2y - x}$$

Method 2 In this approach we must *first work separately* in the numerator and denominator to form single fractions.

Make sure you understand the steps in forming a single fraction in the numerator and denominator.

$$\frac{1 + \dfrac{2x}{y}}{2 - \dfrac{x}{y}} = \frac{\dfrac{y}{y} + \dfrac{2x}{y}}{\dfrac{2y}{y} - \dfrac{x}{y}} = \frac{\dfrac{y + 2x}{y}}{\dfrac{2y - x}{y}}$$

Invert the divisor and multiply.

$$= \frac{y + 2x}{y} \cdot \frac{y}{2y - x}$$

$$= \frac{y + 2x}{2y - x}$$

Check Yourself 1

Simplify

$$\frac{\dfrac{x}{y} - 1}{\dfrac{2x}{y} + 2}$$

Again, simplifying a complex fraction means writing an equivalent simple fraction in lowest terms. The following example illustrates.

EXAMPLE 2 Simplifying Complex Fractions

Simplify

$$\frac{1 - \dfrac{2y}{x} + \dfrac{y^2}{x^2}}{1 - \dfrac{y^2}{x^2}}$$

Solution We choose the first method of simplification in this case. The LCD of all the fractions that appear is x^2. So we multiply the numerator and denominator by x^2.

Distribute x^2 over the numerator and denominator, and simplify.

$$\frac{1 - \dfrac{2y}{x} + \dfrac{y^2}{x^2}}{1 - \dfrac{y^2}{x^2}} = \frac{\left(1 - \dfrac{2y}{x} + \dfrac{y^2}{x^2}\right) \cdot x^2}{\left(1 - \dfrac{y^2}{x^2}\right) \cdot x^2}$$

$$= \frac{x^2 - 2xy + y^2}{x^2 - y^2}$$

Factor the numerator and denominator, and divide by the common factor $x - y$.

$$= \frac{(x - y)(x - y)}{(x + y)(x - y)} = \frac{x - y}{x + y}$$

Check Yourself 2

Simplify

$$\dfrac{1 + \dfrac{5}{x} + \dfrac{6}{x^2}}{1 - \dfrac{9}{x^2}}$$

We will illustrate the second method of simplification for purposes of comparison.

EXAMPLE 3 Simplifying Complex Fractions

Simplify

$$\dfrac{1 - \dfrac{1}{x + 2}}{x - \dfrac{2}{x - 1}}$$

Solution Form single fractions in the numerator and denominator:

$$\dfrac{1 - \dfrac{1}{x + 2}}{x - \dfrac{2}{x - 1}} = \dfrac{\dfrac{x + 2}{x + 2} - \dfrac{1}{x + 2}}{\dfrac{x(x - 1)}{x - 1} - \dfrac{2}{x - 1}} = \dfrac{\dfrac{x + 1}{x + 2}}{\dfrac{x^2 - x - 2}{x - 1}}$$

$$= \dfrac{x + 1}{x + 2} \cdot \dfrac{x - 1}{x^2 - x - 2}$$

$$= \dfrac{x + 1}{x + 2} \cdot \dfrac{x - 1}{(x - 2)(x + 1)}$$

$$= \dfrac{x - 1}{(x + 2)(x - 2)}$$

Again, take time to make sure you understand how the numerator and denominator are rewritten as single fractions.

Note Method 2 is probably the more efficient in this case. The LCD of the denominators would be $(x + 2)(x - 1)$, leading to a somewhat more complicated process if method 1 were used.

Check Yourself 3

Simplify

$$\dfrac{2 + \dfrac{5}{x - 3}}{x - \dfrac{1}{2x + 1}}$$

The following algorithm summarizes our work with complex fractions.

SIMPLIFYING COMPLEX FRACTIONS

METHOD 1

1. Multiply the numerator and denominator of the complex fraction by the LCD of all the fractions that appear within the numerator and denominator.

2. Simplify the resulting rational expression, writing the expression in lowest terms.

METHOD 2

1. Write the numerator and denominator of the complex fraction as single fractions, if necessary.

2. Invert the denominator and multiply as before, writing the result in lowest terms.

CHECK YOURSELF ANSWERS

1. $\dfrac{x - y}{2x + 2y}$ **2.** $\dfrac{x + 2}{x - 3}$ **3.** $\dfrac{2x + 1}{(x - 3)(x + 1)}$

6.4 EXERCISES

Build Your Skills

Simplify each of the following complex fractions.

1. $\dfrac{\dfrac{2}{3}}{\dfrac{6}{8}}$

2. $\dfrac{\dfrac{5}{6}}{\dfrac{10}{15}}$

3. $\dfrac{\dfrac{2}{3} + \dfrac{1}{2}}{\dfrac{3}{4} - \dfrac{1}{3}}$

4. $\dfrac{\dfrac{3}{4} + \dfrac{1}{2}}{\dfrac{7}{8} - \dfrac{1}{4}}$

5. $\dfrac{2 + \dfrac{1}{3}}{3 - \dfrac{1}{5}}$

6. $\dfrac{1 + \dfrac{3}{4}}{2 - \dfrac{1}{8}}$

7. $\dfrac{\dfrac{x}{8}}{\dfrac{x^2}{4}}$

8. $\dfrac{\dfrac{a^2}{10}}{\dfrac{a^3}{15}}$

9. $\dfrac{\dfrac{3}{m}}{\dfrac{6}{m^2}}$

10. $\dfrac{\dfrac{15}{x^2}}{\dfrac{20}{x^3}}$

11. $\dfrac{\dfrac{y+1}{y}}{\dfrac{y-1}{2y}}$

12. $\dfrac{\dfrac{x+3}{4x}}{\dfrac{x-3}{2x}}$

13. $\dfrac{\dfrac{a+2b}{3a}}{\dfrac{a^2+2ab}{9b}}$

14. $\dfrac{\dfrac{m-3n}{4m}}{\dfrac{m^2-3mn}{8n}}$

15. $\dfrac{\dfrac{x-2}{x^2-9}}{\dfrac{x^2-4}{x^2+3x}}$

16. $\dfrac{\dfrac{x+5}{x^2-6x}}{\dfrac{x^2-25}{x^2-36}}$

17. $\dfrac{2-\dfrac{1}{x}}{2+\dfrac{1}{x}}$ L.CD

18. $\dfrac{3+\dfrac{1}{b}}{3-\dfrac{1}{b}}$

19. $\dfrac{\dfrac{1}{x}-\dfrac{1}{y}}{\dfrac{1}{xy}}$ L. CD

20. $\dfrac{\dfrac{1}{ab}}{\dfrac{1}{a}+\dfrac{1}{b}}$

21. $\dfrac{\dfrac{x^2}{y^2}-1}{\dfrac{x}{y}+1}$ L.CD

22. $\dfrac{\dfrac{m}{n}+2}{\dfrac{m^2}{n^2}-4}$

23. $\dfrac{1+\dfrac{3}{a}-\dfrac{4}{a^2}}{1+\dfrac{2}{a}-\dfrac{3}{a^2}}$ L. CD

24. $\dfrac{1-\dfrac{2}{x}-\dfrac{8}{x^2}}{1-\dfrac{1}{x}-\dfrac{6}{x^2}}$

25. $\dfrac{\dfrac{x^2}{y}+2x+y}{\dfrac{1}{y^2}-\dfrac{1}{x^2}}$ L. CD

26. $\dfrac{\dfrac{a}{b}+1-\dfrac{2b}{a}}{\dfrac{1}{b^2}-\dfrac{4}{a^2}}$

27. $\dfrac{1+\dfrac{1}{x-1}}{1-\dfrac{1}{x-1}}$ L CD

28. $\dfrac{2 - \dfrac{1}{m-2}}{2 + \dfrac{1}{m-2}}$ *2nd method*

29. $\dfrac{1 - \dfrac{1}{y-1}}{y - \dfrac{8}{y+2}}$ *2nd method*

30. $\dfrac{1 + \dfrac{1}{x+2}}{x - \dfrac{18}{x-3}}$

31. $\dfrac{\dfrac{1}{x-3} + \dfrac{1}{x+3}}{\dfrac{1}{x-3} - \dfrac{1}{x+3}}$

32. $\dfrac{\dfrac{2}{m-2} + \dfrac{1}{m-3}}{\dfrac{2}{m-2} - \dfrac{1}{m-3}}$

33. $\dfrac{\dfrac{x}{x+1} + \dfrac{1}{x-1}}{\dfrac{x}{x-1} - \dfrac{1}{x+1}}$

34. $\dfrac{\dfrac{y}{y-4} + \dfrac{1}{y+2}}{\dfrac{4}{y-4} - \dfrac{1}{y+2}}$

35. $\dfrac{\dfrac{a+1}{a-1} - \dfrac{a-1}{a+1}}{\dfrac{a+1}{a-1} + \dfrac{a-1}{a+1}}$

36. $\dfrac{\dfrac{x+2}{x-2} - \dfrac{x-2}{x+2}}{\dfrac{x+2}{x-2} + \dfrac{x-2}{x+2}}$

37. $1 + \dfrac{1}{1 + \dfrac{1}{x}}$

38. $1 + \dfrac{1}{1 - \dfrac{1}{y}}$

Transcribe Your Skills

39. Outline the two different methods used to simplify a complex fraction. What are the advantages of each method?

40. Can the expression $\dfrac{x^{-1} + y^{-1}}{x^{-2} + y^{-2}}$ be written as $\dfrac{x^2 + y^2}{x + y}$? If not, what is the correct simplified form?

Think About These

41. $1 + \dfrac{1}{1 + \dfrac{1}{1 + \dfrac{1}{x}}}$

42. Extend the "continued fraction" patterns of Exercises 37 and 41 to write the next complex fraction.

43. Simplify the complex fraction of Exercise 42.

44. Compare your results in Exercises 37, 41, and 43. Could you have predicted the result?

Suppose you drive at 40 mi/h from city A to city B. You then return along the same route from city B to city A at 50 mi/h. What is your average rate for the round trip? Your obvious guess would be 45 mi/h, but you are in for a surprise.

Suppose that the cities are 200 mi apart. Your time from city A to city B is the distance divided by the rate, or

$$\frac{200 \text{ mi}}{40 \text{ mi/h}} = 5 \text{ h}$$

Similarly, your time from city B to city A is

$$\frac{200 \text{ mi}}{50 \text{ mi/h}} = 4 \text{ h}$$

The total time is then 9 h, and now using *rate equals distance divided by time,* we have

$$\frac{400 \text{ mi}}{9 \text{ h}} = \frac{400}{9} \text{ mi/h} = 44\frac{4}{9} \text{ mi/h}$$

Note that the rate for the round trip is independent of the distance involved. For instance, try the same computations above if cities A and B are 400 mi apart.

The answer to the problem above is the complex fraction

$$R = \frac{2}{\dfrac{1}{R_1} + \dfrac{1}{R_2}}$$

where R_1 is the rate going, R_2 is the rate returning, and R is the rate for the round trip.

45. Verify that if $R_1 = 40$ mi/h and $R_2 = 50$ mi/h, then $R = 44\dfrac{4}{9}$ mi/h, by simplifying the complex fraction *after* substituting those values.

46. Simplify the given complex fraction first. *Then* substitute 40 for R_1 and 50 for R_2 to calculate R.

47. Repeat Exercise 45 where $R_1 = 50$ mi/h and $R_2 = 60$ mi/h.

48. Use the procedure of Exercise 46 with the above values for R_1 and R_2.

49. If the expression $\dfrac{x}{(x + 3)^2}$ represents the rate of soil accumulation for a given area of cropland and the expression $\dfrac{x^2}{x^2 - 9}$ represents the rate of erosion for the same area of cropland, the ratio of soil gain to soil loss is given by the expression

$$\dfrac{\dfrac{x}{(x + 3)^2}}{\dfrac{x^2}{x^2 - 9}}$$

Simplify this expression.

50. If the expression $\dfrac{x^2}{x^2 - 16}$ represents the rate of soil accumulation for a given area of cropland and the expression $\dfrac{x}{(x - 4)^2}$ represents the rate of erosion for the same area of cropland, the ratio of soil gain to soil loss is given by the expression

$$\dfrac{\dfrac{x^2}{x^2 - 16}}{\dfrac{x}{(x - 4)^2}}$$

Simplify this expression.

Skillscan (Section 1.3)

Evaluate the following:

a. 2^2 **b.** 3^3 **c.** -5^2 **d.** $(-5)^2$ **e.** 4^3
f. $(-2)^5$

SUMMARY

Simplification of Rational Expressions [6.1]

Rational expressions have the form

$$\dfrac{P}{Q}$$

where P and Q are polynomials and Q cannot have the value 0

$\dfrac{x^2 - 5x}{x - 3}$ is a rational expression. The variable x cannot have the value 3.

Fundamental Principle of Fractions

For polynomials P, Q, and R,

$$\dfrac{P}{Q} = \dfrac{PR}{QR}$$

where $Q \neq 0$ and $R \neq 0$

This principle can be used in two ways. We can multiply or divide the numerator and denominator of a rational expression by the same nonzero polynomial.

This uses the fact that

$$\dfrac{R}{R} = 1$$

where $R \neq 0$.

Simplifying Rational Expressions

To simplify a rational expression, use the following algorithm.

$$\frac{x^2 - 4}{x^2 - 2x - 8}$$

$$= \frac{(x - 2)(x + 2)}{(x - 4)(x + 2)}$$

$$= \frac{x - 2}{x - 4}$$

1. Completely factor both the numerator and denominator of the expression.

2. Divide the numerator and denominator by *all* common factors.

3. The resulting expression will be in simplest form (or in lowest terms).

Multiplication and Division of Rational Expressions [6.2]

Multiplying Rational Expressions

In symbols, the multiplication pattern is as follows: For polynomials P, Q, R, and S,

$$\frac{P}{Q} \cdot \frac{R}{S} = \frac{PR}{QS}$$

where $Q \neq 0$ and $S \neq 0$.

$$\frac{2x - 6}{x^2 - 9} \cdot \frac{x^2 + 3x}{6x + 24}$$

$$= \frac{2(x - 3)}{(x - 3)(x + 3)} \cdot \frac{x(x + 3)}{6(x + 4)}$$

$$= \frac{x}{3(x + 4)}$$

In practice, we apply the following algorithm to multiply two rational expressions.

1. Write each numerator and denominator in completely factored form.

2. Divide by any common factors appearing in both a numerator and a denominator.

3. Multiply as needed to form the desired product.

Dividing Rational Expressions

In symbols, the division pattern is

$$\frac{P}{Q} \div \frac{R}{S} = \frac{P}{Q} \cdot \frac{S}{R} = \frac{PS}{QR}$$

where $Q \neq 0$, $R \neq 0$, and $S \neq 0$.

To divide two rational expressions, you can apply the following algorithm.

$$\frac{5y}{2y - 8} \div \frac{10y^2}{y^2 - y - 12}$$

$$= \frac{5y}{2y - 8} \cdot \frac{y^2 - y - 12}{10y^2}$$

$$= \frac{5y}{2(y - 4)} \cdot \frac{(y - 4)(y + 3)}{10y^2}$$

$$= \frac{y + 3}{4y}$$

1. Invert the divisor (the *second* rational expression) to write the problem as one of multiplication.

2. Proceed as in the algorithm for multiplication of rational expressions.

Addition and Subtraction of Rational Expressions [6.3]

To add or subtract rational expressions with the same denominator, add or subtract the numerator and then write that sum over the common denominator. The result should be written in lowest terms.

In symbols,

$$\frac{P}{R} + \frac{Q}{R} = \frac{P + Q}{R}$$

and

$$\frac{P}{R} - \frac{Q}{R} = \frac{P - Q}{R}$$

where $R \neq 0$.

$$\frac{5w}{w^2 - 16} - \frac{20}{w^2 - 16}$$

$$= \frac{5w - 20}{w^2 - 16}$$

$$= \frac{5(w - 4)}{(w + 4)(w - 4)}$$

$$= \frac{5}{w + 4}$$

Least Common Denominator

The *least common denominator* (LCD) of a group of rational expressions is the simplest polynomial that is divisible by each of the individual denominators of the rational expressions. To find the LCD, you can use the following algorithm.

1. Write each of the denominators in completely factored form.

2. Write the LCD as the product of each prime factor, to the highest power to which it appears in the factored form of any of the denominators.

To find the LCD for

$$\frac{2}{x^2 + 2x + 1} \quad \text{and} \quad \frac{3}{x^2 + x}$$

write

$$x^2 + 2x + 1 = (x + 1)(x + 1)$$

$$x^2 + x = x(x + 1)$$

The LCD is

$$x(x + 1)(x + 1)$$

$$\frac{2}{(x + 1)^2} - \frac{3}{x(x + 1)}$$

Now to add or subtract rational expressions with different denominators, we first find the LCD by the procedure outlined above. We then rewrite each of the rational expressions with that LCD as a common denominator. Then we can add or subtract as before.

$$= \frac{2 \cdot x}{(x + 1)^2 x}$$

$$- \frac{3(x + 1)}{x(x + 1)(x + 1)}$$

$$= \frac{2x - 3(x + 1)}{x(x + 1)(x + 1)}$$

$$= \frac{-x - 3}{x(x + 1)(x + 1)}$$

Complex Fractions [6.4]

Complex fractions are fractions that have a fraction in their numerator or denominator (or both).

There are two commonly used methods for simplifying complex fractions. They are outlined below.

Simplify $\dfrac{1 - \dfrac{2}{x}}{1 - \dfrac{4}{x^2}}$.

Method 1

1. Multiply the numerator and denominator of the complex fraction by the LCD of all the fractions that appear within the numerator and denominator.

2. Simplify the resulting rational expression, writing the result in lowest terms.

Method 1:

$$= \frac{\left(1 - \dfrac{2}{x}\right)x^2}{\left(1 - \dfrac{4}{x^2}\right)x^2}$$

$$= \frac{x^2 - 2x}{x^2 - 4} = \frac{x(x - 2)}{(x + 2)(x - 2)}$$

$$= \frac{x}{x + 2}$$

Method 2:

$$= \frac{\dfrac{x-2}{x}}{\dfrac{x^2-4}{x^2}}$$

$$= \frac{x-2}{x} \cdot \frac{x^2}{x^2-4}$$

$$= \frac{x-2}{x} \cdot \frac{x^2}{(x+2)(x-2)}$$

$$= \frac{x}{x+2}$$

Method 2

1. Write the numerator and denominator of the complex fraction as single fractions, if necessary.

2. Then invert the denominator and multiply as before, writing the result in lowest terms.

SUMMARY EXERCISES

This supplementary exercise set is provided to give you practice with each of the objectives of the chapter. Each exercise is keyed to the appropriate chapter section. The answers are provided in the instructor's manual that accompanies this text. Your instructor will provide guidelines on how best to use these exercises in your instructional program.

[6.1] For what value of the variable will each of the following rational expressions be undefined?

1. $\dfrac{x}{2}$

2. $\dfrac{3}{y}$

3. $\dfrac{2}{x-5}$

4. $\dfrac{3x}{2x+5}$

[6.1] Simplify each of the following rational expressions.

5. $\dfrac{18x^5}{24x^3}$

6. $\dfrac{15m^3n}{-5mn^2}$

7. $\dfrac{8y-64}{y-8}$

8. $\dfrac{5x-20}{x^2-16}$

9. $\dfrac{9-x^2}{x^2+2x-15}$

10. $\dfrac{3w^2+8w-35}{2w^2+13w+15}$

11. $\dfrac{6a^2-ab-b^2}{9a^2-b^2}$

12. $\dfrac{6w-3z}{8w^3-z^3}$

[6.2] Multiply or divide as indicated. Express your results in simplest form.

13. $\dfrac{x^5}{24} \cdot \dfrac{20}{x^3}$

14. $\dfrac{a^3b}{4ab^2} \div \dfrac{ab}{12ab^2}$

15. $\dfrac{6y-18}{9y} \cdot \dfrac{10}{5y-15}$

16. $\dfrac{m^2-3m}{m^2-5m+6} \cdot \dfrac{m^2-4}{m^2+7m+10}$

17. $\dfrac{a^2-2a}{a^2-4} \div \dfrac{2a^2}{3a+6}$

18. $\dfrac{r^2+2rs}{r^3-r^2s} \div \dfrac{5r+10s}{r^2-2rs+s^2}$

19. $\dfrac{x^2-2xy-3y^2}{x^2-xy-2y^2} \cdot \dfrac{x^2-4y^2}{x^2-8xy+15y^2}$

20. $\dfrac{w^3+3w^2+2w+6}{w^4-4} \div (w^3+27)$

[6.3] Perform the indicated operations. Express your results in simplified form.

21. $\dfrac{5x+7}{x+4} - \dfrac{2x-5}{x+4}$

22. $\dfrac{3}{4x^2} + \dfrac{5}{6x}$

23. $\dfrac{2}{x-5} - \dfrac{1}{x}$

24. $\dfrac{2}{y+5} + \dfrac{3}{y+4}$

25. $\dfrac{2}{3m-3} - \dfrac{5}{2m-2}$

26. $\dfrac{7}{x-3} - \dfrac{5}{3-x}$

27. $\dfrac{5}{4x+4} + \dfrac{5}{2x-2}$

28. $\dfrac{2a}{a^2-9a+20} + \dfrac{8}{a-4}$

29. $\dfrac{2}{s-1} - \dfrac{6s}{s^2+s-2}$

30. $\dfrac{4}{x^2-9} - \dfrac{3}{x^2-4x+3}$

31. $\dfrac{x^2-14x-8}{x^2-2x-8} + \dfrac{2x}{x-4} - \dfrac{3}{x+2}$

32. $\dfrac{w^2+2wz+z^2}{w^2-wz-2z^2} \cdot \left(\dfrac{3}{w+z} - \dfrac{1}{w-z}\right)$

[6.4] Simplify each of the following complex fractions.

33. $\dfrac{\dfrac{x^2}{12}}{\dfrac{x^3}{8}}$

34. $\dfrac{\dfrac{y-1}{y^2-4}}{\dfrac{y^2-1}{y^2-y-2}}$

35. $\dfrac{1+\dfrac{a}{b}}{1-\dfrac{a}{b}}$

36. $\dfrac{2-\dfrac{x}{y}}{4-\dfrac{x^2}{y^2}}$

37. $\dfrac{\dfrac{1}{r}-\dfrac{1}{s}}{\dfrac{1}{r^2}-\dfrac{1}{s^2}}$

38. $\dfrac{1-\dfrac{1}{x+2}}{1+\dfrac{1}{x+2}}$

39. $\dfrac{1-\dfrac{2}{x-1}}{x+\dfrac{3}{x-4}}$

40. $\dfrac{\dfrac{w}{w+1}-\dfrac{1}{w-1}}{\dfrac{w}{w-1}+\dfrac{1}{w+1}}$

41. $\dfrac{1}{1-\dfrac{1}{1-\dfrac{1}{y-1}}}$

42. $1-\dfrac{1}{1+\dfrac{1}{1-\dfrac{1}{x}}}$

43. $\dfrac{1-\dfrac{1}{x-1}}{x-\dfrac{8}{x+2}}$

44. $\dfrac{1}{1+\dfrac{1}{1+\dfrac{1}{y+1}}}$

SELF-TEST

The purpose of this self-test is to help you check your progress and to review for a chapter test in class. Allow yourself about an hour to take the test. When you are done, check your answers in the back of the book. If you missed any problems, be sure to go back and review the appropriate sections in the chapter and the exercises that are provided.

Simplify each of the following rational expressions.

1. $\dfrac{-21x^5y^3}{28xy^5}$

2. $\dfrac{3w^2 + w - 2}{3w^2 - 8w + 4}$

3. $\dfrac{x^3 - 2x^2 - 3x}{x^3 + 3x^2 + 2x}$

4. $\dfrac{x^4 - xy^3}{x^3 - xy^2}$

Multiply or divide as indicated.

5. $\dfrac{3ab^2}{5ab^3} \cdot \dfrac{20a^2b}{21b}$

6. $\dfrac{m^2 - 3m}{m^2 - 9} \div \dfrac{4m}{m^2 - m - 12}$

7. $\dfrac{x^2 - 3x}{5x^2} \cdot \dfrac{10x}{x^2 - 4x + 3}$

8. $\dfrac{x^2 + 3xy}{2x^3 - x^2y} \div \dfrac{x^2 + 6xy + 9y^2}{4x^2 - y^2}$

9. $\dfrac{9x^2 + 9x - 4}{6x^2 - 11x + 3} \cdot \dfrac{15 - 10x}{3x + 4}$

10. $\dfrac{x^2 - 7x - 18}{2x^2 + 9x + 10} \div (2x + 5)$

Add or subtract as indicated.

11. $\dfrac{5}{x - 2} - \dfrac{1}{x}$

12. $\dfrac{2}{x + 3} + \dfrac{12}{x^2 - 9}$

13. $\dfrac{6x}{x^2 - x - 2} - \dfrac{2}{x + 1}$

14. $\dfrac{3}{x^2 - 3x - 4} + \dfrac{5}{x^2 - 16}$

15. $\dfrac{1}{x + 2} + \dfrac{1}{x + 1} + \dfrac{1}{x - 3}$

16. $\dfrac{4}{x^2 - 9} - \dfrac{5}{x^2 - 4x + 3}$

Simplify each of the following complex fractions.

17. $\dfrac{3 - \dfrac{x}{y}}{9 - \dfrac{x^2}{y^2}}$

18. $\dfrac{1 - \dfrac{10}{z + 3}}{2 - \dfrac{12}{z - 1}}$

19. $\dfrac{\dfrac{1}{x} + \dfrac{1}{y}}{x^2 - y^2}$

20. $\dfrac{\dfrac{1}{x + y} - \dfrac{1}{x - y}}{\dfrac{2y}{x^2 - y^2}}$

CUMULATIVE REVIEW EXERCISES

This is a review of selected topics from the first six chapters.

Solve each of the following.

1. $4x - 2(x + 1) = 3(5 - x) + 8$

2. $|5x - 2| \le 8$

3. $|3x - 5| = 4$

Graph each of the following.

4. $2x + 3y = 12$

5. $-7(x - 2) \le y$

6. Find the equation of the line that passes through the point $(-1, -2)$ and is perpendicular to the line $4x + 5y = 15$.

7. Solve the following system of equations:

$2x - 5y = 4$
$3x - 2y = -5$

8. Solve the following system of inequalities graphically.

$2x - y < 1$
$3x + y < 6$

Simplify each of the following.

9. $(2x^2 - 3x + 4) - [-3x^2 + 5 + (2x^2 - 7x)]$

10. $(2x + 3)(5x - 4)$

Factor each of the following completely.

11. $3x^3 - x^2 - 2x$

12. $16x^2 - 25y^2$

13. $3x^2 - 3xy + x - y$

Simplify each of the following rational expressions.

14. $\dfrac{2x^2 + 13x + 15}{6x^2 + 7x - 3}$

15. $\dfrac{3}{x - 5} - \dfrac{2}{x - 1}$

16. $\dfrac{2x}{x^2 - 9x + 20} + 8x - 4$

17. $\dfrac{a^2 - 4a}{a^2 - 6a + 8} \cdot \dfrac{a^2 - 4}{2a^2}$

18. Simplify the following complex fraction.

$$\dfrac{\dfrac{x}{x - 1} + \dfrac{1}{x - 1}}{\dfrac{x}{x + 1} + \dfrac{1}{x + 1}}$$

Solve each of the following problems.

19. Leighton has $20,000 to invest. He invests some at 6 percent interest and the rest at 8 percent. His total yearly interest is $1440. How much did Leighton invest at 8 percent?

20. The sum of two numbers is 47. One number is 7 less than 5 times the other. Find the numbers.

DESERTIFICATION Currently, about one-third of the land area on earth can be classified as arid or semiarid. Much of this area is what is commonly referred to as desert. A desert is a region where evaporation exceeds precipitation and the average amount of precipitation is less than 25 centimeters per year (cm/yr). The amount of desert land is constantly changing, due to both natural and human activities.

While natural desert formation can cause problems for humans, it is the human activities which help create deserts that are of greatest concern. Research shows that activities such as overgrazing, improper irrigation, and cultivation of poor soils and steep terrain can increase the rate at which deserts form.

Desertification, *the formation of new deserts*, occurs when rangeland or cropland is converted to desertlike land with a drop in agricultural productivity of 10 percent or more. Land altered by desertification seldom returns to its former productivity. It has been permanently degraded. Mathematical models used to study global climate patterns indicate that increased desertification may lead to a decrease in global rainfall. This decrease in global rainfall, in turn, may lead to increased drought conditions around the world.

Even though the specific regions affected by desertification cannot be predicted, the number of people who have been affected can be determined. By 1989, 230 million people, a number almost equal to the entire U.S. population, were suffering to some degree from loss of productive land to desertification. If the projections for increased desertification are correct, 350 million more people will suffer from the loss of productive lands by the year 2000. Approximately 6 million hectares of new desert are formed each year.

If we wish to slow the expansion of deserts around the world, we need to make some specific changes. Land-use changes need to include removing agriculture from unsuitable soils and terrain, improved irrigation practices, erosion control measures, and a reduction in livestock grazing on much of the world's rangeland. Slowed population growth should be a priority in many of the arid and semiarid regions because many of these areas appear to be approaching the limit to the number of people they can support. ∎

Roots and Radicals **7.1**

TAPE IN16

OBJECTIVES: 1. To introduce the radical notation
2. To evaluate expressions involving radicals
3. To use a scientific calculator to evaluate radical expressions

In Chapter 5 we discussed the properties of integer exponents. Over the next five sections, we will be working toward an extension of those properties. To achieve that objective, we must develop a notation that "reverses" the power process.

A statement such as

$$x^2 = 9$$

is read "x squared equals 9."

In this section we are concerned with the relationship between the base x and the number 9. Equivalently, we can say that "x is the square root of 9."

We know from experience that x must be 3 (since $3^2 = 9$) or -3 [since $(-3)^2 = 9$]. We see that 9 has the two square roots, 3 and -3. In fact, every positive number has *two* square roots, one positive and one negative. In general:

If $x^2 = a$, we say x is a *square root* of a.

We also know that

$$3^3 = 27$$

and similarly we call 3 a *cube root* of 27. Here 3 is the *only* real number with that property. Every real number (positive or negative) has *one* real cube root.

> In general, we can state that if
>
> $x^n = a$
>
> then x is an *nth root* of a.

We are now ready for new notation. The symbol $\sqrt{}$ is called a *radical sign*. We saw above that 3 was the positive square root of 9.

We call 3 the *principal square root* of 9, and we write

$$\sqrt{9} = 3$$

to indicate that 3 is the principal square root of 9.

In some applications we will want to indicate the negative square root; to do so we must write

$$-\sqrt{9} = -3$$

to indicate the negative root.

If both square roots need to be indicated, we can write

$$\pm\sqrt{9} = \pm 3$$

As we will come back to later, a negative number has no *real square roots.*

The symbol $\sqrt{}$ first appeared in print in 1525. In Latin, "radix" means root, and this was contracted to a small r. The present symbol may have been used because it resembled the manuscript form of that small r.

You will see this used later in our work with quadratic equations in Chapter 8.

Every radical expression contains three parts, as shown below. The principal nth root of a is written as

The index 2 for square roots is generally not written. We understand that

$$\sqrt{a}$$

is the principal square root of a.

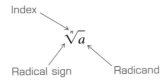

Index

$$\sqrt[n]{a}$$

Radical sign   Radicand

EXAMPLE 1 Evaluating Square Roots

Evaluate, if possible.

(a) $\sqrt{49} = 7$

(b) $-\sqrt{49} = -7$

(c) $\pm\sqrt{49} = \pm7$

(d) $\sqrt{-49}$ is not a real number.

Note that neither 7^2 nor $(-7)^2$ equals -49.

Let's examine (d) more carefully. Suppose that for some real number x,

$$x = \sqrt{-49}$$

By our earlier definition, this means that

$$x^2 = -49$$

We consider imaginary numbers in detail in Section 7.6.

which is impossible. There is no real square root for -49. We call $\sqrt{-49}$ an *imaginary number*.

Check Yourself 1

Evaluate, if possible.

1. $\sqrt{64}$ **2.** $-\sqrt{64}$ **3.** $\pm\sqrt{64}$ **4.** $\sqrt{-64}$

Our next example considers cube roots.

EXAMPLE 2 Evaluating Radicals

Evaluate, if possible.

(a) $\sqrt[3]{64} = 4$ since $4^3 = 64$

(b) $-\sqrt[3]{64} = -4$

(c) $\sqrt[3]{-64} = -4$ since $(-4)^3 = -64$

Note that the cube root of a *negative* number is *negative*.

Check Yourself 2

Evaluate.

1. $\sqrt[3]{125}$ **2.** $-\sqrt[3]{125}$ **3.** $\sqrt[3]{-125}$

Let's consider radicals with other indices in the next example.

The word "indices" is the plural of "index."

EXAMPLE 3 Evaluating Radicals

Evaluate, if possible.

(a) $\sqrt[4]{81} = 3$ since $3^4 = 81$

(b) $\sqrt[4]{-81}$ is not a real number.

(c) $\sqrt[5]{32} = 2$ since $2^5 = 32$

(d) $\sqrt[5]{-32} = -2$ since $(-2)^5 = -32$

In general, an *even* root of a *negative* number is *not real*; it is *imaginary*.

Check Yourself 3

Evaluate, if possible.

1. $\sqrt[4]{16}$ **2.** $\sqrt[5]{243}$ **3.** $\sqrt[4]{-16}$ **4.** $\sqrt[5]{-243}$

All the numbers of our previous examples and exercises were chosen so that the results would be *rational numbers*. That is, the radicands were

Perfect squares: $1, 4, 9, 16, 25, \ldots$

Perfect cubes: $1, 8, 27, 64, 125, \ldots$

and so on.

The square root of a number that is *not* a perfect square (or the cube root of a number that is *not* a perfect cube) is not a rational number.

Expressions such as $\sqrt{2}$, $\sqrt{3}$, and $\sqrt{5}$ are *irrational numbers*. A calculator with a square root key $\boxed{\sqrt{}}$ will give decimal approximations for such numbers.

EXAMPLE 4 Evaluating Radicals with a Calculator

Using a calculator, find the decimal approximation for each of the following. Round all answers to three decimal places.

(a) $\sqrt{17}$

Enter 17 in your calculator and press the $\boxed{\sqrt{}}$ key. The display will read 4.123105626 (if your calculator displays 10 digits). If this is rounded to three decimal places, the result is 4.123.

(b) $\sqrt{28}$

Enter 28 and press the $\boxed{\sqrt{}}$ key. The display reads 5.2915026222. Rounded to three decimal places, the result is 5.292.

(c) $\sqrt{-11}$

Enter -11 by first entering 11 and then pressing the $\boxed{+/-}$ key. Take the square root by pressing the $\boxed{\sqrt{}}$ key. The display will read ERROR. This indicates that -11 does not have a real square root.

 On some calculators, the square root is shown as the "2nd function" or "inverse" of x^2. If that is the case, press the $\boxed{\text{2nd function}}$ key and then the $\boxed{x^2}$ key.

Check Yourself 4

Use a calculator to find the decimal approximation for each of the following. Round each answer to three decimal places.

1. $\sqrt{13}$ 3.60

2. $\sqrt{38}$ 6.10

3. $\sqrt{-21}$ undef 0

 Not all scientific calculators have this key. Most graphing calculators use the up-arrow to indicate exponentiation.

To evaluate roots other than square roots by using scientific calculators, the key marked y^x can be used together with the $\boxed{INV}$ key. (On some calculators, the $\boxed{INV}$ key is $\boxed{2nd\ F}$.)

EXAMPLE 5 Evaluating Radicals with a Calculator

Using a calculator, find a decimal approximation for each of the following. Round each answer to three decimal places.

(a) $\sqrt[4]{12}$

Enter 12 and press $\boxed{INV}$ $\boxed{y^x}$. Then enter 4 and press $\boxed{=}$. The display will read 1.861209718. Rounded to three decimal places, the result is 1.861.

On a graphing calculator, you would enter

12 $\boxed{\wedge}$ $\boxed{(\!(}$ 1 $\boxed{\div}$ 4 $\boxed{)\!)}$

This is further explained in Section 7.5.

(b) $\sqrt[5]{27}$

Enter 27 and press $\boxed{INV}$ $\boxed{y^x}$. Then enter 5 and press $\boxed{=}$. The display will read 1.933182045. Rounded to three decimal places, the result is 1.933.

Check Yourself 5

Use a calculator to find the decimal approximation for each of the following. Round each answer to three decimal places.

1. $\sqrt[4]{35}$

2. $\sqrt[5]{29}$

A certain amount of caution should be exercised in dealing with principal even roots. For example, consider the statement

$$\sqrt{x^2} = x \tag{1}$$

Since $x = 2$,

$\sqrt{x^2} = x$

First, let $x = 2$ in Equation (1).

$$\sqrt{2^2} = \sqrt{4} = 2 \tag{2}$$

Since here $x = -2$,

$\sqrt{x^2} \neq x$

Now let $x = -2$.

$$\sqrt{(-2)^2} = \sqrt{4} = 2 \tag{3}$$

We see that statement (1) is not true when x is negative, but we can write

$$\sqrt{x^2} = \begin{cases} x & \text{where } x \geq 0 \\ -x & \text{where } x < 0 \end{cases}$$

From your earlier work with absolute values you will remember that

$$|x| = \begin{cases} x & \text{where } x \geq 0 \\ -x & \text{where } x < 0 \end{cases}$$

and we can summarize the discussion by writing

$$\sqrt{x^2} = |x| \qquad (4)$$

Statement (4) can be extended to

$$\sqrt[n]{x^n} = |x|$$

when n is *even*.

EXAMPLE 6 Evaluating Radicals

Evaluate.

(a) $\sqrt{5^2} = 5$

(b) $\sqrt{(-4)^2} = |-4| = 4$

(c) $\sqrt[4]{2^4} = 2$

(d) $\sqrt[4]{(-3)^4} = |-3| = 3$

Note Alternately we could write

$$\sqrt{(-4)^2} = \sqrt{16} = 4$$

Check Yourself 6

Evaluate.

1. $\sqrt{6^2}$ **2.** $\sqrt{(-6)^2}$ **3.** $\sqrt[4]{3^4}$ **4.** $\sqrt[4]{(-3)^4}$

The case for roots with indices that are odd does *not* require the use of absolute value, as illustrated in Example 6. For instance,

$$\sqrt[3]{3^3} = \sqrt[3]{27} = 3$$

$$\sqrt[3]{(-3)^3} = \sqrt[3]{-27} = -3$$

and we see that

$$\sqrt[n]{x^n} = x \qquad \text{where } n \text{ is odd}$$

To summarize, we can write

$$\sqrt[n]{x^n} = \begin{cases} |x| & n \text{ is even} \\ x & n \text{ is odd} \end{cases}$$

Let's turn now to a final example in which variables are involved in the radicand.

EXAMPLE 7 Simplifying Radicals

Simplify the following.

(a) $\sqrt[3]{a^3} = a$

(b) $\sqrt{16m^2} = 4|m|$

Note that we can determine the power of the variable in the root by dividing the power in the radical by the index. In Example 7*d*, 8 ÷ 4 = 2.

(c) $\sqrt[5]{32x^5} = 2x$

(d) $\sqrt[4]{x^8} = x^2$ since $(x^2)^4 = x^8$

(e) $\sqrt[3]{27y^6} = 3y^2$ Do you see why?

Check Yourself 7

Simplify.

1. $\sqrt[4]{x^4}$

2. $\sqrt{49w^2}$

3. $\sqrt[5]{a^{10}}$

4. $\sqrt[3]{8y^9}$

CHECK YOURSELF ANSWERS

1. (1) 8; (2) −8; (3) ±8; (4) not a real number   **2.** (1) 5; (2) −5; (3) −5
3. (1) 2; (2) 3; (3) not a real number; (4) −3   **4.** (1) 3.606; (2) 6.164; (3) not a
real number   **5.** (1) 2.432; (2) 1.961   **6.** (1) 6; (2) 6; (3) 3; (4) 3
7. (1) $|x|$; (2) $7|w|$; (3) a^2; (4) $2y^3$

7.1 EXERCISES

Build Your Skills

Evaluate each of the following roots where possible.

1. $\sqrt{49}$

2. $\sqrt{36}$

3. $-\sqrt{36}$

4. $-\sqrt{81}$

5. $\pm\sqrt{81}$

6. $\pm\sqrt{49}$

7. $\sqrt{-49}$

8. $\sqrt{-25}$

9. $\sqrt[3]{27}$

10. $\sqrt[3]{64}$

11. $\sqrt[3]{-64}$

12. $-\sqrt[3]{125}$

13. $-\sqrt[3]{216}$

14. $\sqrt[3]{-27}$

15. $\sqrt[4]{81}$

16. $\sqrt[5]{32}$

17. $\sqrt[5]{-32}$

18. $\sqrt[4]{-81}$

19. $-\sqrt[4]{16}$

20. $\sqrt[5]{-243}$

21. $\sqrt[4]{-16}$

22. $-\sqrt[5]{32}$

23. $-\sqrt[5]{243}$

24. $-\sqrt[4]{625}$

25. $\sqrt{\dfrac{4}{9}}$

26. $\sqrt{\dfrac{9}{25}}$

27. $\sqrt[3]{\dfrac{8}{27}}$

28. $\sqrt[3]{-\dfrac{27}{64}}$

29. $\sqrt{6^2}$

30. $\sqrt{9^2}$

31. $\sqrt{(-3)^2}$

32. $\sqrt{(-5)^2}$

33. $\sqrt[3]{4^3}$ |4|

34. $\sqrt[3]{(-5)^3}$

35. $\sqrt[4]{3^4}$

36. $\sqrt[4]{(-2)^4}$

Simplify each of the following roots.

37. $\sqrt{x^2}$

38. $\sqrt[3]{w^3}$

39. $\sqrt[5]{y^5}$

40. $\sqrt[7]{z^7}$

41. $\sqrt{9x^2}$

42. $\sqrt{81y^2}$ 9|y|

43. $\sqrt{a^4b^6}$ |a²b³|

44. $\sqrt{w^6z^{10}}$ w

45. $\sqrt{16x^4}$ 4x

46. $\sqrt{49y^6}$ 7y³

47. $\sqrt[4]{y^{20}}$ y⁵

48. $\sqrt[3]{a^{18}}$ a⁶

49. $\sqrt[4]{m^8n^{12}}$ m²n³

50. $\sqrt[3]{a^6b^9}$ a²b³

51. $\sqrt[3]{125a^3}$ 5a

52. $\sqrt[3]{-27x^3}$ 3x

53. $\sqrt[5]{32x^5y^{15}}$ 2·xx³

54. $\sqrt[5]{-32m^{10}n^5}$ -2 m²n

 Using a calculator, evaluate the following. Round each answer to three decimal places.

55. $\sqrt{15}$

56. $\sqrt{29}$

57. $\sqrt{156}$

58. $\sqrt{213}$

59. $\sqrt{-15}$

60. $\sqrt{-79}$

61. $\sqrt[3]{83}$

62. $\sqrt[3]{97}$

63. $\sqrt[5]{123}$

64. $\sqrt[5]{283}$

65. $\sqrt[3]{-15}$

66. $\sqrt[3]{-29}$

Transcribe Your Skills

67. If you use a calculator to find the square root of 3, you get the display 1.7320508. If you enter this number into the calculator and square it, you don't get exactly 3. Why not?

68. Explain why $\sqrt[6]{x^{12}}$ will always be positive.

Think About These

Label each of the following statements true or false. For statements which are false, give an example that shows the statement is not true.

69. $\sqrt{16x^{16}} = 4x^4$

70. $\sqrt{36c^2} = 6c$

71. $\sqrt[3]{(4x^6y^9)^3} = 4x^6y^9$

72. $\sqrt[4]{(x-4)^4} = x - 4$

73. $\sqrt{x^4 + 16} = x^2 + 4$

74. $\sqrt{x^8 + 27} = x^2 + 3$

75. $\sqrt{16x^{-4}y^{-4}}$ is not a real number.

76. $\sqrt[3]{-8x^6y^6}$ is not a real number.

 77. A large water tank in the shape of a cube has a volume of 343 m³. What is the length of the edge of the tank?

 78. A cubical grain storage bin has a volume of 512 m³. How tall is the bin?

Skillscan (Section 5.1)

Find each of the following products.

a. $(9x^2)(2x)$ **b.** $(4a^4)(5a)$ **c.** $(25m^2)(3m)$

d. $(8r^3)(2r)$ **e.** $(64y^6)(2y)$ **f.** $(16w^4)(w^3)$

g. $(49a^6)(2a)$ **h.** $(100s^4)(2s)$

Simplification of Radical Expressions   **7.2**

OBJECTIVES: 1. To simplify radical expressions by using the product property
            2. To simplify radical expressions by using the quotient property

TAPE IN16

In the last section, we introduced the radical notation. For some applications, we will want to make sure that all radical expressions are written in *simplified form*. To accomplish this objective, we will need two basic properties. In stating these

A precise set of conditions for a radical to be in simplified form will follow in this section.

properties, and in our subsequent examples, we again assume that all variables represent positive real numbers whenever the index of a radical is even. To develop our first property, consider an expression such as

$$\sqrt{25 \cdot 4}$$

One approach to simplify the expression would be

$$\sqrt{25 \cdot 4} = \sqrt{100} = 10$$

Now what happens if we separate the original radical as follows?

$$\sqrt{25 \cdot 4} = \sqrt{25} \cdot \sqrt{4}$$
$$= 5 \cdot 2 = 10$$

The result in either case is the same, and this suggests our first property for radicals.

As we stated in the first paragraph, a and b are positive real numbers and n is an even integer.

PRODUCT PROPERTY FOR RADICALS

$$\sqrt[n]{ab} = \sqrt[n]{a} \cdot \sqrt[n]{b}$$

In words, the radical of a product is equal to the product of the radicals.

The second property we will need is similar.

To convince yourself that this must be the case, at least for square roots, let $a = 100$ and $b = 4$ and evaluate both sides of the equation.

QUOTIENT PROPERTY FOR RADICALS

$$\sqrt[n]{\frac{a}{b}} = \frac{\sqrt[n]{a}}{\sqrt[n]{b}}$$

In words, the radical of a quotient is the quotient of the radicals.

CAUTION

Be Careful! Students sometimes assume that since

$$\sqrt{ab} = \sqrt{a} \cdot \sqrt{b}$$

it should also be true that

$$\sqrt{a + b} \stackrel{?}{=} \sqrt{a} + \sqrt{b}$$

You can easily see that this is *not* true. Let $a = 9$ and $b = 16$ in the statement.

With these two properties, we are now ready to define the simplified form for a radical expression. A radical is in simplified form if the following three conditions are satisfied.

SIMPLIFIED FORM FOR A RADICAL EXPRESSION

1. **The radicand has no factor raised to a power greater than or equal to the index.**
2. **No fraction appears in the radical.**
3. **No radical appears in a denominator.**

Our initial example deals with satisfying condition 1 above. Essentially, we want to find the largest perfect-square factor (in the case of a square root) in the radicand and then apply the product property to simplify the expression.

EXAMPLE 1 Simplifying Radicals

Write each expression in simplified form.

(a) $\sqrt{18} = \sqrt{9 \cdot 2}$
$= \sqrt{9} \cdot \sqrt{2}$
$= 3\sqrt{2}$

The largest perfect-square factor of 18 is 9.

(b) $\sqrt{75} = \sqrt{25 \cdot 3}$
$= \sqrt{25} \cdot \sqrt{3}$
$= 5\sqrt{3}$

The largest perfect-square factor of 75 is 25.

(c) $\sqrt{27x^3} = \sqrt{9x^2 \cdot 3x}$
$= \sqrt{9x^2} \cdot \sqrt{3x} = 3x\sqrt{3x}$

(d) $\sqrt{72a^3b^4} = \sqrt{36a^2b^4 \cdot 2a}$
$= \sqrt{36a^2b^4} \cdot \sqrt{2a}$
$= 6ab^2\sqrt{2a}$

The largest perfect-square factor of $27x^3$ is $9x^2$. Note that the exponent must be *even* in a perfect square.

Check Yourself 1

Write each expression in simplified form.

1. $\sqrt{45}$ $\ 3\sqrt{5}$ **2.** $\sqrt{200}$ $\ 2\sqrt{10}$ **3.** $\sqrt{75p^5}$ $\ 5p^2\sqrt{3p}$ **4.** $\sqrt{98m^3n^4}$

$5p^2\sqrt{3p}$

$49\,m\,n^2$
$7mn^2\sqrt{2n}$

   Writing a cube root in simplified form involves finding factors of the radicand that are perfect cubes, as illustrated in Example 2. The process illustrated in this example is extended in an identical fashion to simplify radical expressions with any index.

EXAMPLE 2 Simplifying Radicals

Write each expression in simplified form.

(a) $\sqrt[3]{48} = \sqrt[3]{8 \cdot 6}$
$= \sqrt[3]{8} \cdot \sqrt[3]{6} = 2\sqrt[3]{6}$

(b) $\sqrt[3]{24x^4} = \sqrt[3]{8x^3 \cdot 3x}$
$= \sqrt[3]{8x^3} \cdot \sqrt[3]{3x} = 2x\sqrt[3]{3x}$

In a perfect cube, the exponent must be a *multiple of 3*.

(c) $\sqrt[3]{54a^7b^4} = \sqrt[3]{27a^6b^3 \cdot 2ab}$
$= \sqrt[3]{27a^6b^3} \cdot \sqrt[3]{2ab} = 3a^2b\sqrt[3]{2ab}$

Check Yourself 2

Write each expression in simplified form.

1. $\sqrt[3]{128w^4}$ **2.** $\sqrt[3]{40x^5y^7}$ **3.** $\sqrt[4]{48a^8b^5}$

Satisfying our second condition for a radical to be in simplified form (no fractions should appear inside the radical) requires the second property for radicals. Consider the following example.

EXAMPLE 3 Simplifying Radicals That Contain Fractions

Write each expression in simplified form.

Apply the quotient property.

(a) $\sqrt{\dfrac{5}{9}} = \dfrac{\sqrt{5}}{\sqrt{9}}$

$\qquad = \dfrac{\sqrt{5}}{3}$

(b) $\sqrt{\dfrac{a^4}{25}} = \dfrac{\sqrt{a^4}}{\sqrt{25}} = \dfrac{a^2}{5}$

(c) $\sqrt[3]{\dfrac{5x^2}{8}} = \dfrac{\sqrt[3]{5x^2}}{\sqrt[3]{8}} = \dfrac{\sqrt[3]{5x^2}}{2}$

Check Yourself 3

Write each expression in simplified form.

1. $\sqrt{\dfrac{7}{16}}$

2. $\sqrt{\dfrac{3}{25a^2}}$

3. $\sqrt[3]{\dfrac{5x}{27}}$

Our next example also begins with the application of the quotient property for radicals. However, an additional step is required because, as we will see, condition 3 (no radicals can appear in a denominator) must also be satisfied during the process.

EXAMPLE 4 Simplifying Radicals That Contain Fractions

Write $\sqrt{\dfrac{3}{5}}$ in simplified form.

$\sqrt{\dfrac{3}{5}} = \dfrac{\sqrt{3}}{\sqrt{5}}$

The application of the quotient property satisfies condition 2—there are now no fractions *inside* a radical. However, we now have a radical in the denominator, violating condition 3. The expression will not be simplified until that radical is removed.

To remove the radical in the denominator, we multiply the numerator and denominator by the *same* expression, here $\sqrt{5}$. This is called *rationalizing the denominator*.

$$\frac{\sqrt{3}}{\sqrt{5}} = \frac{\sqrt{3} \cdot \sqrt{5}}{\sqrt{5} \cdot \sqrt{5}}$$

$$= \frac{\sqrt{15}}{\sqrt{25}} = \frac{\sqrt{15}}{5}$$

The value of the expression is *not* changed as we multiply by $\frac{\sqrt{5}}{\sqrt{5}}$, or 1.

The point here is to arrive at a perfect square inside the radical in the denominator. This is done by multiplying the numerator and the denominator by $\sqrt{5}$ since

$$\sqrt{5} \cdot \sqrt{5} = \sqrt{5^2} = \sqrt{25}$$

Check Yourself 4

Simplify $\sqrt{\dfrac{3}{7}}$. $\frac{21}{49}$

Let's look at some further examples that involve rationalizing the denominator of an expression.

EXAMPLE 5 Rationalizing the Denominator

Write each expression in simplified form.

(a) $\dfrac{3}{\sqrt{8}} = \dfrac{3 \cdot \sqrt{2}}{\sqrt{8} \cdot \sqrt{2}}$

$$= \frac{3\sqrt{2}}{\sqrt{16}} = \frac{3\sqrt{2}}{4}$$

(b) $\sqrt[3]{\dfrac{5}{4}} = \dfrac{\sqrt[3]{5}}{\sqrt[3]{4}}$

We multiply numerator and denominator by $\sqrt{2}$. Why did we choose $\sqrt{2}$? Note that

$$\sqrt{8} = \sqrt{2^3}$$

so

$$\sqrt{8}\sqrt{2} = \sqrt{2^3}\sqrt{2}$$
$$= \sqrt{2^4}$$

Now note that

$$\sqrt[3]{4} \cdot \sqrt[3]{2} = \sqrt[3]{8} = 2$$

so multiplying the numerator and denominator by $\sqrt[3]{2}$ will produce a perfect cube inside the radical in the denominator. Continuing, we have

$$\frac{\sqrt[3]{5}}{\sqrt[3]{4}} = \frac{\sqrt[3]{5} \cdot \sqrt[3]{2}}{\sqrt[3]{4} \cdot \sqrt[3]{2}}$$

$$= \frac{\sqrt[3]{10}}{\sqrt[3]{8}} = \frac{\sqrt[3]{10}}{2}$$

Why did we use $\sqrt[3]{2}$? Note that

$$\sqrt[3]{4} \cdot \sqrt[3]{2} = \sqrt[3]{2^2} \cdot \sqrt[3]{2}$$
$$= \sqrt[3]{2^3}$$

and the exponent is a multiple of 3.

Check Yourself 5

Simplify each expression.

1. $\dfrac{5}{\sqrt{12}}$ $5 = 60$

2. $\sqrt[3]{\dfrac{2}{9}}$ $\frac{3}{3}$ $\frac{6}{9}$

As our final example, we illustrate the process of rationalizing a denominator when variables are involved in a rational expression.

EXAMPLE 6 Rationalizing the Denominator

Simplify each expression.

(a) $\sqrt{\dfrac{8x^3}{3y}}$

By the quotient property we have

$$\sqrt{\dfrac{8x^3}{3y}} = \dfrac{\sqrt{8x^3}}{\sqrt{3y}}$$

Since the numerator can be simplified in this case, let's start with that procedure.

$$\dfrac{\sqrt{8x^3}}{\sqrt{3y}} = \dfrac{\sqrt{4x^2} \cdot \sqrt{2x}}{\sqrt{3y}} = \dfrac{2x\sqrt{2x}}{\sqrt{3y}}$$

Multiplying the numerator and denominator by $\sqrt{3y}$ will rationalize the denominator.

$$\dfrac{2x\sqrt{2x} \cdot \sqrt{3y}}{\sqrt{3y} \cdot \sqrt{3y}} = \dfrac{2x\sqrt{6xy}}{\sqrt{9y^2}} = \dfrac{2x\sqrt{6xy}}{3y}$$

(b) $\dfrac{2}{\sqrt[3]{3x}}$

To satisfy condition 3, we must remove the radical from the denominator. For this we need a perfect cube inside the radical in the denominator. Multiplying the numerator and denominator by $\sqrt[3]{9x^2}$ will provide the perfect cube. So

$$\dfrac{2\sqrt[3]{9x^2}}{\sqrt[3]{3x} \cdot \sqrt[3]{9x^2}} = \dfrac{2\sqrt[3]{9x^2}}{\sqrt[3]{27x^3}}$$

$$= \dfrac{2\sqrt[3]{9x^2}}{3x}$$

Note

$\sqrt[3]{9x^2} = \sqrt[3]{3^2x^2}$

so

$\sqrt[3]{3x} \cdot \sqrt[3]{9x^2} = \sqrt[3]{3^3x^3}$

and each exponent is a multiple of 3.

Check Yourself 6

Simplify each expression.

1. $\sqrt{\dfrac{12a^3}{5b}}$

2. $\dfrac{3}{\sqrt[3]{2w^2}}$

The following algorithm summarizes our work in simplifying radical expressions.

SIMPLIFYING RADICAL EXPRESSIONS

STEP 1 To satisfy condition 1: Determine the largest perfect-square factor of the radicand. Apply the product property to "remove" that factor from inside the radical.

STEP 2 To satisfy condition 2: Use the quotient property to write the expression in the form

$$\frac{\sqrt{a}}{\sqrt{b}}$$

If b is a perfect square, remove the radical in the denominator. If that is not the case, proceed to step 3.

STEP 3 Multiply the numerator and denominator of the radical expression by an appropriate radical to remove the radical in the denominator. Simplify the resulting expression where necessary.

In the case of a cube root, steps 1 and 2 would refer to perfect cubes, etc.

CHECK YOURSELF ANSWERS

1. (1) $3\sqrt{5}$; (2) $10\sqrt{2}$; (3) $5p^2\sqrt{3p}$; (4) $7mn^2\sqrt{2m}$ **2.** (1) $4w\sqrt[3]{2w}$;
(2) $2xy^2\sqrt[3]{5x^2y}$; (3) $2a^2b\sqrt[4]{3b}$ **3.** (1) $\dfrac{\sqrt{7}}{4}$; (2) $\dfrac{\sqrt{3}}{5a}$; (3) $\dfrac{\sqrt[3]{5x}}{3}$ **4.** $\dfrac{\sqrt{21}}{7}$

5. (1) $\dfrac{5\sqrt{3}}{6}$; (2) $\dfrac{\sqrt[3]{6}}{3}$ **6.** (1) $\dfrac{2a\sqrt{15ab}}{5b}$; (2) $\dfrac{3\sqrt[3]{4w}}{2w}$

7.2 EXERCISES

Build Your Skills

Use the product property to write each expression in simplified form.

1. $\sqrt{12}$

2. $\sqrt{24}$

3. $\sqrt{50}$

4. $\sqrt{28}$

5. $-\sqrt{108}$

6. $\sqrt{32}$

7. $\sqrt{52}$

8. $-\sqrt{96}$

9. $\sqrt{60}$

10. $\sqrt{150}$

11. $-\sqrt{125}$

12. $\sqrt{128}$

13. $\sqrt{288}$

14. $-\sqrt{300}$

15. $\sqrt{450}$

16. $\sqrt{432}$

17. $\sqrt[3]{16}$

18. $\sqrt[3]{-54}$

19. $\sqrt[3]{-48}$

20. $\sqrt[3]{250}$

21. $\sqrt[3]{135}$

22. $\sqrt[3]{-160}$

23. $\sqrt[3]{32}$

24. $\sqrt[4]{96}$

25. $\sqrt[4]{243}$

26. $\sqrt[4]{1250}$

Use the product property to write each expression in simplified form. Assume that all variables represent positive real numbers.

27. $\sqrt{18z^2}$

28. $\sqrt{45a^2}$

29. $\sqrt{63x^4}$

30. $\sqrt{54w^4}$

31. $\sqrt{98m^3}$

32. $\sqrt{75a^5}$

33. $\sqrt{80x^2y^3}$

34. $\sqrt{108p^5q^2}$

35. $\sqrt[3]{40b^3}$

36. $\sqrt[3]{16x^3}$

37. $\sqrt[3]{48p^9}$

38. $\sqrt[3]{-80a^6}$

39. $\sqrt[3]{54m^7}$

40. $\sqrt[3]{250x^{13}}$

41. $\sqrt[3]{-24a^5b^4}$

42. $\sqrt[3]{128r^6s^2}$

43. $\sqrt[3]{56x^6y^5z^4}$

44. $-\sqrt[3]{250a^4b^{15}c^9}$

45. $\sqrt[4]{32x^8}$

46. $\sqrt[4]{162y^{12}}$

47. $\sqrt[4]{243a^{15}}$

48. $\sqrt[4]{80p^{11}}$

49. $\sqrt[4]{96w^5z^{13}}$

50. $\sqrt[4]{128a^{12}b^{17}}$

51. $\sqrt[5]{64w^{10}}$

52. $\sqrt[5]{96a^5b^{12}}$

Use the quotient property to write each expression in simplified form. Assume that all variables represent positive real numbers.

53. $\sqrt{\dfrac{5}{16}}$

54. $\sqrt{\dfrac{11}{36}}$

55. $\sqrt{\dfrac{x^4}{25}}$

56. $\sqrt{\dfrac{a^6}{49}}$

57. $\sqrt{\dfrac{5}{9y^4}}$

58. $\sqrt{\dfrac{7}{25x^2}}$

59. $\sqrt[3]{\dfrac{5}{8}}$

60. $\sqrt[3]{\dfrac{3}{64}}$

61. $\sqrt[3]{\dfrac{4x^2}{27}}$

62. $\sqrt[4]{\dfrac{5x^3}{16}}$

63. $\sqrt[4]{\dfrac{3}{81a^8}}$

64. $\sqrt[3]{\dfrac{7}{8y^6}}$

Write each expression in simplified form. Assume that all variables represent positive real numbers.

65. $\sqrt{\dfrac{4}{5}}$

66. $\sqrt{\dfrac{7}{3}}$

67. $\dfrac{3}{\sqrt{10}}$

68. $\dfrac{5}{\sqrt{7}}$

69. $\sqrt{\dfrac{5}{8}}$

70. $\dfrac{7}{\sqrt{12}}$

71. $\dfrac{\sqrt{6}}{\sqrt{7}}$

72. $\dfrac{\sqrt{5}}{\sqrt{11}}$

73. $\dfrac{2\sqrt{3}}{\sqrt{10}}$

74. $\dfrac{3\sqrt{5}}{\sqrt{3}}$

75. $\sqrt[3]{\dfrac{7}{4}}$

76. $\sqrt[3]{\dfrac{5}{9}}$

77. $\dfrac{5}{\sqrt[3]{16}}$

78. $\dfrac{\sqrt[3]{3}}{\sqrt[3]{4}}$

79. $\sqrt{\dfrac{3}{x}}$

80. $\sqrt{\dfrac{7}{y}}$

81. $\sqrt{\dfrac{12}{w}}$

82. $\dfrac{\sqrt{18}}{\sqrt{a}}$

83. $\dfrac{\sqrt{8m^3}}{\sqrt{5n}}$

84. $\sqrt{\dfrac{24x^5}{7y}}$

85. $\sqrt[3]{\dfrac{5}{y}}$

86. $\sqrt[3]{\dfrac{7}{x^2}}$

87. $\dfrac{3}{\sqrt[3]{2x}}$

88. $\dfrac{5}{\sqrt[3]{3a}}$

89. $\sqrt[3]{\dfrac{2}{5x^2}}$

90. $\sqrt[3]{\dfrac{5}{7w^2}}$

91. $\sqrt[3]{\dfrac{a^5}{b^7}}$

92. $\dfrac{\sqrt[3]{2}}{\sqrt[3]{9m^2}}$

Transcribe Your Skills

93. In this section we encountered two new properties. What is a property?

94. Describe a radical expression that can be simplified.

Think About These

Label each of the following statements as true or false.

95. $\sqrt{16x^{16}} = 4x^4$

96. $\sqrt{x^2 + y^2} = x + y$

97. $\dfrac{\sqrt{x^2 - 25}}{\sqrt{x - 5}} = \sqrt{x + 5}$

98. $\sqrt[3]{x^6} \cdot \sqrt[3]{x^3 - 1} = x^2 \sqrt[3]{x - 1}$

99. $\sqrt[3]{(8b^6)^2} = (\sqrt[3]{8b^6})^2$

100. $\dfrac{\sqrt[3]{8x^3}}{\sqrt[3]{2x}} = \sqrt[3]{4x^2}$

Simplify.

101. $\dfrac{7\sqrt{x^2y^4} \cdot \sqrt{36xy}}{6\sqrt{x^{-6}y^{-2}} \cdot \sqrt{49x^{-1}y^{-3}}}$

102. $\dfrac{3\sqrt[3]{32c^{12}d^2} \cdot \sqrt[3]{2c^5d^4}}{4\sqrt[3]{9c^8d^{-2}} \cdot \sqrt[3]{3c^{-3}d^{-4}}}$

 103. The study of water transport around obstacles generates the following equation.

$$x = \sqrt{1 - \dfrac{k^2}{16\pi^2}}$$

Simplify the expression for x (eliminate the fraction in the radicand).

104. The study of soil mechanics generates the following formula.

$$t = C\sqrt[3]{\dfrac{L}{mx^2}}$$

Simplify the expression for t by rationalizing the denominator.

Skillscan (Section 1.2)

Use the distributive property to combine like terms in each of the following expressions.

a. $7a + 6a$ **b.** $9x - 3x$ **c.** $10y - 15y$ **d.** $8w + 12w$ **e.** $8x + 5x - 12x$ **f.** $8b - 10b + 3b$ **g.** $10m + 5n - 5m$ **h.** $8r + 7s - 4s$

7.3 Addition and Subtraction of Radical Expressions

TAPE IN16

OBJECTIVES: 1. To add radical expressions
2. To subtract radical expressions

The addition and subtraction of radical expressions exactly parallel our earlier work with polynomials containing like terms. Let's review for a moment.

To add $3x^2 + 4x^2$, we have

This uses the distributive property.

$$3x^2 + 4x^2 = (3 + 4)x^2$$
$$= 7x^2$$

Keep in mind that we were able to simplify or combine the above expressions because of like terms in x^2. (Recall that like terms have the same variable factor raised to the same power.)

We *cannot* combine terms such as

$$4a^3 + 3a^2 \qquad \text{or} \qquad 3x - 5y$$

By extending these ideas, radical expressions can be combined *only* if they are *similar,* that is, if the expressions contain the same radicand with the same index. This is illustrated in the first example.

EXAMPLE 1 Adding and Subtracting Radical Expressions

Add or subtract as indicated.

Apply the distributive property again.

(a) $3\sqrt{7} + 2\sqrt{7} = (3 + 2)\sqrt{7}$
$\qquad\qquad = 5\sqrt{7}$

(b) $7\sqrt{3} - 4\sqrt{3} = (7 - 4)\sqrt{3} = 3\sqrt{3}$

(c) $5\sqrt{10} - 3\sqrt{10} + 2\sqrt{10} = (5 - 3 + 2)\sqrt{10}$
$\qquad\qquad\qquad\qquad\qquad = 4\sqrt{10}$

The expressions have different radicands, $\sqrt{5}$ and $\sqrt{3}$.

(d) $2\sqrt{5} + 3\sqrt{3}$ cannot be combined or simplified.

The expressions have different indices, 2 and 3.

(e) $\sqrt{7} + \sqrt[3]{7}$ cannot be simplified.

(f) $5\sqrt{x} + 2\sqrt{x} = (5 + 2)\sqrt{x}$
$\qquad\qquad = 7\sqrt{x}$

(g) $5\sqrt{3ab} - 2\sqrt{3ab} + 3\sqrt{3ab} = (5 - 2 + 3)\sqrt{3ab} = 6\sqrt{3ab}$

The radicands are not the same.

(h) $\sqrt[3]{3x^2} + \sqrt[3]{3x}$ cannot be simplified.

Check Yourself 1

Add or subtract as indicated.

1. $5\sqrt{3} + 2\sqrt{3}$

2. $7\sqrt{5} - 2\sqrt{5} + 3\sqrt{5}$

3. $2\sqrt{3} + 3\sqrt{2}$

4. $\sqrt{2y} + 5\sqrt{2y} - 3\sqrt{2y}$

5. $2\sqrt[3]{3m} - 5\sqrt[3]{3m}$

6. $\sqrt{5x} - \sqrt[3]{5x}$ cann

Often it is necessary to simplify radical expressions by the methods of Section 7.2 before they can be combined. The following example illustrates how the product property is applied.

EXAMPLE 2 Adding and Subtracting Radical Expressions

Add or subtract as indicated.

(a) $\sqrt{48} + 2\sqrt{3}$

In this form, the radicals cannot be combined. However, note that the first radical can be simplified by our earlier methods because 48 has the perfect-square factor 16.

$$\sqrt{48} = \sqrt{16 \cdot 3} = 4\sqrt{3}$$

With this result we can proceed as before.

$$\sqrt{48} + 2\sqrt{3} = 4\sqrt{3} + 2\sqrt{3}$$
$$= (4 + 2)\sqrt{3} = 6\sqrt{3}$$

— how did I get 2

(b) $\sqrt{50} - \sqrt{32} + \sqrt{98} = 5\sqrt{2} - 4\sqrt{2} + 7\sqrt{2}$
$$= (5 - 4 + 7)\sqrt{2} = 8\sqrt{2}$$

Note that each of the radicands has a perfect-square factor. The reader should provide the details for the simplification of each radical.

(c) $x\sqrt{2x} + 3\sqrt{8x^3}$

Note that

$$3\sqrt{8x^3} = 3\sqrt{4x^2 \cdot 2x}$$
$$= 3\sqrt{4x^2} \cdot \sqrt{2x}$$
$$= 3 \cdot 2x\sqrt{2x} = 6x\sqrt{2x}$$

So

$$x\sqrt{2x} + 3\sqrt{8x^3} = x\sqrt{2x} + 6x\sqrt{2x}$$
$$= (x + 6x)\sqrt{2x} = 7x\sqrt{2x}$$

We can now combine the similar radicals.

(d) $\sqrt[3]{2a} - \sqrt[3]{16a} + \sqrt[3]{54a} = \sqrt[3]{2a} - 2\sqrt[3]{2a} + 3\sqrt[3]{2a}$
$$= (1 - 2 + 3)\sqrt[3]{2a}$$
$$= 2\sqrt[3]{2a}$$

Check Yourself 2

Add or subtract as indicated.

but you can simply like term

1. $\sqrt{125} + 3\sqrt{5}$ *can't add*

2. $\sqrt{75} - \sqrt{27} + \sqrt{48}$

3. $5\sqrt{24y^3} - y\sqrt{6y}$ *(now)*

4. $\sqrt[3]{81x} - \sqrt[3]{3x} + \sqrt[3]{24x}$

It may also be necessary to apply the quotient property before combining rational expressions. Consider the following example.

EXAMPLE 3 Adding and Subtracting Radical Expressions

Add or subtract as indicated.

(a) $2\sqrt{6} + \sqrt{\dfrac{2}{3}}$

We apply the quotient property to the *second term* and rationalize the denominator.

Multiply by $\dfrac{\sqrt{3}}{\sqrt{3}}$, or 1.

$$\sqrt{\frac{2}{3}} = \frac{\sqrt{2}}{\sqrt{3}} = \frac{\sqrt{2} \cdot \sqrt{3}}{\sqrt{3} \cdot \sqrt{3}} = \frac{\sqrt{6}}{3}$$

So

$$2\sqrt{6} + \sqrt{\frac{2}{3}} = 2\sqrt{6} + \frac{\sqrt{6}}{3}$$

Note that $\dfrac{\sqrt{6}}{3}$ and $\dfrac{1}{3}\sqrt{6}$ are equivalent.

$$= \left(2 + \frac{1}{3}\right)\sqrt{6} = \frac{7}{3}\sqrt{6}$$

(b) $\sqrt{20x} - \sqrt{\dfrac{x}{5}}$

Again we first simplify the two expressions. So

$$\sqrt{20x} - \sqrt{\frac{x}{5}} = 2\sqrt{5x} - \frac{\sqrt{x} \cdot \sqrt{5}}{\sqrt{5} \cdot \sqrt{5}}$$

$$= 2\sqrt{5x} - \frac{\sqrt{5x}}{5}$$

$$= \left(2 - \frac{1}{5}\right)\sqrt{5x} = \frac{9}{5}\sqrt{5x}$$

Check Yourself 3

Add or subtract as indicated.

1. $3\sqrt{7} + \sqrt{\dfrac{1}{7}}$

2. $\sqrt{40x} - \sqrt{\dfrac{2x}{5}}$

Our final example illustrates how our earlier methods for adding fractions may have to be applied in working with radical expressions.

EXAMPLE 4 Adding and Subtracting Radical Expressions

Add $\dfrac{\sqrt{5}}{3} + \dfrac{2}{\sqrt{5}}$.

Solution Our first step will be to rationalize the denominator of the second fraction, to write the sum as

$$\frac{\sqrt{5}}{3} + \frac{2\sqrt{5}}{\sqrt{5} \cdot \sqrt{5}}$$

or

$$\frac{\sqrt{5}}{3} + \frac{2\sqrt{5}}{5}$$

The LCD of the fractions is 15, and rewriting each fraction with that denominator, we have

$$\frac{\sqrt{5} \cdot 5}{3 \cdot 5} + \frac{2\sqrt{5} \cdot 3}{5 \cdot 3} = \frac{5\sqrt{5} + 6\sqrt{5}}{15}$$

$$= \frac{11\sqrt{5}}{15}$$

Check Yourself 4

Subtract $\dfrac{3}{\sqrt{10}} - \dfrac{\sqrt{10}}{5}$.

CHECK YOURSELF ANSWERS

1. (1) $7\sqrt{3}$; (2) $8\sqrt{5}$; (3) cannot be simplified; (4) $3\sqrt{2y}$; (5) $-3\sqrt[3]{3m}$; (6) cannot be simplified   **2.** (1) $8\sqrt{5}$; (2) $6\sqrt{3}$; (3) $9y\sqrt{6y}$; (4) $4\sqrt[3]{3x}$

3. (1) $\dfrac{22}{7}\sqrt{7}$; (2) $\dfrac{9}{5}\sqrt{10x}$ **4.** $\dfrac{\sqrt{10}}{10}$

7.3 EXERCISES

Build Your Skills

Add or subtract as indicated. Assume that all variables represent positive real numbers.

1. $3\sqrt{5} + 4\sqrt{5}$

2. $5\sqrt{6} + 3\sqrt{6}$

3. $7\sqrt{7} - 4\sqrt{7}$

4. $12\sqrt{3} - 9\sqrt{3}$

5. $3\sqrt{5} + 4\sqrt{3}$

6. $3\sqrt{7} - 7\sqrt{3}$

7. $5\sqrt{x} + 3\sqrt{x}$

8. $9\sqrt{y} - 5\sqrt{y}$

9. $11\sqrt{3a} - 8\sqrt{3a}$

10. $2\sqrt{5w} + 3\sqrt{5w}$

11. $2x\sqrt{5} + 3x\sqrt{5}$

12. $7y\sqrt{10} - 2y\sqrt{10}$

13. $7\sqrt{m} + 6\sqrt{n}$

14. $8\sqrt{a} - 6\sqrt{b}$

15. $2\sqrt[3]{2} + 7\sqrt[3]{2}$

16. $5\sqrt[4]{3} - 2\sqrt[4]{3}$

17. $8\sqrt{6} - 2\sqrt{6} + 3\sqrt{6}$

18. $8\sqrt{3} + 2\sqrt{3} - 7\sqrt{3}$

17- 57

Simplify the radical expressions where necessary. Then add or subtract as indicated. Again, assume that all variables represent positive real numbers.

19. $\sqrt{20} + \sqrt{5}$

20. $\sqrt{27} + \sqrt{3}$

21. $\sqrt{18} + \sqrt{50}$

22. $\sqrt{28} + \sqrt{63}$

23. $\sqrt{54} - \sqrt{24}$

24. $\sqrt{72} - \sqrt{18}$

25. $3\sqrt{3} + \sqrt{75}$

26. $\sqrt{125} - 2\sqrt{5}$

27. $4\sqrt{28} - \sqrt{63}$

28. $2\sqrt{40} + \sqrt{90}$

29. $\sqrt{98} - \sqrt{18} + \sqrt{8}$

30. $\sqrt{108} - \sqrt{27} + \sqrt{75}$

31. $\sqrt[3]{81} + \sqrt[3]{3}$

32. $\sqrt[3]{16} - \sqrt[3]{2}$

33. $2\sqrt[3]{128} - 3\sqrt[3]{2}$

34. $3\sqrt[3]{81} - 2\sqrt[3]{3}$

35. $\sqrt{5x} + \sqrt{20x}$

36. $\sqrt{63a} - \sqrt{7a}$

37. $\sqrt{54w} - \sqrt{24w}$

38. $\sqrt{27p} + \sqrt{75p}$

39. $\sqrt{18x^3} + \sqrt{8x^3}$

40. $\sqrt{125y^3} - \sqrt{20y^3}$

41. $\sqrt{75a^3} - a\sqrt{27a}$

42. $w\sqrt{63w} + \sqrt{28w^3}$

43. $\sqrt[3]{81m^2} - \sqrt[3]{24m^2}$

44. $\sqrt[3]{16b^2} + \sqrt[3]{54b^2}$

45. $\sqrt[4]{162x} - \sqrt[4]{32x}$

46. $\sqrt[4]{48s^3} + \sqrt[4]{243s^3}$

47. $\sqrt[3]{54x^4} - \sqrt[3]{16x^4} + \sqrt[3]{128x^4}$

48. $\sqrt[3]{81a^5} + \sqrt[3]{24a^5} - \sqrt[3]{192a^5}$

49. $\sqrt[3]{16w^5} + 2w\sqrt[3]{2w^2} - \sqrt[3]{2w^5}$

50. $\sqrt[4]{2z^7} - z\sqrt[4]{32z^3} + \sqrt[4]{162z^7}$

51. $\sqrt{3} + \sqrt{\dfrac{1}{3}}$

52. $\sqrt{6} - \sqrt{\dfrac{1}{6}}$

53. $\sqrt{6} - \sqrt{\dfrac{2}{3}}$

54. $\sqrt{10} + \sqrt{\dfrac{2}{5}}$

55. $\sqrt{84} + \sqrt{\dfrac{3}{7}}$

56. $\sqrt{120} - \sqrt{\dfrac{5}{6}}$

57. $\sqrt{5x} - \sqrt{\dfrac{x}{5}}$

58. $\sqrt{3x} + \sqrt{\dfrac{x}{3}}$

59. $\sqrt{60x} + \sqrt{\dfrac{3x}{5}}$

60. $\sqrt{24a} - \sqrt{\dfrac{2a}{3}}$

61. $\sqrt[3]{48} - \sqrt[3]{\dfrac{3}{4}}$

62. $\sqrt[3]{96} + \sqrt[3]{\dfrac{4}{9}}$

63. $\dfrac{\sqrt{12}}{3} - \dfrac{1}{\sqrt{3}}$

64. $\dfrac{\sqrt{20}}{5} + \dfrac{2}{\sqrt{5}}$

65. $\dfrac{1}{\sqrt{a}} + \dfrac{1}{\sqrt{b}}$

66. $\dfrac{1}{\sqrt{x}} - \dfrac{1}{\sqrt{y}}$

Transcribe Your Skills

67. Is there any prime number whose square root is an integer? Explain your answer.

68. Explain why $\sqrt{3 + 5x + y}$ is considered a single term but $3 + 5x + y$ is considered the sum of three terms.

Think About These

Simplify each of the following radical expressions.

69. $x\sqrt[3]{8x^4} + 4\sqrt[3]{27x^7}$

70. $\sqrt[3]{8x^2} - \sqrt[3]{27x^2}$

71. $\sqrt[4]{16a^5} + \sqrt[4]{81a^5}$

72. $\sqrt[4]{256x^7} - 2\sqrt[4]{81x^7}$

Label each of the following as true or false.

73. $\sqrt{16 + x^2} = 4 + x$

74. $\sqrt{2} + \sqrt[3]{6} = \sqrt[5]{8}$

75. $\dfrac{5}{\sqrt{xy}} - \dfrac{4}{\sqrt{x}} = \dfrac{5\sqrt{xy} - 4y\sqrt{x}}{xy}$

76. $\sqrt{8y - 8} + \sqrt{2y - 2} = 3\sqrt{2y - 2}$

77. $\sqrt{7x^2} + \sqrt{9x^6} = 4x^4$

78. $\dfrac{1}{\sqrt{5}} + \dfrac{3}{\sqrt{10}} = \dfrac{4}{\sqrt{15}}$

79. A survey crew is trying to determine the extent of desert encroachment on nearby farmland. Two adjacent, square pieces of property are surveyed. One piece of property measures 4000 m², and the second piece of property measures 6250 m². Find the distance across the two pieces of property in simplified radical form.

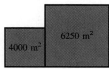

80. Three square pieces of property share the same irrigation ditch, as shown below. How long is the irrigation ditch in simplified radical form?

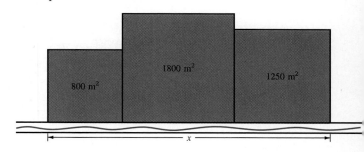

Skillscan (Section 5.3)

Perform the indicated multiplication.

a. $3(a + 6)$　　　**b.** $5(x - 4)$　　　**c.** $y(y - 7)$

d. $m(m + 9)$　　　**e.** $(s + 3)(s - 3)$　　　**f.** $(x - 5)(x + 5)$

g. $(w - z)(w - z)$　　　**h.** $(m + 8)(m + 8)$

Multiplication and Division of Radical Expressions　7.4

OBJECTIVES: 1. To multiply radical expressions
　　　　　　　　 2. To divide radical expressions

TAPE IN16

In Section 7.2 we introduced the product and quotient properties for radical expressions. At that time they were used for simplifying radicals.

　If we turn those properties around, we have our rules for the multiplication and division of radical expressions. For multiplication:

MULTIPLYING RADICAL EXPRESSIONS

$\sqrt[n]{a} \cdot \sqrt[n]{b} = \sqrt[n]{a \cdot b}$

In words, the product of two roots is the root of the product of the radicands.

The use of this multiplication rule is illustrated in our first example. Again, we assume that all variables represent positive real numbers.

EXAMPLE 1 Multiplying Rational Expressions

Multiply.

Just multiply the radicands.

(a) $\sqrt{7} \cdot \sqrt{5} = \sqrt{7 \cdot 5} = \sqrt{35}$

(b) $\sqrt{3x} \cdot \sqrt{10y} = \sqrt{3x \cdot 10y}$
$= \sqrt{30xy}$

(c) $\sqrt[3]{4x} \cdot \sqrt[3]{7x} = \sqrt[3]{4x \cdot 7x}$
$= \sqrt[3]{28x^2}$

Check Yourself 1

Multiply.

1. $\sqrt{6} \cdot \sqrt{7}$ $\sqrt{42}$

2. $\sqrt{5a} \cdot \sqrt{11b}$ $55ab$

3. $\sqrt[3]{3y} \cdot \sqrt[3]{5y}$ $15y^2$

Keep in mind that all radical expressions should be written in simplified form. Often we have to apply the methods of Section 7.2 to simplify a product once it has been formed.

EXAMPLE 2 Simplifying Rational Products

Multiply and simplify.

Now, $\sqrt{18}$ is not in simplified form. And 9 is a perfect-square factor of 18.

(a) $\sqrt{3} \cdot \sqrt{6} = \sqrt{18}$
$= \sqrt{9 \cdot 2} = \sqrt{9}\sqrt{2}$
$= 3\sqrt{2}$

(b) $\sqrt{5x} \cdot \sqrt{15x} = 75x^2$
$= \sqrt{25x^2 \cdot 3} = \sqrt{25x^2} \cdot \sqrt{3}$
$= 5x\sqrt{3}$

Now we want a factor that is a perfect cube.

(c) $\sqrt[3]{4a^2b} \cdot \sqrt[3]{10a^2b^2} = \sqrt[3]{40a^4b^3} = \sqrt[3]{8a^3b^3 \cdot 5a}$
$= \sqrt[3]{8a^3b^3} \cdot \sqrt[3]{5a} = 2ab\sqrt[3]{5a}$

Check Yourself 2

Multiply and simplify.

1. $\sqrt{10} \cdot \sqrt{20}$ $\sqrt{200}$ $10\sqrt{2}$

2. $\sqrt{6x} \cdot \sqrt{15x}$

3. $\sqrt[3]{9p^2q^2} \cdot \sqrt[3]{6pq^2}$

We are now ready to combine our work of this section with the techniques of Section 7.3 for the addition and subtraction of radicals. This will allow us to multiply radical expressions with more than one term. Consider the following examples.

EXAMPLE 3 Simplifying Rational Products

Multiply and simplify.

(a) $\sqrt{2}(\sqrt{5} + \sqrt{7})$

Distributing $\sqrt{2}$, we have

$\sqrt{2} \cdot \sqrt{5} + \sqrt{2} \cdot \sqrt{7} = \sqrt{10} + \sqrt{14}$

The expression cannot be simplified further.

(b) $\sqrt{3}(\sqrt{6} + 2\sqrt{15}) = \sqrt{3} \cdot \sqrt{6} + \sqrt{3} \cdot 2\sqrt{15}$
$= \sqrt{18} + 2\sqrt{45}$
$= 3\sqrt{2} + 6\sqrt{5}$

(c) $\sqrt{x}(\sqrt{2x} + \sqrt{8x}) = \sqrt{x} \cdot \sqrt{2x} + \sqrt{x} \cdot \sqrt{8x}$
$= \sqrt{2x^2} + \sqrt{8x^2}$
$= x\sqrt{2} + 2x\sqrt{2} = 3x\sqrt{2}$

> We distribute $\sqrt{2}$ over the sum $\sqrt{5} + \sqrt{7}$ to multiply.

> Distribute $\sqrt{3}$.

> Alternately we could choose to simplify $\sqrt{8x}$ in the original expression as our first step. We leave it to the reader to verify that the result would be the same.

Check Yourself 3

Multiply and simplify.

1. $\sqrt{3}(\sqrt{10} + \sqrt{2})$ **2.** $\sqrt{2}(3 + 2\sqrt{6})$ **3.** $\sqrt{a}(\sqrt{3a} + \sqrt{12a})$

If both of the radical expressions involved in a multiplication statement have two terms, we must apply the patterns for multiplying polynomials developed in Chapter 5. The following example illustrates.

EXAMPLE 4 Simplifying Rational Products

Multiply and simplify.

(a) $(\sqrt{3} + 1)(\sqrt{3} + 5)$

To write the desired product, we use the FOIL pattern for multiplying binomials.

$(\sqrt{3} + 1)(\sqrt{3} + 5) = \overset{\text{First}}{\sqrt{3} \cdot \sqrt{3}} + \overset{\text{Outer}}{5 \cdot \sqrt{3}} + \overset{\text{Inner}}{1 \cdot \sqrt{3}} + \overset{\text{Last}}{1 \cdot 5}$
$= 3 + 6\sqrt{3} + 5$
$= 8 + 6\sqrt{3}$

> Combine the outer and inner terms.

> Combine the first and last terms.

(b) $(\sqrt{6} + \sqrt{2})(\sqrt{6} - \sqrt{2})$

Multiplying as before, we have

$\sqrt{6} \cdot \sqrt{6} - \sqrt{6} \cdot \sqrt{2} + \sqrt{6} \cdot \sqrt{2} - \sqrt{2} \cdot \sqrt{2} = 6 - 2 = 4$

> Note that sum of the outer and inner terms $-\sqrt{12} + \sqrt{12}$ is 0.

Two binomial radical expressions that differ *only* in the sign of the second term are called *conjugates* of each other. So

Note the form of the product

$(a + b)(a - b)$

which gives

$a^2 - b^2$

when *a* and *b* are square roots. The product will be rational.

$$\sqrt{6} + \sqrt{2} \quad \text{and} \quad \sqrt{6} - \sqrt{2}$$

are conjugates, and their product does *not* contain a radical—the product is a rational number. That will always be the case with two conjugates. This will have particular significance later in this section.

(c) $(\sqrt{2} + \sqrt{5})^2 = (\sqrt{2} + \sqrt{5})(\sqrt{2} + \sqrt{5})$

We write the square as a product and apply the multiplication pattern for binomials.

Multiplying as before, we have

$$(\sqrt{2} + \sqrt{5})^2 = \sqrt{2} \cdot \sqrt{2} + \sqrt{2} \cdot \sqrt{5} + \sqrt{2} \cdot \sqrt{5} + \sqrt{5} \cdot \sqrt{5}$$
$$= 2 + \sqrt{10} + \sqrt{10} + 5$$
$$= 7 + 2\sqrt{10}$$

This square can also be evaluated by using our earlier formula for the square of a binomial

$$(a + b)^2 = a^2 + 2ab + b^2$$

where $a = \sqrt{2}$ and $b = \sqrt{5}$.

Check Yourself 4

Multiply and simplify.

1. $(\sqrt{2} + 3)(\sqrt{2} + 5)$　　　　　　**2.** $(\sqrt{5} - \sqrt{3})(\sqrt{5} + \sqrt{3})$

3. $(\sqrt{7} - \sqrt{3})^2$

We are now ready to state our basic rule for the division of radical expressions. Again, it is simply a restatement of our earlier quotient property.

DIVIDING RADICAL EXPRESSIONS

$$\frac{\sqrt[n]{a}}{\sqrt[n]{b}} = \sqrt[n]{\frac{a}{b}}$$

In words, the quotient of two roots is the root of the quotient of the radicands.

Although we illustrate the use of this property in one of the examples that follow, the division of rational expressions is most often carried out by rationalizing the denominator. This process can be divided into two types of problems, those with a monomial divisor and those with binomial divisors. The following series of examples will illustrate.

EXAMPLE 5 Simplifying Rational Quotients

Simplify each expression. Again assume that all variables represent positive real numbers.

(a) $\dfrac{3}{\sqrt{5}} = \dfrac{3 \cdot \sqrt{5}}{\sqrt{5} \cdot \sqrt{5}} = \dfrac{3\sqrt{5}}{5}$

We multiply numerator and denominator by $\sqrt{5}$ to rationalize the denominator.

Use

$\dfrac{\sqrt{5}}{\sqrt{5}} = 1$

(b) $\dfrac{\sqrt{7x}}{\sqrt{10y}} = \dfrac{\sqrt{7x} \cdot \sqrt{10y}}{\sqrt{10y} \cdot \sqrt{10y}}$

$= \dfrac{\sqrt{70xy}}{10y}$

(c) $\dfrac{3}{\sqrt[3]{2}} = \dfrac{3\sqrt[3]{4}}{\sqrt[3]{2} \cdot \sqrt[3]{4}}$

In this case we want a perfect cube in the denominator, and so we multiply numerator and denominator by $\sqrt[3]{4}$.

$= \dfrac{3\sqrt[3]{4}}{2}$

Note that

$\sqrt[3]{2} \cdot \sqrt[3]{4} = \sqrt[3]{2} \cdot \sqrt[3]{2^2}$

$= \sqrt[3]{2^3}$

$= 2$

These division problems are identical to those we saw earlier in Section 7.1 when we were simplifying radical expressions. They are shown here to illustrate this case of division with radicals.

Check Yourself 5

Simplify each expression.

1. $\dfrac{5}{\sqrt{7}}$

2. $\dfrac{\sqrt{3a}}{\sqrt{5b}}$

3. $\dfrac{5}{\sqrt[3]{9}}$

Our division rule can be particularly useful if the radicands in the numerator and denominator have common factors. Consider the following example.

EXAMPLE 6 Simplifying Rational Quotients

Simplify

$\dfrac{\sqrt{10}}{\sqrt{15a}}$

Note the common factor 5 in the radicand of the numerator and denominator.

Solution We apply the division rule so that the radicand can be reduced as a fraction:

$\dfrac{\sqrt{10}}{\sqrt{15a}} = \sqrt{\dfrac{10}{15a}} = \sqrt{\dfrac{2}{3a}}$

Now we use the quotient property and rationalize the denominator:

$\sqrt{\dfrac{2}{3a}} = \dfrac{\sqrt{2}}{\sqrt{3a}} = \dfrac{\sqrt{2} \cdot \sqrt{3a}}{\sqrt{3a} \cdot \sqrt{3a}}$

$= \dfrac{\sqrt{6a}}{3a}$

(handwritten: $\frac{3x}{6x}$)

Check Yourself 6

Simplify $\dfrac{\sqrt{15}}{\sqrt{18x}}$.

(handwritten work)

We now turn our attention to a second type of division problem involving radical expressions. Here the divisors (the denominators) are binomials. This will use the idea of conjugates that we saw in Example 4, earlier in this section. The following example illustrates.

EXAMPLE 7 Rationalizing Denominators

Rationalize each denominator.

(a) $\dfrac{6}{\sqrt{6} + \sqrt{2}}$

> *If a radical expression has a sum or difference in the denominator, multiply the numerator and denominator by the conjugate of the denominator to rationalize.*
>
> *See Example 4b for the details of the multiplication in the denominator.*

Recall that $\sqrt{6} - \sqrt{2}$ is the conjugate of $\sqrt{6} + \sqrt{2}$, and the product of conjugates is *always a rational number*. Therefore, to rationalize the denominator, we multiply by $\sqrt{6} - \sqrt{2}$.

$$\dfrac{6}{\sqrt{6} + \sqrt{2}} = \dfrac{6(\sqrt{6} - \sqrt{2})}{(\sqrt{6} + \sqrt{2})(\sqrt{6} - \sqrt{2})}$$

$$= \dfrac{6(\sqrt{6} - \sqrt{2})}{4}$$

$$= \dfrac{3(\sqrt{6} - \sqrt{2})}{2}$$

(b) $\dfrac{\sqrt{5} + \sqrt{3}}{\sqrt{5} - \sqrt{3}}$

> *Combine like terms, factor, and divide the numerator and denominator by 2 to simplify.*

Multiply numerator and denominator by $\sqrt{5} + \sqrt{3}$:

$$\dfrac{(\sqrt{5} + \sqrt{3})(\sqrt{5} + \sqrt{3})}{(\sqrt{5} - \sqrt{3})(\sqrt{5} + \sqrt{3})} = \dfrac{5 + \sqrt{15} + \sqrt{15} + 3}{5 - 3}$$

$$= \dfrac{8 + 2\sqrt{15}}{2} = \dfrac{2(4 + \sqrt{15})}{2}$$

$$= 4 + \sqrt{15}$$

Check Yourself 7

Rationalize the denominator.

1. $\dfrac{4}{\sqrt{3} - \sqrt{2}}$

2. $\dfrac{\sqrt{6} + \sqrt{3}}{\sqrt{6} - \sqrt{3}}$

CHECK YOURSELF ANSWERS

1. (1) $\sqrt{42}$; (2) $\sqrt{55ab}$; (3) $\sqrt[3]{15y^2}$ **2.** (1) $10\sqrt{2}$; (2) $3x\sqrt{10}$; (3) $3pq\sqrt[3]{2q}$
3. (1) $\sqrt{30} + \sqrt{6}$; (2) $3\sqrt{2} + 4\sqrt{3}$; (3) $3a\sqrt{3}$ **4.** (1) $17 + 8\sqrt{2}$; (2) 2;
(3) $10 - 2\sqrt{21}$ **5.** (1) $\dfrac{5\sqrt{7}}{7}$; (2) $\dfrac{\sqrt{15ab}}{5b}$; (3) $\dfrac{5\sqrt[3]{3}}{3}$ **6.** $\dfrac{\sqrt{30x}}{6x}$
7. (1) $4(\sqrt{3} + \sqrt{2})$; (2) $3 + 2\sqrt{2}$

Build Your Skills

Multiply each of the following expressions.

1. $\sqrt{7} \cdot \sqrt{6}$

2. $\sqrt{3} \cdot \sqrt{10}$

3. $\sqrt{a} \cdot \sqrt{11}$

4. $\sqrt{10} \cdot \sqrt{w}$

5. $\sqrt{7x} \cdot \sqrt{13}$

6. $\sqrt{17} \cdot \sqrt{2b}$

7. $\sqrt{3a} \cdot \sqrt{11b}$

8. $\sqrt{5x} \cdot \sqrt{17y}$

9. $\sqrt{3} \cdot \sqrt{7} \cdot \sqrt{2}$

10. $\sqrt{5} \cdot \sqrt{7} \cdot \sqrt{3}$

11. $\sqrt[3]{4} \cdot \sqrt[3]{9}$

12. $\sqrt[3]{5} \cdot \sqrt[3]{7}$

13. $\sqrt[3]{5a} \cdot \sqrt[3]{9a}$

14. $\sqrt[3]{4y} \cdot \sqrt[3]{7y}$

15. $\sqrt[4]{8} \cdot \sqrt[4]{7}$

16. $\sqrt[5]{10} \cdot \sqrt[5]{8}$

Multiply and simplify each of the following expressions.

17. $\sqrt{3} \cdot \sqrt{12}$

18. $\sqrt{5} \cdot \sqrt{20}$

19. $\sqrt{7} \cdot \sqrt{7}$

20. $\sqrt{10} \cdot \sqrt{10}$

21. $\sqrt{5x} \cdot \sqrt{15x}$

22. $\sqrt{3a} \cdot \sqrt{15a}$

23. $\sqrt{8x^2} \cdot \sqrt{4x}$

24. $\sqrt{12w^3} \cdot \sqrt{6w}$

25. $\sqrt{3} \cdot \sqrt{3} \cdot \sqrt{3}$

26. $\sqrt{5} \cdot \sqrt{5} \cdot \sqrt{5}$

27. $\sqrt[3]{4} \cdot \sqrt[3]{6}$

28. $\sqrt[3]{9} \cdot \sqrt[3]{15}$

29. $\sqrt[3]{9p^2} \cdot \sqrt[3]{6p}$

30. $\sqrt[3]{25x^2} \cdot \sqrt[3]{10x^2}$

31. $\sqrt[3]{4x^2y} \cdot \sqrt[3]{10xy^3}$

32. $\sqrt[3]{18r^2s^2} \cdot \sqrt[3]{9r^2s}$

33. $\sqrt[4]{8} \cdot \sqrt[4]{6}$

34. $\sqrt[4]{27} \cdot \sqrt[4]{15}$

35. $\sqrt[4]{9a^3} \cdot \sqrt[4]{18a^2}$

36. $\sqrt[4]{8x^2} \cdot \sqrt[4]{10x^3}$

37. $\sqrt{2}(\sqrt{3} + 5)$

38. $\sqrt{3}(\sqrt{5} - 7)$

39. $\sqrt{5}(\sqrt{3} - \sqrt{2})$

40. $\sqrt{7}(\sqrt{5} + \sqrt{2})$

41. $\sqrt{3}(\sqrt{6} + \sqrt{5})$

42. $\sqrt{5}(\sqrt{10} - \sqrt{2})$

43. $\sqrt{5}(\sqrt{2} + 2\sqrt{3})$

44. $\sqrt{7}(\sqrt{3} - 3\sqrt{2})$

45. $\sqrt{3}(5\sqrt{2} - \sqrt{18})$

46. $\sqrt{2}(2\sqrt{10} + \sqrt{40})$

47. $\sqrt{x}(\sqrt{3x} + \sqrt{27x})$

48. $\sqrt{y}(\sqrt{8y} - \sqrt{2y})$

49. $\sqrt{a}(\sqrt{a} + \sqrt{ab})$

50. $\sqrt{w}(\sqrt{w} - \sqrt{wz})$

51. $\sqrt[3]{4}(\sqrt[3]{4} + \sqrt[3]{32})$

52. $\sqrt[3]{6}(\sqrt[3]{32} - \sqrt[3]{4})$

53. $\sqrt[3]{xy}(\sqrt[3]{x^2y} - \sqrt[3]{xy^2})$

54. $\sqrt[3]{r^2s}(\sqrt[3]{rs^5} + \sqrt[3]{r^4s^2})$

55. $(\sqrt{2} + 3)(\sqrt{2} - 4)$

56. $(\sqrt{3} - 1)(\sqrt{3} + 5)$

57. $(\sqrt{3} - 2)(\sqrt{3} - 5)$

58. $(\sqrt{5} + 1)(\sqrt{5} + 4)$

59. $(\sqrt{5} - \sqrt{3})(\sqrt{5} + \sqrt{2})$

60. $(\sqrt{6} + \sqrt{2})(\sqrt{5} - \sqrt{2})$

61. $(\sqrt{x} - 3)(\sqrt{x} + 2)$

62. $(\sqrt{w} + 5)(\sqrt{w} + 2)$

63. $(\sqrt{2} + 3\sqrt{5})(\sqrt{2} - 2\sqrt{5})$

64. $(\sqrt{6} - 2\sqrt{3})(\sqrt{6} - 3\sqrt{3})$

65. $(\sqrt{5} + 3)(\sqrt{5} - 3)$

66. $(\sqrt{10} + 2)(\sqrt{10} - 2)$

67. $(\sqrt{6} + \sqrt{3})(\sqrt{6} - \sqrt{3})$

68. $(\sqrt{7} + \sqrt{5})(\sqrt{7} - \sqrt{5})$

69. $(\sqrt{15} + 2\sqrt{3})(\sqrt{15} - 2\sqrt{3})$

70. $(3\sqrt{2} + 2)(3\sqrt{2} - 2)$

71. $(\sqrt{x} + 3)(\sqrt{x} - 3)$

72. $(\sqrt{y} + 5)(\sqrt{y} - 5)$

73. $(\sqrt{a} + \sqrt{3})(\sqrt{a} - \sqrt{3})$

74. $(\sqrt{m} + \sqrt{7})(\sqrt{m} - \sqrt{7})$

75. $(\sqrt{3} - 5)^2$

76. $(\sqrt{5} + \sqrt{2})^2$

77. $(\sqrt{a} + 3)^2$

78. $(\sqrt{x} - 4)^2$

79. $(\sqrt{x} + \sqrt{y})^2$

80. $(\sqrt{r} - \sqrt{s})^2$

Rationalize the denominator in each of the following expressions. Simplify where necessary.

81. $\dfrac{\sqrt{3}}{\sqrt{7}}$

82. $\dfrac{\sqrt{5}}{\sqrt{3}}$

83. $\dfrac{\sqrt{2a}}{\sqrt{3b}}$

84. $\dfrac{\sqrt{5x}}{\sqrt{6y}}$

85. $\dfrac{3}{\sqrt[3]{4}}$

86. $\dfrac{2}{\sqrt[3]{9}}$

87. $\dfrac{\sqrt{15}}{\sqrt{20x}}$

88. $\dfrac{\sqrt{7}}{\sqrt{28w}}$

89. $\dfrac{1}{2 + \sqrt{3}}$

90. $\dfrac{2}{3 - \sqrt{2}}$

91. $\dfrac{8}{3 - \sqrt{5}}$

92. $\dfrac{20}{4 + \sqrt{6}}$

93. $\dfrac{12}{\sqrt{6} - \sqrt{2}}$

94. $\dfrac{8}{\sqrt{7} + \sqrt{3}}$

95. $\dfrac{\sqrt{5}}{\sqrt{5} + \sqrt{3}}$

96. $\dfrac{\sqrt{6}}{\sqrt{6} - \sqrt{3}}$

97. $\dfrac{\sqrt{7} - 2}{\sqrt{7} + 2}$

98. $\dfrac{\sqrt{5} + 3}{\sqrt{5} - 3}$

99. $\dfrac{\sqrt{6} + \sqrt{3}}{\sqrt{6} - \sqrt{3}}$

100. $\dfrac{\sqrt{7} - \sqrt{5}}{\sqrt{7} + \sqrt{5}}$

101. $\dfrac{\sqrt{a}}{\sqrt{a} + 2}$

102. $\dfrac{\sqrt{x}}{\sqrt{x} - 5}$

103. $\dfrac{\sqrt{w} + 3}{\sqrt{w} - 3}$

104. $\dfrac{\sqrt{x} - 5}{\sqrt{x} + 5}$

105. $\dfrac{\sqrt{x} - \sqrt{y}}{\sqrt{x} + \sqrt{y}}$

106. $\dfrac{\sqrt{m} + \sqrt{n}}{\sqrt{m} - \sqrt{n}}$

Transcribe Your Skills

107. Why is the product of conjugates always equal to a rational number?

108. What does the term "square root" have to do with the shape of a square? How is it related to the area of a square?

Think About These

Simplify each of the following expressions. Assume that values for the variables do not result in negative radicands.

109. $\sqrt{x} \cdot \sqrt{x^2 + x}$

110. $\sqrt{x^2 - 4} \cdot \sqrt{x^2 + 3x + 2}$

111. $\sqrt{x} \cdot \sqrt{x^2 - 2x} \cdot \sqrt{x - 2}$

112. $\sqrt{x^2 + 3x + 2} \cdot \sqrt{x^2 - 1} \cdot \sqrt{x^2 + x - 2}$

113. $\dfrac{\sqrt{2x^2 + 3x}}{\sqrt{x}}$

114. $\dfrac{\sqrt{x^2 - 9}}{\sqrt{x + 3}}$

115. $\dfrac{\sqrt{x + 2}}{\sqrt{x + 1}}$

116. $\dfrac{\sqrt{2x + 1}}{\sqrt{x - 1}}$

117. $\dfrac{\sqrt{2x^2 + 5x - 3}}{\sqrt{x + 3}}$

118. $\dfrac{\sqrt{3x^2 - 8x - 3}}{\sqrt{x^2 - 9}}$

119. $\dfrac{\sqrt{x^2 + 4x + 3}}{\sqrt{x^2 + 3x + 2}}$

120. $\dfrac{\sqrt{x^3 + 7x^2 + 10x}}{\sqrt{2x^2 + 4x}}$

 121. In the discussion of water flow through an irrigation pipe, the following expression arises:

$$\frac{x - y}{(\sqrt{x} + \sqrt{y})\sqrt{2z}}$$

Rationalize this expression.

 122. A surveyor determines the area of a parcel of farmland lost to desertification to be $\dfrac{1}{2}\sqrt{a}(\sqrt{a} + \sqrt{3a})$. Write this expression in simplified form.

Skillscan (Section 7.1)

Evaluate each of the following expressions.

a. $\sqrt{16}$ **b.** $\sqrt[3]{27}$ **c.** $\sqrt[3]{-125}$ **d.** $(\sqrt{9})^3$
e. $(\sqrt[3]{8})^4$ **f.** $(\sqrt[4]{16})^3$ **g.** $(\sqrt{4})^{-5}$ **h.** $(\sqrt[3]{-64})^{-2}$

Rational Exponents   **7.5**

OBJECTIVES: 1. To define the meaning of a rational exponent
2. To simplify expressions involving rational exponents

TAPE IN17

In Section 7.1 we introduced the radical notation, along with the concept of roots. In this section, we use that concept to develop a new notation, using exponents that will provide an alternate way of writing these roots.

That new notation will involve *rational numbers as exponents*. To start the development, we will extend all the previous properties of exponents to include rational exponents.

Given that extension, suppose that

$a = 4^{1/2}$ (1)

Squaring both sides of the equation yields

$$a^2 = (4^{1/2})^2$$

or

We will see later in this chapter that the property $(x^m)^n = x^{mn}$ holds for rational numbers m and n.

$$a^2 = 4^{(1/2)(2)}$$

$$a^2 = 4^1$$

$$a^2 = 4 \tag{2}$$

From Equation (2) we see that a is the number whose square is 4; that is, a is the principal square root of 4. Using our earlier notation, we can write

$$a = \sqrt{4}$$

But from (1)

$$a = 4^{1/2}$$

and to be consistent, we must have

$4^{1/2}$ indicates the principal *square root* of 4.

$$4^{1/2} = \sqrt{4}$$

This argument can be repeated for any exponent of the form $\dfrac{1}{n}$, so it seems reasonable to make the following definition.

> **If a is any real number and n is a positive integer ($n > 1$), then**
>
> $$a^{1/n} = \sqrt[n]{a}$$
>
> **We restrict a so that a is nonnegative when n is even.**
>
> **In words, $a^{1/n}$ indicates the principal nth root of a.**

The following example illustrates the use of rational exponents to represent roots.

EXAMPLE 1 Writing Expressions in Radical Form

Write each expression in radical form and then simplify.

(a) $25^{1/2} = \sqrt{25} = 5$

$27^{1/3}$ is the *cube root* of 27.

(b) $27^{1/3} = \sqrt[3]{27} = 3$

(c) $-36^{1/2} = -\sqrt{36} = -6$

(d) $(-36)^{1/2} = \sqrt{-36}$ is not a real number.

$32^{1/5}$ is the *fifth root* of 32.

(e) $32^{1/5} = \sqrt[5]{32} = 2$

Check Yourself 1

Write each expression in radical form and simplify.

1. $8^{1/3}$ 2

2. $-64^{1/2}$ ~ 8

3. $81^{1/4}$ 3

We are now ready to extend our exponent notation to allow *any* rational exponent, again assuming that our previous exponent properties must still be valid. Note that

$$a^{m/n} = (a^{1/n})^m = (a^m)^{1/n}$$

This is because

$$\frac{m}{n} = (m)\left(\frac{1}{n}\right) = \left(\frac{1}{n}\right)(m)$$

From our earlier work, we know that $a^{1/n} = \sqrt[n]{a}$, and combining this with the above observation, we offer the following definition for $a^{m/n}$.

> **For any real number a and positive integers m and n with $n > 1$,**
> $$a^{m/n} = (\sqrt[n]{a})^m = \sqrt[n]{a^m}$$

The two radical forms for $a^{m/n}$ are equivalent, and the choice of which form to use generally depends on whether we are evaluating numerical expressions or rewriting expressions containing variables in radical form.

This new extension of our rational exponent notation is applied in the following example.

EXAMPLE 2 Simplifying Expressions with Rational Exponents

Simplify each expression.

(*a*) $9^{3/2} = (\sqrt{9})^3$
$$= 3^3 = 27$$

(*b*) $\left(\dfrac{16}{81}\right)^{3/4} = \left(\sqrt[4]{\dfrac{16}{81}}\right)^3$
$$= \left(\dfrac{2}{3}\right)^3 = \dfrac{8}{27}$$

(*c*) $(-8)^{2/3} = (\sqrt[3]{-8})^2$
$$= (-2)^2 = 4$$

In (*a*) we could also have evaluated the expression as

$$9^{3/2} = \sqrt{9^3} = \sqrt{729}$$
$$= 27$$

This illustrates why we use $(\sqrt[n]{a})^m$ for $a^{m/n}$ when evaluating numerical expressions. The numbers involved will be smaller and easier to work with.

Check Yourself 2

Simplify each expression.

1. $16^{3/4}$ **2.** $\left(\dfrac{8}{27}\right)^{2/3}$ **3.** $(-32)^{3/5}$

Now we want to extend our rational exponent notation. Using the definition of negative exponents, we can write

$$a^{-m/n} = \frac{1}{a^{m/n}}$$

The following example illustrates the use of negative rational exponents.

EXAMPLE 3 Simplifying Expressions with Rational Exponents

Simplify each expression.

(a) $16^{-1/2} = \dfrac{1}{16^{1/2}} = \dfrac{1}{4}$

(b) $27^{-2/3} = \dfrac{1}{27^{2/3}} = \dfrac{1}{(\sqrt[3]{27})^2} = \dfrac{1}{3^2} = \dfrac{1}{9}$

Check Yourself 3

Simplify each expression.

1. $16^{-1/4}$ **2.** $81^{-3/4}$

Calculators can be used to evaluate expressions that contain rational exponents by using the $\boxed{y^x}$ key and the parenthesis keys.

EXAMPLE 4 Estimating Powers Using a Calculator

Using a graphing calculator, evaluate each of the following. Round all answers to three decimal places.

(a) $45^{2/5}$

 If you are using a scientific calculator, try using the $\boxed{y^x}$ key in place of the $\boxed{\wedge}$ key.

Enter 45 and press the $\boxed{\wedge}$ key. Then use the following keystrokes:

$\boxed{(}\ \ 2\ \ \boxed{\div}\ \ 5\ \ \boxed{)}$

Press $\boxed{=}$, and the display will read 4.584426407. Rounded to three decimal places, the result is 4.584.

(b) $38^{-2/3}$

Enter 38 and press the $\boxed{\wedge}$ key. Then use the following keystrokes:

$\boxed{(}\ \ \boxed{(-)}\ \ 2\ \ \boxed{\div}\ \ 3\ \ \boxed{)}$

The $\boxed{(-)}$ key changes the sign of the exponent to minus.

Press $\boxed{=}$, and the display will read 0.088473037. Rounded to three decimal places, the result is 0.088.

Check Yourself 4

Evaluate each of the following by using a scientific calculator. Round each answer to three decimal places.

1. $23^{3/5}$ **2.** $18^{-4/7}$

As we mentioned earlier, we assume that all our previous exponent properties will continue to hold for rational exponents. Those properties are restated here for reference.

PROPERTIES OF EXPONENTS

For any nonzero real numbers *a* and *b* and rational numbers *m* and *n*,

1. $a^m \cdot a^n = a^{m+n}$

2. $\dfrac{a^m}{a^n} = a^{m-n}$

3. $(a^m)^n = a^{mn}$

4. $(ab)^m = a^m b^m$

5. $\left(\dfrac{a}{b}\right)^m = \dfrac{a^m}{b^m}$

We restrict *a* and *b* to being nonnegative real numbers when *m* or *n* indicates an even root.

The following example illustrates the use of our extended properties to simplify expressions involving rational exponents. Here we assume that all variables represent positive real numbers.

EXAMPLE 5 Simplifying Expressions

Simplify each expression.

(a) $x^{2/3} \cdot x^{1/2} = x^{2/3+1/2}$
$$= x^{4/6+3/6} = x^{7/6}$$

Property 1—add the exponents.

(b) $\dfrac{w^{3/4}}{w^{1/2}} = w^{3/4-1/2}$
$$= w^{3/4-2/4} = w^{1/4}$$

Property 2—subtract the exponents.

(c) $(a^{2/3})^{3/4} = a^{(2/3)(3/4)}$
$$= a^{1/2}$$

Property 3—multiply the exponents.

Check Yourself 5

Simplify each expression.

1. $z^{3/4} \cdot z^{1/2}$ **2.** $\dfrac{x^{5/6}}{x^{1/3}}$ **3.** $(b^{5/6})^{2/5}$

As you would expect from your previous experience with exponents, simplifying expressions often involves using several exponent properties.

EXAMPLE 6 Simplifying Expressions

Simplify each expression.

(a) $(x^{2/3} \cdot y^{5/6})^{3/2}$
$= (x^{2/3})^{3/2} \cdot (y^{5/6})^{3/2}$ Property 4
$= x^{(2/3)(3/2)} \cdot y^{(5/6)(3/2)} = xy^{5/4}$ Property 3

(b) $\left(\dfrac{r^{-1/2}}{s^{1/3}}\right)^6 = \dfrac{(r^{-1/2})^6}{(s^{1/3})^6}$ Property 5

$= \dfrac{r^{-3}}{s^2} = \dfrac{1}{r^3 s^2}$ Property 3

We simplify *inside the parentheses* as the first step.

(c) $\left(\dfrac{4a^{-2/3} \cdot b^2}{a^{1/3} \cdot b^{-4}}\right)^{1/2} = \left(\dfrac{4b^2 \cdot b^4}{a^{1/3} \cdot a^{2/3}}\right)^{1/2} = \left(\dfrac{4b^6}{a}\right)^{1/2}$

$= \dfrac{(4b^6)^{1/2}}{a^{1/2}} = \dfrac{4^{1/2}(b^6)^{1/2}}{a^{1/2}}$

$= \dfrac{2b^3}{a^{1/2}}$

Check Yourself 6

Simplify each expression.

1. $(a^{3/4} \cdot b^{1/2})^{2/3}$ **2.** $\left(\dfrac{w^{1/2}}{z^{-1/4}}\right)^4$ **3.** $\left(\dfrac{8x^{-3/4}y}{x^{1/4} \cdot y^{-5}}\right)^{1/3}$

Note Here we use $a^{m/n} = \sqrt[n]{a^m}$, which is generally the preferred form in this situation.

We can also use the relationships between rational exponents and radicals to write expressions involving rational exponents as radicals and vice versa.

EXAMPLE 7 Writing Expressions in Radical Form

Write each expression in radical form.

(a) $a^{3/5} = \sqrt[5]{a^3}$

(b) $(mn)^{3/4} = \sqrt[4]{(mn)^3}$
$= \sqrt[4]{m^3 n^3}$

Note that the exponent applies *only* to the variable y.

(c) $2y^{5/6} = 2\sqrt[6]{y^5}$

Now the exponent applies to $2y$ because of the parentheses.

(d) $(2y)^{5/6} = \sqrt[6]{(2y)^5}$
$= \sqrt[6]{32y^5}$

Check Yourself 7

Write each expression in radical form.

1. $(ab)^{2/3}$ **2.** $3x^{3/4}$ **3.** $(3x)^{3/4}$

EXAMPLE 8 Writing Expressions in Exponential Form

Using rational exponents, write each expression and simplify.

(a) $\sqrt[3]{5x} = (5x)^{1/3}$

(b) $\sqrt{9a^2b^4} = (9a^2b^4)^{1/2}$
$$= 9^{1/2}(a^2)^{1/2}(b^4)^{1/2} = 3ab^2$$

(c) $\sqrt[4]{16w^{12}z^8} = (16w^{12}z^8)^{1/4}$
$$= 16^{1/4}(w^{12})^{1/4}(z^8)^{1/4} = 2w^3z^2$$

Check Yourself 8

Using rational exponents, write each expression and simplify.

1. $\sqrt{7a}$ **2.** $\sqrt[3]{27p^6q^9}$ **3.** $\sqrt[4]{81x^8y^{16}}$

CHECK YOURSELF ANSWERS

1. (1) 2; (2) -8; (3) 3   **2.** (1) 8; (2) $\dfrac{4}{9}$; (3) -8 **3.** (1) $\dfrac{1}{2}$; (2) $\dfrac{1}{27}$

4. (1) 6.562; (2) 0.192   **5.** (1) $z^{5/4}$; (2) $x^{1/2}$; (3) $b^{1/3}$ **6.** (1) $a^{1/2}b^{1/3}$;

(2) w^2z; (3) $\dfrac{2y^2}{x^{1/3}}$ **7.** (1) $\sqrt[3]{a^2b^2}$; (2) $3\sqrt[4]{x^3}$; (3) $\sqrt[4]{27x^3}$ **8.** (1) $(7a)^{1/2}$;

(2) $3p^2q^3$; (3) $3x^2y^4$

7.5 EXERCISES

Build Your Skills

Use the definition of $a^{1/n}$ to evaluate each of the following expressions.

1. $36^{1/2}$

2. $100^{1/2}$

3. $-25^{1/2}$

4. $(-64)^{1/2}$

5. $(-49)^{1/2}$

6. $-49^{1/2}$

7. $27^{1/3}$

8. $(-64)^{1/3}$

9. $81^{1/4}$

10. $-32^{1/5}$

11. $\left(\dfrac{4}{9}\right)^{1/2}$

12. $\left(\dfrac{27}{8}\right)^{1/3}$

Use the definition of $a^{m/n}$ to evaluate each of the following expressions. Use your calculator to check each answer.

13. $27^{2/3}$

14. $16^{3/2}$

15. $(-8)^{4/3}$

16. $125^{2/3}$

17. $32^{2/5}$

18. $-81^{3/4}$

19. $81^{3/2}$

$\left(\sqrt[4]{81}\right)^2$

3

1-31 -33 -85

every other

odd

20. $(-243)^{3/5}$

21. $\left(\dfrac{8}{27}\right)^{2/3}$

22. $\left(\dfrac{9}{4}\right)^{3/2}$

 Use the definition of $a^{-m/n}$ to evaluate each of the following expressions. Use your calculator to check each answer.

23. $25^{-1/2}$

24. $27^{-1/3}$

25. $81^{-1/4}$

26. $121^{-1/2}$

27. $9^{-3/2}$

28. $16^{-3/4}$

29. $64^{-5/6}$

30. $16^{-3/2}$

31. $\left(\dfrac{4}{25}\right)^{-1/2}$

32. $\left(\dfrac{27}{8}\right)^{-2/3}$

Use the properties of exponents to simplify each expression. Assume all variables represent positive real numbers.

33. $x^{1/2} \cdot x^{1/2}$

34. $a^{2/3} \cdot a^{1/3}$

35. $y^{3/5} \cdot y^{1/5}$

36. $m^{1/4} \cdot m^{5/4}$

37. $b^{2/3} \cdot b^{3/2}$

38. $p^{5/6} \cdot p^{2/3}$

39. $\dfrac{x^{2/3}}{x^{1/3}}$

40. $\dfrac{a^{5/6}}{a^{1/6}}$

41. $\dfrac{s^{7/5}}{s^{2/5}}$

42. $\dfrac{z^{9/2}}{z^{3/2}}$

43. $\dfrac{w^{5/4}}{w^{1/2}}$

44. $\dfrac{b^{7/6}}{b^{2/3}}$

45. $(x^{3/4})^{4/3}$

46. $(y^{4/3})^{3/4}$

47. $(a^{2/5})^{3/2}$

48. $(p^{3/4})^{2/3}$

49. $(y^{-3/4})^{8}$

50. $(w^{-2/3})^{6}$

51. $(a^{2/3} \cdot b^{3/2})^{6}$

52. $(p^{3/4} \cdot q^{5/2})^{4}$

53. $(2x^{1/5} \cdot y^{3/5})^{5}$

54. $(3m^{3/4} \cdot n^{5/4})^{4}$

55. $(s^{3/4} \cdot t^{1/4})^{4/3}$

56. $(x^{5/2} \cdot y^{5/7})^{2/5}$

57. $(8p^{3/2} \cdot q^{5/2})^{2/3}$

58. $(16a^{1/3} \cdot b^{2/3})^{3/4}$

59. $(x^{3/5} \cdot y^{3/4} \cdot z^{3/2})^{2/3}$

60. $(p^{5/6} \cdot q^{2/3} \cdot r^{5/3})^{3/5}$

61. $\dfrac{a^{5/6} \cdot b^{3/4}}{a^{1/3} \cdot b^{1/2}}$

62. $\dfrac{x^{2/3} \cdot y^{3/4}}{x^{1/2} \cdot y^{1/2}}$

63. $\dfrac{(r^{-1} \cdot s^{1/2})^{3}}{r \cdot s^{-1/2}}$

64. $\dfrac{(w^{-2} \cdot z^{-1/4})^{6}}{w^{-8} z^{1/2}}$

65. $\left(\dfrac{x^{12}}{y^{8}}\right)^{1/4}$

66. $\left(\dfrac{p^{9}}{q^{6}}\right)^{1/3}$

67. $\left(\dfrac{m^{-1/4}}{n^{1/2}}\right)^{4}$

68. $\left(\dfrac{r^{1/5}}{s^{-1/2}}\right)^{10}$

69. $\left(\dfrac{r^{-1/2} \cdot s^{3/4}}{t^{1/4}}\right)^{4}$

70. $\left(\dfrac{a^{1/3} \cdot b^{-1/6}}{c^{-1/6}}\right)^{6}$

71. $\left(\dfrac{8x^{3} \cdot y^{-6}}{z^{-9}}\right)^{1/3}$

72. $\left(\dfrac{16p^{-4} \cdot q^6}{r^2}\right)^{-1/2}$

73. $\left(\dfrac{16m^{-3/5} \cdot n^2}{m^{1/5} \cdot n^{-2}}\right)^{1/4}$

74. $\left(\dfrac{27x^{5/6} \cdot y^{-4/3}}{x^{-7/6} \cdot y^{5/3}}\right)^{1/3}$

75. $\left(\dfrac{x^{3/2} \cdot y^{1/2}}{z^2}\right)^{1/2}\left(\dfrac{x^{3/4} \cdot y^{3/2}}{z^{-3}}\right)^{1/3}$

76. $\left(\dfrac{p^{1/2} \cdot q^{4/3}}{r^{-4}}\right)^{3/4}\left(\dfrac{p^{15/8} \cdot q^{-3}}{r^6}\right)^{1/3}$

Write each of the following expressions in radical form. Do not simplify.

77. $a^{3/4}$

78. $m^{5/6}$

79. $2x^{2/3}$

80. $3m^{-2/5}$

81. $3x^{2/5}$

82. $2y^{-3/4}$

83. $(3x)^{2/5}$

84. $(2y)^{-3/4}$

Write each of the following expressions, using rational exponents, and simplify where necessary.

85. $\sqrt{7a}$

86. $\sqrt{25w^4}$

87. $\sqrt[3]{8m^6n^9}$

88. $\sqrt[5]{32r^{10}s^{15}}$

Evaluate each of the following, using a scientific calculator. Round each answer to three decimal places.

89. $46^{3/5}$

90. $23^{2/7}$

91. $12^{-2/5}$

92. $36^{-3/4}$

Transcribe Your Skills

93. Describe the difference between x^{-2} and $x^{1/2}$.

94. Some rational exponents, like $\dfrac{1}{2}$, can easily be rewritten as terminating decimals (0.5). Others, like $\dfrac{1}{3}$, cannot. What is it that determines which rational numbers can be rewritten as terminating decimals?

Think About These

Apply the appropriate multiplication patterns. Then simplify your result.

95. $a^{1/2}(a^{3/2} + a^{3/4})$

96. $2x^{1/4}(3x^{3/4} - 5x^{-1/4})$

97. $(a^{1/2} + 2)(a^{1/2} - 2)$

98. $(w^{1/3} - 3)(w^{1/3} + 3)$

99. $(m^{1/2} + n^{1/2})(m^{1/2} - n^{1/2})$

100. $(x^{1/3} + y^{1/3})(x^{1/3} - y^{1/3})$

101. $(x^{1/2} + 2)^2$

102. $(a^{1/3} - 3)^2$

103. $(r^{1/2} + s^{1/2})^2$

104. $(p^{1/2} - q^{1/2})^2$

As is suggested by several of the preceding exercises, certain expressions containing rational exponents are factorable. For instance, to factor $x^{2/3} - x^{1/3} - 6$, let $u = x^{1/3}$. Note that $x^{2/3} = (x^{1/3})^2 = u^2$.

Substituting, we have $u^2 - u - 6$, and factoring yields $(u - 3)(u + 2)$ or $(x^{1/3} - 3)(x^{1/3} + 2)$.

Use this technique to factor each of the following expressions.

105. $x^{2/3} + 4x^{1/3} + 3$

106. $y^{2/5} - 2y^{1/5} - 8$

107. $a^{4/5} - 7a^{2/5} + 12$

108. $w^{4/3} + 3w^{2/3} - 10$

109. $x^{4/3} - 4$

110. $x^{2/5} - 16$

Perform the indicated operations. Assume that n represents a positive integer and that the denominators are not zero.

111. $x^{3n} \cdot x^{2n}$

112. $p^{1-n} \cdot p^{n+3}$

113. $(y^2)^{2n}$

114. $(a^{3n})^3$

115. $\dfrac{r^{n+2}}{r^n}$

116. $\dfrac{w^n}{w^{n-3}}$

117. $(a^3 \cdot b^2)^{2n}$

118. $(c^4 \cdot d^2)^{3m}$

119. $\left(\dfrac{x^{n+2}}{x^n}\right)^{1/2}$

120. $\left(\dfrac{b^n}{b^{n-3}}\right)^{1/3}$

Write each of the following expressions in exponent form, simplify, and give the result as a single radical.

121. $\sqrt{\sqrt{x}}$

122. $\sqrt{\sqrt[3]{a}}$

123. $\sqrt[4]{\sqrt{y}}$

124. $\sqrt{\sqrt[3]{w}}$

Simplify each of the following expressions. Write your answer in scientific notation.

125. $(4 \times 10^8)^{1/2}$

126. $(8 \times 10^6)^{1/3}$

127. $(16 \times 10^{-12})^{1/4}$

128. $(9 \times 10^{-4})^{1/2}$

129. $(8 \times 10^{-6})^{2/3}$

130. $(16 \times 10^{-8})^{3/4}$

 131. While investigating rainfall runoff in a region of semi-arid farmland, a researcher encounters the following expression:

$$t = C\left(\frac{L}{xy^2}\right)^{1/3}$$

Evaluate t when $C = 20$, $L = 600$, $x = 3$, and $y = 5$.

 132. The average velocity of water in an open irrigation ditch is given by the formula

$$V = \frac{1.5x^{2/3}y^{1/2}}{z}$$

Evaluate V when $x = 27$, $y = 16$, and $z = 12$.

Skillscan (Section 5.3)

Multiply.

a. $2x(3 + 5x)$ **b.** $(-3x)(2 - 5x)$ **c.** $(3 + 2x)(3 - x)$
d. $(3 - 4w)(5 - 2w)$ **e.** $(7 - 2a)(7 + 2a)$
f. $(5 + 3m)(5 - 3m)$ **g.** $(5 + y)^2$ **h.** $(4 - 3r)^2$

7.6 Complex Numbers

TAPE IN17

OBJECTIVES: 1. To define a complex number
2. To perform operations with complex numbers

Radicals such as

$$\sqrt{-4} \quad \text{and} \quad \sqrt{-49}$$

are *not* real numbers since no real number squared produces a negative number. Our work in this section will extend our number system to include these *imaginary numbers*. This will allow us to consider radicals such as $\sqrt{-4}$.

First we offer a definition:

$i = \sqrt{-1}$ is called the *imaginary unit*.

> **The number _i_ is defined as**
> $$i = \sqrt{-1}$$
> **Note that this means that**
> $$i^2 = -1$$

This definition of the number *i* gives us an alternate means of indicating the square root of a negative number.

> When *a* is a positive real number,
> $$\sqrt{-a} = \sqrt{a}\,i \quad \text{or} \quad i\sqrt{a}$$

EXAMPLE 1 Using the Number *i*

Write each expression as a multiple of *i*.

(a) $\sqrt{-4} = \sqrt{4}\,i = 2i$

(b) $-\sqrt{-9} = -\sqrt{9}\,i = -3i$

(c) $\sqrt{-8} = \sqrt{8}\,i = 2\sqrt{2}\,i$ or $2i\sqrt{2}$

(d) $\sqrt{-7} = \sqrt{7}\,i$ or $i\sqrt{7}$

We simplify $\sqrt{8}$ as $2\sqrt{2}$. Note that we write *i* in front of the radical to make it clear that *i* is not part of the radicand.

Check Yourself 1

Write each radical as a multiple of *i*.

1. $\sqrt{-25}$ $\sqrt{25}\,i = \sqrt{5}\,i$

2. $\sqrt{-24}$ $\sqrt{-24}\,i$ $2i\sqrt{6}$

We are now ready to define complex numbers in terms of the number *i*.

> A *complex number* is any number that can be written in the form
> *a + bi*
> where *a* and *b* are real numbers and
> $$i = \sqrt{-1}$$

The term "imaginary number" was introduced by René Descartes in 1637. Euler used *i* to indicate $\sqrt{-1}$ in 1748, but it was not until 1832 that Gauss used the term "complex number."

The first application of these numbers was made by Charles Steinmetz (1865–1923) in explaining the behavior of electric circuits.

The form *a + bi* is called the *standard form* of a complex number. We call *a* the *real part* of the complex number and *b* the *imaginary part*.

$3 + 7i$ is an example of a complex number with real part 3 and imaginary part 7.

$5i$ is also a complex number since it can be written as $0 + 5i$.

-3 is a complex number since it can be written as $-3 + 0i$.

Also, 5*i* is called a *pure imaginary* number.

The real numbers can be considered a subset of the set of complex numbers.

The basic operations (addition, subtraction, multiplication, and division) on complex numbers are defined below. We start with addition and subtraction.

ADDITION AND SUBTRACTION OF COMPLEX NUMBERS

For the complex numbers $a + bi$ and $c + di$,

$(a + bi) + (c + di) = (a + c) + (b + d)i$

$(a + bi) - (c + di) = (a - c) + (b - d)i$

In words, we add or subtract the real parts and the imaginary parts of the complex numbers.

The following example illustrates the use of these definitions.

EXAMPLE 2 Adding Complex Numbers

Perform the indicated operations.

(a) $(5 + 3i) + (6 - 7i) = (5 + 6) + (3 - 7)i$
$$= 11 - 4i$$

(b) $5 + (7 - 5i) = (5 + 7) + (-5i)$
$$= 12 - 5i$$

(c) $(8 - 2i) - (3 - 4i) = (8 - 3) + [-2 - (-4)]i$
$$= 5 + 2i$$

Check Yourself 2

Perform the indicated operations.

1. $(4 - 7i) + (3 - 2i)$ $7 - 9i$ **2.** $-7 + (-2 + 3i)$

3. $(-4 + 3i) - (-2 - i)$ $-2 + 4i$ $-7x$ $-9 + 3i$

Since complex numbers are binomial in form, the product of two complex numbers is found by applying our earlier multiplication pattern for binomials. The following example illustrates.

EXAMPLE 3 Multiplying Complex Numbers

Multiply.

$(2 + 3i)(3 - 4i)$

       First   Outer   Inner   Last

$= 2(3) + 2(-4i) + (3i)(3) + (3i)(-4i)$

We can replace i^2 with -1 because of the definition of i, and we usually do so because of the resulting simplification.

$= 6 - 8i + 9i - 12i^2$

$= 6 + i - 12(-1)$

$= 6 + i + 12 = 18 + i$

Check Yourself 3

Multiply $(2 - 5i)(3 - 2i)$.

$6 + 4i - 15i + 10i^2$

$6 - 19i + 10(i^2)$

Since

$$(a + bi)(c + di) = ac + adi + bci + bdi^2$$
$$= ac + adi + bci - bd$$
$$= (ac - bd) + (ad + bc)i$$

we can state a formula for the general product of two complex numbers. However, you will find it much easier to get used to the multiplication pattern as it is applied to complex numbers than to memorize this formula.

There is one particular product form that will seem very familiar. We call $a + bi$ and $a - bi$ *complex conjugates.* For instance,

$$3 + 2i \qquad \text{and} \qquad 3 - 2i$$

are complex conjugates.

Consider the product

$$(3 + 2i)(3 - 2i) = 3^2 - (2i)^2$$
$$= 9 - 4i^2 = 9 - 4(-1)$$
$$= 9 + 4 = 13$$

The product of $3 + 2i$ and $3 - 2i$ is a real number. In general, we can write the product of two complex conjugates as

$$(a + bi)(a - bi) = a^2 + b^2$$

The fact that this product is always a real number will be very useful when we consider the division of complex numbers later in this section.

> Note that this is the same form as the conjugate for radical binomials shown in Section 7.4.

> The multiplication pattern is also identical to that for conjugates of radicals.

EXAMPLE 4 Multiplying Complex Numbers

Multiply.

$$(7 - 4i)(7 + 4i) = 7^2 - (4i)^2$$
$$= 7^2 - 4^2(-1)$$
$$= 7^2 + 4^2$$
$$= 49 + 16 = 65$$

> We could get the same result by applying the formula above with $a = 7$ and $b = 4$.

Check Yourself 4

Multiply $(5 + 3i)(5 - 3i)$.

$5^2 + 3i^2 \qquad 25 + 9(i) = 34$

We are now ready to discuss the division of complex numbers. Generally we find the quotient by multiplying the numerator and denominator by the conjugate of the denominator. The following example illustrates.

EXAMPLE 5 Dividing Complex Numbers

Divide.

Think of $3i$ as $0 + 3i$ and of its conjugate as $0 - 3i$, or $-3i$.

(a) $\dfrac{6 + 9i}{3i}$

The conjugate of $3i$ is $-3i$, and so we multiply the numerator and denominator by $-3i$:

$$\frac{6 + 9i}{3i} = \frac{(6 + 9i)(-3i)}{(3i)(-3i)}$$

Note Multiplying the numerator and denominator by i would accomplish the same result in this case. Try it yourself.

$$= \frac{-18i - 27i^2}{-9i^2}$$

$$= \frac{-18i - 27(-1)}{(-9)(-1)}$$

$$= \frac{27 - 18i}{9} = 3 - 2i$$

We multiply by $\dfrac{3 - 2i}{3 - 2i}$, or 1.

(b) $\dfrac{3 - i}{3 + 2i} = \dfrac{(3 - i)(3 - 2i)}{(3 + 2i)(3 - 2i)}$ $9 - 6i + 6i - 4i$

$$= \frac{9 - 6i - 3i + 2i^2}{9 - 4i^2}$$

$$= \frac{9 - 9i - 2\,(1)}{9 + 4}$$

To write a complex number in standard form, we separate the real component from the imaginary.

$$= \frac{7 - 9i}{13} = \frac{7}{13} - \frac{9}{13}i$$

(c) $\dfrac{2 + i}{4 - 5i} = \dfrac{(2 + i)(4 + 5i)}{(4 - 5i)(4 + 5i)}$

$$= \frac{8 + 4i + 10i + 5i^2}{16 - 25i^2}$$

$3 + 14i$ $\dfrac{3}{41} = \dfrac{14}{41}\bigg/i$

$$= \frac{8 + 14i - 5}{16 + 25}$$

$\dfrac{}{41}$

$$= \frac{3 + 14i}{41} = \frac{3}{41} + \frac{14}{41}i$$

Check Yourself 5

Divide.

1. $\dfrac{5 + i}{5 - 3i}$

2. $\dfrac{4 + 10i}{2i}$

We conclude this section with the following diagram, which summarizes the structure of the system of complex numbers.

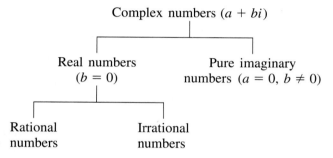

Complex numbers $(a + bi)$

Real numbers $(b = 0)$

Pure imaginary numbers $(a = 0, b \neq 0)$

Rational numbers

Irrational numbers

CHECK YOURSELF ANSWERS

1. (1) $5i$; (2) $2i\sqrt{6}$ **2.** (1) $7 - 9i$; (2) $-9 + 3i$; (3) $-2 + 4i$ **3.** $-4 - 19i$

4. 34   **5.** (1) $\dfrac{11}{17} + \dfrac{10}{17}i$; (2) $5 - 2i$

(9

7.6 EXERCISES

Build Your Skills

Write each of the following roots as a multiple of i. Simplify your results where necessary.

1. $\sqrt{-16}$

2. $\sqrt{-36}$

3. $-\sqrt{-64}$

4. $-\sqrt{-25}$

5. $\sqrt{-21}$

6. $\sqrt{-19}$

7. $\sqrt{-12}$

8. $\sqrt{-24}$

9. $-\sqrt{-108}$

10. $-\sqrt{-192}$

Perform the indicated operations.

11. $(3 + i) + (5 + 2i)$

12. $(2 + 3i) + (4 + 5i)$

13. $(3 - 2i) + (-2 + 7i)$

14. $(-5 - 3i) + (-2 - 4i)$

15. $(5 + 4i) - (3 + 2i)$

1-9 , 15-19

3 5-45

53-61

16. $(7 + 6i) - (3 + 5i)$

17. $(8 - 5i) - (3 + 2i)$

18. $(7 - 3i) - (-2 - 5i)$

19. $(5 + i) + (2 + 3i) + 7i$

20. $(3 - 2i) + (2 + 3i) + 7i$

21. $(2 + 3i) - (3 - 5i) + (4 + 3i)$

22. $(5 - 7i) + (7 + 3i) - (2 - 7i)$

23. $(7 + 3i) - [(3 + i) - (2 - 5i)]$

24. $(8 - 2i) - [(4 + 3i) - (-2 + i)]$

25. $(5 + 3i) + (-5 - 3i)$

26. $(8 - 7i) + (-8 + 7i)$

Find each of the following products. Write your answers in standard form.

27. $3i(3 + 5i)$

28. $2i(7 + 3i)$

29. $4i(3 - 7i)$

30. $2i(6 + 3i)$

31. $-2i(4 - 3i)$

32. $-5i(2 - 7i)$

33. $6i\left(\dfrac{2}{3} + \dfrac{5}{6}i\right)$

34. $4i\left(\dfrac{1}{2} + \dfrac{3}{4}i\right)$

35. $(3 + 2i)(2 + 3i)$

36. $(5 - 2i)(3 - i)$

37. $(4 - 3i)(2 + 5i)$

38. $(7 + 2i)(3 - 2i)$

39. $(-2 - 3i)(-3 + 4i)$

40. $(-5 - i)(-3 - 4i)$

41. $(5 - 2i)^2$

42. $(3 + 7i)^2$

Write the conjugate of each of the following complex numbers. Then find the product of the given number and its conjugate.

43. $3 - 2i$

44. $5 + 2i$

45. $2 + 3i$

46. $7 - i$

47. $-3 - 2i$

48. $-5 - 7i$

49. $5i$

50. $-3i$

Find each of the following quotients, and write your answer in standard form.

51. $\dfrac{3 + 2i}{i}$

52. $\dfrac{5 - 3i}{-i}$

53. $\dfrac{6 - 4i}{2i}$

54. $\dfrac{8 + 12i}{-4i}$

55. $\dfrac{3}{2 + 5i}$

56. $\dfrac{5}{2 - 3i}$

57. $\dfrac{13}{2 + 3i}$

58. $\dfrac{-17}{3 + 5i}$

59. $\dfrac{2 + 3i}{4 + 3i}$

60. $\dfrac{4 - 2i}{5 - 3i}$

61. $\dfrac{3 - 4i}{3 + 4i}$

62. $\dfrac{7 + 2i}{7 - 2i}$

Transcribe Your Skills

63. The first application of complex numbers was suggested by the Norwegian surveyor Caspar Wessel in 1797. He found that complex numbers could be used to represent distance and direction on a two-dimensional grid. Why would a surveyor care about such a thing?

64. To what sets of numbers does 1 belong?

Think About These

In this section, we defined $\sqrt{-4} = \sqrt{4}i = 2i$ in the process of expressing the square root of a negative number as a multiple of i.

Particular care must be taken with products where two negative radicands are involved. For instance,

$$\sqrt{-3} \cdot \sqrt{-12} = (i\sqrt{3})(i\sqrt{12})$$
$$= i^2\sqrt{36} = (-1)\sqrt{36} = -6$$

is correct. However, if we try to apply the product property for radicals, we have

$$\sqrt{-3} \cdot \sqrt{-12} \overset{?}{=} \sqrt{(-3)(-12)} = \sqrt{36} = 6$$

which is *not* correct. The property $\sqrt{a} \cdot \sqrt{b} = \sqrt{ab}$ is *not* applicable in the case where a and b are both negative. Radicals such as $\sqrt{-a}$ must be written in the standard form $i\sqrt{a}$ *before* multiplying, in order to use the rules for real-valued radicals.

Find each of the following products.

65. $\sqrt{-5} \cdot \sqrt{-7}$

66. $\sqrt{-3} \cdot \sqrt{-10}$

67. $\sqrt{-2} \cdot \sqrt{-18}$

68. $\sqrt{-4} \cdot \sqrt{-25}$

69. $\sqrt{-6} \cdot \sqrt{-15}$

70. $\sqrt{-5} \cdot \sqrt{-30}$

71. $\sqrt{-10} \cdot \sqrt{-10}$

72. $\sqrt{-11} \cdot \sqrt{-11}$

Since $i^2 = -1$, the positive integral powers of i form an interesting pattern. Consider the following.

$$i = i \qquad\qquad i^5 = i^4 \cdot i = 1 \cdot i = i$$
$$i^2 = -1 \qquad\qquad i^6 = i^4 \cdot i^2 = 1(-1) = -1$$
$$i^3 = i^2 \cdot i = (-1)i = -i \qquad i^7 = i^4 \cdot i^3 = 1(-i) = -i$$
$$i^4 = i^2 \cdot i^2 = (-1)(-1) = 1 \qquad i^8 = i^4 \cdot i^4 = 1 \cdot 1 = 1$$

Given the pattern above, do you see that any power of i will simplify to i, -1, $-i$, or 1? The easiest approach to simplifying higher powers of i is to write that power in terms of i^4 (because $1^4 = 1$). As an example,

$$i^{18} = i^{16} \cdot i^2 = (i^4)^4 \cdot i^2 = 1^4(-1) = -1$$

Use these comments to simplify each of the following powers of i.

73. i^{10}

74. i^9

75. i^{20}

76. i^{15}

77. i^{38}

78. i^{40}

79. i^{51}

80. i^{61}

Skillscan (Section 5.5)

Factor each of the following completely.

a. $x^2 - 3x$ **b.** $2x^3 - 4x$ **c.** $2x^2 - x - 3$

d. $x^2 - 6x - 9$ **e.** $2x^2 - x - 1$ **f.** $x^2 - 16$

SUMMARY

Roots and Radicals [7.1]

Square Roots Every positive number has two square roots. The positive or principal square root of a number a is denoted by

$$\sqrt{a}$$

The negative square root is written as

$$-\sqrt{a}$$

$\sqrt{25} = 5$

5 is the principal square root of 25 since $5^2 = 25$.

$-\sqrt{49} = -7$

since $(-7)^2 = 49$.

$\sqrt[3]{27} = 3$

$\sqrt[3]{-64} = -4$

$\sqrt[4]{81} = 3$

Higher Roots Cube roots, fourth roots, and so on are denoted by using an index and a radical. The principal *n*th root of *a* is written as

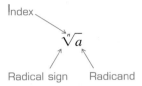

$\sqrt[3]{4^2} = 4$

$\sqrt{(-5)^2} = 5$

$\sqrt[3]{(-3)^3} = -3$

$\sqrt{m^2} = |m|$

$\sqrt[3]{27x^3} = 3x$

Radicals Containing Variables In general,

$$\sqrt[n]{x^n} = \begin{cases} |x| & n \text{ is even} \\ x & n \text{ is odd} \end{cases}$$

Simplification of Radical Expressions [7.2]

Simplifying radical expressions entails applying two properties for radicals.

Product Property

$\sqrt{35} = \sqrt{5 \cdot 7}$

$\quad = \sqrt{5} \cdot \sqrt{7}$

$$\sqrt[n]{ab} = \sqrt[n]{a} \cdot \sqrt[n]{b}$$

Quotient Property

$\sqrt{\dfrac{2}{5}} = \dfrac{\sqrt{2}}{\sqrt{5}}$

$$\sqrt[n]{\frac{a}{b}} = \frac{\sqrt[n]{a}}{\sqrt[n]{b}}, \; b \neq 0$$

Simplified Form for Radicals A radical is in *simplified form* if the following three conditions are satisfied.

$\sqrt{18x^3} = \sqrt{9x^2 \cdot 2x}$

$\quad = \sqrt{9x^2} \cdot \sqrt{2x}$

$\quad = 3x\sqrt{2x}$

1. The radicand has no factor raised to a power greater than or equal to the index.

$\sqrt{\dfrac{5}{9}} = \dfrac{\sqrt{5}}{\sqrt{9}} = \dfrac{\sqrt{5}}{3}$

2. No fraction appears in the radical.

$\sqrt{\dfrac{3}{7x}} = \dfrac{\sqrt{3}}{\sqrt{7x}} = \dfrac{\sqrt{3} \cdot \sqrt{7x}}{\sqrt{7x} \cdot \sqrt{7x}}$

$\quad = \dfrac{\sqrt{21x}}{\sqrt{49x^2}} = \dfrac{\sqrt{21x}}{7x}$

3. No radical appears in a denominator.

Note Satisfying condition 3 may require *rationalizing the denominator*.

Addition and Subtraction of Radical Expressions [7.3]

Radical expressions may be combined only if they are *similar*, that is, if they have the same radicand with the same index.
    Similar radicals are combined by application of the distributive property.

$$8\sqrt{5} + 3\sqrt{5} = (8 + 3)\sqrt{5}$$
$$= 11\sqrt{5}$$

$$2\sqrt{18} - 4\sqrt{2}$$
$$= 2\sqrt{9 \cdot 2} - 4\sqrt{2}$$
$$= 2\sqrt{9} \cdot \sqrt{2} - 4\sqrt{2}$$
$$= 2 \cdot 3\sqrt{2} - 4\sqrt{2}$$
$$= 6\sqrt{2} - 4\sqrt{2} = (6 - 4)\sqrt{2}$$
$$= 2\sqrt{2}$$

Multiplication and Division of Radical Expressions [7.4]

Multiplication

To multiply two monomial expressions, we use

$$\sqrt[n]{a} \cdot \sqrt[n]{b} = \sqrt[n]{ab}$$

and simplify the product.

$$\sqrt{3x} \cdot \sqrt{6x^2} = \sqrt{18x^3}$$
$$= \sqrt{9x^2 \cdot 2x}$$
$$= \sqrt{9x^2} \cdot \sqrt{2x}$$
$$= 3x\sqrt{2x}$$

$$\sqrt{2}(5 + \sqrt{8})$$
$$= \sqrt{2} \cdot 5 + \sqrt{2} \cdot \sqrt{8}$$
$$= 5\sqrt{2} + 4$$

If binomial expressions are involved, we use the distributive property or the FOIL method.

$$(3 + \sqrt{2})(5 - \sqrt{2})$$
$$= 5 - 3\sqrt{2} + 5\sqrt{2} - 2$$
$$= 13 + 2\sqrt{2}$$

Division

To divide two monomial expressions, rationalize the denominator by multiplying the numerator and denominator by the appropriate radical.

$$\frac{5}{\sqrt{8}} = \frac{5 \cdot \sqrt{2}}{\sqrt{8} \cdot \sqrt{2}} = \frac{5\sqrt{2}}{\sqrt{16}}$$
$$= \frac{5\sqrt{2}}{4}$$

If the divisor (the denominator) is a binomial, multiply the numerator and denominator by the conjugate of the denominator.

Note $3 + \sqrt{5}$ is the conjugate of $3 - \sqrt{5}$.

$$\frac{2}{3 - \sqrt{5}}$$
$$= \frac{2(3 + \sqrt{5})}{(3 - \sqrt{5})(3 + \sqrt{5})}$$
$$= \frac{2(3 + \sqrt{5})}{4}$$
$$= \frac{3 + \sqrt{5}}{2}$$

Rational Exponents [7.5]

Rational exponents are an alternate way of indicating roots. We use the following definition:

$36^{1/2} = \sqrt{36} = 6$

$-27^{1/3} = \sqrt[3]{-27} = -3$

$243^{1/5} = \sqrt[5]{243} = 3$

$25^{-1/2} = \dfrac{1}{\sqrt{25}} = \dfrac{1}{5}$

> **If a is any real number and n is a positive integer $(n > 1)$,**
>
> $$a^{1/n} = \sqrt[n]{a}$$

We restrict a so that a is nonnegative when n is even.
Also we can define the following:

$27^{2/3} = (\sqrt[3]{27})^2$

$\qquad = 3^2 = 9$

$(a^4b^8)^{3/4} = \sqrt{(a^4b^8)^3}$

$\qquad = \sqrt[4]{a^{12}b^{24}} = a^3b^6$

> **If a is any real number and m and n are positive integers such that $n > 1$, then**
>
> $$a^{m/n} = (\sqrt[n]{a})^m = \sqrt[n]{a^m}$$

The five properties for exponents, in Section 5.1, continue to hold for rational exponents.

Complex Numbers [7.6]

A *complex number* is any number that can be written in the form

$a + bi$

where a are b and real numbers and

$i = \sqrt{-1}$

The basic operations of addition, subtraction, multiplication, and division are defined for complex numbers as follows.

Addition and Subtraction For complex numbers $a + bi$ and $c + di$,

$(2 + 3i) + (-3 - 5i)$

$= (2 - 3) + (3 - 5)i$

$= -1 - 2i$

$$(a + bi) + (c + di) = (a + c) + (b + d)i$$

and

$(5 - 2i) - (3 - 4i)$

$= (5 - 3) + [-2 - (-4)]i$

$= 2 + 2i$

$$(a + bi) - (c + di) = (a - c) + (b - d)i$$

Multiplication To multiply complex numbers, you can use

$$(a + bi)(c + di) = (ac - bd) + (ad + bc)i$$

Note It is generally easier to use the FOIL multiplication pattern and the definition of i, rather than to apply the above formula.

$(2 + 5i)(3 - 4i)$
$= 6 - 8i + 15i - 20i^2$
$= 6 + 7i - 20(-1)$
$= 26 + 7i$

Division To divide two complex numbers, we multiply the numerator and denominator by the complex conjugate of the denominator and write the result in standard form.

$$\frac{3 + 2i}{3 - 2i} = \frac{(3 + 2i)(3 + 2i)}{(3 - 2i)(3 + 2i)}$$

$$= \frac{9 + 6i + 6i + 4i^2}{9 - 4i^2}$$

$$= \frac{9 + 12i + 4(-1)}{9 - 4(-1)}$$

$$= \frac{5 + 12i}{13} = \frac{5}{13} + \frac{12}{13}i$$

SUMMARY EXERCISES

This summary exercise set is provided to give you practice with each of the objectives of the chapter. Each exercise is keyed to the appropriate chapter section. The answers are provided in the instructor's manual that accompanies this text. Your instructor will provide guidelines on how best to use these exercises in your instructional program.

[7.1] Evaluate each of the following roots over the set of real numbers.

1. $\sqrt{121}$

2. $-\sqrt{64}$

3. $\sqrt{-81}$

4. $\sqrt[3]{64}$

5. $\sqrt[3]{-64}$

6. $\sqrt[4]{81}$

7. $\sqrt{\dfrac{9}{16}}$

8. $\sqrt[3]{-\dfrac{8}{27}}$

9. $\sqrt{8^2}$

[7.1] Simplify each of the following expressions. Assume that all variables represent positive real numbers for all subsequent exercises in this exercise set.

10. $\sqrt{4x^2}$

11. $\sqrt{a^4}$

12. $\sqrt{36y^2}$

13. $\sqrt{49w^4z^6}$

14. $\sqrt[3]{x^9}$

15. $\sqrt[3]{-27b^6}$

16. $\sqrt[3]{8r^3s^9}$

17. $\sqrt[4]{16x^4y^8}$

18. $\sqrt[5]{32p^5q^{15}}$

[7.2] Use the product property to write each of the following expressions in simplified form.

19. $\sqrt{45}$

20. $-\sqrt{75}$

21. $\sqrt{60x^2}$

22. $\sqrt{108a^3}$

23. $\sqrt[3]{32}$

24. $\sqrt[3]{-80w^4z^3}$

[7.2] Use the quotient property to write each of the following expressions in simplified form.

25. $\sqrt{\dfrac{9}{16}}$

26. $\sqrt{\dfrac{7}{36}}$

27. $\sqrt{\dfrac{y^4}{49}}$

28. $\sqrt{\dfrac{2x}{9}}$

29. $\sqrt{\dfrac{5}{16x^2}}$

30. $\sqrt[3]{\dfrac{5a^2}{27}}$

[7.2] Rationalize the denominator, and write each of the following expressions in simplified form.

31. $\sqrt{\dfrac{3}{7}}$

32. $\dfrac{\sqrt{12}}{\sqrt{x}}$

33. $\dfrac{\sqrt{10a}}{\sqrt{5b}}$

34. $\sqrt[3]{\dfrac{3}{a^2}}$

35. $\dfrac{2}{\sqrt[3]{3x}}$

36. $\dfrac{\sqrt[3]{x^2}}{\sqrt[3]{y^5}}$

[7.3] Simplify each of the following expressions where necessary. Then add or subtract as indicated.

37. $7\sqrt{10} + 4\sqrt{10}$

38. $5\sqrt{3x} - 2\sqrt{3x}$

39. $7\sqrt[3]{2x} + 3\sqrt[3]{2x}$

40. $8\sqrt{10} - 3\sqrt{10} + 2\sqrt{10}$

41. $\sqrt{72} + \sqrt{50}$

42. $\sqrt{54} - \sqrt{24}$

43. $9\sqrt{7} - 2\sqrt{63}$

44. $\sqrt{20} - \sqrt{45} + 2\sqrt{125}$

45. $2\sqrt[3]{16} + 3\sqrt[3]{54}$

46. $\sqrt{27w^3} - w\sqrt{12w}$

47. $\sqrt[3]{128a^5} + 6a\sqrt[3]{2a^2}$

48. $\sqrt{20} + \dfrac{3}{\sqrt{5}}$

49. $\sqrt{72x} - \sqrt{\dfrac{x}{2}}$

50. $\sqrt[3]{81a^4} - a\sqrt[3]{\dfrac{a}{9}}$

51. $\dfrac{\sqrt{15}}{3} - \dfrac{1}{\sqrt{15}}$

[7.4] Multiply and simplify each of the following expressions.

52. $\sqrt{3x} \cdot \sqrt{7y}$

53. $\sqrt{6x^2} \cdot \sqrt{18}$

54. $\sqrt[3]{4a^2b} \cdot \sqrt[3]{ab^2}$

55. $\sqrt{5}(\sqrt{3} + 2)$

56. $\sqrt{6}(\sqrt{8} - \sqrt{2})$

57. $\sqrt{a}(\sqrt{5a} + \sqrt{125a})$

58. $(\sqrt{3} + 5)(\sqrt{3} - 7)$

59. $(\sqrt{7} - \sqrt{2})(\sqrt{7} + \sqrt{3})$

60. $(\sqrt{5} - 2)(\sqrt{5} + 2)$

61. $(\sqrt{7} - \sqrt{3})(\sqrt{7} + \sqrt{3})$

62. $(2 + \sqrt{3})^2$

63. $(\sqrt{5} - \sqrt{2})^2$

[7.4] Divide and simplify each of the following expressions.

64. $\dfrac{1}{3 + \sqrt{2}}$

65. $\dfrac{11}{5 - \sqrt{3}}$

66. $\dfrac{\sqrt{5} - 2}{\sqrt{5} + 2}$

67. $\dfrac{\sqrt{x} - 3}{\sqrt{x} + 3}$

[7.4] Evaluate each of the following expressions.

68. $49^{1/2}$

69. $-100^{1/2}$

70. $(-27)^{1/3}$

71. $16^{1/4}$

72. $64^{2/3}$

73. $25^{3/2}$

74. $\left(\dfrac{4}{9}\right)^{3/2}$

75. $49^{-1/2}$

76. $81^{-3/4}$

[7.5] Use the properties of exponents to simplify each of the following expressions.

77. $x^{3/2} \cdot x^{5/2}$

78. $b^{2/3} \cdot b^{3/2}$

79. $\dfrac{r^{8/5}}{r^{3/5}}$

80. $\dfrac{a^{5/4}}{a^{1/2}}$

81. $(x^{3/5})^{2/3}$

82. $(y^{-4/3})^6$

83. $(x^{4/5}y^{3/2})^{10}$

84. $(16x^{1/3} \cdot y^{2/3})^{3/4}$

85. $\left(\dfrac{x^{-2}y^{-1/6}}{x^{-4}y}\right)^3$

86. $\left(\dfrac{27y^3z^{-6}}{x^{-3}}\right)^{1/3}$

[7.5] Write each of the following expressions in radical form.

87. $x^{3/4}$

88. $(w^2z)^{2/5}$

89. $3a^{2/3}$

90. $(3a)^{2/3}$

[7.5] Write each of the following expressions, using rational exponents, and simplify where necessary.

91. $\sqrt[5]{7x}$

92. $\sqrt{16w^4}$

93. $\sqrt[3]{27p^3q^9}$

94. $\sqrt[4]{16a^8b^{16}}$

[7.6] Write each of the following roots as a multiple of i. Simplify your result.

95. $\sqrt{-49}$

96. $\sqrt{-13}$

97. $-\sqrt{-60}$

[7.6] Perform the indicated operations.

98. $(2 + 3i) + (3 - 5i)$

99. $(7 - 3i) + (-3 - 2i)$

100. $(5 - 3i) - (2 + 5i)$

101. $(-4 + 2i) - (-1 - 3i)$

[7.6] Find each of the following products.

102. $4i(7 - 2i)$

103. $(5 - 2i)(3 + 4i)$

104. $(3 - 4i)^2$

105. $(2 - 3i)(2 + 3i)$

[7.6] Find each of the following quotients, and write your answer in standard form.

106. $\dfrac{5 - 15i}{5i}$

107. $\dfrac{10}{3 - 4i}$

108. $\dfrac{3 - 2i}{3 + 2i}$

109. $\dfrac{5 + 10i}{2 + i}$

SELF-TEST

The purpose of this self-test is to help you check your progress and to review for a chapter test in class. Allow yourself about an hour to take the test. When you are done, check your answers in the back of the book. If you missed any problems, be sure to go back and review the appropriate sections in the chapter and the exercises that are provided.

Simplify each expression. Assume that all variables represent positive real numbers in all subsequent problems.

1. $\sqrt{49a^4}$

2. $\sqrt[3]{-27w^6z^9}$

Use the product or quotient properties to write each expression in simplified form.

3. $\sqrt[3]{9p^7q^5}$

4. $\dfrac{7x}{\sqrt{64y^2}}$

Rationalize the denominator, and write each expression in simplified form.

5. $\sqrt{\dfrac{5x}{8y}}$

6. $\dfrac{3}{\sqrt[3]{9x}}$

Simplify each expression where necessary. Then add or subtract as indicated.

7. $\sqrt{3x^3} + x\sqrt{75x} - \sqrt{27x^3}$

8. $\sqrt[3]{54m^4} + m\sqrt[3]{16m}$

Multiply or divide as indicated. Then simplify your result.

9. $\sqrt{6x}(\sqrt{18x} - \sqrt{2x})$

10. $\dfrac{\sqrt{6} - \sqrt{3}}{\sqrt{6} + \sqrt{3}}$

Use the properties of exponents to simplify each expression.

11. $(16x^4)^{-3/2}$

12. $(27m^{3/2}n^{-6})^{2/3}$

13. $\left(\dfrac{16r^{-1/3}s^{5/3}}{rs^{-7/3}}\right)^{3/4}$

Write the expression in radical form and simplify.

14. $(a^7b^3)^{2/5}$

Write the expression, using rational exponents. Then simplify.

15. $\sqrt[3]{125p^9q^6}$

Perform the indicated operations.

16. $(-2 + 3i) - (-5 - 7i)$

17. $(5 - 3i)(-4 + 2i)$

18. $\dfrac{10 - 20i}{3 - i}$

Evaluate each expression.

19. $(25)^{-3/2}$

20. $(16)^{3/4}$

CUMULATIVE REVIEW EXERCISES

Here is a review of selected topics for the first seven chapters.

Solve each of the following.

1. $5x - 3(2x + 6) = 4 - (3x - 2)$

2. $4x - 2(x - 5) \le -2$

3. $|3x - 2| \le 6$

4. $|8 - 3x| = 5$

5. Find the equation of the line that is parallel to the line $6x + 7y = 42$ and has a y intercept of -3.

6. Solve the following system of equations.

$2x + 5y = 9$

$4x - 3y = 5$

Perform the indicated operations.

7. $3x - 2[x - (3x - 1)] + 6x(x - 2)$

8. $x(2x - 1)(x + 3)$

Factor each of the following completely.

9. $6x^3 + 7x^2 - 3x$

10. $16x^{16} - 9y^8$

Simplify each of the following rational expressions.

11. $\dfrac{5}{x - 1} - \dfrac{2x + 6}{x^2 + 2x - 3}$

12. $\dfrac{x + 1}{x^2 - 5x - 6} \div \dfrac{x^2 - 1}{x - 6}$

13. $\dfrac{1 - \dfrac{3}{x + 3}}{\dfrac{1}{x^2 - 9}}$

Simplify each of the following radical expressions.

14. $\sqrt{147} - \sqrt{75} + 2\sqrt{27}$

15. $(\sqrt{3} - 5)(\sqrt{2} + 3)$

Graph each equation.

16. $y = 3x - 5$

17. $x = -5$

18. $2x - 3y = 12$

Solve the following word problems.

19. In a right triangle, one leg is 4 cm longer than the other leg. If the hypotenuse is 5 cm, find the length of the longer leg.

20. One number is 4 more than a second number and 9 less than a third number. Find the three numbers if their sum is 50.

QUADRATIC EQUATIONS, FUNCTIONS, AND INEQUALITIES

WETLANDS Wetlands are land areas covered by water all or part of the year (excluding lakes, reservoirs, and streams). Generally classified as either coastal (estuarine) or inland wetlands, these broad categories contain many specific types of wetlands: marshes, bogs, prairie potholes, swamps, mud flats, moors, floodplains, and wet arctic tundra.

Wetlands were once thought of as wasteland, to be converted to better waterways, filled to create more land, or chosen as a good site for landfills. We are finally beginning to realize how important and productive wetlands are. Coastal wetlands are some of the most biologically productive habitats on earth. They are the feeding grounds, spawning areas, and nurseries for the majority of saltwater fish and shellfish species. One-third of all North American bird species rely on wetlands at some point in their lives.

Wetlands are also major water storage areas. They store water during heavy rains and release it slowly, helping to prevent flooding and erosion problems. They help improve water quality by filtering, diluting, and degrading various pollutants. Wetlands help recharge groundwater systems by holding the water and releasing it slowly into the land surface.

Wetlands also act as sewage treatment facilities. At least 31 cities and towns in North America use wetland sewage treatment, which is cheaper and easier to maintain than traditional sewage treatment plants.

Wetland losses are a major environmental problem. Tropical wetlands in Asia and South America are being degraded by expanding agricultural and chemical pollution. About 90 percent of the Tule marshes in California have been destroyed or degraded by chemical runoff from nearby agricultural lands. Inland wetlands account for approximately 95 percent of the remaining U.S. wetlands. The yearly loss of these wetlands is approximately 121,000 hectares.

In the mid-1980s, the U.S. government began to take steps to stop the loss of inland wetlands. These steps have helped set a national policy of no net loss of wetlands by mandating the replacement of any wetlands destroyed, but newly created wetlands can never equal the complex ecosystems that evolve naturally over time. Instead of replacement, our goal should be preservation and restoration of the existing natural systems. ■

8.1 Solving Quadratic Equations by Factoring

TAPE IN18

OBJECTIVES: 1. To solve quadratic equations by factoring
2. To find the zeros of quadratic functions

The factoring techniques we considered in Chapter 5 provide us with tools for solving equations that can be written in the form

This is a quadratic equation in one variable, here x. You can recognize such a *quadratic* equation by the fact that the highest power of the variable x is the second power.

$$ax^2 + bx + c = 0 \qquad a \neq 0$$

where a, b, and c are constants. This is called a *quadratic equation in standard form*. Using factoring to solve quadratic equations requires the *zero-product principle*. It says that if the product of two factors is 0, then one or both of the factors must be equal to 0. In symbols:

> ### ZERO-PRODUCT PRINCIPLE
> If $a \cdot b = 0$, then $a = 0$ or $b = 0$ or $a = b = 0$.

Let's see how the principle is applied to solving quadratic equations.

EXAMPLE 1 Solving Equations by Factoring

Solve

To use the zero-product principle, 0 must be on one side of the equation.

$$x^2 - 2x - 15 = 0$$

Solution Factoring on the left, we have

$$(x - 5)(x + 3) = 0$$

By the zero-product principle, we know that one or both of the factors must be zero. We can then write

$$x - 5 = 0 \qquad \text{or} \qquad x + 3 = 0$$

Solving each equation gives

$$x = 5 \qquad \text{or} \qquad x = -3$$

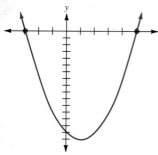

Graph the function
$y = x^2 - 2x - 15$
on your graphing calculator. The solutions to the equation $0 = x^2 - 2x - 15$ will be those values on the curve at which $y = 0$. Those are the points at which the graph intercepts the x axis.

The two solutions are 5 and -3. These are sometimes called the *zeros*, or *roots*, of the equation.

Quadratic equations can be checked in the same way as linear equations were checked, by substitution. For instance, if $x = 5$, we have

$$5^2 - 2 \cdot 5 - 15 \stackrel{?}{=} 0$$
$$25 - 10 - 15 \stackrel{?}{=} 0$$
$$0 = 0$$

which is a true statement. We leave it to you to check the solution of -3.

Check Yourself 1

Solve $x^2 - 9x + 20 = 0$.

Other factoring techniques are also used in solving quadratic equations. The following examples illustrate.

EXAMPLE 2 Solving Equations by Factoring

(*a*) Solve $x^2 - 5x = 0$.

Again, factor the left side of the equation and apply the zero-product principle:

$$x(x - 5) = 0$$

Now

$$x = 0 \quad \text{or} \quad x - 5 = 0$$
$$x = 5$$

The two solutions are 0 and 5.

(*b*) Solve $x^2 - 9 = 0$.

Factoring yields

$$(x + 3)(x - 3) = 0$$

$$x + 3 = 0 \quad \text{or} \quad x - 3 = 0$$
$$x = -3 \qquad\qquad x = 3$$

The two solutions are -3 and 3.

CAUTION

Note A *common mistake* is to forget the statement $x = 0$ when you are solving equations of this type. Be sure to include the *two* solutions.

 The graph of $y = x^2 - 9$ is

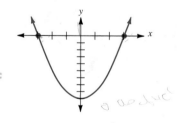

Check Yourself 2

Solve by factoring.

1. $x^2 + 8x = 0$ **2.** $x^2 - 16 = 0$

Example 3 illustrates a crucial point. Our solution technique depends on the zero-product principle. This means that the product of factors *must be equal to 0*. The importance of this is shown now.

EXAMPLE 3 Solving Equations by Factoring

Solve $2x^2 - x = 3$.

Solution The first step in the solution is to write the equation in standard form (that is, one side of the equation is 0). So start by adding -3 to both sides of the equation. Then

$$2x^2 - x - 3 = 0$$

Make sure all terms are on one side of the equation. The other side will be 0.

You can now factor and solve by using the zero-product principle.

$$(2x - 3)(x + 1) = 0$$

$$2x - 3 = 0 \qquad \text{or} \qquad x + 1 = 0$$

$$2x = 3 \qquad\qquad\qquad x = -1$$

$$x = \frac{3}{2}$$

The two solutions are $\frac{3}{2}$ and -1.

CAUTION
Be Careful! Consider the equation

$x(2x - 1) = 3$

Students are sometimes tempted to now write

$x = 3$ or $2x - 1 = 3$

That is *not correct*. The fact that the product of factors is 3 does *not* ensure that one or both of the factors are equal to 3.
   Subtract 3 from both sides of the equation *as the first step*, to write

$x^2 - 2x - 3 = 0$

in standard form. Only *now* can you factor and proceed as before.

Check Yourself 3

Solve $3x^2 = 5x + 2$.

In all the examples above, the quadratic equations have had two distinct real number solutions. That may or may not always be the case, as we shall see.

EXAMPLE 4 Solving Equations by Factoring

Solve $x^2 - 6x + 9 = 0$.

Solution Factoring, we have

$$(x - 3)(x - 3) = 0$$

and

$$x - 3 = 0 \qquad \text{or} \qquad x - 3 = 0$$

$$x = 3 \qquad\qquad\qquad x = 3$$

Note that the graph of $y = x^2 - 6x + 9$ touches the x axis at only the point (3, 0).

The solution is 3. A quadratic (or second-degree) equation always has *two* solutions. When an equation such as this one has two solutions which are the same number, we call 3 the *repeated* (or *double*) *solution* of the equation.

Even though a quadratic equation will always have two solutions, they may not always be real numbers. More about this in the next section.

Check Yourself 4

Solve $x^2 + 6x + 9 = 0$.

Always examine the quadratic member of an equation for common factors. It will make your work much easier, as the next example illustrates.

EXAMPLE 5 Solving Equations by Factoring

Solve $3x^2 - 3x - 60 = 0$.

Solution First, note the common factor 3 in the quadratic member of the equation. Factoring out the 3, we have

$3(x^2 - x - 20) = 0$

Now divide both sides of the equation by 3:

$$\frac{3}{3}(x^2 - x - 20) = \frac{0}{3}$$

or

Note the advantage of dividing both members by 3. The coefficients in the quadratic member become smaller, and that member is much easier to factor.

$x^2 - x - 20 = 0$

We can now factor and solve as before:

$(x - 5)(x + 4) = 0$

$x - 5 = 0$ or $x + 4 = 0$

$x = 5$ $x = -4$

The two solutions are 5 and -4.

Check Yourself 5

Solve $2x^2 - 10x - 48 = 0$.

In Chapter 3, we introduced the concept of a function and expressed the equation of a line in function form. Another type of function is called a quadratic function.

> A *quadratic function* is a function that can be written in the form
>
> $f(x) = ax^2 + bx + c$
>
> where a, b, and c are real numbers and $a \neq 0$.

For example, $f(x) = 3x^2 - 2x - 1$ and $g(x) = x^2 - 2$ are quadratic functions. In working with functions, we often want to find the values of x for which $f(x) = 0$. These values are called the *zeros* of the function. They represent the points where the graph of the function crosses the x axis. To find the zeros of a quadratic function, a quadratic equation must be solved.

EXAMPLE 6 Finding the Zeros of a Function

Find the zeros of $f(x) = x^2 - x - 2$.

Solution To find the zeros of $f(x) = x^2 - x - 2$, we must solve the quadratic equation $f(x) = 0$.

$$f(x) = 0$$
$$x^2 - x - 2 = 0$$
$$(x - 2)(x + 1) = 0$$
$$x - 2 = 0 \quad \text{or} \quad x + 1 = 0$$
$$x = 2 \qquad\qquad x = -1$$

The zeros of the function are -1 and 2.

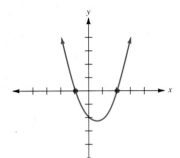

The graph of $f(x) = x^2 - x - 2$ intercepts the x axis at the points $(2, 0)$ and $(-1, 0)$, so these are the *zeros* of the equation.

Check Yourself 6

Find the zeros of $f(x) = 2x^2 - x - 3$.

CHECK YOURSELF ANSWERS

1. $4, 5$ **2.** (1) $0, -8$; (2) $4, -4$ **3.** $-\dfrac{1}{3}, 2$ **4.** -3 **5.** $-3, 8$

6. $-1, \dfrac{3}{2}$

8.1 EXERCISES

Build Your Skills

Solve each of the following quadratic equations by factoring.

1. $x^2 + 4x + 3 = 0$

2. $x^2 - 5x + 4 = 0$

3. $x^2 - 2x - 15 = 0$

4. $x^2 + 4x - 32 = 0$

5. $x^2 - 11x + 30 = 0$

6. $x^2 + 14x + 48 = 0$

7. $x^2 - 4x - 21 = 0$

8. $x^2 + 5x - 36 = 0$

9. $x^2 - 5x = 50$

10. $x^2 + 14x = -33$

11. $x^2 = 2x + 35$

12. $x^2 = 6x + 27$

13. $x^2 - 8x = 0$

14. $x^2 + 7x = 0$

15. $x^2 + 10x = 0$

16. $x^2 - 9x = 0$

17. $x^2 = 5x$

18. $4x = x^2$

19. $x^2 - 25 = 0$

20. $x^2 - 49 = 0$

21. $x^2 = 64$

22. $x^2 = 36$

23. $4x^2 + 12x + 9 = 0$

24. $9x^2 - 30x + 25 = 0$

25. $2x^2 - 17x + 36 = 0$

26. $5x^2 + 17x - 12 = 0$

27. $3x^2 - x = 4$

28. $6x^2 = 13x - 6$

29. $6x^2 = 7x - 2$

30. $4x^2 - 3 = x$

31. $2m^2 = 12m + 54$

32. $5x^2 - 55x = 60$

33. $4x^2 - 24x = 0$

34. $6x^2 - 9x = 0$

35. $5x^2 = 15x$

36. $7x^2 = -49x$

37. $x(x - 2) = 15$

38. $x(x + 3) = 28$ $x(x+3) = 28$

39. $x(2x - 3) = 9$

40. $x(3x + 1) = 52$

41. $2x(3x + 1) = 28$

42. $3x(2x - 1) = 30$

43. $(x - 3)(x - 1) = 15$

44. $(x + 4)(x + 1) = 18$

45. $(2x + 1)(x - 4) = 11$

46. $(3x - 5)(x + 2) = 14$

multi li first
startand form factor

Find the zeros of the following functions.

47. $f(x) = 3x^2 - 24x + 36$

48. $f(x) = 2x^2 - 6x - 56$

49. $f(x) = 4x^2 + 16x - 20$

50. $f(x) = 3x^2 - 33x + 54$

multi first.
starclared form factori

Transcribe Your Skills

51. Explain the differences between solving the equations $3(x - 2)(x + 5) = 0$ and $3x(x - 2)(x + 5) = 0$.

52. How can a graphing calculator be used to determine the zeros of a quadratic function?

Think About These

Write an equation that has the following solutions. *Hint:* Write the binomial factors and then the quadratic member of the equation.

53. $2, -3$

54. $0, 5$

55. $6, 2$

56. $-4, 4$

The zero-product rule can be extended to three or more factors. If $a \cdot b \cdot c = 0$, then at least one of these factors is 0. Use this information to solve the following equations.

57. $x^3 - 3x^2 - 10x = 0$

58. $x^3 + 8x^2 + 15x = 0$

59. $x^3 - 9x = 0$

60. $x^3 = 16x$

Extend the ideas in the previous exercises to find solutions for the following equations. *Hint:* Apply factoring by grouping.

61. $x^3 + x^2 - 4x - 4 = 0$

62. $x^3 - 5x^2 - x + 5 = 0$

63. $x^4 - 10x^2 + 9 = 0$

64. $x^4 - 5x^2 + 4 = 0$

The net productivity of a forested wetland as related to the amount of water moving through the wetland can be expressed by a quadratic equation.

If y represents the amount of wood produced, in grams per square meter, and x represents the amount of water present, in centimeters, determine where the productivity is zero in each of the wetlands represented by the equations below.

65. $y = -3x^2 + 300x$

66. $y = -4x^2 + 500x$

67. $y = -6x^2 + 792x$

68. $y = -7x^2 + 1022x$

9-31

43 - 44

69. If the lower x values in the above exercises represent too little water present in the wetland, what do you think the higher values represent?

 Graphing Utility

Use your graphing utility to graph $f(x)$.

70. $f(x) = 3x^2 - 24x + 36$ Note the x values at which the graph crosses the x axis. Compare your answer to the solution for Exercise 47.

71. $f(x) = 2x^2 - 6x - 56$ Note the x values at which the graph crosses the x axis. Compare your answer to the solution for Exercise 48.

Skillscan (Section 7.2)

Simplify each of the following expressions.

a. $\sqrt{25}$ **b.** $\sqrt{121}$ **c.** $\sqrt{32}$ **d.** $\sqrt{50}$ **e.** $\dfrac{4 + \sqrt{48}}{4}$

f. $\dfrac{3 + \sqrt{45}}{3}$

8.2   Solving Quadratic Equations by Completing the Square

TAPE IN19

OBJECTIVES: 1. To solve quadratic equations by the square root property
2. To solve quadratic equations by completing the square

In Section 8.1 we identified all equations of the form

$$ax^2 + bx + c = 0 \qquad a \neq 0$$

as quadratic equations in standard form. In that section, we discussed solving such equations whenever the quadratic member of the equation was factorable. In this section, we want to extend our equation-solving techniques so that we can find solutions for all such quadratic equations.

Let's use the factoring method of solution for a special type of quadratic equation.

EXAMPLE 1 Solving Equations by Factoring

Solve the quadratic equation $x^2 = 16$ by factoring.

Solution We write the equation in standard form:

$$x^2 - 16 = 0$$

Here we factor the quadratic member of the equation as a difference of squares.

Factoring, we have

$$(x + 4)(x - 4) = 0$$

and finally, the solutions are

$$x = -4 \qquad \text{or} \qquad x = 4$$

Check Yourself 1

Solve each of the following quadratic equations.

1. $5x^2 = 180$ **2.** $x^2 = 25$

The equation of Example 1 could have been solved in an alternative fashion. We could have used what is called the *square root method.* Again, given the equation

$$x^2 = 16$$

we can write the equivalent statement

$$x = \sqrt{16} \qquad \text{or} \qquad x = -\sqrt{16}$$

This yields the solutions

$$x = 4 \qquad \text{or} \qquad x = -4$$

The discussion above leads us to the following general result.

Note: Be sure to include *both* the positive and the negative square roots when you use the square root method.

SQUARE ROOT PROPERTY

If $x^2 = k$, where k is a complex number, then

$$x = \sqrt{k} \qquad \text{or} \qquad x = -\sqrt{k}$$

The following example further illustrates the use of this property.

This follows from our earlier definition of $\sqrt{x^2}$ as $|x|$.

$$x^2 = k$$
$$\sqrt{x^2} = \sqrt{k}$$

and since $\sqrt{x^2} = |x|$,

$$|x| = \sqrt{k}$$

which means

$$x = -\sqrt{k} \qquad \text{or} \qquad x = \sqrt{k}$$

EXAMPLE 2 Using the Square Root Method

Solve each equation by using the square root method.

(*a*) $x^2 = 9$

By the square root property,

$$x = \sqrt{9} \qquad \text{or} \qquad x = -\sqrt{9}$$
$$= 3 \qquad\qquad\qquad = -3$$

This is often abbreviated as

$$x = \pm 3$$

(*b*) $x^2 - 17 = 0$

Add 17 to both sides of the equation:

$$x^2 = 17$$
$$x = \pm\sqrt{17}$$

 Graph the function related to each part of Example 2. In part *a*, the graph intercepts the *x* axis at (3, 0) and (−3, 0). In parts *b* and *c* we can approximate the solution by looking at the points at which the graph intercepts the *x* axis. In part *d*, where there are no **real** solutions, the graph never intercepts the *x* axis.

If a calculator were used, $\sqrt{17} = 4.123$ (rounded to three decimal places).

(*c*) $2x^2 - 3 = 0$
$$2x^2 = 3$$
$$x^2 = \frac{3}{2}$$
$$x = \pm\sqrt{\frac{3}{2}}$$
$$x = \pm\frac{\sqrt{6}}{2}$$

(*d*) $x^2 + 1 = 0$
$$x^2 = -1$$
$$x = \pm\sqrt{-1}$$
$$x = \pm i$$

In Example 2*d* we see that complex-number solutions may result.

Check Yourself 2

Solve each equation.

1. $x^2 = 5$ $\pm\sqrt{5}$ or $i\sqrt{5}$ **2.** $x^2 - 2 = 0$ $x\sqrt{2}$

3. $3x^2 - 8 = 0$ **4.** $x^2 + 9 = 0$ $\sqrt{3}i - \sqrt{3}$

We can also use the approach of Example 2 to solve an equation of the form

$$(x + 3)^2 = 16$$

As before, by the square root property we have

$$x + 3 = \pm 4$$

Solving for x yields

Subtract 3 from both sides of
the equations.

$$x = -3 \pm 4$$

which means that there are two solutions:

$$x = -3 + 4 \qquad \text{or} \qquad x = -3 - 4$$
$$= 1 \qquad\qquad\qquad = -7$$

EXAMPLE 3 Using the Square Root Method

Use the square root method to solve each equation.

The two solutions $5 + \sqrt{5}$ and
$5 - \sqrt{5}$ are abbreviated as
$5 \pm \sqrt{5}$.

(a) $(x - 5)^2 - 5 = 0$
$$(x - 5)^2 = 5$$
$$x - 5 = \pm\sqrt{5}$$
$$x = 5 \pm \sqrt{5}$$

(b) $3(y + 1)^2 - 2 = 0$
$$3(y + 1)^2 = 2$$
$$(y + 1)^2 = \frac{2}{3}$$

We have solved for y and
rationalized the denominator.

$$\sqrt{\frac{2}{3}} = \frac{\sqrt{2}}{\sqrt{3}} = \frac{\sqrt{2} \cdot \sqrt{3}}{\sqrt{3} \cdot \sqrt{3}} = \frac{\sqrt{6}}{3}$$

Then we combine the terms on
the right, using the common
denominator of 3.

$$y + 1 = \pm\sqrt{\frac{2}{3}}$$

$$y = -1 \pm \frac{\sqrt{6}}{3}$$

$$= \frac{-3 \pm \sqrt{6}}{3}$$

Check Yourself 3

Using the square root method, solve each equation.

1. $(x - 2)^2 - 3 = 0$ **2.** $2(x - 1)^2 = 1$

We have now seen two solution methods for quadratic equations—factoring the quadratic member of an equation in standard form and the method of square roots. Both methods are relatively easy to apply. Unfortunately, not all quadratic equations can be solved directly with these techniques. We must extend our solution techniques.

The square root method is useful in this process because any quadratic equation can be written in the form

$$(x + h)^2 = k$$

which yields the solution

$$x = -h \pm \sqrt{k}$$

If $(x + h)^2 = k$, then

$x + h = \pm\sqrt{k}$ and

$x = -h \pm \sqrt{k}$

The process of changing an equation in standard form

$$ax^2 + bx + c = 0$$

to the form

$$(x + h)^2 = k$$

is called the method of (completing the square, and it is based on the relationship between the middle term and the last term of any perfect-square trinomial.

Let's look at three perfect-square trinomials to see whether we can detect a pattern:

$$x^2 + 4x + \quad 4 = (x + 2)^2 \tag{1}$$
$$x^2 - 6x + \quad 9 = (x - 3)^2 \tag{2}$$
$$x^2 + 8x + 16 = (x + 4)^2 \tag{3}$$

Note that in each case the last (or constant) term is the square of one-half of the coefficient of x in the middle (or linear) term.

Note that this relationship is true *only* if the leading, or x^2, coefficient is 1. That will be important later.

In Equation (2),

$$x^2 - 6x + 9 = (x - 3)^2$$

$\dfrac{1}{2}$ of this coefficient is -3, and $(-3)^2 = 9$, the constant.

Verify this relationship for yourself in statement (3). To summarize, in perfect-square trinomials, the constant is always the square of one-half the coefficient of x.

We are now ready to use the above observation in the solution of quadratic equations by completing the square. Consider the following example.

EXAMPLE 4 Completing the Square to Solve an Equation

Solve $x^2 + 8x - 7 = 0$ by completing the square.

Solution First, we rewrite the equation with the constant on the *right-hand side:*

$$x^2 + 8x = 7$$

$\frac{1}{2} \cdot 8 = 4$ and $4^2 = 16$

Our objective is to have a perfect-square trinomial on the left-hand side. We know that we must add the square of one-half of the x coefficient to complete the square. In this case, that value is 16, so now we add 16 to each side of the equation:

$$x^2 + 8x + 16 = 7 + 16$$

This can be expressed as

Factor the perfect-square trinomial on the left, and combine like terms on the right. Use the fact that if $(x + h)^2 = k$, then $x = -h \pm \sqrt{k}$.

$$(x + 4)^2 = 23$$

Now the square root property yields

$$x + 4 = \pm\sqrt{23}$$

Subtracting 4 from both sides of the equation gives

When you graph the related function, $y = x^2 + 8x - 7$, you will note that the x values for the x intercepts are just below 1 and just above -9. Be certain that you see how these points relate to the exact solutions, $-4 + \sqrt{23}$ and $-4 - \sqrt{23}$.

$$x = -4 \pm \sqrt{23}$$

Check Yourself 4

Solve $x^2 - 6x - 2 = 0$ by completing the square.

EXAMPLE 5 Completing the Square to Solve an Equation

Solve $x^2 + 5x - 3 = 0$ by completing the square.

$$x^2 + 5x - 3 = 0$$

$$x^2 + 5x = 3 \qquad \text{Add 3 to both sides.}$$

Add the square of one-half of the x coefficient to both sides of the equation. Note that

$$\frac{1}{2} \cdot 5 = \frac{5}{2}$$

$$x^2 + 5x + \left(\frac{5}{2}\right)^2 = 3 + \left(\frac{5}{2}\right)^2 \qquad \text{Make the left-hand side a perfect square.}$$

$$\left(x + \frac{5}{2}\right)^2 = \frac{37}{4}$$

$$x + \frac{5}{2} = \pm\frac{\sqrt{37}}{2} \qquad \text{Take the square root of both sides.}$$

$$x = \frac{-5 \pm \sqrt{37}}{2} \qquad \text{Solve for } x.$$

Check Yourself 5

Solve $x^2 + 3x - 7 = 0$ by completing the square.

Some equations have non-real complex solutions, as the next example illustrates.

EXAMPLE 6 Completing the Square to Solve an Equation

Solve $x^2 + 4x + 13 = 0$ by completing the square.

$x^2 + 4x + 13 = 0$ Subtract 13 from both sides.

$x^2 + 4x = -13$ Add $\left[\dfrac{1}{2}(4)\right]^2$ to both sides.

$x^2 + 4x + 4 = -13 + 4$ Factor the left-hand side.

$(x + 2)^2 = -9$ Take the square root of both sides.

$x + 2 = \pm\sqrt{-9}$ Simplify the radical.

$x + 2 = \pm\sqrt{9}i$

$x + 2 = \pm 3i$

$x = -2 \pm 3i$

Note that the graph of $y = x^2 + 4x + 13$ does not intercept the x axis.

Check Yourself 6

Solve $x^2 + 10x + 41 = 0$.

Our next example illustrates the situation where the leading coefficient of the quadratic member is not equal to 1. As you will see, an extra step will be required.

EXAMPLE 7 Completing the Square to Solve an Equation

Solve $3x^2 + 6x - 7 = 0$ by completing the square.

$3x^2 + 6x = 7$ Add 7 to both sides.

$x^2 + 2x = \dfrac{7}{3}$ Divide both sides by 3.

$x^2 + 2x + 1 = \dfrac{7}{3} + 1$

$(x + 1)^2 = \dfrac{10}{3}$

$x + 1 = \pm\sqrt{\dfrac{10}{3}}$

$x = -1 \pm \sqrt{\dfrac{10}{3}}$

$= \dfrac{-3 \pm \sqrt{30}}{3}$

Be Careful! Before you can complete the square on the left, the coefficient of x^2 must be equal to 1. Otherwise, we must *divide* both sides of the equation by that coefficient.

We have rationalized the denominator and combined the terms on the right side.

don't complete the square.

Check Yourself 7

Solve $2x^2 - 8x + 3 = 0$ by completing the square.

The following algorithm summarizes our work in this section with solving quadratic equations by completing the square.

COMPLETING THE SQUARE

STEP 1   Isolate the constant on the right side of the equation.

STEP 2   Divide both sides of the equation by the coefficient of the x^2 term if that coefficient is not equal to 1.

STEP 3   Add the square of one-half of the coefficient of the linear term to both sides of the equation. This will give a perfect-square trinomial on the left side of the equation.

STEP 4   Write the left side of the equation as the square of a binomial, and simplify on the right side.

STEP 5   Use the square root property, and then solve the resulting linear equations.

CHECK YOURSELF ANSWERS

1. (1) $-6, 6$; (2) $-5, 5$ **2.** (1) $\sqrt{5}, -\sqrt{5}$; (2) $\sqrt{2}, -\sqrt{2}$; (3) $\dfrac{2\sqrt{6}}{3}, -\dfrac{2\sqrt{6}}{3}$;

(4) $3i, -3i$ **3.** (1) $2 \pm \sqrt{3}$; (2) $\dfrac{2 \pm \sqrt{2}}{2}$ **4.** $3 \pm \sqrt{11}$ **5.** $\dfrac{-3 \pm \sqrt{37}}{2}$

6. $-5 \pm 4i$ **7.** $\dfrac{4 \pm \sqrt{10}}{2}$

8.2 EXERCISES

Build Your Skills

Solve each of the following by factoring or completing the square.

1. $x^2 + 6x + 5 = 0$

2. $x^2 + 5x + 6 = 0$

3. $z^2 - 2z - 35 = 0$

4. $q^2 - 5q - 24 = 0$

5. $2x^2 - 5x - 3 = 0$

6. $3x^2 + 10x - 8 = 0$

7. $6y^2 - y - 2 = 0$

8. $10z^2 + 3z - 1 = 0$

Use the square root method to find solutions for each of the following equations.

9. $x^2 = 36$

10. $x^2 = 144$

11. $y^2 = 7$

12. $p^2 = 18$

13. $2x^2 - 12 = 0$

14. $3x^2 - 66 = 0$

15. $2t^2 + 12 = 4$

16. $3u^2 - 5 = -32$

17. $(x + 1)^2 = 12$

18. $(2x - 3)^2 = 5$

19. $(2z + 1)^2 - 3 = 0$

20. $(3p - 4)^2 + 9 = 0$

Find the constant that must be added to each of the following binomial expressions to form a perfect-square trinomial.

21. $x^2 + 12x$

22. $r^2 - 14r$

23. $y^2 - 8y$

24. $w^2 + 16w$

25. $x^2 - 3x$

26. $z^2 + 5z$

27. $n^2 + n$

28. $x^2 - x$

29. $x^2 + \dfrac{1}{2}x$

30. $x^2 - \dfrac{1}{3}x$

31. $x^2 + 2ax$

32. $y^2 - 4ay$

Solve each of the following equations by completing the square.

33. $x^2 + 12x - 2 = 0$

34. $x^2 - 14x - 7 = 0$

35. $y^2 - 2y = 8$

36. $z^2 + 4z - 72 = 0$

37. $x^2 - 2x - 5 = 0$

38. $x^2 - 2x = 3$

39. $x^2 + 10x + 13 = 0$

40. $x^2 + 3x - 17 = 0$

41. $z^2 - 5z - 7 = 0$

42. $q^2 - 8q + 20 = 0$

43. $m^2 - m - 3 = 0$

44. $y^2 + y - 5 = 0$

45. $x^2 + \dfrac{1}{2}x = 1$

46. $x^2 - \dfrac{1}{3}x = 2$

47. $2x^2 + 2x - 1 = 0$

48. $3x^2 - 3x = 1$

49. $3x^2 - 6x = 2$

50. $4x^2 + 8x - 1 = 0$

51. $3x^2 - 2x + 12 = 0$

52. $7y^2 - 2y + 3 = 0$

53. $x^2 + 8x + 20 = 0$

54. $x^2 - 2x + 10 = 0$

Transcribe Your Skills

55. Why must the leading coefficient of the quadratic member be set equal to 1 before using the technique of completing the square?

56. What relationship exists between the solution of a quadratic equation and the graph of a quadratic function?

Think About These

Find the constant that must be added to each binomial to form a perfect-square trinomial. Let x be the variable; other letters represent constants.

57. $x^2 + 2ax$

58. $x^2 + 2abx$

59. $x^2 + 3ax$

60. $x^2 + abx$

61. $a^2x^2 + 2ax$

62. $a^2x^2 + 4abx$

Solve each of the following equations by completing the square.

63. $x^2 + 2ax = 4$

64. $x^2 + 2ax - 8 = 0$

Graphing Utility

Use your graphing utility to find the graph for each of the following. Approximate the x intercepts for each graph. (You may have to adjust the viewing window to see both intercepts.) Compare your answers to those of Exercises 33 to 36.

65. $y = x^2 + 12x - 2$

66. $y = x^2 - 14x - 7$

67. $y = x^2 - 2x - 8$

68. $y = x^2 + 4x - 72$

Skillscan (Section 1.3)

Evaluate the expression $b^2 - 4ac$ for each of the following sets of values for a, b, and c.

a. $a = 2, \quad b = 3, \quad c = 4$ **b.** $a = 2, \quad b = -3, \quad c = 4$
c. $a = 1, \quad b = -3, \quad c = -2$ **d.** $a = 1, \quad b = 6, \quad c = 9$
e. $a = -1, b = 2, c = 4$ **f.** $a = -1, b = -3, c = -4$

8.3   Solving Quadratic Equations by Using the Quadratic Formula

TAPE IN19

OBJECTIVES: 1. To solve quadratic equations by using the quadratic formula
2. To determine the nature of the solutions of a quadratic equation by using the discriminant

As we pointed out earlier, the factoring and square root methods are limited to certain types of quadratic equations. From the last section, we know that *any* quadratic equation can be solved by the method of completing the square. However, that method can become involved and time-consuming. Therefore, we now want to derive a general formula that will provide solutions for any quadratic equation. That formula is found by using the algorithm for completing the square on the standard form of the quadratic equation

$$ax^2 + bx + c = 0 \qquad a \neq 0$$

DERIVING THE QUADRATIC FORMULA

STEP 1   Isolate the constant on the right side of the equation.
$$ax^2 + bx = -c$$

STEP 2   Divide both sides by the coefficient of the x^2 term.
$$x^2 + \frac{b}{a}x = -\frac{c}{a}$$

STEP 3   Add the square of one-half the x coefficient to both sides.
$$x^2 + \frac{b}{a}x + \frac{b^2}{4a^2} = -\frac{c}{a} + \frac{b^2}{4a^2}$$

STEP 4   Factor the left side as a perfect-square binomial. Then apply the square root property.
$$\left(x + \frac{b}{2a}\right)^2 = \frac{-4ac + b^2}{4a^2}$$
$$x + \frac{b}{2a} = \pm\sqrt{\frac{b^2 - 4ac}{4a^2}}$$

STEP 5   Solve the resulting linear equations.
$$x = -\frac{b}{2a} \pm \frac{\sqrt{b^2 - 4ac}}{2a}$$

STEP 6   Simplify.
$$= \frac{-b \pm \sqrt{b^2 - 4ac}}{2a}$$

We now use the result derived above to state the *quadratic formula,* a formula that allows us to find the solutions for any quadratic equation.

THE QUADRATIC FORMULA

Given any quadratic equation in the form

$ax^2 + bx + c = 0$ where $a \neq 0$

the two solutions to the equation are

$$x = \frac{-b \pm \sqrt{b^2 - 4ac}}{2a}$$

EXAMPLE 1 Using the Quadratic Formula

Solve, using the quadratic formula:

$6x^2 - 7x - 3 = 0$

Note that the equation is in standard form.

Solution First, we determine the values for a, b, and c. Here

$a = 6$ $b = -7$ $c = -3$

Substituting those values into the quadratic formula, we have

$$x = \frac{-(-7) \pm \sqrt{(-7)^2 - 4(6)(-3)}}{2(6)}$$

Since $b^2 - 4ac = 121$, a perfect square, the two solutions in this case are rational numbers.

Now simplifying inside the radical gives

$$x = \frac{7 \pm \sqrt{121}}{12}$$

$$= \frac{7 \pm 11}{12}$$

This gives us the solutions

$$x = \frac{3}{2} \quad \text{or} \quad x = -\frac{1}{3}$$

 Compare these solutions to the graph of $y = 6x^2 - 7x - 3$

Note that since the solutions for the equation of this example were rational, the original equation could have been solved by our earlier method of factoring.

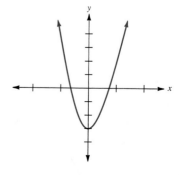

Check Yourself 1

Solve, using the quadratic formula.

$3x^2 + 2x - 8 = 0$

To use the quadratic formula, first the equation must be written in standard form.

EXAMPLE 2 Using the Quadratic Formula

Solve by using the quadratic formula:

The equation *must be in standard form* to determine *a*, *b*, and *c*.

$$9x^2 = 12x - 4$$

Solution First, we must write the equation in standard form:

Subtract 12*x*, and add 4 to both sides. The resulting left member will be *equal to 0*.

$$9x^2 - 12x + 4 = 0$$

Second, we find the values of *a*, *b*, and *c*. Here

The graph of $y = 9x^2 - 12x + 4$ intercepts the *x* axis only at the point (2/3, 0)

$$a = 9 \qquad b = -12 \qquad c = 4$$

Substituting these values into the quadratic formula, we find

$$x = \frac{-(-12) \pm \sqrt{(-12)^2 - 4(9)(4)}}{2(9)}$$

$$= \frac{12 + \sqrt{0}}{18}$$

and simplifying yields

The quadratic equation has but one real solution. This is sometimes called a *repeated* or *double solution*.

$$x = \frac{2}{3}$$

Check Yourself 2

Using the quadratic formula, solve the equation.

$$4x^2 - 4x = -1$$

Thus far our examples and exercises have led to rational solutions. That is not always the case, as the next example illustrates.

EXAMPLE 3 Using the Quadratic Formula

Using the quadratic formula, solve

$$x^2 - 3x = 5$$

Solution Once again, to use the quadratic formula, we write the equation in standard form:

$$x^2 - 3x - 5 = 0$$

We now determine values for a, b, and c and substitute.

$$x = \frac{-(-3) \pm \sqrt{(-3)^2 - 4(1)(-5)}}{2(1)}$$

Simplifying as before, we have

$$x = \frac{3 \pm \sqrt{29}}{2}$$

Check Yourself 3

Using the quadratic equation, solve $2x^2 = x + 7$.

The following example will require some special care in simplifying the solution.

EXAMPLE 4 Using the Quadratic Formula

Using the quadratic formula, solve

$$3x^2 - 6x + 2 = 0$$

Solution Here we have $a = 3$, $b = -6$, and $c = 2$. Substituting gives

$$x = \frac{-(-6) \pm \sqrt{(-6)^2 - 4(3)(2)}}{2(3)}$$

$$= \frac{6 \pm \sqrt{12}}{6}$$

We now look for the largest perfect-square factor of 12, the radicand.

Simplifying, we note that $\sqrt{12}$ is equal to $\sqrt{4 \cdot 3}$, or $2\sqrt{3}$. We can then write the solutions as

$$x = \frac{6 \pm 2\sqrt{3}}{6} = \frac{2(3 \pm \sqrt{3})}{6} = \frac{3 \pm \sqrt{3}}{3}$$

CAUTION

Be Careful! Students are sometimes tempted to reduce this result to

$$\frac{6 \pm 2\sqrt{3}}{6} \stackrel{?}{=} 1 \pm 2\sqrt{3}$$

This is *not a valid step.* We must divide *each of the terms* in the numerator by 2 when simplifying the expression.

Check Yourself 4

Solve by using the quadratic formula.

$$x^2 - 4x = 6$$

Let's examine a case in which the solutions are non-real complex numbers.

EXAMPLE 5 Using the Quadratic Formula

Solve by using the quadratic formula:

$$x^2 - 2x = -2$$

Solution Rewriting in standard form, we have

$$x^2 - 2x + 2 = 0$$

Labeling the coefficients, we find that

$$a = 1 \qquad b = -2 \qquad c = 2$$

Applying the quadratic formula, we have

The solutions will be complex any time $b^2 - 4ac$ is negative.

$$x = \frac{2 \pm \sqrt{-4}}{2}$$

  The graph of $y = x^2 - 2x + 2$ does not intercept the x axis, so there are no real solutions.

and noting that $\sqrt{-4}$ is $2i$, we can simplify to

$$x = 1 \pm i$$

Check Yourself 5

Solve by using the quadratic formula.

$$x^2 - 4x + 6 = 0$$

We summarize the discussion thus far: In attempting to solve a quadratic equation, you should first try the factoring method. If this method doesn't work, you can apply the quadratic formula or the square root method to find the solution. The following algorithm outlines the steps.

SOLVING A QUADRATIC EQUATION BY USING THE QUADRATIC FORMULA

STEP 1 Write the equation in standard form (one side is equal to 0).

STEP 2 Determine the values for a, b, and c.

STEP 3 Substitute those values into the quadratic formula.

STEP 4 Simplify.

Given a quadratic equation, the radicand $b^2 - 4ac$ determines the number of real solutions. Because of this, we call the result of substituting a, b, and c into that part of the quadratic formula the *discriminant*.

Although not necessarily distinct or real, every second-degree equation has two solutions.

Because the discriminant is a real number, there are three possibilities (the *trichotomy property*):

  Graphically, we can see the number of real solutions as the number of times the related quadratic function intercepts the x axis.

$$\text{If } b^2 - 4ac \begin{cases} < 0 & \text{there are } \textit{no real solutions, } \text{but two complex} \\ & \text{solutions} \\ = 0 & \text{there is } \textit{one real solution } \text{(a double solution)} \\ > 0 & \text{there are } \textit{two distinct real solutions} \end{cases}$$

EXAMPLE 6 Analyzing the Discriminant

How many real solutions are there for each of the following quadratic equations?

(a) $x^2 + 7x - 15 = 0$

The discriminant $[49 - 4(1)(-15)]$ is 109. This indicates there are two real solutions.

(b) $3x^2 - 5x + 7 = 0$

The discriminant is negative. There are no real solutions.

(c) $9x^2 - 12x + 4 = 0$

The discriminant is 0. There is exactly one real solution (a double solution).

[handwritten annotations:]

$c1$

$b^2 - 4ac$

$49 - 4(6) - 15)$

$49 + 60 = 109$

$A\ 3 \quad b -5 \quad c\ 7$

$25 - 4(3)(7)$

$12\times7 - 84$

25

We could find two complex solutions by using the quadratic formula, or by completing the square.

Check Yourself 6

How many real solutions are there for each of the following quadratic equations?

1. $2x^2 - 3x + 2 = 0$ *one*

2. $3x^2 + x - 11 = 0$ *two*

3. $4x^2 - 4x + 1 = 0$ *none*

4. $x^2 = -5x - 7$ *1*

Frequently (see Examples 3 and 4, for instance) the solutions of a quadratic equation involve square roots. When we are solving algebraic equations, it is generally best to leave solutions in this form. However, if an equation resulting from an application has been solved by the use of the quadratic formula, we will often estimate the root and sometimes accept only positive solutions. Consider the following application that can be solved by using the quadratic formula.

EXAMPLE 7 Solving a Thrown-Ball Application

If a ball is thrown upward from the ground, the equation to find the height h of such a ball thrown with an initial velocity of 80 ft/s is

$$h = 80t - 16t^2$$

Find the time it takes the ball to reach a height of 48 ft.

Solution First we substitute 48 for h, and then we rewrite the equation in standard form.

$$16t^2 - 80t + 48 = 0$$

To simplify the computation, we divide both sides of the equation by the common factor 16. This yields

$$t^2 - 5t + 3 = 0$$

We solve for t as before, using the quadratic equation, with the result

$$t = \frac{5 \pm \sqrt{13}}{2}$$

Here h measures the height above the ground, in feet, t seconds (s) after the ball is thrown upward.

Note that the result of dividing by 16

$$\frac{0}{16} = 0$$

is 0 on the right.

There are two solutions because the ball reaches the height *twice*, once on the way up and once on the way down.

This gives us two solutions, $\dfrac{5 + \sqrt{13}}{2}$ and $\dfrac{5 - \sqrt{13}}{2}$. But because we have specified units of time, we generally estimate the answer to the nearest tenth or hundredth of a second.

In this case, estimating to the nearest tenth of a second gives solutions of 0.7 and 4.3 s.

Check Yourself 7

The equation to find the height h of a ball thrown with an initial velocity of 64 ft/s is

$$h = 64t - 16t^2$$

Find the time it takes the ball to reach a height of 32 ft.

EXAMPLE 8 Solving a Thrown-Ball Application

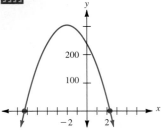

The graph of $h = 240 - 64t - 16t^2$ shows the height, h, at any time t.

The height h of a ball thrown downward from the top of a 240-ft building with an initial velocity of 64 ft/s is given by

$$h = 240 - 64t - 16t^2$$

At what time will the ball reach a height of 176 ft?

Solution Let $h = 176$, and write the equation in standard form.

$$16t^2 + 64t - 64 = 0$$

or

$$t^2 + 4t - 4 = 0$$

Again we divide both sides of the equation by 16 to simplify the computation.

and the quadratic formula with $a = 1$, $b = 4$, and $c = -4$ yields

$$t = -2 \pm 2\sqrt{2}$$

The ball has a height of 64 ft approximately 0.8 s.

Estimating these solutions, we have $t = -4.8$ and $t = 0.8$ s, but of these two values only the *positive value* makes any sense. (To accept the negative solution would be to say that the ball reached the specified height before it was thrown.)

Check Yourself 8

The height h of a ball thrown upward from the top of a 96-ft building with an initial velocity of 16 ft/s is given by

$$h = 96 + 16t - 16t^2$$

When will the ball have a height of 32 ft? (Estimate your answer to the nearest tenth of a second.)

CHECK YOURSELF ANSWERS

1. -2, $\dfrac{4}{3}$ **2.** $\dfrac{1}{2}$ **3.** $\dfrac{1 \pm \sqrt{57}}{4}$ **4.** $2 \pm \sqrt{10}$ **5.** $2 \pm i\sqrt{2}$

6. (1) None; (2) two; (3) one; (4) none **7.** 0.6 and 3.4 s **8.** 2.6 s

Factoring the easyest way.

8.3 EXERCISES

Build Your Skills

Solve each of the following quadratic equations by first factoring and then using the quadratic formula.

1. $x^2 - 5x - 14 = 0$

2. $x^2 + 7x - 18 = 0$

3. $t^2 + 8t - 65 = 0$

4. $q^2 + 3q - 130 = 0$

5. $5x^2 + 4x - 1 = 0$

6. $3x^2 + 2x - 1 = 0$

7. $16t^2 - 24t + 9 = 0$

8. $6m^2 - 23m + 10 = 0$

Solve each of the following quadratic equations by (a) completing the square and (b) using the quadratic formula.

9. $x^2 - 2x - 5 = 0$ *whort ever square*

10. $x^2 + 6x - 1 = 0$

11. $x^2 + 3x - 27 = 0$ *Faberce*

12. $t^2 + 4t - 7 = 0$

13. $2x^2 - 6x - 3 = 0$ *doesn't factor*

14. $2x^2 - 6x + 1 = 0$

15. $2q^2 - 4q + 1 = 0$ — *don't complet the squar*

16. $4r^2 - 2r + 1 = 0$

17. $3x^2 - x - 2 = 0$ —

18. $2x^2 - 8x + 3 = 0$

19. $2y^2 - y - 5 = 0$ ~

20. $3m^2 + 2m - 1 = 0$

Solve each of the following equations by using the quadratic formula.

True to factor

21. $x^2 - 4x + 3 = 0$

22. $x^2 - 7x + 3 = 0$

23. $p^2 - 8p + 16 = 0$ *-square*

24. $u^2 + 7u - 30 = 0$

25. $2x^2 - 2x - 3 = 0$ *from ku*

26. $2x^2 - 3x - 7 = 0$

27. $-3s^2 + 2s - 1 = 0$ *sumul un.*

28. $5t^2 - 2t - 2 = 0$

Hint: Clear each of following equations of fractions or grouping symbols first.

29. $2x^2 - \dfrac{1}{2}x - 5 = 0$

30. $3x^2 + \dfrac{1}{3}x - 3 = 0$

31. $5t^2 - 2t - \dfrac{2}{3} = 0$

32. $3y^2 + 2y + \dfrac{3}{4} = 0$

33. $(x - 2)(x + 3) = 4$

34. $(x + 1)(x - 8) = 3$

35. $(t + 1)(2t - 4) - 7 = 0$

36. $(2w + 1)(3w - 2) = 1$

37. $3x - 5 = \dfrac{1}{x}$

38. $x + 3 = \dfrac{1}{x}$

39. $2t - \dfrac{3}{t} = 3$

40. $4p - \dfrac{1}{p} = 6$

Factor first

9 - 27/81

13 - 49

41. $\dfrac{5}{y^2} + \dfrac{2}{y} - 1 = 0$

42. $\dfrac{6}{x^2} - \dfrac{2}{x} = 1$

For each quadratic equation, find the value of the discriminant and give the number of real solutions.

43. $2x^2 - 5x = 0$ $2x^2 - 5x = 0$
                   $a\ 2$

44. $3x^2 + 8x = 0$ $b\ -5$

45. $m^2 - 8m + 16 = 0$ $c\ -0$ $b^2 - 4ac$ discrimb Folulmer

46. $4p^2 + 12p + 9 = 0$

47. $3x^2 - 7x + 1 = 0$ $25 - 4(2)(0)$

48. $2x^2 - x + 5 = 0$ $25 - 0$

49. $2w^2 - 5w + 11 = 0$ $2\ solution.$

50. $7q^2 - 3q + 1 = 0$

Find all the solutions of each of the following quadratic equations. Use any applicable method.

51. $x^2 - 8x + 16 = 0$

52. $4x^2 + 12x + 9 = 0$

53. $3t^2 - 7t + 1 = 0$

54. $2z^2 - z + 5 = 0$

55. $5y^2 - 2y = 0$

56. $7z^2 - 6z - 2 = 0$

57. $(x - 1)(2x + 7) = -6$

58. $4x^2 - 3 = 0$

59. $x^2 + 9 = 0$

60. $(4x - 5)(x + 2) = 1$

61. $x - 3 - \dfrac{10}{x} = 0$

62. $1 + \dfrac{2}{x} + \dfrac{2}{x^2} = 0$

The equation

$$h(t) = 112t - 16t^2$$

is the equation for the height of an arrow, shot upward from the ground with an initial velocity of 112 ft/s, where t is the time, in seconds, after the arrow leaves the ground. Your answers to Exercises 63 and 64 should be to the nearest tenth of a second.

63. Find the time it takes for the arrow to reach a height of 112 ft.

64. Find the time it takes for the arrow to reach a height of 144 ft.

The equation

$$h(t) = 320 - 32t - 16t^2$$

is the equation for the height of a ball, thrown downward from the top of a 320-ft building with an initial velocity of 32 ft/s, where t is the time, to the nearest tenth of a second, after the ball is thrown down from the top of the building.

65. Find the time it takes for the ball to reach a height of 240 ft.

66. Find the time it takes for the ball to reach a height of 96 ft.

Transcribe Your Skills

67. Can the solution of a quadratic equation with integer coefficients include one real and one imaginary number? Justify your answer.

68. Explain how the discriminant is used to predict the nature of the solutions of a quadratic equation.

Think About These

Solve each of the following equations for x.

69. $x^2 + y^2 = z^2$

70. $2x^2y^2z^2 = 1$

71. $x^2 - 36a^2 = 0$

72. $ax^2 - 9b^2 = 0$

73. $2x^2 + 5ax - 3a^2 = 0$

74. $3x^2 - 16bx + 5b^2 = 0$

75. $2x^2 + ax - 2a^2 = 0$

76. $3x^2 - 2bx - 2b^2 = 0$

77. Given that the polynomial $x^3 - 3x^2 - 15x + 25 = 0$ has as one of its solutions $x = 5$, find the other two solutions. (*Hint:* If you divide the given polynomial by $x - 5$, the quotient will be a quadratic equation. The remaining two solutions will be the solutions for *that* equation.)

78. Given that $2x^3 + 2x^2 - 5x - 2 = 0$ has as one of its solutions $x = -2$, find the other two solutions. (*Hint:* In this case, divide the original polynomial by $x + 2$.)

79. Find all the zeros of the function $f(x) = x^3 + 1$.

80. Find the zeros of the function $f(x) = x^2 + x + 1$.

81. Find all *six* solutions to the equation $x^6 - 1 = 0$. (*Hint:* Factor the left-hand side of the equation first as the difference of squares, then as the sum and difference of cubes.)

82. Find all six solutions to $x^6 = 64$.

The number of plant and animal species in a forested wetland shows a quadratic relationship with the amount of
water, in centimeters, in the wetland.

For a specific wetland,

$$y = -0.05x^2 + 7x - 215$$

expresses this relationship, where y is the number of species
and x is the amount of water, in centimeters.

 83. Find the amount of water necessary to support 20 species of plants and animals. (Express your answer to the
nearest tenth of a centimeter.)

84. Find the amount of water necessary to support 15 species of plants and animals. (Express your answer to the
nearest tenth of a centimeter.)

 Graphing Utility

Use your graphing utility to find the graph for each of the
following. Approximate the x intercepts for each graph.
(You may have to adjust the viewing window to see both

intercepts.) Compare your answers to those for Exercises 9
to 12.

85. $y = x^2 - 2x - 5$

86. $y = x^2 + 6x - 1$

87. $y = x^2 + 3x - 27$

88. $y = x^2 + 4x - 7$

Skillscan (Section 2.4)

Graph each of the following inequalities.

a. $x \geq 2$ **b.** $x > 3$ **c.** $x < -4$ **d.** $x \leq -5$

e. $1 < x < 5$ **f.** $-2 \leq x \leq 3$

g. $x < -3$ or $x > 2$

h. $x \leq -4$ or $x \geq 5$

Quadratic Inequalities   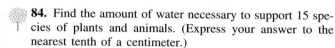 **8.4**

OBJECTIVE: To solve quadratic inequalities and graph the solution sets

TAPE IN18

In Section 2.4, we discussed solutions for linear inequalities. Here we expand upon
the ideas of that section to solve quadratic inequalities.

> A *quadratic inequality* is an inequality that can be written in the form
> $$ax^2 + bx + c < 0 \qquad \text{where } a \neq 0$$
> **Note** The inequality symbol $<$ can be replaced by the symbol $>$, $\leq$, or $\geq$ in
> the above definition.

Before we discuss the solution of quadratic inequalities, we want to take time
to review some properties of multiplication over the real numbers. Recall that

1. The product of two positive numbers is always positive.

2. The product of two negative numbers is always positive.

3. The product of a positive number and a negative number is always negative.

EXAMPLE 1 Evaluating a Product over an Interval

For each interval, state whether the product is always positive or always negative.

Try substituting a number in the interval, like 0, to see what the sign of the expression is.

(a) $(t + 1)(t + 5)$ where $-1 < t < 3$

Since both factors are positive over the interval, the product is *positive* over that interval.

(b) $(x - 3)(x + 2)$ where $-2 < x < 3$

Notice that, over the stated interval, the factor $x - 3$ is always negative and the factor $x + 2$ is always positive. As a result, the product is *negative* over the entire interval.

Check Yourself 1

State whether each of the following products is negative or positive over the given interval.

1. $(t - 3)(t - 9)$ where $-1 < t < 3$

2. $(x + 1)(x - 2)$ where $-1 < x < 2$

Products such as the ones discussed in Example 1 will occur frequently in the process of solving quadratic inequalities.

When solving quadratic inequalities, we can use sign graphs to determine the solution. The following example illustrates.

If we expand the binomial, we get
$x^2 - 2x - 3 < 0$
Look at the graph of
$y = x^2 - 2x - 3$

EXAMPLE 2 Solving a Quadratic Inequality

Solve $(x - 3)(x + 1) < 0$.

Solution We start by finding the solutions of the corresponding quadratic equation. So

$$(x - 3)(x + 1) = 0$$

has solutions 3 and -1. These are the points where one of the factors has the value 0, and they are called the *critical points*.

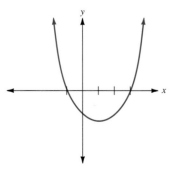

Where is y less than 0 on the graph?

Our solution process depends on determining where each of the factors is positive or negative. To help visualize that process, we start with a number line and label it as shown below. We begin with our first critical point of -1.

$x + 1$ is negative if x is less than -1.

$x + 1$ is positive if x is greater than -1.

Sign of $x + 1$ $\quad - - - -|+ + + + + + + + + +$

-1

We now continue in the same manner with the second critical point, 3.

Sign of $x - 3$ $- - - - - - - - - - - | + + + +$

3

$x - 3$ is negative if x is less than 3.

$x - 3$ is positive if x is greater than 3.

In practice, we combine the two steps above for the following result.

Sign of $x + 1$ $- - - - | + + + + + + + | + + + +$
Sign of $x - 3$ $- - - - | - - - - - - - | + + + +$

-1 3

Sign of product $+ + + + - - - - - - - + + + +$

Examining the signs of the factors, we see that in this case

For any x less than -1, the product is positive.

For any x between -1 and 3, the product is negative.

For any x greater than 3, the product is again positive.

We return to the original inequality:

$(x - 3)(x + 1) < 0$

We can see that this is true only between -1 and 3. In set notation, the solution can be written as

$\{x | -1 < x < 3\}$

On a number line the graph of the solution is

-1 3

Both factors are negative.

The factors have opposite signs.

Both factors are positive.

The product of the two binomials must be negative.

Check Yourself 2

Solve and graph the solution set.

$(x - 2)(x + 4) < 0$

 Examine the graph of $y = x^2 - 5x + 4$. For what values of x is y (the graph) greater than zero?

To use this method, the quadratic member of the inequality must be factored.

EXAMPLE 3 Solving a Quadratic Inequality

Solve $x^2 - 5x + 4 > 0$.

Solution Factoring the quadratic member, we have

$(x - 1)(x - 4) > 0$

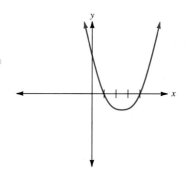

The critical points are 1 and 4, and we form the sign graph as before.

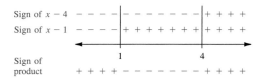

In this case we want those values of x for which the product is *positive,* and we can see from the sign graph above that the solution is

$$\{x \mid x < 1 \text{ or } x > 4\}$$

The graph of the solution set is shown below.

CAUTION

Be Careful! This method works *only* when one side of the inequality is factorable and the other is 0. It is sometimes necessary to rewrite the inequality in an equivalent form in order to attain that form. The following example illustrates.

Check Yourself 3

Solve and graph the solution set.

$$2x^2 - x - 3 > 0$$

Thus far we have considered only inequalities that have been related to 0.

EXAMPLE 4 Solving a Quadratic Inequality

Solve $(x + 1)(x - 4) \geq 6$.

Solution First, we multiply to clear the parentheses.

$$x^2 - 3x - 4 \geq 6$$

Now we subtract 6 from both sides so that the inequality is *related to 0:*

$$x^2 - 3x - 10 \geq 0$$

Factoring the quadratic member, we have

$$(x - 5)(x + 2) \geq 0$$

We can now proceed with the sign graph method as before.

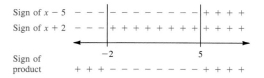

From the graph we see that the solution is

$\{x | x \leq -2 \text{ or } x \geq 5\}$

The graph is shown below.

-2   5

Both factors are negative if x is less than or equal to -2.

Both factors are positive if x is greater than or equal to 5.

Check Yourself 4

Solve and graph the solution set.

$(x - 5)(x + 7) \leq -11$

CHECK YOURSELF ANSWERS

1. (1) Positive; (2) negative

2. ←———○———————○———→ $\{x | -4 < x < 2\}$
        -4   2

3. ←———○—————————○———→ $\left\{x | x < -1 \text{ or } x > \dfrac{3}{2}\right\}$
        -1 $\frac{3}{2}$

4. ←———●—————————●———→ $\{x | -6 \leq x \leq 4\}$
        -6   4

Build Your Skills

Solve each of the following inequalities, and graph the solution set.

1. $(x - 3)(x + 4) < 0$

2. $(x - 2)(x + 5) > 0$

3. $(x - 3)(x + 4) > 0$

4. $(x - 2)(x + 5) < 0$

5. $(x - 3)(x + 4) \leq 0$

6. $(x - 2)(x + 5) \geq 0$

7. $(x - 3)(x + 4) \geq 0$

8. $(x - 2)(x + 5) \leq 0$

Solve each of the following inequalities, and graph the solution set.

9. $x^2 - 3x - 4 > 0$

10. $x^2 - 2x - 8 < 0$

11. $x^2 + x - 12 \leq 0$

12. $x^2 - 2x - 15 \geq 0$

13. $x^2 - 5x + 6 \geq 0$

14. $x^2 + 7x + 10 \leq 0$

15. $x^2 + 2x \leq 24$

16. $x^2 - 3x > 18$

17. $x^2 > 27 - 6x$

18. $x^2 \leq 7x - 12$

19. $2x^2 + x - 6 \leq 0$

20. $3x^2 - 10x - 8 < 0$

21. $4x^2 + x < 3$

22. $5x^2 - 13x \geq 6$

23. $x^2 - 16 \leq 0$

24. $x^2 - 9 > 0$

25. $x^2 \geq 25$

26. $x^2 < 49$

27. $4 - x^2 < 0$

28. $36 - x^2 \geq 0$

29. $x^2 - 4x \leq 0$

30. $x^2 + 5x > 0$

31. $x^2 \geq 6x$

32. $x^2 < 3x$

33. $4x > x^2$

34. $6x \leq x^2$

35. $x^2 - 4x + 4 \leq 0$

36. $x^2 + 6x + 9 \geq 0$

37. $(x + 3)(x - 6) \leq 10$

38. $(x + 4)(x - 5) > 22$

Transcribe Your Skills

39. Can a quadratic inequality be solved if the quadratic member of the inequality is not factorable? If so, explain how the solution can be found. If not, explain why not.

40. Is it necessary to relate a quadratic inequality to 0 in order to solve it? Why or why not?

Think About These

An inequality of the form

$$(x - a)(x - b)(x - c) < 0$$

can be solved by using a sign graph to consider the signs of *all three factors*. Use this suggestion to solve each of the following inequalities. Then graph the solution set.

41. $x(x - 2)(x + 1) < 0$

42. $x(x + 3)(x - 2) \geq 0$

43. $(x - 3)(x + 2)(x - 1) \geq 0$

44. $(x - 5)(x + 1)(x - 4) < 0$

45. $x^3 - 2x^2 - 15x \leq 0$

46. $x^3 + 2x^2 - 24x > 0$

Solve each of the following exercises.

47. A small manufacturer's weekly profit is given by

$$P(x) = -2x^2 + 220x$$

where x is the number of items manufactured and sold. Find the number of items that must be manufactured and sold if the profit is to be greater than or equal to $6000.

48. Suppose that a company's profit is given by

$$P(x) = -2x^2 + 360x$$

How many items must be produced and sold so that the profit will be at least $16,000?

49. If a ball is thrown vertically upward from the ground with an initial velocity of 80 ft/s, its approximate height is given by

$$h(t) = -16t^2 + 80t$$

where t is the time (in seconds) after the ball was released. When will the ball have a height of at least 96 ft?

50. Suppose a ball's height (in meters) is given by

$$h(t) = -5t^2 + 20t$$

When will the ball have a height of at least 15 m?

Graphing Utility

Use your graphing utility to find the graph for each of the following equations. Use that graph to estimate the solution of the related inequality.

51. $y = x^2 + 6x;\ x^2 + 6x < 0$

52. $y = x^2 - 49;\ x^2 - 49 \geq 0$

53. $y = x^2 - 5x - 6;\ x^2 - 5x - 6 \leq 0$

54. $y = 2x^2 + 7x - 15;\ 2x^2 + 7x - 15 > 0$

Skillscan (Section 2.3)

Solve each of the following applications. Be sure to show the equation used for the solution.

a. One number is 4 more than another. If the sum of the two numbers is 24, find the two numbers.   **b.** One number is 2 less than 3 times another. If the sum of the two numbers is 26, find the two numbers.   **c.** The sum of two consecutive odd integers is 64. Find the two integers.   **d.** If the sum of three consecutive even integers is 72, what are the three integers?   **e.** The length of a rectangle is 5 cm more than its width. If the perimeter of the rectangle is 46 cm, find the dimensions of the rectangle.   **f.** The length of a rectangle is 3 in more than twice its width. If the perimeter of the rectangle is 42 in, find the dimensions of the rectangle.

Applications and Problem Solving   8.5

OBJECTIVE: To use quadratic equations in the solution of applications

TAPE IN18

With the techniques introduced in this chapter for solving quadratic equations by factoring, we can now examine a new group of applications that lead to quadratic equations. Whether or not a problem leads to a quadratic equation, the key to problem solving lies in a step-by-step, organized approach to the process. You might want to take time now to review the five-step process introduced in Section 2.3 since all the examples of this section are built on that same model.

Step 5, the process of verifying or checking your solution, is particularly important when an application leads to a quadratic equation. By verifying solutions you may find that both, only one, or none of the derived solutions satisfy the physical conditions stated in the original problem.

> Of course, we can expect more than one solution to check where models lead to quadratic equations. Remember that quadratic equations may have two real number solutions.

As an example, consider the following: The product of a whole number and 1 more than twice that number is 55. What is the number?

As you will see, this problem leads to the equation

$$2x^2 + x - 55 = 0 \qquad \text{or} \qquad (2x + 11)(x - 5) = 0$$

> This is a result of the equation $x(2x + 1) = 55$.

The equation has solutions of $-\dfrac{11}{2}$ and 5.

However, since the problem asks for a *whole* number, we must reject the solution $-\dfrac{11}{2}$ because it does not meet the original conditions of the problem. Similarly, we reject other values that are not meaningful, such as a negative solution in a problem asking for a length or other dimension.

EXAMPLE 1 A Number Application

One integer is 3 less than twice another. If their product is 35, find the two integers.

Step 1 The unknowns are the two integers.

Step 2 Let x represent the first integer. Then

represents the second.

Step 3 Form an equation:

$$x(2x - 3) = 35$$

Product of
the two integers

Step 4 Remove the parentheses and solve as before.

$$2x^2 - 3x = 35$$
$$2x^2 - 3x - 35 = 0$$

Factor on the left:

$(2x + 7)(x - 5) = 0$

$2x + 7 = 0$ or $x - 5 = 0$

$2x = -7$ $\qquad\qquad x = 5$

This represents the solutions to the equation but not the answer to the original problem.

$x = -\dfrac{7}{2}$

Step 5 Solving in step 4, we have the two solutions $-\dfrac{7}{2}$ and 5. Since the original problem asks for *integers,* we must consider only 5 as a solution.

The desired integers, x and $2x - 3$, are then 5 and 7.

To verify, their product is 35.

Check Yourself 1

One integer is 2 more than 3 times another. If their product is 56, what are the two integers?

Problems involving consecutive integers may also lead to quadratic equations. Recall that consecutive integers can be represented by x, $x + 1$, $x + 2$, and so on. Consecutive even (or odd) integers are represented by x, $x + 2$, $x + 4$, and so on.

EXAMPLE 2 A Number Application

The sum of the squares of two consecutive integers is 85. What are the two integers?

Step 1 The unknowns are the two consecutive integers.

Step 2 Let x be the first integer and $x + 1$ the second integer.

Step 3 Form the equation.

$$\underbrace{x^2 + (x + 1)^2}_{\substack{\text{Sum of}\\\text{squares}}} = 85$$

Step 4 Solve.

$x^2 + (x + 1)^2 = 85$

$x^2 + x^2 + 2x + 1 = 85$ $\qquad$ Remove parentheses.

$2x^2 + 2x - 84 = 0$ $\qquad$ Note common factor 2.

$x^2 + x - 42 = 0$ $\qquad$ Divide both members of equation by 2.

$(x + 7)(x - 6) = 0$

$x + 7 = 0$ or $x - 6 = 0$

$x = -7$ $\qquad\qquad x = 6$

Step 5 The solution in step 4 leads to two possibilities, -7 or 6. Since *both numbers are integers,* both meet the conditions of the original problem. The consecutive integers are

-7 (x) or   6 (x)

-6 $(x + 1)$ 7 $(x + 1)$

The two consecutive integers we are looking for are

$\quad -7 \quad$ and $\quad -6$

or $\quad 6 \quad$ and $\quad 7$

You should verify that in both cases the sum of the squares is 85.

Check Yourself 2

The sum of the squares of two consecutive even numbers is 100. Find the two integers.

Let's proceed now to applications involving geometry.

EXAMPLE 3 A Geometric Application

The length of a rectangle is 3 cm greater than its width. If the area of the rectangle is 108 cm^2, what are the dimensions of the rectangle?

Step 1 You are asked to find the dimensions (the length and the width) of the rectangle.

Step 2 Whenever geometric figures are involved in an application, start by drawing, and *then labeling,* a sketch of the problem. Letting x represent the width and $x + 3$ the length, we have

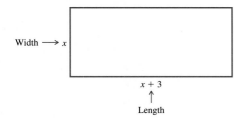

Width $\longrightarrow x$

$x + 3$

↑

Length

Step 3 Once the drawing is correctly labeled, this step should be easy. The area of a rectangle is the product of its length and width, so

$x(x + 3) = 108$

Step 4

$$x(x + 3) = 108$$

$$x^2 + 3x - 108 = 0$$

$$(x + 12)(x - 9) = 0$$

$$x + 12 = 0 \quad \text{or} \quad x - 9 = 0$$

$$x = -12 \qquad\qquad x = 9$$

Multiply and write in standard form.

Factor and solve as before.

Step 5 We reject -12 (cm) as a solution. A length cannot be negative, and so we must consider only 9 (cm) in finding the required dimensions.

The width x is 9 cm, and the length $x + 3$ is 12 cm. Since this gives a rectangle of area 108 cm^2, the solution is verified.

Check Yourself 3

In a triangle, the base is 4 in less than its height. If its area is 30 in^2, find the length of the base and the height of the triangle.

EXAMPLE 4 A Rectangular Box Application

An open box is formed from a rectangular piece of cardboard, whose length is 2 in more than its width, by cutting 2-in squares from each corner and folding up the sides. If the volume of the box is to be 96 in^3, what must be the size of the original piece of cardboard?

Step 1 We are asked for the dimensions of the sheet of cardboard.

Step 2 Again, sketch the problem.

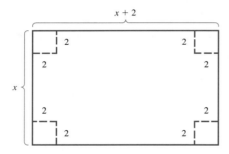

Step 3 To form an equation for volume, we sketch the completed box.

Note The original width of the cardboard was x. Removing two 2-in squares leaves $x - 4$ for the width of the box. Similarly, the length of the box is $x - 2$. Do you see why?

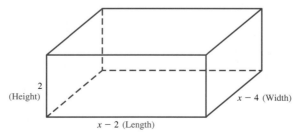

Since volume is the product of height, length, and width,

$$2(x - 2)(x - 4) = 96$$

Step 4

$$2(x - 2)(x - 4) = 96$$ Divide both sides by 2, multiply
on the left, and write in standard
$$(x - 2)(x - 4) = 48$$ form. Then solve as before.

$$x^2 - 6x + 8 = 48$$

$$x^2 - 6x - 40 = 0$$

$$(x - 10)(x + 4) = 0$$

$$x = 10 \text{ in} \qquad \text{or} \qquad x = -4 \text{ in}$$

Step 5 Again, we need consider only the positive solution. The width x of the original piece of cardboard is 10 in, and its length $x + 2$ is 12 in. The dimensions of the completed box will be 6 by 8 by 2 in, which gives the required volume of 96 in^3.

Check Yourself 4

A similar box is to be made by cutting 3-in squares from a piece of cardboard that is 4 in longer than it is wide. If the required volume is 180 in^3, find the dimensions of the original square of cardboard.

Another geometric result that generates quadratic equations in applications is the *Pythagorean theorem*. Recall that the theorem gives an important relationship between the lengths of the sides of a right triangle (a triangle with a 90° angle).

THE PYTHAGOREAN THEOREM

In any right triangle, the square of the longest side (the hypotenuse) is equal to the sum of the squares of the two shorter sides (the legs).

$$c^2 = a^2 + b^2$$

Let's look at an application of the theorem.

EXAMPLE 5 A Triangular Application

One leg of a right triangle is 2 in longer than the other. If the hypotenuse is 4 in longer than the shortest side, find the lengths of the three sides of the triangle.

Step 1 You must find the lengths of the three sides of the triangle.

Step 2 Since the legs can be represented by x and $x + 2$ and the hypotenuse by $x + 4$, a sketch of the problem should look like this:

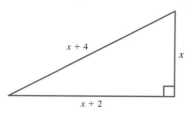

Step 3 Now apply the Pythagorean theorem to write an equation:

$$\underbrace{x^2 + (x + 2)^2}_{\text{Sum of squares of legs}} = \underbrace{(x + 4)^2}_{\text{Square of hypotenuse}}$$

Step 4 Clearing the parentheses, we have

$$x^2 + x^2 + 4x + 4 = x^2 + 8x + 16$$

or

Write in standard form.

$$x^2 - 4x - 12 = 0$$

$$(x - 6)(x + 2) = 0$$

$$x = 6 \quad \text{or} \quad x = -2$$

Step 5 Again, since a length cannot be negative, we reject the solution of -2. The legs then are 6 and 8 in long, and the hypotenuse has length 10 in.

Check Yourself 5

One leg of a right triangle is 7 cm greater than the other. The hypotenuse is 8 cm greater than the length of the shorter leg. Find the lengths of the three sides of the triangle.

In the next example, the solution of the quadratic equation contains radicals. Substituting a pair of solutions such as $\dfrac{3 \pm \sqrt{5}}{2}$ is a very difficult process. The emphasis is on checking the "reasonableness" of the answer.

EXAMPLE 6 A Triangular Application

One leg of a right triangle is 4 cm longer than the other leg. The length of the hypotenuse of the triangle is 12 cm. Find the length of the two legs.

Step 1 We wish to find the length of the two legs of the right triangle.

Step 2 We assign variable x to the shorter leg and $x + 4$ to the other leg.

As in any geometric problem, a sketch of the information will help us visualize.

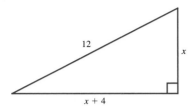

Step 3 Now we apply the Pythagorean theorem to write an equation for the solution.

$$x^2 + (x + 4)^2 = (12)^2$$

The sum of the squares of the legs of the triangle is equal to the square of the hypotenuse.

Step 4 From step 3 we have the equation

$$x^2 + x^2 + 8x + 16 = 144$$

or

$$2x^2 + 8x - 128 = 0$$

Dividing both sides by 2, we have the equivalent equation

$$x^2 + 4x - 64 = 0$$

Dividing both sides of a quadratic equation by a common factor is always a prudent step. It simplifies your work with the quadratic formula.

Using the quadratic formula, we get

$$x = -2 + 2\sqrt{17} \qquad \text{or} \qquad x = -2 - 2\sqrt{17}$$

Step 5 We can reject $-2 - 2\sqrt{17}$ (do you see why?), but we should still check the reasonableness of the value $-2 + 2\sqrt{17}$.

We could substitute $-2 + 2\sqrt{17}$ into the original equation, but it seems more prudent to simply check that it "makes sense" as a solution.

Remembering that $\sqrt{16} = 4$, we will estimate $-2 + 2\sqrt{17}$ as

$\sqrt{17}$ is just slightly more than $\sqrt{16}$, or 4.

$$-2 + 2(4) = 6$$

Our equation of step 3,

$$x^2 + (x + 4)^2 = (12)^2$$

where x equals 6, becomes

$$36 + 100 \approx 144$$

This indicates that our answer is reasonable at least.

Check Yourself 6

One leg of a right triangle is 2 cm longer than the other. The hypotenuse is 1 cm less than twice the length of the shorter leg. Find the length of each side of the triangle.

As we continue looking at examples that demonstrate applications of quadratic equations, we will use the five-step problem-solving process, but we will not always label the steps. However, be sure to think about those steps as you read through the examples.

EXAMPLE 7 A Path (or Frame) Application

Bill has a swimming pool that is 15 ft wide and 25 ft long. He is going to pour a concrete patio of uniform width around the pool.

If he has enough concrete to cover 400 ft^2, how wide will the patio be?

Solution We will first draw a picture of the problem, letting the patio width be represented by x.

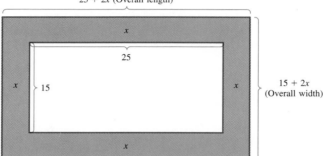

One approach to the problem is to think of the area of the patio as the total area (patio and pool) minus the area of the pool.

From our sketch, we see the total region has width $15 + 2x$ and length $25 + 2x$. Its area is then given by

$(25 + 2x)(15 + 2x)$

while the area of the pool is $25 \cdot 15$, or 375 ft^2.

Let's translate those last two paragraphs to an algebraic sentence:

$(25 + 2x)(15 + 2x) - 375 = 400$

Now we want to solve the above equation. We remove the parentheses, and after simplifying on the left, we have

$4x^2 + 80x = 400$

or, in standard form,

$4x^2 + 80x - 400 = 0$

We divide both sides of the equation by 4, so

$x^2 + 20x - 100 = 0$

and applying the quadratic formula, we have

$x = -10 \pm 10\sqrt{2}$

Because the width must be positive, we need only verify the value $-10 + 10\sqrt{2}$.

Not only would this make for a difficult substitution, but also it is not the kind of answer that Bill is looking for. (Can you imagine calling Bill to tell him that the patio should be exactly $-10 + 10\sqrt{2}$ ft wide?) This leads us to a second reason for estimating our solutions in these application problems—it gives us answers that fit our real-world experience.

In this case, we could say that $\sqrt{2}$ is approximately equal to 1.4. That makes the patio 4 ft wide, an answer that works well when we substitute it into the equation derived above.

$x \approx -10 + 10(1.4)$

$= -10 + 14$

$= 4$

Check Yourself 7

Assume (from Example 7) that Bill has enough concrete for 800 ft². How wide would the patio be, given this amount of concrete?

Note It would *not* be twice as wide.

Let's turn to another field for an application that leads to solving a quadratic equation. Many equations of motion in physics involve quadratic equations.

EXAMPLE 8 A Thrown Ball Application

If a ball is thrown vertically upward from the ground, with an initial velocity of 48 ft/s, its height h, in feet, above the ground after t seconds is given by

$$h(t) = 48t - 16t^2$$

This equation does not take into account any air resistance.

(*a*) How long does it take the ball to return to the ground?

Since $h = 0$ represents a height of 0 (the ground), we simply substitute 0 for h and solve for t:

$0 = 48t - 16t^2$

$\quad = 16t(3 - t)$

So

$16t = 0 \qquad \text{or} \qquad 3 - t = 0$

$\quad t = 0 \qquad\qquad\qquad t = 3$

 Graph the function $y = 48t - 16t^2$. This curve now represents the relationship between the y value (the height of the ball) and the x value (the number of seconds since the ball was thrown).

The solution $t = 0$ represents the starting point (on the ground), so it takes the ball 3 s to return to the ground.

(*b*) How long will it take the ball to reach a height of 32 ft on the way up?

In this case we want the time t to reach a height of 32 ft, so we substitute 32 for h and again solve for t. Letting h be 32, we have

$$32 = 48t - 16t^2$$

Writing the equation in standard form yields

$$16t^2 - 48t + 32 = 0$$

$$t^2 - 3t + 2 = 0$$

$$(t - 1)(t - 2) = 0$$

$$t = 1 \text{ s} \qquad \text{or} \qquad t = 2 \text{ s}$$

Note that we have chosen to write the quadratic member *on the left* so that the coefficient of t^2 (16) is positive. That will simplify the later factoring step. We then divide both sides by 16 and factor as before.

Which solution do we want? The following figure will help.

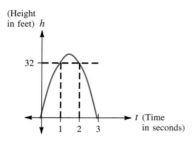

It takes 2 s to reach 32 ft on the way down.

Since we want the time t to reach the height of 32 ft on the way *up*, the desired solution is 1 s.

Check Yourself 8

The height of a ball thrown upward from the ground with a velocity 80 ft/s is given by

$$h(t) = 80t - 16t^2$$

where h represents the height above the ground, in feet.

1. How long will it take the ball to return to the ground?

2. How long will it take the ball to pass through a height of 96 ft on the way back *down* to the ground?

EXAMPLE 9 A Thrown Ball Application

If a ball is thrown vertically upward from the roof of a building 70 m tall with an initial velocity of 15 m/s, its approximate height h, after t seconds, is given by

$$h(t) = 70 + 15t - 5t^2$$

Note that when $t = 0$, the height is 70 m.

(*a*) How long does it take the ball to reach the ground?

Substituting 0 for h, we have a quadratic equation in t:

$$0 = 70 + 15t - 5t^2$$

Dividing both sides of the equation by 5 and writing the quadratic member on the left, we have

$$t^2 - 3t - 14 = 0$$

Using the quadratic formula yields

$$t = \frac{3 \pm \sqrt{65}}{2}$$

Since $\sqrt{65}$ is only slightly more than 8, we can estimate the two solutions for t as

$$t = -\frac{5}{2} \text{ s} \qquad \text{or} \qquad t = \frac{11}{2} \text{ s}$$

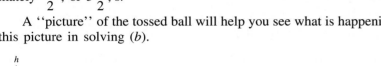

Our values for t are approximately

$$\frac{3 + 8}{2} \qquad \text{or} \qquad \frac{3 - 8}{2}$$

The solution must be positive (why?), so the ball falls to the ground after approximately $\frac{11}{2}$, or $5\frac{1}{2}$, s.

A "picture" of the tossed ball will help you see what is happening. Try to use this picture in solving (b).

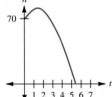

(b) At what two times will the ball reach a height of 75 m?

We substitute 75 for h in the original equation.

$$75 = 70 + 15t - 5t^2$$

or

$$5t^2 - 15t + 5 = 0$$

Dividing by 5, we have the equivalent equation

$$t^2 - 3t + 1 = 0$$

Using the quadratic formula yields

$$t = \frac{3 \pm \sqrt{5}}{2}$$

If we use 2.2 as an estimate for $\sqrt{5}$, we find that

$$t = 0.4 \text{ s} \qquad \text{or} \qquad t = 2.6 \text{ s}$$

Another look at the sketch of the problem above will help us to interpret these results.

Check Yourself 9

A ball is thrown vertically upward from the top of a building 100 m high with an initial velocity of 25 m/s. After t seconds, the height h is given by

$$h(t) = 100 + 25t - 5t^2$$

1. How long does it take the ball to reach the ground?

2. How long does it take the ball to reach a height of 80 m?

An important economic application involves supply and demand. Our next example illustrates that application.

EXAMPLE 10 An Economic Application

The number of intermediate algebra workbooks that a publisher is willing to produce is determined by the supply curve

$$S(p) = -p^2 + 30p - 180 \qquad \text{where } p \text{ is the unit price in dollars}$$

The demand for these workbooks is determined by the equation

$$D(p) = -10p + 130$$

Find the equilibrium price (the price at which supply and demand are equal).

Solution Because supply equals demand ($S = D$), we can write

$$-p^2 + 30p - 180 = -10p + 130$$

Rewriting this statement as a quadratic equation in standard form yields

$$p^2 - 40p + 310 = 0$$

When we apply the quadratic formula, we find the solutions

$$p = 20 \pm 3\sqrt{10}$$

If we approximate $\sqrt{10}$ as 3.2, we have

$$p \approx 10.40 \qquad \text{or} \qquad p \approx 29.60$$

Although you might assume that the publisher will choose the higher price, it will, in fact, choose $10.40. If you want to discover why, try substituting the two solutions into the original demand equation.

Check Yourself 10

The demand equation for floppy disks that accompany a text is predicted to be

$$D = -6p + 30 \qquad \text{where } p \text{ is the unit price in dollars}$$

The supply equation is predicted to be

$$S = -p^2 + 12p - 20$$

Find the equilibrium price.

CHECK YOURSELF ANSWERS

1. 4, 14   **2.** $-8, -6$, or 6, 8   **3.** Base 6 in, height 10 in   **4.** 12 in by 16 in
5. 5 cm, 12 cm, 13 cm   **6.** Approximately 4.3, 6.3, and 7.6 cm   **7.** $-10 +$

$10\sqrt{3}$, approximately 7.3 ft   **8.** (1) 5 s; (2) 3 s   **9.** (1) $\dfrac{5 + \sqrt{105}}{2} \approx 7.6$ s;

(2) $\dfrac{5 + \sqrt{41}}{2} \approx 5.7$ s   **10.** Approximately \$3.43

Build Your Skills

1. One integer is 3 more than twice another. If the product of those integers is 65, find the two integers.

2. One positive integer is 5 less than 3 times another, and their product is 78. What are the two integers?

3. The sum of two integers is 10, and their product is 24. Find the two integers.

4. The sum of two integers is 12. If the product of the two integers is 27, what are the two integers?

5. The product of two consecutive integers is 72. What are the two integers?

6. If the product of two consecutive odd integers is 63, find the two integers.

7. The sum of the squares of two consecutive whole numbers is 61. Find the two whole numbers.

8. If the sum of the squares of two consecutive even integers is 100, what are the two integers?

9. The sum of two integers is 9, and the sum of the squares of those two integers is 41. Find the two integers.

10. The sum of two natural numbers is 12. If the sum of the squares of those numbers is 74, what are the two numbers?

11. The sum of the squares of three consecutive integers is 50. Find the three integers.

12. If the sum of the squares of three consecutive odd positive integers is 83, what are the three integers?

13. Twice the square of a positive integer is 12 more than 5 times that integer. What is the integer?

14. Find an integer such that if 10 is added to the integer's square, the result is 40 more than that integer.

15. The width of a rectangle is 3 ft less than its length. If the area of the rectangle is 70 ft², what are the dimensions of the rectangle?

16. The length of a rectangle is 5 cm more than its width. If the area of the rectangle is 84 cm², find the dimensions of the rectangle.

17. The length of a rectangle is 2 cm more than 3 times its width. If the area of the rectangle is 85 cm², find the dimensions of the rectangle.

18. If the length of a rectangle is 3 ft less than twice its width and the area of the rectangle is 54 ft², what are the dimensions of the rectangle?

19. The length of a rectangle is 1 cm more than its width. If the length of the rectangle is doubled, the area of the rectangle is increased by 30 cm². What were the dimensions of the original rectangle?

20. The height of a triangle is 2 in more than the length of the base. If the base is tripled in length, the area of the new triangle is 48 in² more than the original. Find the height and base of the original triangle.

21. One leg of a right triangle is twice the length of the other. The hypotenuse is 6 m long. Find the length of each leg.

22. One leg of a right triangle is 2 ft longer than the shorter side. If the length of the hypotenuse is 14 ft, how long is each leg?

23. One leg of a right triangle is 1 in shorter than the other leg. The hypotenuse is 3 in longer than the shorter side. Find the length of each side.

24. The hypotenuse of a given right triangle is 5 cm longer than the shorter leg. The length of the shorter leg is 2 cm less than that of the longer leg. Find the length of each side.

25. The sum of the lengths of the two legs of a right triangle is 25 m. The hypotenuse is 22 m long. Find the length of the two legs.

26. The sum of the lengths of one side of a right triangle and the hypotenuse is 15 cm. The other leg is 5 cm shorter than the hypotenuse. Find the length of each side.

27. A box is to be made from a rectangular piece of tin that is twice as long as it is wide. To accomplish this, a 10-cm square is cut from each corner, and the sides are folded up. The volume of the finished box is to be 5000 cm³. Find the dimensions of the original piece of tin.

Hint 1: To solve this equation, you will want to use the following sketch of the piece of tin. Note that the original dimensions are represented by x and $2x$. Do you see why? Also recall that the volume of the resulting box will be the product of the length, width, and height.

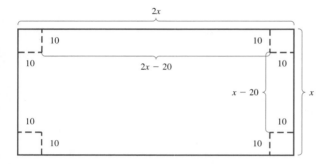

Hint 2: From this sketch, you can see that the equation that results from $V = LWH$ will be

$(2x - 20)(x - 20)10 = 5000$

28. An open box is formed from a square piece of material by cutting 2-in squares from each corner of the material and folding up the sides. If the volume of the box that is formed is to be 72 in³, what was the size of the original piece of material?

29. An open carton is formed from a rectangular piece of cardboard that is 4 ft longer than it is wide, by removing 1-ft squares from each corner and folding up the sides. If the volume of the carton is then 32 ft³, what were the dimensions of the original piece of cardboard?

30. A box that has a volume of 2000 in³ was made from a square piece of tin. The square piece cut from each corner had sides of length 4 in. What were the original dimensions of the square?

31. A square piece of cardboard is to be formed into a box. After 5-cm squares are cut from each corner and the sides are folded up, the resulting box will have a volume of 400 cm³. Find the length of a side of the original piece of cardboard.

32. A rectangular piece of cardboard has a length that is 2 cm longer than twice its width. If 2-cm squares are cut

from each of its corners, it can be folded into a box that has a volume of 250 cm³. What were the original dimensions of the piece of cardboard?

33. Shirley has a swimming pool that is 20 ft wide and 30 ft long. She wishes to pour a concrete patio of uniform width around the pool. If she has enough concrete to cover 400 ft², find the maximum width of the patio. (*Hint:* See Example 7 in this section.)

34. Repeat Exercise 33 to find the width of the patio if Shirley has enough concrete to cover 500 ft².

35. Jason has a greenhouse that is 20 ft long and 12 ft wide. He wishes to pour a concrete sidewalk all the way around the greenhouse. If he has enough concrete to cover 160 ft², what will the width of the sidewalk be?

36. In Exercise 35, if Jason doubles the amount of concrete available, what will the maximum sidewalk width become?

37. A rectangular lawn measures 30 by 35 m. It is surrounded by a sidewalk of uniform width. The total area of the property, including both lawn and sidewalk, is 1200 m². Find the width of the sidewalk.

38. Christine owns a rectangular lot that has a total area of 2000 ft². The lot consists of a garden surrounded by a path of uniform width. If the garden is 60 by 30 ft, find the width of the path.

39. A rectangular mirror measures 40 by 60 cm. It is surrounded by a wooden frame of uniform width. The total area of the frame and mirror is 3000 cm². Find the width of the frame.

40. A rectangular mirror is 25 by 40 in. The total area of the mirror and its frame is 1500 in². Find the width of the frame.

41. Jacob wishes to create a cross path through his garden, as indicated by the sketch below. The dimensions of the garden are 4 by 5 m. Find the width of the path that would take up one-fourth of the garden area. (*Hint:* The area of the path must be $5x + 4x - x^2$. Do you see why?)

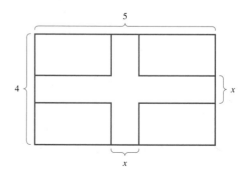

42. In Exercise 41, find the width of the path that would take up one-fifth of the area of Jacob's garden.

43. Having seen the path in Jacob's garden, Gretchen would like to do the same in her garden. If her garden is 7 m long and 5 m wide, what is the width of the path that would take up one-fifth of the garden area?

44. What is the maximum width of the path that would take up no more than one-seventh of the area of Gretchen's garden?

45. One leg of a right triangle is 3 cm longer than the other. The hypotenuse of the triangle is 6 cm longer than the shorter leg. Find the lengths of the three sides of the triangle.

46. The lengths of the shorter leg, the longer leg, and the hypotenuse of a right triangle can be represented by consecutive even integers. Find the lengths of the three sides of the triangle.

47. The length of a rectangle is 7 in more than its width. If the diagonal of the rectangle is 8 in longer than the width, find the dimensions of the rectangle.

48. The length of a rectangle is 7 ft longer than its width, and the diagonal is 9 ft longer than its width. Find the length of the diagonal of the rectangle.

49. The diagonal of a square is 5 cm long. Find the length of one side of the square.

50. Find the length of one side of a square that has a diagonal $7\sqrt{5}$ in long.

51. The diagonal of a square is 3 in longer than one of the sides. Find the length of the sides.

52. The length of the diagonal of a square is 2 cm less than the sum of the lengths of the four sides. Find the length of the sides.

53. If a ball is thrown vertically upward from the ground, its height h after t seconds is given by

$$h(t) = 64t - 16t^2$$

(a) How long does it take the ball to return to the ground? [*Hint:* Let $h(t) = 0$.]

(b) How long does it take the ball to reach a height of 48 ft on the way up?

54. If a ball is thrown vertically upward from the ground, its height h after t seconds is given by

$$h(t) = 96t - 16t^2$$

(a) How long does it take the ball to return to the ground?

(b) How long does it take the ball to pass through a height of 128 ft on the way back down to the ground?

55. Suppose that the cost $C(x)$, in dollars, of producing x chairs is given by

$$C(x) = 2400 - 40x + 2x^2$$

How many chairs can be produced for $5400?

56. Suppose that the profit $T(x)$, in dollars, of producing and selling x appliances is given by

$$T(x) = -3x^2 + 240x - 1800$$

How many appliances must be produced and sold to achieve a profit of $3000?

If a ball is thrown vertically upward from the roof of a building 32 ft high with an initial velocity of 64 ft/s, its approximate height h after t seconds is given by

$$h(t) = 32 + 64t - 16t^2$$

Note The difference between this equation and the one we used in Example 9 has to do with the units used. When we used meters, the t^2 coefficient was -5 (from the fact that the acceleration due to gravity is approximately 10 m/s²). When we use feet as the unit for height, the t^2 coefficient is -16 (that same acceleration becomes 32 ft/s²).

57. How long does it take the ball to fall back to the ground?

58. When will the ball reach a height of 64 ft?

Changing the initial velocity to 96 ft/s will change only the t coefficient (although this is a velocity that would require an arm like Dwight Gooden's). Our new equation becomes

$$h(t) = 32 + 96t - 16t^2$$

59. How long will it take the ball to return to the thrower?

60. When will the ball reach a height of 144 ft?

The only part of the height equation that we have not discussed is the constant. You have probably noticed that the constant is always equal to the initial height of the ball (50 ft in our previous problems). Now, let's have *you* develop an equation.

A ball is thrown upward from the roof of a 100-m building with an initial velocity of 20 m/s.

61. Find the equation for the height of the ball h after t seconds.

62. How long will it take the ball to fall back to the ground?

63. When will the ball reach a height of 75 m?

64. Will the ball ever reach a height of 125 m? (*Hint:* Check the discriminant.)

A ball is thrown upward from the roof of a 100-ft building with an initial velocity of 20 ft/s.

65. Find the height of the ball h after t seconds.

66. How long will it take the ball to fall back to the ground?

67. When will the ball reach a height of 80 ft?

68. Will the ball ever reach a height of 120 ft? Explain.

69. A small manufacturer's weekly profit in dollars is given by

$$P(x) = -3x^2 + 270x$$

Find the number of items x that must be produced to realize a profit of $5100.

70. Suppose the profit in dollars is given by

$$P(x) = -2x^2 + 240x$$

Now how many items must be sold to realize a profit of $5100?

71. The demand equation for a certain computer chip is given by

$$D = -2p + 14$$

The supply equation is predicted to be

$$S = -p^2 + 16p - 2$$

Find the equilibrium price.

72. The demand equation for a certain type of printer is predicted to be

$$D = -200p + 36,000$$

The supply equation is predicted to be

$$S = -p^2 + 400p - 24,000$$

Find the equilibrium price.

Skillscan (Section 1.2)

Simplify each of the following

a. $3\left(\dfrac{2}{3}x + 4\right)$ **b.** $4\left(\dfrac{3x}{4} + \dfrac{1}{2}\right)$ **c.** $2x(x^2 + 3)$

d. $3x\left(\dfrac{x}{3} + 2\right)$ **e.** $-7(x - 3)$ **f.** $-2(-x - 1)$

SUMMARY

Solving Quadratic Equations by Factoring [8.1]

To solve:

$2x^2 - x = 15$

To Solve a Quadratic Equation by Factoring You can use the following procedure.

1. Write the equation in standard form:

$2x^2 - x - 15 = 0$

$$ax^2 + bx + c = 0 \qquad a \neq 0$$

where a, b, and c are constants.

$(2x + 5)(x - 3) = 0$

$2x + 5 = 0 \quad$ or $\quad x - 3 = 0$

$2x = -5 \qquad\qquad x = 3$

$x = -\dfrac{5}{2}$

2. Factor the polynomial on the left as a product of linear factors.

3. Use the zero-product principle to set each factor equal to zero.

4. Solve the resulting linear equations to obtain solutions for the original quadratic equation.

Solving Quadratic Equations by Completing the Square [8.2]

To solve:

$(x - 3)^2 = 5$

$x - 3 = \pm\sqrt{5}$

or $\quad x = 3 \pm \sqrt{5}$

Square Root Property If $x^2 = k$, where k is a complex number, then

$$x = \sqrt{k} \qquad \text{or} \qquad x = -\sqrt{k}$$

Completing the Square

1. Isolate the constant on the right side of the equation.

2. Divide both sides of the equation by the coefficient of the x^2 term if that coefficient is not equal to 1.

3. Add the square of one-half of the coefficient of the linear term to both sides of the equation. This will give a perfect-square trinomial on the left side of the equation.

4. Write the left side of the equation as the square of a binomial, and simplify on the right side.

5. Use the square root property, and then solve the resulting linear equations.

To solve:

$$2x^2 + 2x - 1 = 0$$

$$2x^2 + 2x = 1$$

$$x^2 + x = \frac{1}{2}$$

$$x^2 + x + \left(\frac{1}{2}\right)^2 = \frac{1}{2} + \left(\frac{1}{2}\right)^2$$

$$x^2 + x + \frac{1}{4} = \frac{1}{2} + \frac{1}{4}$$

$$\left(x + \frac{1}{2}\right)^2 = \frac{3}{4}$$

$$x + \frac{1}{2} = \pm\sqrt{\frac{3}{4}}$$

or

$$x + \frac{1}{2} = \pm\frac{\sqrt{3}}{2}$$

$$x = \frac{-1 \pm \sqrt{3}}{2}$$

Solving Quadratic Equations by Using the Quadratic Formula [8.3]

Any quadratic equation can be solved by using the following algorithm.

1. Write the equation in standard form (set it equal to 0).

   $$ax^2 + bx + c = 0$$

2. Determine the values for a, b, and c.

3. Substitute those values into the quadratic formula

   $$x = \frac{-b \pm \sqrt{b^2 - 4ac}}{2a}$$

4. Write the solutions in simplest form.

To solve

$$x^2 - 2x = 4$$

Write the equation as

$$x^2 - 2x - 4 = 0$$

$$a = 1 \quad b = -2 \quad c = -4$$

$$x = \frac{-(-2) \pm \sqrt{(-2)^2 - 4(1)(-4)}}{2 \cdot 1}$$

$$= \frac{2 \pm \sqrt{20}}{2}$$

$$= \frac{2 \pm 2\sqrt{5}}{2}$$

$$= 1 \pm \sqrt{5}$$

The Discriminant The expression $b^2 - 4ac$ is called the *discriminant* for a quadratic equation. There are three possibilities:

1. If $b^2 - 4ac < 0$, there are no real solutions (but two complex solutions).

2. If $b^2 - 4ac = 0$, there is one real solution (a double solution).

3. If $b^2 - 4ac > 0$, there are two distinct real solutions.

Finding the Zeros of a Quadratic Function [8.1–8.3]

Find the zeros of $f(x) = 3x^2 + x - 2$. Set

$$f(x) = 0$$
$$3x^2 + x - 2 = 0$$
$$(3x - 2)(x + 1) = 0$$
$$3x - 2 = 0 \quad \text{or} \quad x + 1 = 0$$
$$x = \frac{2}{3} \quad \text{or} \quad x = -1$$

The zeros of $f(x)$ are $\frac{2}{3}$ and -1.

To find the zeros of a quadratic function, set the function equal to 0 and solve the resulting quadratic equation.

Quadratic Inequalities [8.4]

To solve

$$x^2 - 3x < 18$$
$$x^2 - 3x - 18 < 0$$
$$(x - 6)(x + 3) < 0$$

Critical points are

$$-3 \quad \text{and} \quad 6$$

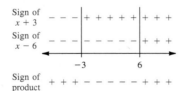

The solution set is

$$\{x \mid -3 < x < 6\}$$

Quadratic inequalities may be solved by first relating the quadratic member to 0. The quadratic member is then factored.

Critical points are determined by noting where the factors are equal to 0, and a sign graph is drawn to indicate where the factors are negative or positive. From this graph and the sense of the original inequality (the quadratic member will be positive or negative), the solution set is determined.

Applications and Problem Solving [8.5]

To solve applications involving quadratic equations, use the five-step process introduced in Chapter 2.

Step 1. Read the problem carefully to determine the unknown quantities.

Step 2. Choose a variable to represent the unknown or unknowns.

Step 3. Translate the problem to the language of algebra, to form an equation.

Step 4. Solve the equation.

Step 5. Verify your solution by returning to the original problem.

The process of verifying the "reasonableness" of your solution is of particular importance when an application leads to a quadratic equation. Both, only one, or possibly none of the derived solutions may satisfy the physical conditions stated in the original problem.

When the use of the quadratic formula or the completing-the-square method leads to irrational solutions, it is generally best to employ approximations or estimations in checking the validity of your results.

SUMMARY EXERCISES

This summary exercise set is provided to give you practice with each of the objectives of the chapter. Each exercise is keyed to the appropriate chapter section. The answers are provided in the instructor's manual that accompanies this text. Your instructor will provide guidelines on how best to use these exercises in your instructional program.

[8.1] Solve each of the following equations, using the zero-product rule.

1. $(x - 4)(x + 5) = 0$

2. $2w(w - 7) = 0$

3. $x^2 - 11x + 30 = 0$

4. $x^2 - 3x = 54$

5. $3y^2 + 5y - 2 = 0$

6. $w^2 - 64 = 0$

7. $4m^2 = 9$

8. $5x^2 = 12x$

[8.1] Solve each of the following quadratic equations by factoring.

9. $x^2 + 5x - 6 = 0$

10. $x^2 - 2x - 8 = 0$

11. $x^2 + 7x = 30$

12. $x^2 - 6x = 40$

13. $x^2 = 11x - 24$

14. $x^2 = 28 - 3x$

15. $x^2 - 10x = 0$

16. $x^2 = 12x$

17. $x^2 - 25 = 0$

18. $x^2 = 144$

19. $2x^2 - x - 3 = 0$

20. $3x^2 - 4x = 15$

21. $3x^2 + 9x - 30 = 0$

22. $4x^2 + 24x = -32$

23. $x(x - 3) = 18$

24. $(x - 2)(2x + 1) = 33$

[8.1] Solve each of the following equations by factoring.

25. $x^3 - 2x^2 - 15x = 0$

26. $x^3 + x^2 - 4x - 4 = 0$

[8.2] Solve each of the following equations, using the square root method.

27. $x^2 - 8 = 0$

28. $3y^2 - 15 = 0$

29. $(x - 2)^2 = 20$

30. $(2x + 1)^2 - 10 = 0$

[8.2] Find the constant that must be added to each of the following binomials to form a perfect-square trinomial.

31. $x^2 - 12x$

32. $y^2 + 3y$

[8.2] Solve the following equations by completing the square.

33. $x^2 - 4x - 5 = 0$

34. $x^2 + 8x + 12 = 0$

35. $w^2 - 10w - 3 = 0$

36. $y^2 + 3y - 1 = 0$

37. $2x^2 - 8x - 5 = 0$

38. $3x^2 + 3x - 1 = 0$

[8.3] Solve each of the following equations by using the quadratic formula.

39. $x^2 - 5x - 24 = 0$

40. $w^2 + 10w + 25 = 0$

41. $x^2 = 3x + 3$

42. $2y^2 - 5y + 2 = 0$

43. $3y^2 + 4y = 1$

44. $2y^2 + 5y + 4 = 0$

45. $(x - 5)(x + 3) = 13$

46. $\dfrac{1}{x^2} - \dfrac{4}{x} + 1 = 0$

47. $3x^2 + 2x + 5 = 0$

48. $(x - 1)(2x + 3) = -5$

[8.3] For each of the following quadratic equations, use the discriminant to determine the number of real solutions.

49. $x^2 - 3x + 3 = 0$

50. $x^2 + 4x = 2$

51. $4x^2 - 12x + 9 = 0$

52. $2x^2 + 3 = 3x$

[8.1–8.3]

Find the zeros of the following functions.

53. $f(x) = x^2 - x + 2$

54. $f(x) = 6x^2 + 7x + 2$

55. $f(x) = -2x^2 - 7x - 6$

56. $f(x) = -x^2 - 1$

[8.4] Solve each of the following inequalities, and graph the solution set.

57. $(x - 2)(x + 5) > 0$

58. $(x - 1)(x - 6) < 0$

59. $(x + 1)(x + 3) \leq 0$

60. $(x + 4)(x - 5) \geq 0$

61. $x^2 - 5x - 24 \leq 0$

62. $x^2 + 4x \geq 21$

63. $x^2 \geq 64$

64. $x^2 + 5x \geq 0$

65. $(x + 2)(x - 6) < 9$

66. $(x - 1)(x + 2) \geq 4$

67. $(x - 4)(x - 1)(x + 3) < 0$

68. $x^3 - 2x - 8x \geq 0$

[8.5] Solve.

69. The sum of two integers is 12, and their product is 32. Find the two integers.

70. The product of two consecutive positive even integers is 80. What are the two integers?

71. Twice the square of a positive integer is 10 more than 8 times that integer. Find the integer.

72. The length of a rectangle is 2 ft more than its width. If the area of the rectangle is 80 ft², what are the dimensions of the rectangle?

73. The length of a rectangle is 3 cm less than twice its width. The area of the rectangle is 35 cm². Find the length and width of the rectangle.

74. An open box is formed by cutting 3-in squares from each corner of a rectangular piece of cardboard which is 3 in longer than it is wide. If the box is to have volume 120 in³, what must be the size of the original piece of cardboard?

75. A rectangular garden, measuring 12 by 16 yd, has its area increased by 128 yd² when a walk of uniform width is added to all sides of the garden. What is the width of the walk?

76. Suppose that a manufacturer's weekly profit P is given by

$$P(x) = -3x^2 + 240x$$

where x is the number of items manufactured and sold. Find the number of items that must be manufactured and sold if the profit is to be at least $4500.

77. If a ball is thrown vertically upward from the ground with an initial velocity of 64 ft/s, its approximate height is given by

$$h(t) = -16t^2 + 64t$$

When will the ball reach a height of at least 48 ft?

78. The length of a rectangle is 1 cm more than twice its width. If the length is doubled, the area of the new rectangle is 36 cm² more than that of the old. Find the dimensions of the original rectangle.

79. One leg of a right triangle is 4 in longer than the other. The hypotenuse of the triangle is 8 in longer than the shorter leg. What are the lengths of the three sides of the triangle?

80. The diagonal of a rectangle is 9 ft longer than the width of the rectangle, and the length is 7 ft more than its width. Find the dimensions of the rectangle.

81. If a ball is thrown vertically upward from the ground, the height h after t seconds is given by

$$h(t) = 128t - 16t^2$$

(a) How long does it take the ball to return to the ground?
(b) How long does it take the ball to reach a height of 240 ft on the way up?

82. Suppose that the cost, in dollars, of producing x stereo systems is given by

$$C(x) = 3000 - 60x + 3x^2$$

How many systems can be produced for $7500?

Solve each of the following applications. Where appropriate, give your answer to the nearest tenth of a unit.

83. One leg of a right triangle is 2 m longer than the other. If the length of the hypotenuse is 8 m, find the length of the other two legs.

84. A square piece of cardboard is to be formed into a box. After squares with sides of length 3 cm are cut

from each corner, the resulting box will have a volume of 150 cm^3. Find the dimensions of the original piece of cardboard.

85. In a rectangular yard 40 ft long and 30 ft wide, Mark wants to form a path of uniform width around a garden area. If Mark wants the remaining garden area to be 1000 ft^2, how wide should he make the path?

86. Suppose that the height (in meters) of a golf ball, hit off a raised tee, is approximated by

$$h(t) = -5t^2 + 10t + 10$$

t seconds after the ball is hit. When will the ball hit the ground?

87. The demand equation for a certain type of computer paper is predicted to be

$$D = -3p + 60$$

The supply equation is predicted to be

$$S = -p^2 + 24p - 3$$

Find the equilibrium price.

The purpose of this self-test is to help you check your progress and to review for a chapter test in class. Allow yourself about an hour to take the test. When you are done, check your answers in the back of the book. If you missed any problems, be sure to go back and review the appropriate sections in the chapter and the exercises that are provided.

Solve each of the following equations, using the zero-product rule.

1. $x^2 - 9x - 36 = 0$

2. $9y^2 = 25$

Solve each of the following equations by factoring.

3. $2x^2 + 7x + 3 = 0$

4. $6x^2 = 10 - 11x$

5. $4x^3 - 9x = 0$

Solve each of the following equations, using the square root method.

6. $4w^2 - 20 = 0$

7. $(x - 1)^2 = 10$

8. $4(x - 1)^2 = 23$

Solve each of the following equations by completing the square.

9. $m^2 + 3m - 1 = 0$

10. $2x^2 - 10x + 3 = 0$

Solve each of the following equations, using the quadratic formula.

11. $x^2 - 5x - 3 = 0$

12. $x^2 + 4x = 7$

13. Find the zeros of the function $f(x) = 3x^2 - 10x - 8$.

Solve each of the following inequalities, and graph the solution set.

14. $x^2 + 5x - 14 < 0$

15. $x^2 - 3x \geq 18$

16. $3x^2 + x - 10 \leq 0$

Solve.

17. The product of two consecutive positive odd integers is 63. Find the two integers.

18. The length of a rectangle is 2 cm more than 3 times its width. If the area of the rectangle is 85 cm², what are the dimensions of the rectangle?

Solve each of the following applications. Approximate your answer to the nearest tenth of a unit.

19. One leg of a right triangle is 1 in longer than the other. If the length of the hypotenuse is 7 in, find the lengths of the other two legs.

20. Suppose that the height (in feet) of a ball thrown upward from a raised platform is approximated by

$$h(t) = -16t^2 + 32t + 32$$

t seconds after the ball has been released. How long will it take the ball to hit the ground?

CUMULATIVE REVIEW EXERCISES

Here is a review of selected topics from the first eight chapters.

Graph each of the following equations.

1. $2x - 3y = 6$

2. $y = -\dfrac{1}{4}x - 2$

3. $y = 4$

Find the slope of the line determined by each set of points.

4. $(-4, 7)$ and $(-3, 4)$

5. $(-2, 3)$ and $(-5, -1)$

Find the distance between each pair of points.

6. $(2, 3)$ and $(-7, 3)$

7. $(2, -5)$ and $(4, 6)$

8. $(-3, 5)$ and $(-6, 9)$

Find the midpoint of the segment defined by each pair of points.

9. $(-4, 3)$ and $(5, 7)$

10. $(-5, -1)$ and $(3, -6)$

Solve each equation.

11. $2x - 7 = 0$

12. $3x - 5 = 5x + 3$

13. $0 = (x - 3)(x + 5)$

14. $x^2 - 3x + 2 = 0$

15. $x^2 + 7x - 30 = 0$

16. $x^2 - 3x - 3 = 0$

17. $(x - 3)^2 = 5$

Solve the following word problems. Show the equation used for the solution.

18. Five times a number decreased by 7 is -72. Find the number.

19. One leg of a right triangle 4 ft longer than the shorter leg. If the hypotenuse is 28 ft, how long is each leg?

20. Suppose that a manufacturer's weekly profit P is given by

$$P(x) = -4x^2 + 320x$$

where x is the number of units manufactured and sold. Find the number of items that must be manufactured and sold to guarantee a profit of at least $4956.

OTHER TYPES OF EQUATIONS, FUNCTIONS, AND INEQUALITIES

FOSSIL FUELS The fossil fuels of oil, coal, and natural gas are the resources most responsible for the world's economic growth since World War II. They have provided the energy to fuel worldwide industrial expansion, driven much of the worldwide expansion in food production, and provided the basis for the increased standard of living in the industrialized nations.

These same fossil fuels have also caused many problems. Their production has contributed to air and water pollution and fouled many coastlines, harming not only the plants and animals but also the local human populations.

Another problem with fossil fuels that has been acknowledged is that they are finite resources. There is a limit to the amount of these resources available for our use. If we continue using these resources at current rates, the oil and natural gas will be exhausted within 50 to 100 years and the coal will be gone in 300 to 500 years.

Fossil fuels are also not evenly distributed around the world. This makes the extraction, production, and sale of oil a very volatile political issue. The United States has only 3 percent of the world's oil resources but uses nearly 30 percent of the oil extracted each year. This imbalance between production and use may not continue in the future.

The fossil fuel resources are vital to the world economy. Knowing these resources will run out someday means we need to begin looking for ways to replace or extend them. Current research shows the most immediate and cost-effective alternative is conservation.

Improving energy conservation measures in the United States will produce three times as much energy by 2020 for less money than finding and developing all new oil and gas deposits believed to exist in the United States. Of the energy consumed in the United States, 43 percent is wasted. Most of this energy comes from fossil fuels. The largest supply of energy in the United States, and most other countries, lies in its energy-wasting buildings, factories, and vehicles. Conservation measures and using fossil fuels only in those areas where other resources cannot be substituted should allow time for the world economy to evolve to a point where fossil fuels play a less vital role.

9.1 Equations and Inequalities Involving Rational Expressions

TAPE IN20

OBJECTIVES:
1. To solve equations containing rational expressions
2. To solve literal equations involving rational expressions for a specified variable
3. To find the zeros of rational functions
4. To solve inequalities involving rational expressions

Applications of your work in algebra will often result in equations involving rational expressions. Our objective in this section is to develop methods to find the solution of such equations.

The usual solution technique is to multiply both sides of the equation by the lowest common denominator (LCD) of all the rational expressions appearing in the equation. The resulting equation will be cleared of fractions, and we can then proceed to solve the equation as before. The following example illustrates the process.

This is justified by the multiplication property of equality.

The resulting equation will be equivalent to the original equation unless a solution results that makes the denominator of the original equation 0. More about this later.

EXAMPLE 1 Clearing Equations of Fractions

Solve

$$\frac{2x}{3} + \frac{x}{5} = 13$$

Solution The LCD for 3 and 5 is 15. Multiplying both sides of the equation by 15, we have

$$15\left(\frac{2x}{3} + \frac{x}{5}\right) = 15 \cdot 13 \qquad \text{Distribute 15 on the left.}$$

$$15 \cdot \frac{2x}{3} + 15 \cdot \frac{x}{5} = 15 \cdot 13$$

Simplify. The equation is now cleared of fractions.

$$10x + 3x = 195$$

$$13x = 195$$

$$x = 15$$

To check, substitute 15 in the original equation.

$$\frac{2 \cdot 15}{3} + \frac{15}{5} \stackrel{?}{=} 13$$

$$10 + 3 \stackrel{?}{=} 13$$

$$13 = 13 \qquad \text{A true statement}$$

So 15 is the solution for the equation.

Be Careful! A common mistake is to confuse an *equation* such as

CAUTION

$$\frac{2x}{3} + \frac{x}{5} = 13$$

and an *expression* such as

$$\frac{2x}{3} + \frac{x}{5}$$

Let's compare.

Equation: $\dfrac{2x}{3} + \dfrac{x}{5} = 13$

Here we want to *solve the equation for x,* as in the example above. We multiply both sides by the LCD to clear fractions and proceed as before.

Expression: $\dfrac{2x}{3} + \dfrac{x}{5}$

Here we want to find *a third fraction* that is equivalent to the given expression. We write each fraction as an equivalent fraction with the LCD as a common denominator.

$$\frac{2x}{3} + \frac{x}{5} = \frac{2x \cdot 5}{3 \cdot 5} + \frac{x \cdot 3}{5 \cdot 3}$$

$$= \frac{10x}{15} + \frac{3x}{15} = \frac{10x + 3x}{15}$$

$$= \frac{13x}{15}$$

Check Yourself 1

Solve $\dfrac{3x}{2} - \dfrac{x}{3} = 7$.

The process is similar when variables are in the denominators. Consider the following example.

EXAMPLE 2 Solving an Equation Involving Rational Expressions

Solve

$$\frac{7}{4x} - \frac{3}{x^2} = \frac{1}{2x^2}$$

We assume that *x cannot* have the value 0. Do you see why?

Solution The LCD of $4x$, x^2, and $2x^2$ is $4x^2$. So multiplying both sides by $4x^2$, we have

$$4x^2\left(\frac{7}{4x} - \frac{3}{x^2}\right) = 4x^2 \cdot \frac{1}{2x^2} \qquad \text{Distribute } 4x^2 \text{ on the left side.}$$

$$4x^2 \cdot \frac{7}{4x} - 4x^2 \cdot \frac{3}{x^2} = 4x^2 \cdot \frac{1}{2x^2} \qquad \text{Simplify.}$$

$$7x - 12 = 2$$

$$7x = 14$$

$$x = 2$$

We will leave the check of the solution, $x = 2$, to you. Be sure to return to the original equation and substitute 2 for x.

Check Yourself 2

Solve $\dfrac{5}{2x} - \dfrac{4}{x^2} = \dfrac{7}{2x^2}$.

The next example illustrates the same solution process when there are binomials in the denominators.

EXAMPLE 3 Solving an Equation Involving Rational Expressions

Solve

Here we assume that x cannot have the value −2 or 3.

$$\frac{4}{x + 2} + 3 = \frac{3x}{x - 3}$$

Solution The LCD is $(x + 2)(x - 3)$. Multiplying by that LCD, we have

Note that multiplying each term by the LCD is the same as multiplying both sides of the equation by the LCD.

$$(x + 2)(x - 3)\left(\frac{4}{x + 2}\right) + (x + 2)(x - 3)(3) = (x + 2)(x - 3)\left(\frac{3x}{x - 3}\right)$$

or, simplifying each term, we have

$$4(x - 3) + 3(x + 2)(x - 3) = 3x(x + 2)$$

We now clear the parentheses and proceed as before.

$$4x - 12 + 3x^2 - 3x - 18 = 3x^2 + 6x$$

$$3x^2 + x - 30 = 3x^2 + 6x$$

$$x - 30 = 6x$$

$$-5x = 30$$

$$x = -6$$

Again we will leave the check of this solution to you.

Check Yourself 3

Solve $\dfrac{5}{x-4} + 2 = \dfrac{2x}{x-3}$.

Factoring plays an important role in solving equations containing rational expressions.

EXAMPLE 4 Solving an Equation Involving Rational Expressions

Solve

$$\frac{3}{x-3} - \frac{7}{x+3} = \frac{2}{x^2 - 9}$$

Solution In factored form, the denominator on the right side is $(x-3)(x+3)$, which forms the LCD, and we multiply each term by that LCD.

$$(x-3)(x+3)\left(\frac{3}{x-3}\right) - (x-3)(x+3)\left(\frac{7}{x+3}\right)$$

$$= (x-3)(x+3)\left[\frac{2}{(x-3)(x+3)}\right]$$

Again, simplifying each term on the right and left sides, we have

$$3(x+3) - 7(x-3) = 2$$

$$3x + 9 - 7x + 21 = 2$$

$$-4x = -28$$

$$x = 7$$

Be sure to check this result by substitution in the original equation.

Check Yourself 4

Solve $\dfrac{4}{x-4} - \dfrac{3}{x+1} = \dfrac{5}{x^2 - 3x - 4}$.

Whenever we multiply both sides of an equation by an expression containing a variable, there is the possibility that a proposed solution may make that multiplier 0. As we pointed out earlier, multiplying by 0 does not give an equivalent equation, and therefore verifying solutions by substitution serves not only as a check of our work but also as a check for extraneous solutions. Consider the following example.

EXAMPLE 5 Solving an Equation Involving Rational Expressions

Solve

Note that we must assume that $x \neq 2$.

$$\frac{x}{x - 2} - 7 = \frac{2}{x - 2}$$

Solution The LCD is $x - 2$, and multiplying, we have

Note that each of the three terms gets multiplied by the $x - 2$.

$$\left(\frac{x}{x - 2}\right)(x - 2) - 7(x - 2) = \left(\frac{2}{x - 2}\right)(x - 2)$$

Simplifying yields

$$x - 7(x - 2) = 2$$

$$x - 7x + 14 = 2$$

$$-6x = -12$$

$$x = 2$$

Now trying to check this result, by substituting 2 for x, we have

$$\frac{2}{2 - 2} - 7 \overset{?}{=} \frac{2}{2 - 2}$$

$$\frac{2}{0} - 7 \overset{?}{=} \frac{2}{0}$$

CAUTION
Because division by 0 is undefined, we conclude that 2 is *not a solution* for the original equation. It is an extraneous solution. The original equation has no solution.

Check Yourself 5

Solve $\dfrac{x - 3}{x - 4} = 4 + \dfrac{1}{x - 4}$.

Equations involving rational expressions may also lead to quadratic equations, as illustrated in this example.

EXAMPLE 6 Solving an Equation Involving Rational Expressions

Solve

Assume $x \neq 3$ and $x \neq 4$.

$$\frac{x}{x - 4} = \frac{15}{x - 3} - \frac{2x}{x^2 - 7x + 12}$$

Solution After factoring the trinomial denominator on the right, the LCD of $x - 4$, $x - 3$, and $x^2 - 7x + 12$ is $(x - 4)(x - 3)$. Multiplying by that LCD, we have

$$(x - 3)(x - 4)\left(\frac{x}{x - 4}\right) = (x - 3)(x - 4)\left(\frac{15}{x - 3}\right)$$

$$- (x - 3)(x - 4)\left[\frac{2x}{(x - 3)(x - 4)}\right]$$

Simplifying yields

$x(x - 3) = 15(x - 4) - 2x$

$x^2 - 3x = 15x - 60 - 2x$ Remove the parentheses.

$x^2 - 16x + 60 = 0$ Write in standard form and factor.

$(x - 6)(x - 10) = 0$

So

$x = 6$ or $x = 10$

Verify that 6 and 10 are both solutions for the original equation.

Check Yourself 6

Solve $\dfrac{3x}{x + 2} - \dfrac{2}{x + 3} = \dfrac{36}{x^2 + 5x + 6}$.

The following algorithm summarizes our work in solving equations containing rational expressions.

> **SOLVING EQUATIONS CONTAINING RATIONAL EXPRESSIONS**
>
> **STEP 1** Clear the equation of fractions by multiplying both sides of the equation by the LCD of all the fractions that appear.
>
> **STEP 2** Solve the equation resulting from step 1.
>
> **STEP 3** Check all solutions by substitution in the original equation.

The methods of this section may also be used to solve certain literal equations for a specified variable. Consider the following.

EXAMPLE 7 Solving a Literal Equation

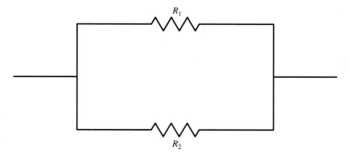

A parallel electric circuit. The symbol for a resistor is ─∿∿─ .

If two resistors with resistances R_1 and R_2 are connected in parallel, the combined resistance R can be found from

$$\frac{1}{R} = \frac{1}{R_1} + \frac{1}{R_2}$$

The numbers 1 and 2 are *subscripts*. We read R_1 as "R sub 1" and R_2 as "R sub 2."

Solve the formula for R.

Solution First, the LCD is RR_1R_2, and we multiply:

$$RR_1R_2 \cdot \frac{1}{R} = RR_1R_2 \cdot \frac{1}{R_1} + RR_1R_2 \cdot \frac{1}{R_2}$$

Simplifying yields

$$R_1R_2 = RR_2 + RR_1$$

Factor out R on the right.　　$R_1R_2 = R(R_2 + R_1)$

Divide by $R_2 + R_1$ to isolate R.　　$\dfrac{R_1R_2}{R_2 + R_1} = R$

or

Symmetric property of equality.　　$R = \dfrac{R_1R_2}{R_1 + R_2}$

Check Yourself 7

Solve for D_1.

$$\frac{1}{F} = \frac{1}{D_1} + \frac{1}{D_2}$$

Note This formula involves the focal length of a convex lens.

The techniques we have just discussed can also be used to find the zeros of rational functions. Remember that a zero of a function is a value of x for which $f(x) = 0$.

EXAMPLE 8 Finding the Zeros of a Function

Find the zeros of

$$f(x) = \frac{1}{x} - \frac{3}{7x} - \frac{8}{21}$$

Solution Set the function equal to 0, and solve the resulting equation for x.

$$f(x) = \frac{1}{x} - \frac{3}{7x} - \frac{4}{21} = 0$$

The LCD for x, $7x$, and 21 is $21x$. So multiplying both sides by $21x$, we have

$$21x\left(\frac{1}{x} - \frac{3}{7x} - \frac{4}{21}\right) = 21x \cdot 0$$

$$21 - 9 - 4x = 0 \qquad \text{Distribute } 21x \text{ on the left side.}$$

$$12 - 4x = 0 \qquad \text{Simplify.}$$

$$12 = 4x$$

$$3 = x$$

So 3 is the value of x for which $f(x) = 0$, that is, 3 is a zero of $f(x)$.

Check Yourself 8

Find the zeros of the function

$$f(x) = \frac{5x + 2}{x - 6} - \frac{11}{4}$$

RATIONAL INEQUALITIES

To solve inequalities involving rational expressions, we need some properties of division over the real numbers. Recall that

1. The quotient of two positive numbers is always positive.

2. The quotient of two negative numbers is always positive.

3. The quotient of a positive number and a negative number is always negative.

As with quadratic inequalities, we solve rational inequalities by using sign graphs. The next example illustrates this technique.

EXAMPLE 9 Solving a Rational Inequality

Solve

$$\frac{x - 3}{x + 2} < 0$$

 The graph of

$$y = \frac{x - 3}{x + 2}$$

is

Solution The inequality states that the quotient of $x - 3$ and $x + 2$ must be negative (less than 0). This means that the numerator and denominator must have opposite signs.

We start by finding the critical points. These are points where either the numerator or denominator is 0. In this case the critical points are 3 and -2.

The solution depends on determining whether the numerator and denominator are positive or negative. To visualize the process, start with a number line and label it as shown below.

Sign of $x - 3$ $- \ - \ -$|$- \ - \ - \ - \ - \ - \ - \ -$|$+ + + + +$

Sign of $x + 2$ $- \ - \ -$|$+ + + + + + + + +$|$+ + + + +$

Sign of quotient $+ + +$|$- \ - \ - \ - \ - \ - \ - \ -$|$+ + + + +$

$(x - 3)(x + 2)$ -2 3

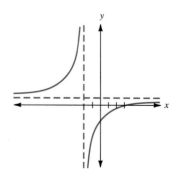

Where is y less than 0?

Examining the sign of the numerator and denominator, we see that

For any x less than -2, the quotient is positive (quotient of two negatives).

For any x between -2 and 3, the quotient is negative (quotient of a negative and a positive).

For any x greater than 3, the quotient is positive (quotient of two positives).

We return to the original inequality

$$\frac{x - 3}{x + 2} < 0$$

This inequality is true only when the quotient is negative, that is, when x is between -2 and 3. This solution can be written as $\{x \mid -2 < x < 3\}$ and represented on a graph as follows.

$$-2 \qquad 3$$

Check Yourself 9

Solve and graph the solution set.

$$\frac{x - 4}{x + 2} > 0$$

As with quadratic inequalities, the solution process illustrated in Example 9 is valid only when the rational expression is isolated on one side of the inequality and is related to 0. If this is not the case, we must write an equivalent inequality as the first step. The following example illustrates.

EXAMPLE 10 Solving a Rational Inequality

Solve

$$\frac{2x - 3}{x + 1} \geq 1$$

Solution Since the rational expression is not related to 0, we use the following procedure.

We have subtracted 1 from both sides.

$$\frac{2x - 3}{x + 1} - 1 \geq 0 \qquad \text{Form a common denominator on the left side.}$$

$$\frac{2x - 3}{x + 1} - \frac{x + 1}{x + 1} \geq 0 \qquad \text{Combine the expressions on the left side.}$$

$$\frac{2x - 3 - (x + 1)}{x + 1} \geq 0 \qquad \text{Simplify.}$$

$$\frac{x - 4}{x + 1} \geq 0$$

We can now proceed as before since the rational expression is related to 0. The critical points are 4 and -1, and the sign graph is formed as shown below.

```
Sign of x − 4        − − −|− − − − − − − −|+ + + +
Sign of x + 1        − − −|+ + + + + + + +|+ + + +

Sign of quotient   + + +|− − − − − − − −|+ + + +
(x − 4)(x + 1)           −1              4
```

From the sign graph the solution is

$\{x \mid x < -1 \text{ or } x \geq 4\}$

The graph is shown below.

We want a *positive* quotient.

Note that 4 is included but −1 cannot be included in the solution set. Why?

Check Yourself 10

Solve and graph the solution set.

$$\frac{2x - 3}{x - 2} \leq 1$$

CHECK YOURSELF ANSWERS

1. 6   **2.** 3   **3.** 9   **4.** −11   **5.** No solution   **6.** $-5, \dfrac{8}{3}$ **7.** $\dfrac{FD_2}{D_2 - F}$

8. $\dfrac{74}{9}$ **9.** $\{x \mid x < -2 \text{ or } x > 4\}$ ←———○———————○———→
                                                              −2   4

10. $\{x \mid 1 \leq x < 2\}$ ←———●————————○———→
                                    1   2

9.1 EXERCISES

Build Your Skills

Decide whether each of the following is an expression or an equation. If it is an equation, find a solution. If it is an expression, rewrite it as a single fraction.

1. $\dfrac{x}{2} - \dfrac{x}{3} = 6$

2. $\dfrac{x}{4} - \dfrac{x}{7} = 3$

3. $\dfrac{x}{2} - \dfrac{x}{5}$

4. $\dfrac{x}{6} - \dfrac{x}{8}$

5. $\dfrac{3x + 1}{4} = x - 1$

6. $\dfrac{3x - 1}{2} - \dfrac{x}{5} - \dfrac{x + 3}{4}$

7. $\dfrac{x}{4} = \dfrac{x}{12} + \dfrac{1}{2}$

8. $\dfrac{2x + 1}{3} + \dfrac{x}{2}$

Solve each of the following equations.

9. $\dfrac{x}{3} + \dfrac{3}{2} = \dfrac{x}{6} + \dfrac{7}{3}$

10. $\dfrac{x}{10} - \dfrac{1}{5} = \dfrac{x}{5} + \dfrac{1}{2}$

11. $\dfrac{4}{x} + \dfrac{3}{4} = \dfrac{10}{x}$

12. $\dfrac{3}{x} = \dfrac{5}{3} - \dfrac{7}{x}$

13. $\dfrac{5}{4x} - \dfrac{1}{2} = \dfrac{1}{2x}$

14. $\dfrac{7}{6x} - \dfrac{1}{3} = \dfrac{1}{2x}$

15. $\dfrac{3}{x + 4} = \dfrac{2}{x + 3}$

16. $\dfrac{5}{x-2} = \dfrac{4}{x-1}$

17. $\dfrac{9}{x} + 2 = \dfrac{2x}{x+3}$

18. $\dfrac{6}{x} + 3 = \dfrac{3x}{x+1}$

19. $\dfrac{3}{x+2} + \dfrac{5}{x} = \dfrac{13}{x+2}$

20. $\dfrac{7}{x} - \dfrac{2}{x-3} = \dfrac{6}{x}$

21. $\dfrac{3}{2} + \dfrac{2}{2x-4} = \dfrac{1}{x-2}$

22. $\dfrac{2}{x-1} + \dfrac{5}{2x-2} = \dfrac{3}{4}$

23. $\dfrac{x}{3x+12} + \dfrac{x-1}{x+4} = \dfrac{5}{3}$

24. $\dfrac{x}{4x-12} - \dfrac{x-4}{x-3} = \dfrac{1}{8}$

25. $\dfrac{x-1}{x+3} - \dfrac{x-3}{x} = \dfrac{3}{x^2+3x}$

26. $\dfrac{x+1}{x-2} - \dfrac{x+3}{x} = \dfrac{6}{x^2-2x}$

27. $\dfrac{1}{x-2} - \dfrac{2}{x+2} = \dfrac{2}{x^2-4}$

28. $\dfrac{1}{x+4} + \dfrac{1}{x-4} = \dfrac{12}{x^2-16}$

29. $\dfrac{7}{x+5} - \dfrac{1}{x-5} = \dfrac{x}{x^2-25}$

30. $\dfrac{2}{x-2} = \dfrac{3}{x+2} + \dfrac{x}{x^2-4}$

31. $\dfrac{11}{x+2} - \dfrac{5}{x^2-x-6} = \dfrac{1}{x-3}$

32. $\dfrac{5}{x-4} = \dfrac{1}{x+2} - \dfrac{2}{x^2-2x-8}$

33. $\dfrac{5}{x-2} - \dfrac{3}{x+3} = \dfrac{24}{x^2+x-6}$

34. $\dfrac{3}{x+1} - \dfrac{5}{x+6} = \dfrac{2}{x^2+7x+6}$

35. $\dfrac{x}{x-3} - 2 = \dfrac{3}{x-3}$

36. $\dfrac{x}{x-5} + 2 = \dfrac{5}{x-5}$

37. $\dfrac{2}{x^2-3x} - \dfrac{1}{x^2+2x} = \dfrac{2}{x^2-x-6}$

38. $\dfrac{2}{x^2-x} - \dfrac{4}{x^2+5x-6} = \dfrac{3}{x^2+6x}$

39. $\dfrac{2}{x^2-4x+3} - \dfrac{3}{x^2-9} = \dfrac{2}{x^2+2x-3}$

40. $\dfrac{2}{x^2-4} - \dfrac{1}{x^2+x-2} = \dfrac{3}{x^2-3x+2}$

41. $2 - \dfrac{6}{x^2} = \dfrac{1}{x}$

42. $3 - \dfrac{7}{x} - \dfrac{6}{x^2} = 0$

43. $1 - \dfrac{7}{x-2} + \dfrac{12}{(x-2)^2} = 0$

44. $1 + \dfrac{3}{x+1} = \dfrac{10}{(x+1)^2}$

45. $1 + \dfrac{3}{x^2-9} = \dfrac{10}{x+3}$

46. $3 - \dfrac{7}{x^2-x-6} = \dfrac{5}{x-3}$

47. $\dfrac{2x}{x-3} + \dfrac{2}{x-5} = \dfrac{3x}{x^2-8x+15}$

48. $\dfrac{x}{x-4} = \dfrac{5x}{x^2-x-12} - \dfrac{3}{x+3}$

49. $\dfrac{2x}{x+2} = \dfrac{5}{x^2-x-6} - \dfrac{1}{x-3}$

50. $\dfrac{3x}{x-1} = \dfrac{2}{x-2} - \dfrac{2}{x^2-3x+2}$

Solve each equation for the indicated variable.

51. $\dfrac{1}{x} = \dfrac{1}{a} - \dfrac{1}{b}$ for x

52. $\dfrac{1}{x} = \dfrac{1}{a} + \dfrac{1}{b}$ for a

53. $\dfrac{1}{R} = \dfrac{1}{R_1} + \dfrac{1}{R_2}$ for R_1

54. $\dfrac{1}{F} = \dfrac{1}{D_1} + \dfrac{1}{D_2}$ for D_2

55. $y = \dfrac{x+1}{x-1}$ for x

56. $y = \dfrac{x-3}{x-2}$ for x

57. $I = \dfrac{A - P}{Pr}$ for P

58. $I = \dfrac{nE}{R + nr}$ for n

Solve each of the following inequalities.

59. $\dfrac{x - 2}{x + 1} < 0$

60. $\dfrac{x + 3}{x - 2} > 0$

61. $\dfrac{x - 4}{x - 2} > 0$

62. $\dfrac{x + 6}{x + 3} < 0$

63. $\dfrac{x - 5}{x + 3} \leq 0$

64. $\dfrac{x + 3}{x - 2} \leq 0$

65. $\dfrac{2x - 1}{x + 3} \geq 0$

66. $\dfrac{3x - 2}{x - 4} \leq 0$

67. $\dfrac{x}{x - 3} + \dfrac{2}{x - 3} \leq 0$

68. $\dfrac{x}{x + 5} - \dfrac{3}{x + 5} > 0$

69. $\dfrac{x}{x + 3} \leq \dfrac{4}{x + 3}$

70. $\dfrac{x}{x - 5} \geq \dfrac{2}{x - 5}$

71. $\dfrac{2x - 5}{x - 2} > 1$

72. $\dfrac{2x + 3}{x + 4} \geq 1$

Find the zeros of each of the following functions.

73. $f(x) = \dfrac{x}{10} - \dfrac{12}{5}$

74. $f(x) = \dfrac{4x}{3} - \dfrac{x}{6}$

75. $f(x) = \dfrac{12}{x + 5} - \dfrac{5}{x}$

76. $f(x) = \dfrac{7}{8} - \dfrac{16}{x - 2} - \dfrac{3}{x}$

77. $f(x) = \dfrac{1}{x - 3} + \dfrac{2}{x} - \dfrac{5}{3x}$

78. $f(x) = \dfrac{2}{x} - \dfrac{1}{x + 1} - \dfrac{3}{x^2 + x}$

79. $f(x) = 1 + \dfrac{39}{x^2} - \dfrac{16}{x}$

80. $f(x) = x - \dfrac{72}{x} + 1$

Transcribe Your Skills

81. What special considerations must be made when an equation contains rational expressions with variables in the denominator?

82. In solving the inequality $\dfrac{x}{x - 1} > 5$, is it incorrect to find the solution by multiplying both sides by $x - 1$? Why? What technique should be used?

Think About These

In Exercises 6.4 we introduced the equation (which is now given in a simplified form)

$$R = \dfrac{2R_1 R_2}{R_1 + R_2}$$

where R was the rate for a round trip, R_1 was the rate going, and R_2 was the rate returning.

83. If Hans travels at an average rate of 40 mi/h to his destination, at what rate will he have to travel for his return trip to average 48 mi/h for the round trip? *Hint:* Substitute 48 for R and 40 for R_1, and solve the resulting equation for R_2.

84. Suppose Olga averages 50 mi/h when she leaves her home for a conference. At what rate will she have to travel on the return trip to average 55 mi/h for the round trip?

85. The focal length F for a combination of two convex lenses with focal lengths F_1 and F_2 is given by

$$\dfrac{1}{F} = \dfrac{1}{F_1} + \dfrac{1}{F_2} - \dfrac{d}{F_1 F_2}$$

where d is the distance between the two lenses. Solve for d in terms of F, F_1, and F_2.

86. Assume that two balls of mass M_1 and mass M_2, both traveling with velocity V, collide. The velocity V_1 of the smaller ball with mass M_1 after the collision is given by

$$V_1 = \dfrac{V(M_2 - M_1)}{M_2 + M_1} - \dfrac{2VM_1}{M_2 + M_1}$$

Solve the equation for V.

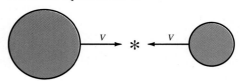

Graphing Utility

Problems such as Exercise 35 can be solved (or at least closely approximated) graphically. If we graph, on the same set of coordinate axes, the expression on the left side of the equation as

$$y_1 = \frac{x}{x - 3} - 2$$

and the expression on the right side of the equation as

$$y_2 = \frac{3}{x - 3}$$

then the points of intersection of the two graphs represent the solutions of the equation

$$\frac{x}{x - 3} - 2 = \frac{3}{x - 3}$$

Use this method to approximate solutions for each of the following.

87. $2 - \dfrac{6}{x^2} = \dfrac{1}{x}$

88. $3 - \dfrac{7}{x} = \dfrac{6}{x^2}$

Skillscan (Section 2.3)

Write the equation necessary for the solution of each of the following problems. Then solve the equation.

a. One number is 5 less than another. If 3 times the first number is 9 more than the second number, find the two numbers.   **b.** One number is 3 more than another. If twice the smaller number is 5 less than the larger number, what are the two numbers?   **c.** The sum of an integer and 3 times the next consecutive integer is 47. What are the two integers?   **d.** Five times an integer is 1 more than 4 times the next consecutive odd integer. Find the two integers.   **e.** Claudia rowed upstream for 3 h. In rowing downstream, her speed was 2 mi/h faster, and the return trip took 2 h. What was her speed each way?   **f.** José left the city at 11 A.M., heading west at a rate of 55 mi/h. An hour later, Jeanine headed east at a rate of 45 mi/h. At what time will they be 305 mi apart?

9.2   Applications and Problem Solving

TAPE IN20

OBJECTIVE: To solve applications leading to equations involving rational expressions

As we pointed out earlier, many problems in algebra lead to equations involving rational expressions. Our five-step problem-solving algorithm will, of course, remain the same as that introduced earlier. We will illustrate the applications of this section in the same step-by-step fashion.

Let's begin with an example involving numbers that will lead to a fractional equation.

EXAMPLE 1 A Number Application

One number is twice another. If the sum of the reciprocals of the numbers is $\dfrac{3}{10}$, what are the two numbers?

Step 1 You are asked to find the two unknown numbers.

Step 2 Let x represent the first number. Then

$2x$ Twice the first number

represents the second number.

Step 3 We now form the equation.

$$\frac{1}{x} + \frac{1}{2x} = \frac{3}{10}$$

Reciprocal of first number   Reciprocal of second number   Sum

Step 4 We now solve the equation as before. The LCD of x, $2x$, and 10 is $10x$. First we multiply both sides of the equation by that LCD.

$$10x\left(\frac{1}{x}\right) + 10x\left(\frac{1}{2x}\right) = 10x\left(\frac{3}{10}\right)$$

Simplifying, we have

$$10 + 5 = 3x$$

or

$$3x = 15$$

$$x = 5$$

and

$$2x = 10$$

The desired numbers are 5 and 10.

Step 5 To check our solutions, we verify that 5 and 10 meet the conditions of the original problem.

$$\frac{1}{5} + \frac{1}{10} \overset{?}{=} \frac{3}{10}$$

$$\frac{2}{10} + \frac{1}{10} \overset{?}{=} \frac{3}{10}$$

$$\frac{3}{10} = \frac{3}{10} \qquad \text{A true statement}$$

and our solutions are verified.

Check Yourself 1

One number is 3 times another. If the sum of their reciprocals is $\frac{2}{9}$, find the two numbers.

Problems involving consecutive integers may also lead to equations involving rational expressions. Consider the following example.

EXAMPLE 2 A Number Application

Find two consecutive odd integers so that the sum of the reciprocal of the first integer and the reciprocal of the second integer is equal to 8 times the reciprocal of the product of the integers.

Step 1 We want to find two consecutive odd integers that satisfy the given conditions.

Step 2 Let x represent the first odd integer. Then $x + 2$ represents the second odd integer.

The reciprocal of the product of the integers is

$$\frac{1}{x(x + 2)}$$

so 8 times that product is

$$\frac{8}{x(x + 2)}$$

Step 3 We can write our equation as follows.

$$\frac{1}{x} + \frac{1}{x + 2} = \frac{8}{x(x + 2)}$$

Step 4 The LCD for the equation of step 3 is $x(x + 2)$, and so multiplying, we have

$$x(x + 2)\left(\frac{1}{x}\right) + x(x + 2)\left(\frac{1}{x + 2}\right) = x(x + 2)\left[\frac{8}{x(x + 2)}\right]$$

Simplifying, we have

$$x + 2 + x = 8$$
$$2x = 6$$
$$x = 3$$

and $\quad x + 2 = 5$

The desired integers are 3 and 5.

Step 5 We leave it to you to verify these solutions.

Check Yourself 2

The sum of the reciprocals of two consecutive integers is 9 times the reciprocal of the product of those integers. What are the two integers?

The solution of many uniform-motion problems will also lead to equations involving rational expressions. Remember that the key equation for solving all motion problems relates the distance traveled, the rate, and the time.

$$d = r \cdot t$$

Distance   Rate   Time

In many applications, we use this equation in different forms by solving for r or for t. So

$$r = \frac{d}{t} \qquad \text{or} \qquad t = \frac{d}{r}$$

The following example illustrates.

EXAMPLE 3 A Distance-Rate-Time Application

Members of a rowing crew know that they can row 15 mi/h in still water. If they can travel 2 mi upstream in the time it takes to travel 3 mi downstream, what is the rate of the river's current?

Step 1 We want to know the rate of the current.

Step 2 Let r represent that rate.

Upstream, the rate is slowed by the current; downstream it is increased by the current.

Note Upstream, the crew will then be able to travel $(15 - r)$ mi/h. Downstream, they can travel $(15 + r)$ mi/h.

Step 3 A table of the given information is often an effective part of any problem-solving strategy. For instance, here we have

	Distance	Rate	Time
Upstream	2	$15 - r$	$\dfrac{2}{15 - r}$
Downstream	3	$15 + r$	$\dfrac{3}{15 + r}$

In completing the time column of the table, we note that

$$t = \frac{d}{r}$$

and we apply the information from the first two columns.

From the original problem we know that the time upstream *is the same as* the time downstream, and so using the table above, we can write

$$\frac{2}{15 - r} = \frac{3}{15 + r}$$

We set the two expressions for time equal to each other.

Step 4 The LCD is $(15 - r)(15 + r)$, and multiplying, we have

$$(15 - r)(15 + r)\left(\frac{2}{15 - r}\right) = (15 - r)(15 + r)\left(\frac{3}{15 + r}\right)$$

Simplifying yields

$$2(15 + r) = 3(15 - r)$$
$$30 + 2r = 45 - 3r$$
$$5r = 15$$
$$r = 3$$

The rate of the current is 3 mi/h.

Step 5 We leave the check of this result to you.

Check Yourself 3

A light plane's speed is 140 mi/h in still air. With a steady tailwind, the plane can fly 480 mi in the same time it takes to fly 360 mi against the wind. Find the rate of the wind.

Let's consider a related example in which we are led to a quadratic equation.

EXAMPLE 4 A Distance-Rate-Time Application

A motorboat makes a 48-mi trip downstream and then returns to its starting point. If the rate of the current was 4 mi/h and the entire trip took 5 h, find the boat's rate in still water.

Step 1 We are asked to find the boat's rate in still water.

Step 2 Let r represent the boat's rate.

Note Here the rate upstream will be $r - 4$. The rate downstream will be $r + 4$.

Step 3 Again, we use a table to organize the given information.

Once more, to complete the time column, use

$$t = \frac{d}{r}$$

	Distance	Rate	Time
Upstream	48	$r - 4$	$\dfrac{48}{r - 4}$
Downstream	48	$r + 4$	$\dfrac{48}{r + 4}$

The total time (the time upstream plus the time downstream) is 5 h. Considering the time column of our table, we can write

$$\frac{48}{r - 4} + \frac{48}{r + 4} = 5$$

Step 4 The LCD for the equation of step 3 is $(r - 4)(r + 4)$, so multiplying, we have

$$(r - 4)(r + 4)\left(\frac{48}{r - 4}\right) + (r - 4)(r + 4)\left(\frac{48}{r + 4}\right) = 5(r - 4)(r + 4)$$

Then

$$48(r + 4) + 48(r - 4) = 5(r - 4)(r + 4)$$
$$96r = 5r^2 - 80$$

Write in standard form.

$$5r^2 - 96r - 80 = 0$$

Factor.

$$(r - 20)(5r + 4) = 0$$

So

Apply the zero-product rule. $$r = 20 \quad \text{or} \quad r = -\frac{4}{5}$$

The boat's rate cannot be negative!

We reject the negative solution, and the rate of the boat in still water is 20 mi/h.

Step 5 You might want to check this result.

Check Yourself 4

A small commuter plane makes a 360-mi trip with a steady tailwind of 30 mi/h. The plane then returns along the same route against the wind. If the entire flight took 5 h, find the rate of the plane in still air.

In the previous two examples, the rate (of the current or the boat) was the unknown quantity. We can use a very similar approach when time is the unknown quantity. Consider the following example.

EXAMPLE 5 A Distance-Rate-Time Application

Pierre took 1 h longer on a trip of 180 mi than he did on a trip of 135 mi. If his rate was the same for both trips, how long did each trip take?

Step 1 We want to know the time for each of the trips.

Step 2 Let t be the time for the 135-mi trip. Then

1 h longer

$t + 1$

is the time for the 180-mi trip.

Step 3 The following table can be used.

	Distance	Time	Rate
135-mi trip	135	t	$\dfrac{135}{t}$
180-mi trip	180	$t + 1$	$\dfrac{180}{t + 1}$

Here, to complete the rate column, we use

$$r = \frac{d}{t}$$

From the original problem we know that the rate was the same for each trip, so we equate the rates shown above.

$$\frac{135}{t} = \frac{180}{t + 1}$$

Step 4 Multiplying by $t(t + 1)$ gives

$$t(t + 1)\left(\frac{135}{t}\right) = t(t + 1)\left(\frac{180}{t + 1}\right)$$

$$135(t + 1) = 180t$$
$$135t + 135 = 180t$$
$$135 = 45t$$
$$t = 3$$

and $\qquad t + 1 = 4$

The 135-mi trip took 3 h, and the 180-mi trip took 4 h.

Step 5 Again you should verify these results.

Check Yourself 5

Cynthia took 2 h longer to bicycle 120 mi than she took for a trip of 72 mi. If her rate was the same for both trips, find the time of each trip.

Our next example introduces a new group of applications that lead to fractional equations. In these applications, we assume that a person (or machine) works at a constant rate throughout a job. We use that assumption in the following way: Suppose that it takes a person 8 h to complete a job. Then the portion of the job completed in 1 h is $\frac{1}{8}$. Similarly, if it takes x hours to complete the job, the portion of the job completed will be $\frac{1}{x}$. Let's see how this is used in the following example.

EXAMPLE 6 A Work Problem

Suppose that one computer printer can complete a company's year-end report in 8 h. A newer-model printer can do the same job in 5 h. How long will the job take if both printers are used?

Solution Let x be the number of hours it will take to print the report with the printers working together. From the discussion above, in *1 h* we have

Portion of job done by first printer		Portion done by new printer		Portion done with printers working together
$\dfrac{1}{8}$	$+$	$\dfrac{1}{5}$	$=$	$\dfrac{1}{x}$

The LCD is $40x$, and we again multiply.

$$40x\left(\frac{1}{8}\right) + 40x\left(\frac{1}{5}\right) = 40x\left(\frac{1}{x}\right)$$

Simplifying yields

$$5x + 8x = 40$$
$$13x = 40$$
$$x = \frac{40}{13}$$

The job will take $\frac{40}{13}$ or $3\frac{1}{13}$ h to complete with the two printers working together.

Check Yourself 6

One inlet pipe can fill a water tank in 30 h; a second pipe can fill the same tank in 20 h. How long will it take to fill the tank if both inlet pipes are opened?

Let's look at another example involving work, this time with a slight variation.

EXAMPLE 7 A Work Problem

Camillo and Jeremy, working together, can install the wallboard for a new home in 6 h. If Jeremy, working alone, could complete the job in 9 h, how long would it take Camillo, working alone?

Solution In this example, our unknown is the time it would take Camillo, working alone, to complete the wallboard installation.

Therefore we let x represent that time.

As in the previous example, we can now write that for 1 h

Portion of job done by Jeremy		Portion of job done by Camillo		Portion done by working together
$\dfrac{1}{9}$	$+$	$\dfrac{1}{x}$	$=$	$\dfrac{1}{6}$

Multiplying to clear fractions yields

$$18x\left(\frac{1}{9}\right) + 18x\left(\frac{1}{x}\right) = 18x\left(\frac{1}{6}\right)$$

$$2x + 18 = 3x$$

$$x = 18$$

It would take Camillo 18 h, working alone, to complete the wallboard installation.

Check Yourself 7

Maria and Nathan, working together, can trim a hedge around their property in 10 h. Maria could do the job in 16 h by herself. How long would it take Nathan?

CHECK YOURSELF ANSWERS

1. 6, 18   **2.** 4, 5   **3.** 20 mi/h   **4.** 150 mi/h   **5.** 3 h, 5 h   **6.** 12 h

7. $\dfrac{80}{3}$ or $26\dfrac{2}{3}$ h

9.2 EXERCISES

Build Your Skills

1. One number is 3 times another number. If the sum of the reciprocals of the numbers is $\dfrac{1}{6}$, what are the two numbers?

2. One number is 4 times another number. The sum of the reciprocals of the numbers is $\dfrac{5}{24}$. Find the two numbers.

3. The sum of the reciprocals of two consecutive integers is equal to 11 times the reciprocal of the product of those integers. What are the two integers?

4. The sum of the reciprocals of two consecutive even integers is equal to 10 times the reciprocal of the product of those integers. Find the two integers.

5. If the same number is added to the numerator and denominator of $\dfrac{2}{5}$, the result is $\dfrac{4}{5}$. What is that number?

6. If the same number is subtracted from the numerator and denominator of $\dfrac{11}{15}$, the result is $\dfrac{1}{3}$. Find that number.

7. The numerator of $\dfrac{4}{7}$ is multiplied by a number. That same number is added to the denominator, and the result is $\dfrac{6}{5}$. What is that number?

8. The numerator of $\dfrac{8}{9}$ is multiplied by a number. That same number is subtracted from the denominator, and the result is 10. What is that number?

9. If the reciprocal of 5 times a number is subtracted from the reciprocal of that number, the result is $\dfrac{4}{25}$. What is the number?

10. If the reciprocal of a number is added to 4 times the reciprocal of that number, the result is $\dfrac{5}{9}$. Find that number.

11. One positive number is 2 more than another. If the sum of the reciprocals of the two numbers is $\dfrac{7}{24}$, what are those numbers?

12. One integer is 3 less than another. If the sum of the reciprocals of the two numbers is $\dfrac{7}{10}$, find the two integers.

13. The sum of the reciprocals of two consecutive integers is $\dfrac{9}{20}$. What are the two integers?

14. If the difference of the reciprocals of two consecutive positive integers is $\dfrac{1}{42}$, find those integers.

15. A motorboat can travel 20 mi/h in still water. If the boat can travel 3 mi downstream on a river in the same time it takes to travel 2 mi/h upstream, what is the rate of the river's current?

16. A small jet has an airspeed (the rate in still air) of 300 mi/h. During one day's flights, the pilot noted that the plane could fly 85 mi with a tailwind in the same time it took to fly 65 mi against that same wind. What was the rate of the wind?

17. A plane flew 720 mi with a steady 30-mi/h tailwind. The pilot then returned to the starting point, flying against that same wind. If the round-trip flight took 10 h, what was the plane's airspeed?

18. Janet and Michael took a canoeing trip, traveling 6 mi upstream along a river against a 2 mi/h current. They then returned downstream to the starting point of their trip. If their entire trip took 4 h, what was their rate in still water?

19. Po Ling can bicycle 75 mi in the same time it takes her to drive 165 mi. If her driving rate is 30 mi/h faster than her rate on the bicycle, find each rate.

20. A passenger train can travel 275 mi in the same time a freight train takes to travel 225 mi. If the speed of the passenger train is 10 mi/h more than that of the freight train, find the speed of each train.

21. A light plane took 1 h longer to fly 540 mi on the first portion of a trip than to fly 360 mi on the second portion. If the rate was the same for each portion, what was the flying time for each leg of the trip?

22. Gilbert took 2 h longer to drive 240 mi on the first day of a business trip than to drive 144 mi on the second day. If his rate was the same both days, what was his driving time for each day?

23. An express train and a passenger bus leave the same city, at the same time, for a destination 350 mi away. The rate of the train is 20 mi/h faster than the rate of the bus. If the train arrives at its destination 2 h ahead of the bus, find each rate.

24. A private plane and a commercial plane take off from an airport at the same time for a city 720 mi away. The rate of the private plane is 180 mi/h less than that of the commercial plane. If the commercial plane arrives 2 h ahead of the private plane, find each plane's rate.

25. One road crew can pave a section of highway in 15 h. A second crew, working with newer equipment, can do the same job in 10 h. How long would it take to pave that same section of highway if both crews worked together?

26. One computer printer can print a company's weekly payroll checks in 60 min. A second printer would take 90 min to complete the job. How long would it take the two printers, operating together, to print the checks?

27. An inlet pipe can fill a tank in 10 h. An outlet pipe can drain that same tank in 30 h. The inlet valve is opened, but the outlet valve is accidentally left open. How long will it take to fill the tank with both valves open?

28. A bathtub can be filled in 8 min. It takes 12 min for the bathtub to drain. If the faucet is turned on but the drain is also left open, how long will it take to fill the tub?

29. An electrician can wire a house in 20 h. If she works with an apprentice, the same job can be completed in 12 h. How long would it take the apprentice, working alone, to wire the house?

30. A landscaper can prepare and seed a new lawn in 12 h. If he works with an assistant, the job takes 8 h. How long would it take the assistant, working alone, to complete the job?

31. An experienced roofer can work twice as fast as her helper. Working together, they can shingle a new section of roof in 4 h. How long would it take the experienced roofer, working alone, to complete the same job?

32. One model copier operates at 3 times the speed of another. Working together, the copiers can copy a report in 8 min. The faster model breaks down, and the other model must be used. How long will the job take with the machine that is available?

33. A college uses two optical scanners to grade multiple-choice tests. One model takes 12 min longer to complete the scoring of a test than the other model. If by both models working together the test can be scored in 8 min, how long would each model, used by itself, take to score the same test?

34. Virginia can complete her company's monthly report in 5 h less than Carl. If they work together, the report will take them 6 h to finish. How long would it take Virginia, working alone?

35. An oil tanker can be filled from three inlet pipes. The first inlet pipe takes 8 h to fill the tank; the second, 12 h; and the third, 24 h. How long will it take to fill the tanker if all three pipes are opened?

36. Three natural gas lines lead to a storage tank. The first requires 10 min to fill the tank; the second, 12 min; and the third, 15 min. How long will it take to fill the tank with all three lines open?

37. If the United States depended solely on coal for heating homes (we will not discuss the pollution problem at this time), the supply would be exhausted in 50 yr. If we depended solely on oil, the supply would be exhausted in 30 yr. If both resources are used, how long will we have before the supply is depleted? *Hint:* Do not simply plug numbers into an equation; think about the problem first.

38. Carla's van has the same size fuel tank as Van's car, but its gas mileage is 4 mi/gal less. If the car can travel 620 mi on a tank of gas and the van can travel 540 mi on a tank of gas, what is the gas mileage of each? *Hint:* Distance = mileage × gallons.

39. Mack's truck has the same size fuel tank as Austin's car, but its gas mileage is 6 mi/gal less. If the car can travel 432 mi on a tank of gas and the truck can travel 336 mi on a tank of gas, what is the gas mileage of each?

40. Two cars get the same gas mileage, but one car has a gas tank 5 gal larger than the other. If the smaller car can travel 384 mi on a tank of gas and the larger car can travel 544 mi on a tank of gas, how much fuel can each car's tank hold?

41. Jeremiah used 6 gal more fuel to travel 405 mi than he did on a 243-mi trip. If his mileage was the same on both trips, how much fuel did he use on each trip?

Transcribe Your Skills

42. Describe the five-step process that is used to solve an application.

43. In some application problems, a valid solution to the equation describing the problem will not be a solution to the original problem itself. What circumstances will cause this to occur?

Skillscan (Section 1.4)

Evaluate each expression if $p = 3$, $q = -4$, and $r = 5$.

a. $\sqrt{4r + 5}$ **b.** $\sqrt{10p + 6}$ **c.** $\sqrt{20p + 4}$

d. $\sqrt{5 - 5q}$ **e.** $5 - \sqrt{p + 1}$ **f.** $10 + \sqrt{r + 4}$

g. $12 + \sqrt{1 - 2q}$ **h.** $8 - \sqrt{6p - 2}$

Equations Involving Radicals   9.3

OBJECTIVE: To solve equations involving radicals

In this section, we wish to establish procedures for solving equations involving radicals. The basic technique we will use involves raising both sides of an equation to some power. However, doing so requires some caution.

For example, let's begin with the equation $x = 1$. Squaring both sides gives us $x^2 = 1$, which has two solutions, 1 and -1. Clearly -1 is not a solution to the original equation. We refer to -1 as an *extraneous solution*.

We must be aware of the possibility of extraneous solutions any time we raise both sides of an equation to any *even power*. Having said that, we are now prepared to introduce the power property of equality.

Note

$x^2 = 1$

$x^2 - 1 = 0$

$(x + 1)(x - 1) = 0$

so the solutions are 1 and -1.

THE POWER PROPERTY OF EQUALITY

Given any two expressions a and b and any positive integer n:

If $a = b$, then $a^n = b^n$.

Note that although in applying the power property you will never lose a solution, you will often find an extraneous one as a result of raising both sides of an equation to some power. Because of this, it is very important that you *check all solutions.*

EXAMPLE 1 Solving a Radical Equation

Solve $\sqrt{x + 2} = 3$.

Solution Squaring each side, we have

Note that

$(\sqrt{x + 2})^2 = x + 2$

That is why squaring both sides of the equation removes the radical.

$$(\sqrt{x + 2})^2 = 3^2$$
$$x + 2 = 9$$
$$x = 7$$

Substituting 7 into the original equation, we find

$$\sqrt{7 + 2} \overset{?}{=} 3$$
$$\sqrt{9} \overset{?}{=} 3$$
$$3 = 3$$

Since this is a true statement, we have found the solution for the equation, $x = 7$.

Check Yourself 1

Solve $\sqrt{x - 5} = 4$.

EXAMPLE 2

Solve $\sqrt{4x + 5} + 1 = 0$.

Solution We must *first isolate the radical* on the left side:

Applying the power property will only remove the radical if that radical is isolated on one side of the equation.

$$\sqrt{4x + 5} = -1$$

Then, squaring both sides, we have

Note that on the right

$(-1)^2 = 1$

$$(\sqrt{4x + 5})^2 = (-1)^2$$
$$4x + 5 = 1$$

and solving for x, we find that

$$x = -1$$

Now we will check the solution by substituting -1 for x in the original equation:

Note $\sqrt{1} = 1$, the principal root.

This is clearly a false statement, so -1 is *not* a solution for the original equation.

$$\sqrt{4(-1) + 5} + 1 \overset{?}{=} 0$$
$$\sqrt{1} + 1 \overset{?}{=} 0$$
and $$2 \neq 0$$

Since -1 is an extraneous solution, there are *no solutions* to the original equation.

Check Yourself 2

Solve $\sqrt{3x - 2} + 2 = 0$.

Let's consider an example in which the procedure we have described will involve squaring a binomial.

EXAMPLE 3 Solving a Radical Equation

Solve $\sqrt{x + 3} = x + 1$.

Solution We can square each side, as before.

$$(\sqrt{x + 3})^2 = (x + 1)^2$$

$$x + 3 = x^2 + 2x + 1$$

Simplifying gives us the quadratic equation

$$x^2 + x - 2 = 0$$

Factoring, we have

$$(x - 1)(x + 2) = 0$$

which gives us the possible solutions

$$x = 1 \quad \text{or} \quad x = -2$$

Now we check for extraneous solutions, and we find that $x = 1$ is a valid solution but that $x = -2$ does not yield a true statement.

These problems can also be solved graphically. With a graphing utility, plot the two graphs $y_1 = \sqrt{x + 3}$ and $y_2 = x + 1$. Note that the graphs have two points of intersection. One is where $x = 1$, the other where $x = -2$.

Verify this for yourself by substituting 1 and then -2 for x in the original equation.

Check Yourself 3

Solve $\sqrt{x - 5} = x - 7$.

CAUTION

Be Careful! Sometimes (as in Example 3), one side of the equation contains a binomial. In that case, we must remember the middle term when we square the binomial. The square of a binomial *is always a trinomial.*

It is not always the case that one of the solutions is extraneous. We may have zero, one, or two valid solutions when we generate a quadratic from a radical equation.

In the following example we see a case in which both of the solutions derived will satisfy the equation.

EXAMPLE 4 Solving a Radical Equation

Solve $\sqrt{7x + 1} - 1 = 2x$.

Solution First, *we must isolate the term involving the radical.*

$$\sqrt{7x + 1} = 2x + 1$$

We can now square both sides of the equation.

$$7x + 1 = 4x^2 + 4x + 1$$

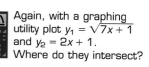 Again, with a graphing utility plot $y_1 = \sqrt{7x + 1}$ and $y_2 = 2x + 1$. Where do they intersect?

Now we write the quadratic equation in standard form.

$$4x^2 - 3x = 0$$

Factoring, we have

$$x(4x - 3) = 0$$

which yields two possible solutions

$$x = 0 \qquad \text{or} \qquad x = \frac{3}{4}$$

Checking the solutions by substitution, we find that both values for x give true statements, as follows.

Letting x be 0, we have

$$\sqrt{7(0) + 1} - 1 \overset{?}{=} 2(0)$$
$$\sqrt{1} - 1 \overset{?}{=} 0$$

or $\qquad\qquad 0 = 0 \qquad$ A true statement

Letting x be $\dfrac{3}{4}$, we have

$$\sqrt{7\left(\frac{3}{4}\right) + 1} - 1 \overset{?}{=} 2\left(\frac{3}{4}\right)$$

$$\sqrt{\frac{25}{4}} - 1 \overset{?}{=} \frac{3}{2}$$

$$\frac{5}{2} - 1 \overset{?}{=} \frac{3}{2}$$

$$\frac{3}{2} = \frac{3}{2} \qquad \text{Again, a true statement}$$

Check Yourself 4

Solve $\sqrt{5x + 1} - 1 = 3x$.

Sometimes when an equation involves more than one radical, we must apply the power property more than once. In such a case, it is generally best to avoid having to work with two radicals on the same side of the equation. The following example illustrates one approach to the solution of such equations.

Note $1 + \sqrt{2x - 6}$ is a binomial of the form $a + b$, where a is 1 and b is $\sqrt{2x - 6}$.

EXAMPLE 5 Solving a Radical Equation

Solve $\sqrt{x - 2} - \sqrt{2x - 6} = 1$.

Solution First we isolate $\sqrt{x - 2}$ by adding $\sqrt{2x - 6}$ to both sides of the equation. This gives

$$\sqrt{x - 2} = 1 + \sqrt{2x - 6}$$

Then squaring each side, we have

$$x - 2 = 1 + 2\sqrt{2x - 6} + 2x - 6$$

The square on the right then has the form $a^2 + 2ab + b^2$.

We now isolate the radical that remains on the right side.

$$-x + 3 = 2\sqrt{2x - 6}$$

We must square again to remove that radical.

$$x^2 - 6x + 9 = 4(2x - 6)$$

Now we solve the quadratic equation that results.

$$x^2 - 14x + 33 = 0$$

$$(x - 3)(x - 11) = 0$$

The equation is now written in standard form.

So

$$x = 3 \qquad \text{and} \qquad x = 11$$

are the possible solutions.

Checking the possible solutions, you will find that $x = 3$ yields the only valid solution. You should verify that for yourself.

Check Yourself 5

Solve $\sqrt{x + 3} - \sqrt{2x + 4} + 1 = 0$.

Earlier in this section, we noted that extraneous roots were possible whenever we raised both sides of the equation to an *even power*. In the following example, we will raise both sides of the equation to an odd power. We will still check the solutions, but in this case it will simply be a check of our work and not a search for extraneous solutions.

EXAMPLE 6 Solving a Radical Equation

Solve $\sqrt[3]{x^2 + 23} = 3$.

Solution Cubing each side, we have

$$x^2 + 23 = 27$$

Since a *cube root* is involved, we *cube* both sides to remove the radical.

which results in the quadratic equation

$$x^2 - 4 = 0$$

This has two solutions

$$x = 2 \qquad \text{or} \qquad x = -2$$

Checking the solutions, we find that both result in true statements. Again, you should verify this result.

Check Yourself 6

Solve $\sqrt[3]{x^2 - 8} - 2 = 0$.

We summarize our work in this section in the following algorithm for solving equations involving radicals.

SOLVING EQUATIONS INVOLVING RADICALS

STEP 1 Isolate the radical on one side of the equation.

STEP 2 Raise each side of the equation to the smallest power that will eliminate the isolated radical.

STEP 3 If any radicals remain in the equation derived in step 2, return to step 1 and continue the solution process.

STEP 4 Solve the resulting equation to determine any possible solutions.

STEP 5 Check all solutions to determine whether extraneous solutions may have resulted from step 2.

CHECK YOURSELF ANSWERS

1. 21   **2.** No solution   **3.** 9   **4.** $0, -\dfrac{1}{9}$ **5.** 6   **6.** 4, -4

9.3 EXERCISES

Build Your Skills

Solve each of the following equations. Be sure to check your solutions.

1. $\sqrt{x} = 2$

2. $\sqrt{x} - 3 = 0$

3. $2\sqrt{y} - 1 = 0$

4. $3\sqrt{2z} = 9$

5. $\sqrt{m + 5} = 3$

6. $\sqrt{y + 7} = 5$

7. $\sqrt{2x + 4} - 4 = 0$

8. $\sqrt{3x + 3} - 6 = 0$

9. $\sqrt{3x - 2} + 2 = 0$

10. $\sqrt{4x + 1} + 3 = 0$

11. $\sqrt{x - 1} = \sqrt{1 - x}$

12. $\sqrt{x + 1} = \sqrt{1 + x}$

13. $\sqrt{w + 3} = \sqrt{3 + w}$

14. $\sqrt{w - 3} = \sqrt{3 - w}$

15. $\sqrt{2x - 3} + 1 = 3$

16. $\sqrt{3x + 1} - 2 = -1$

17. $2\sqrt{3z + 2} - 1 = 5$

18. $3\sqrt{4q - 1} - 2 = 7$

19. $\sqrt{15 - 2x} = x$

20. $\sqrt{48 - 2y} = y$

21. $\sqrt{x + 5} = x - 1$

22. $\sqrt{2x - 1} = x - 8$

23. $\sqrt{3m - 2} + m = 10$

24. $\sqrt{2x + 1} + x = 7$

25. $\sqrt{t + 9} + 3 = t$

26. $\sqrt{2y + 7} + 4 = y$

27. $\sqrt{6x + 1} - 1 = 2x$

28. $\sqrt{7x + 1} - 1 = 3x$

29. $\sqrt[3]{x - 5} = 3$

30. $\sqrt[3]{x + 6} = 2$

31. $\sqrt[3]{x^2 - 1} = 2$

32. $\sqrt[3]{x^2 + 11} = 3$

Solve each of the following equations. Be sure to check your solutions.

33. $\sqrt{2x} = \sqrt{x + 1}$

34. $\sqrt{3x} = \sqrt{5x - 1}$

35. $2\sqrt{3r} = \sqrt{r + 11}$

36. $5\sqrt{2q - 7} = \sqrt{15q}$

37. $\sqrt{x + 2} + 1 = \sqrt{x + 4}$

38. $\sqrt{x + 5} - 1 = \sqrt{x + 3}$

39. $\sqrt{4m - 3} - 2 = \sqrt{2m - 5}$

40. $\sqrt{2c - 1} = \sqrt{3c + 1} - 1$

41. $\sqrt{x + 1} + \sqrt{x} = 1$

42. $\sqrt{z - 1} - \sqrt{6 - z} = 1$

43. $\sqrt{5x + 6} - \sqrt{x + 3} = 3$

44. $\sqrt{5y + 6} - \sqrt{3y + 4} = 2$

45. $\sqrt{y^2 + 12y} - 3\sqrt{5} = 0$

46. $\sqrt{x^2 + 2x} - 2\sqrt{6} = 0$

47. $\sqrt{\dfrac{x - 3}{x + 2}} = \dfrac{2}{3}$

48. $\dfrac{\sqrt{x - 2}}{x - 2} = \dfrac{x - 5}{\sqrt{x - 2}}$

49. $\sqrt{\sqrt{t} + 5} = 3$

50. $\sqrt{\sqrt{s} - 1} = \sqrt{s - 7}$

51. For what values of x is $\sqrt{(x - 1)^2} = x - 1$ a true statement?

52. For what values of x is $\sqrt[3]{(x - 1)^3} = x - 1$ a true statement?

Transcribe Your Skills

53. Why is it necessary to check all solutions that are found for an equation that contains radicals?

54. A radical equation has been solved incorrectly as follows.

$$\sqrt{2x - 1} = x - 8$$
$$2x - 1 = x^2 - 64$$
$$0 = x^2 - 2x - 63$$
$$0 = (x - 9)(x + 7)$$
$$x - 9 = 0 \qquad x + 7 = 0$$
$$x = 9 \qquad\quad x = -1$$

Find the error in the solution process.

Think About These

Solve for the indicated variable.

55. $h = \sqrt{pq}$ for q

56. $c = \sqrt{a^2 + b^2}$ for a

57. $v = \sqrt{2gR}$ for R

58. $v = \sqrt{2gR}$ for g

59. $r = \sqrt{\dfrac{S}{2\pi}}$ for S

60. $r = \sqrt{\dfrac{3V}{4\pi}}$ for V

61. $r = \sqrt{\dfrac{2V}{\pi h}}$ for V

62. $r = \sqrt{\dfrac{2V}{\pi h}}$ for h

63. $d = \sqrt{(x - 1)^2 + (y - 2)^2}$ for x

64. $d = \sqrt{(x - 1)^2 + (y - 2)^2}$ for y

A weight suspended on the end of a string is a *pendulum.* The most common example of a pendulum (this side of Edgar Allen Poe) is the kind found in many clocks.

The regular back-and-forth motion of the pendulum is *periodic,* and one such cycle of motion is called a *period.* The time, in seconds, that it takes for one period is given by the radical equation

$$t = 2\pi\sqrt{\dfrac{l}{g}}$$

where g is the force of gravity (10 m/s^2) and l is the length of the pendulum.

65. Find the period (to the nearest hundredth of a second) if the pendulum is 0.9 m long.

66. Find the period if the pendulum is 0.049 m long.

67. Solve the equation for length l.

68. How long would the pendulum be if the period were exactly 1 s?

Solve each of the following applications.

69. The sum of an integer and its square root is 12. Find the integer.

70. The difference between an integer and its square root is 12. What is the integer?

71. The sum of an integer and twice its square root is 24. What is the integer?

72. The sum of an integer and 3 times its square root is 40. Find the integer.

 Graphing Utility

To solve equations such as

$$\sqrt{6x + 1} - 1 = 2x$$

graphically, we graph both

$$y = \sqrt{6x + 1} - 1 \quad \text{and} \quad y = 2x$$

on the same set of axes.

The solutions are the points of intersection. Use this method to find the *number of solutions* for each of the following equations.

73. $\sqrt{6x + 1} + 1 = 2x$

74. $\sqrt{x + 9} = x - 3$

75. $\sqrt{5x - 6} - \sqrt{x + 3} = 3$

76. $\sqrt{x^2 + 12x} = 3\sqrt{5}$

Skillscan (Section 5.5)

Factor each of the following trinomials completely.

a. $x^2 - 5x + 6$ **b.** $x^2 - 2x - 15$ **c.** $x^4 - 5x^2 + 6$
d. $x^4 - 2x^2 - 15$ **e.** $x^2 - 13x + 36$ **f.** $x^2 + 4x - 5$
g. $x^4 - 13x^2 + 36$ **h.** $x^4 + 4x^2 - 5$

9.4   Equations That Are Quadratic in Form

TAPE IN21

OBJECTIVE: To solve equations that are quadratic in form

Consider the following equations:

$$2x - 5\sqrt{x} + 3 = 0 \tag{1}$$

$$x^4 - 4x^2 + 3 = 0 \tag{2}$$

$$(x^2 - x)^2 - 8(x^2 - x) + 12 = 0 \tag{3}$$

None of these equations are quadratic, yet each can be readily solved by using quadratic methods.

Compare the following quadratic equations to the original three equations.

Let $u = \sqrt{x}$ in (1).

$$2u^2 - 5u + 3 = 0 \tag{4}$$

Let $u = x^2$ in (2).

$$u^2 - 4u + 3 = 0 \tag{5}$$

Let $u = x^2 - x$ in (3).

$$u^2 - 8u + 12 = 0 \tag{6}$$

In each case, a simple substitution has been made that resulted in a quadratic equation. Equations that can be rewritten in this manner are said to be *equations in quadratic form*.

EXAMPLE 1 Letting $\sqrt{x} = u$

Solve

$$2x - 5\sqrt{x} + 3 = 0$$

Solution By substituting u for $\sqrt{x}$, we have

$2u^2 - 5u + 3 = 0$

Note that $u^2 = x$ since $u = \sqrt{x}$.

Factoring yields

$(2u - 3)(u - 1) = 0$

which gives the intermediate solutions

$u = \dfrac{3}{2}$ or $u = 1$

By intermediate solutions we mean values for u rather than for the original variable x.

We must now solve for x and then check our solutions. Since $\sqrt{x} = u$, we can write

$\sqrt{x} = \dfrac{3}{2}$ or $\sqrt{x} = 1$

$x = \dfrac{9}{4}$ $\qquad$ $x = 1$

We square both sides in each equation.

To check these solutions, we again simply substitute these values into the original equation. You should verify that each is a valid solution.

Check Yourself 1

Solve $3x - 8\sqrt{x} + 4 = 0$.

For certain equations in quadratic form, we can either solve by substitution (as we have done above) or solve directly by treating the equation as quadratic in some other power of the variable (in the case of the equation of the following example, x^2). In our next example, we show both methods of solution.

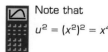

Note that

$u^2 = (x^2)^2 = x^4$

The graph of $y = x^4 - 4x^2 + 3 = 0$ is

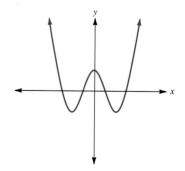

EXAMPLE 2 Letting $x^2 = u$

(a) Solve $x^4 - 4x^2 + 3 = 0$ by substitution.

Let $u = x^2$. Then

$u^2 - 4u + 3 = 0$

Factoring, we have

$(u - 1)(u - 3) = 0$

so

$u = 1$ or $u = 3$

Given these intermediate solutions, since $u = x^2$, we can write

$x^2 = 1$ or $x^2 = 3$

which, by using the square root method, yields the four solutions

$x = \pm 1$ or $x = \pm\sqrt{3}$

It crosses the x axis four times.

There are *four* solutions to the original *fourth-degree* solution.

We can check each of these solutions by substituting into the original equation. When we do so, we find that all four are valid solutions to the original equation.

(b) Solve $x^4 - 4x^2 + 3 = 0$ directly.

By treating the equation as quadratic in x^2, we can factor the left member, to write

$(x^2 - 1)(x^2 - 3) = 0$

This gives us the two equations

$x^2 - 1 = 0 \qquad \text{or} \qquad x^2 - 3 = 0$

Now

$x^2 = 1 \qquad \text{or} \qquad x^2 = 3$

Again, we have the four possible solutions

We apply the square root property.

$x = \pm 1 \qquad \text{or} \qquad x = \pm\sqrt{3}$

All check when they are substituted into the original equation.

Check Yourself 2

Solve $x^4 - 9x^2 + 20 = 0$ by substitution and by factoring directly.

In the following example, a binomial is replaced with u to make it easier to proceed with the solution.

EXAMPLE 3 Setting an Expression Equal to u

Solve

$(x^2 - x)^2 - 8(x^2 - x) + 12 = 0$

Solution Because of the repeated factor $x^2 - x$, we substitute u for $x^2 - x$. Factoring the resulting equation

$u^2 - 8u + 12 = 0$

gives

$(u - 6)(u - 2) = 0$

So

$u = 6 \qquad \text{or} \qquad u = 2$

We now have two intermediate solutions to work with. Since $u = x^2 - x$, we have two cases:

If $u = 6$, then   If $u = 2$, then

$$x^2 - x = 6 \qquad\qquad x^2 - x = 2$$

Write in standard form.

$$x^2 - x - 6 = 0 \qquad\qquad x^2 - x - 2 = 0$$

Factor the quadratic member.

$$(x - 3)(x + 2) = 0 \qquad (x - 2)(x + 1) = 0$$

$$x = 3 \quad \text{or} \quad x = -2 \qquad x = 2 \quad \text{or} \quad x = -1$$

The quadratic equations now yield four solutions that we must check. Substituting into the original equation, you will find that all four are valid solutions.

Check Yourself 3

Solve for x.

$$(x^2 - 2x)^2 - 11(x^2 - 2x) + 24 = 0$$

To summarize our work with equations in quadratic form, two approaches are commonly used. The first involves substitution of a new intermediate variable to make the original equation quadratic. The second solves the original equation directly by treating the equation as quadratic in some other power of the original variable. The following algorithms outline the two approaches.

SOLVING BY SUBSTITUTION

STEP 1 Make an appropriate substitution so that the equation becomes quadratic.

STEP 2 Solve the resulting equation for the intermediate variable.

STEP 3 Use the intermediate values found in step 2 to find possible solutions for the original variable.

STEP 4 Check the solutions of step 3 by substitution into the original equation.

SOLVING BY FACTORING

STEP 1 Treat the original equation as quadratic in some power of the variable, and factor.

STEP 2 Solve the resulting equations.

STEP 3 Check the solutions of step 2 by substitution into the original equation.

CHECK YOURSELF ANSWERS

1. $\dfrac{4}{9}, 4$ **2.** $\pm 2, \pm\sqrt{5}$ **3.** $-1, -2, 3, 4$

9.4 EXERCISES

Build Your Skills

Solve each of the following equations by factoring directly and then applying the zero-product rule.

1. $x^4 - 9x^2 + 20 = 0$

2. $t^4 - 7t^2 + 12 = 0$

3. $x^4 + x^2 - 12 = 0$

4. $x^4 - 7x^2 - 18 = 0$

5. $2w^4 - 9w^2 + 4 = 0$

6. $3x^4 - 5x^2 + 2 = 0$

7. $x^4 - 4x^2 + 4 = 0$

8. $y^4 - 6y^2 + 9 = 0$

9. $3x^4 + 16x^2 - 12 = 0$

10. $2x^4 + 9x^2 - 5 = 0$

11. $2z^4 + 4z^2 - 70 = 0$

12. $3y^4 + 10y^2 - 8 = 0$

13. $4t^4 - 20t^2 = 0$

14. $r^4 - 81 = 0$

Solve each of the following equations by substitution.

15. $x^4 - 9x^2 + 20 = 0$

16. $w^4 + w^2 - 12 = 0$

17. $2y^4 + y^2 - 15 = 0$

18. $4x^4 - 5x^2 - 9 = 0$

19. $b - 20\sqrt{b} + 64 = 0$

20. $z - 6\sqrt{z} + 8 = 0$

21. $t - 8\sqrt{t} - 9 = 0$

22. $y - 24\sqrt{y} - 25 = 0$

23. $(x - 2)^2 - 3(x - 2) - 10 = 0$

24. $(w + 1)^2 - 5(w + 1) + 6 = 0$

25. $(x^2 + 2x)^2 + 3(x^2 + 2x) + 2 = 0$

26. $(x^2 - 4x)^2 - (x^2 - 4x) - 12 = 0$

Solve each of the following equations by any method.

27. $7m - 41\sqrt{m} - 6 = 0$

28. $(x + 1) - 6\sqrt{x + 1} + 8 = 0$

29. $(w - 3)^2 - 2(w - 3) = 15$

30. $(x^2 - 4x)^2 + 7(x^2 - 4x) + 12 = 0$

31. $2y^4 - 5y^2 = 12$

32. $4t^4 - 29t^2 + 25 = 0$

Transcribe Your Skills

33. Describe what is meant by an equation that is quadratic in form.

34. The equation $x^{1/2} = -2$ has no real solution. How can we tell this without actually going through the solution process?

Think About These

An equation involving rational exponents may sometimes be solved by substitution. For instance, to solve an equation of the form $ax^{1/2} + bx^{1/4} + c = 0$, make the substitution $u = x^{1/4}$. Note that $u^2 = (x^{1/4})^2 = x^{2/4} = x^{1/2}$.

Use the suggestion above to solve each of the following equations. Be sure to check your solutions.

35. $x^{1/2} - 4x^{1/4} + 3 = 0$

36. $x^{1/2} - 5x^{1/4} + 6 = 0$

37. $x^{1/2} - x^{1/4} = 2$

38. $2x^{1/2} + x^{1/4} - 1 = 0$

39. $x^{2/3} + 2x^{1/3} - 3 = 0$

40. $x^{2/5} - x^{1/5} = 6$

Certain equations involving rational expressions can also be solved by the method of substitution. For instance, to solve an equation of the form

$$\frac{a}{x^2} + \frac{b}{x} + c = 0$$

make the substitution $u = \dfrac{1}{x}$. Note that $u^2 = \left(\dfrac{1}{x}\right)^2 = \dfrac{1}{x^2}$.

Use the suggestion above to solve the following equations. Then, for purposes of comparison, treat each of the equations as a quadratic equation and solve it.

41. $\dfrac{1}{x^2} - \dfrac{6}{x} + 8 = 0$

42. $\dfrac{2}{x^2} - \dfrac{1}{x} = 3$

43. $\dfrac{3}{x^2} - \dfrac{5}{x} = 2$

44. $\dfrac{1}{(x + 1)^2} - \dfrac{5}{x + 1} + 4 = 0$

45. $\dfrac{1}{(x-2)^2} + \dfrac{1}{x-2} = 6$

46. $\dfrac{8}{(x-3)^2} - \dfrac{2}{x-3} = 1$

Solve each of the following applications.

47. The sum of an integer and twice its square root is 24. What is the integer?

48. The sum of an integer and 3 times its square root is 40. Find the integer.

49. The sum of the reciprocal of an integer and the square of its reciprocal is $\dfrac{3}{4}$. What is the integer?

50. The difference between the reciprocal of an integer and the square of its reciprocal is $\dfrac{2}{9}$. Find the integer.

Skillscan

Find the value of x in each of the following.

a. $x = ky$ if $k = 2$ and $y = 3$ **b.** $2x = 3ky$ if $k = 4$ and $y = 5$ **c.** $x = kt^2$ if $k = 3$ and $t = -2$ **d.** $x = \dfrac{5k}{y}$ if $k = 4$ and $y = 2$ **e.** $x = \dfrac{24k}{5y}$ if $k = 4$ and $y = 2$

f. $3y = \dfrac{x}{2k}$ if $k = 3$ and $y = 4$

Variation 9.5

OBJECTIVE: To solve problems involving direct, inverse, and joint variation

TAPE IN6

We have seen that a function is a means of describing a relationship between two quantities. One particular type of function is so common that there exists a precise terminology to describe the relationship. That type of function is our focus in this section. In each case we will see that a quantity is *varying* with respect to one or more other quantities in a particular fashion.

Considering these special functions is called the study of variation *because the quantity varies.*

Suppose that our quantities are related in a manner such that one is a constant multiple of the other. There are many real-world applications.

The circumference of a circle is a constant multiple of the length of its diameter:

The idea that the circumference is a constant times the diameter was known to the Babylonians as early as 2000 B.C.

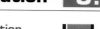

$$C = \pi d$$

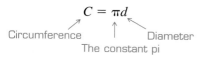
Circumference   Diameter
The constant pi

If a rate or speed is constant, the distance traveled is a constant multiple of time.

$$d = rt$$

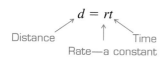
Distance   Time
Rate—a constant

If you earn a fixed hourly pay, your total pay is a constant multiple of the number of hours that you worked.

$$T = ph$$

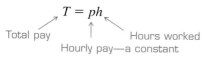
Total pay   Hours worked
Hourly pay—a constant

In all the cases above, the changes in one variable are proportional to the changes in the other. For instance, if the diameter of a circle is *doubled*, its circumference is *doubled*. This leads to our first definition.

> **DIRECT VARIATION**
>
> If y is a constant multiple of x, we write
>
> $y = kx$ k is a constant
>
> and say that y *varies directly* as x, or that y is *directly proportional* to x. The constant k is called the *constant of variation*.

Typically, in a variation problem, you will be given the type of variation involved and related values for the variables. From this information you can determine the constant of variation and, therefore, the equation relating the quantities. The following examples illustrate.

EXAMPLE 1 Solving a Direct Variation Problem

If y varies directly as x and if $y = 40$ when $x = 8$, find the equation relating x and y. Also find the value of y when $x = 10$.

Solution Since y varies directly with x, from our definition we have

As you will see, in direct variation, as the absolute value of one variable *increases*, the absolute value of the other also *increases*.

$y = kx$

We need to determine k, and this is easily done by letting $x = 8$ and $y = 40$.

$40 = k(8)$

$k = 5$

The desired equation relating x and y is then

$y = 5x$

To complete the example, if $x = 10$,

$y = 5(10) = 50$

Check Yourself 1

If y varies directly as x and if $y = 72$ when $x = 9$, find the value of y when $x = 12$.

As we said, problems of variation occur frequently in applications of mathematics to many fields. The following is a typical example.

EXAMPLE 2 An Application Involving a Spring

Example 2 is an application of *Hooke's law,* which states that stress is proportional to strain.

In physics, it is known that the force F needed to stretch a spring x units varies directly with x. If a force of 18 lb stretches a spring 3 in, how far will the same spring be stretched by a force of 30 lb?

Solution From the problem we know that $F = kx$, and letting $F = 18$ and $x = 3$, we have

$18 = k(3)$

$k = 6$

Therefore, $F = 6x$ relates the two variables. Now, to find x when $F = 30$, we write

$30 = 6x$

$x = 5$

So the force of 30 lb will stretch the spring 5 in.

Check Yourself 2

The pressure at a point under water is directly proportional to the depth. If the pressure at a depth of 2 ft is 125 pounds per square foot (lb/ft^2), find the pressure at a depth of 10 ft.

Many applications require that one variable be directly proportional to some power of a second variable. For instance, we might say that y is directly proportional to the square of x and write

$y = kx^2$

Consider the following example.

EXAMPLE 3 An Application Involving Gravity

The distance s that an object will fall from rest (neglecting air resistance) varies directly with the square of the time t. If an object falls 64 ft in 2 s, how far will it fall in 5 s?

Solution The relating equation in this example is

$s = kt^2$

By letting $s = 64$ and $t = 2$, we can determine k:

$64 = k(2)^2$

$64 = 4k$

so

$k = 16$

We now know that the desired equation is $s = 16t^2$, and substituting 5 for t, we have

$s = 16(5)^2 = 400$ ft

Check Yourself 3

The distance that an object falls from rest varies directly with the square of time t. If an object falls 144 ft in 3 s, how far does it fall in 6 s?

Perhaps the most common example is the relationship between rate and time. The faster something travels (rate *increases*), the sooner it arrives (time *decreases*).

If two quantities are related so that an *increase* in the absolute value of the first gives a proportional *decrease* in the absolute value of the second, we say that the variables *vary inversely* with each other. This leads to our second definition.

INVERSE VARIATION

If *y varies inversely* as *x*, we write

$$y = \frac{k}{x} \qquad k \text{ is a constant}$$

We can also say that *y* is *inversely proportional* to *x*.

EXAMPLE 4 Solving an Inverse Variation Problem

If *y* varies inversely as *x* and if $y = 18$ when $x = \frac{1}{2}$, find the equation relating *x* and *y*. Also find the value of *y* when $x = 3$.

Solution Since *y* varies inversely as *x*, we can write

$$y = \frac{k}{x}$$

Now with $y = 18$ and $x = \frac{1}{2}$, we have

$$18 = \frac{k}{\frac{1}{2}}$$

$$9 = k$$

We now have the desired equation relating *x* and *y*:

$$y = \frac{9}{x}$$

and when $x = 3$,

$$y = \frac{9}{3} = 3$$

Check Yourself 4

If w is inversely proportional to z and if $w = 15$ when $z = 5$, find the equation relating z and w. Also find the value of w when $z = 3$.

Let's consider an application which involves the idea of inverse variation.

EXAMPLE 5 An Application Involving Illumination

The intensity of illumination I of a light source varies inversely as the square of the distance d from that source. If the illumination 4 ft from the source is 9 footcandles (fc), find the illumination 6 ft from the source.

Solution From the given information we know that

$$I = \frac{k}{d^2}$$

Letting $d = 4$ and $I = 9$, we first find the constant of variation k:

$$9 = \frac{k}{4^2} \qquad \text{or} \qquad 9 = \frac{k}{16}$$

and

$$k = 144$$

The equation relating I and d is then

$$I = \frac{144}{d^2}$$

and when $d = 6$,

$$I = \frac{144}{6^2} = \frac{144}{36} = 4 \text{ fc}$$

Check Yourself 5

At a constant temperature, the volume of a gas varies inversely as the pressure. If a gas has volume 200 ft^3 under a pressure of 40 pounds per square inch (lb/in^2), what will be its volume under a pressure of 50 lb/in^2?

It is also common for one quantity to depend on *several* others. We can find a familiar example from geometry.

The volume of a cylinder depends on its height and the square of its radius.

This is an example of *joint variation*. We say that the volume V varies jointly with the height h and the square of the radius r. We can write

$$V = khr^2$$

In general:

You probably recognize that k, the constant of variation, is π in this case.

JOINT VARIATION

If **z varies jointly** as x and y, we write

$z = kxy$ k is a constant

The solution techniques for problems involving joint variation are similar to those used earlier, as Example 6 illustrates.

EXAMPLE 6 Solving a Joint Variation Problem

Assume that z varies jointly as x and y. If $z = 100$ when $x = 2$ and $y = 20$, find the value of z if $x = 4$ and $y = 30$.

Solution From the given information we have

$z = kxy$

Letting $z = 100$, $x = 2$, and $y = 20$ gives

$$100 = k(2)(20) \qquad \text{or} \qquad k = \frac{5}{2}$$

The equation relating z with x and y is then

$$z = \frac{5}{2}xy$$

and when $x = 4$ and $y = 30$, by substitution

$$z = \frac{5}{2}(4)(30) = 300$$

Check Yourself 6

Assume that r varies jointly as s and t. If $r = 64$ when $s = 3$ and $t = 8$, find the value of r when $s = 16$ and $t = 12$.

Once again, there are many physical applications of the concept of joint variation. The following is a typical example.

EXAMPLE 7 An Application Involving "Safe Load"

The "safe load" for a wooden rectangular beam varies jointly as its width and the square of its depth.

If the safe load of a beam 2 in wide and 8 in deep is 640 lb, what is the safe load of a beam 4 in wide and 6 in deep?

Solution From the given information,

$$S = kwd^2$$

S is the safe load; w, the width; and d, the depth.

Substituting the given values yields

$$640 = k(2)(8)^2$$

$$k = 5$$

We then have the equation

$$S = 5wd^2$$

and, for the 4-in by 6-in beam,

$$S = 5(4)(6)^2 = 720 \text{ lb}$$

Check Yourself 7

The force of a wind F blowing on a vertical wall varies jointly as the surface area of the wall A and the square of the wind velocity v.

If a wind of 20 mi/h has a force of 100 lb on a wall with an area of 50 ft^2, what force will a wind of 40 mi/h produce on the same wall?

There is one final category of variation problems. This category involves applications in which inverse variation is combined with direct or joint variation in stating the equation relating the variables.

These are called *combined variation* problems. In general, a typical statement form is as follows:

COMBINED VARIATION

If z varies directly as x and inversely as y, we write

$$z = \frac{kx}{y} \quad \text{where } k \text{ is a constant}$$

EXAMPLE 8 Solving a Combined Variation Problem

Assume that w varies directly as x and inversely as the square of y. When $x = 8$ and $y = 4$, $w = 18$. Find w if $x = 4$ and $y = 6$.

Solution From the given information we can write

$$w = \frac{kx}{y^2}$$

Substituting the known values, we have

$$18 = \frac{k \cdot 8}{4^2} \quad \text{or} \quad k = 36$$

We now have the equation

$$w = \frac{36x}{y^2}$$

and letting $x = 4$ and $y = 6$, we get

$$w = \frac{36 \cdot 4}{(6)^2} = 4$$

Check Yourself 8

Ohm's law for an electric circuit states that the current I varies directly as the electromotive force E and inversely as the resistance R.

If the current is 10 amperes (A), the electromotive force is 110 volts (V) and the resistance is 11 ohms (Ω). Find the current for an electromotive force of 220 V and a resistance of 5 Ω.

Translating verbal problems to algebraic equations is the basis for all variation applications. The following table gives some typical examples from our work in this section.

Variation Statement	Algebraic Equation
y varies directly as x.	$y = kx$
y varies inversely as x.	$y = \dfrac{k}{x}$
z varies jointly as x and as y.	$z = kxy$
z varies directly as x and inversely as y.	$z = \dfrac{kx}{y}$

All four basic types of variation problems involve essentially the same solution technique. The following algorithm summarizes the steps involved in all the variation problems that we have considered.

SOLVING PROBLEMS INVOLVING VARIATION

STEP 1   Translate the given problem to an algebraic equation involving the constant of variation k.

STEP 2   Use the given values to find that constant.

STEP 3   Replace k with the value found in step 2 to form the general equation relating the variables.

STEP 4   Substitute the appropriate values of the variables to solve for the corresponding value of the desired unknown quantity.

CHECK YOURSELF ANSWERS

1. $y = 96$ **2.** 625 lb/ft^2 **3.** 576 ft **4.** $w = \dfrac{75}{z}$, $w = 25$ **5.** 160 ft^3

6. 512 **7.** 400 lb **8.** $I = \dfrac{kE}{R}$ where $k = 1$, so $I = 44 \text{ A}$

9.5 EXERCISES

Build Your Skills

Translate each of the following statements of variation to an algebraic equation, using k as the constant of variation.

1. s varies directly as the square of x.

2. z is directly proportional to the square root of w.

3. r is inversely proportional to s.

4. m varies inversely as the cube of n.

5. V varies directly as T and inversely as P.

6. A varies jointly as x and y.

7. V varies jointly as h and the square of r.

8. t is directly proportional to d and inversely proportional to r.

9. w varies jointly as x and y and inversely as the square of z.

10. p varies jointly as r and the square of s and inversely as the cube of t.

Find k, the constant of variation, given each of the following sets of conditions.

11. y varies directly with x; $y = 54$ when $x = 6$

12. m varies inversely with p; $m = 5$ when $p = 3$

13. r is inversely proportional to the square of s; $r = 5$ when $s = 4$

14. u varies directly with the square of w; $u = 75$ when $w = 5$

15. V varies jointly as x and y; $V = 100$ when $x = 5$ and $y = 4$

16. w is directly proportional to u and inversely proportional to v; $w = 20$ when $u = 10$ and $v = 3$

17. z varies directly as the square of x and inversely as y; $z = 20$ when $x = 2$ and $y = 4$

18. p varies jointly as the square of q and r; $p = 144$ when $q = 6$ and $r = 2$

19. m varies jointly as n and the square of p and inversely as r; $m = 40$ when $n = 5$, $p = 2$, and $r = 4$

20. x varies directly as the square of y and inversely as w and z; $x = 8$ when $y = 4$, $w = 3$, and $z = 2$

Solve each of the following variation problems.

21. Let y vary directly with x. If $y = 60$ when $x = 5$, find the value of y when $x = 8$.

22. Suppose that z varies inversely as the square of w and that $z = 3$ when $w = 4$. Find the value of z when $w = 6$.

23. Variable A varies jointly with x and y, and $A = 120$ when $x = 6$ and $y = 5$. Find the value of A when $x = 8$ and $y = 3$.

24. Let p be directly proportional to q and inversely proportional to the square of r. If $p = 3$ when $q = 8$ and $r = 4$, find p when $q = 9$ and $r = 6$.

25. Suppose that s varies directly with r and inversely with the square of t. If $s = 4$ when $r = 12$ and $t = 6$, find the value of s when $r = 8$ and $t = 4$.

26. Variable p varies jointly with the square root of r and the square of q. If $p = 72$ when $r = 16$ and $q = 3$, find the value of p when $r = 25$ and $q = 2$.

Solve each of the following variation problems.

27. The length that a spring will stretch varies directly as the force applied to the spring. If a force of 10 lb will stretch a spring 2 in, what force will stretch the same spring 3 in?

28. If the temperature of a gas is held constant, the volume occupied by that gas varies inversely as the pressure to which the gas is subjected. If the volume of a gas is 8 ft^3 when the pressure is 12 lb/in^2, find the volume of the gas if the pressure is 16 lb/in^2.

29. If the current, in amperes, in an electric circuit is inversely proportional to the resistance, the current is 55 A when the resistance is 2 Ω. Find the current when the resistance is 5 Ω.

30. The distance that a ball rolls down an inclined plane varies directly as the square of the time. If the ball rolls 36 ft in 3 s, how far will it roll in 5 s?

31. The volume of a right circular cone varies jointly as the height and the square of the radius. If the volume of the cone is 15π cm^3 when the height is 5 cm and the radius is 3 cm, find the volume when the height is 6 cm and the radius is 4 cm.

32. The safe load of a rectangular beam varies jointly as its width and the square of its depth. If the safe load of a beam is 1000 lb when the width is 2 in and the depth is 10 in, find its safe load when the width is 4 in and the depth is 8 in.

33. The stopping distance (in feet) of an automobile varies directly as the square of its speed (in miles per hour). If a car can stop in a distance of 80 ft from 40 mi/h, how long will it take to stop from a speed of 60 mi/h?

34. The period (the time required for one complete swing) of a simple pendulum is directly proportional to the square root of its length. If a pendulum with length 9 cm has a period of 3.3 s, find the period of a pendulum with length 16 cm.

35. The distance (in miles) that a person can see to the horizon from a point above the earth's surface is directly proportional to the square root of the height (in feet) of that point. If a person 100 ft above the earth's surface can see 12.5 mi, how far can an observer in a light plane at 3600 ft see to the horizon?

36. The illumination produced by a light source on a surface varies inversely as the square of the distance of that surface from the source. If a light source produces an illumination of 48 fc on a wall 4 ft from the source, what will be the illumination (in footcandles) of a wall 8 ft from the source?

37. The electric resistance of a wire varies directly as its length and inversely as the square of its diameter. If a wire with length 200 ft and diameter 0.1 in has a resistance of 2 Ω, what will be the resistance in a wire of length 400 ft with diameter 0.2 in?

38. The frequency of a guitar string varies directly as the square root of the tension on the string and inversely as the length of the string. If a frequency of 440 cycles per second, or *hertz* (Hz), is produced by a tension of 36 lb on a string of length 60 cm, what frequency (in hertz) will be produced by a tension of 64 lb on a string of length 40 cm?

Transcribe Your Skills

39. Explain what is meant by a constant of variation.

40. If y varies directly as x, we write $y = kx$. Describe the relationship between x and y if k is a negative constant.

Skillscan

Find the value of y in each of the following

a. $y = 2(x - 1)^2 + 3$; $x = 2$ **b.** $y = -3(x + 2)^2 - 2$; $x = 1$ **c.** $y = -2(x + 3)^2$; $x = -3$ **d.** $y = 4(x - 4)^2 + 1$; $x = 4$ **e.** $y = 2x^2 + 8x - 1$; $x = -2$ **f.** $y = -x^2 + 6x + 2$; $x = 3$

SUMMARY

Equations and Inequalities Involving Rational Expressions [9.1]

Solve

$$\frac{3}{x - 3} - \frac{2}{x + 2} = \frac{19}{x^2 - x - 6}$$

Multiply by the LCD
$(x - 3)(x + 2)$:

$3(x + 2) - 2(x - 3) = 19$

$3x + 6 - 2x + 6 = 19$

$x = 7$

To solve an equation involving rational expressions, you should apply the following algorithm.

1. Clear the equation of fractions by multiplying both sides of the equation by the LCD of all the fractions that appear in the equation.

2. Solve the equation resulting from step 1.

3. Check all solutions by substitution in the original equation.

Check:

$$\frac{3}{4} - \frac{2}{9} = \frac{19}{36}$$

$$\frac{19}{36} = \frac{19}{36}$$

Solving Rational Inequalities Inequalities containing rational expressions are solved in a similar fashion to quadratic inequalities. The rational expression must first be related to 0. Critical points in this case occur when the numerator is equal to 0 or when the denominator is equal to 0. A sign graph is used to indicate where the quotient is negative or positive, and the solution set is determined.

To solve

$$\frac{x+2}{x-1} \geq 0$$

Critical points are -2 and 1.

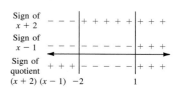

The solution set is

$$\{x \mid x \leq -2 \text{ or } x > 1\}$$

Applications and Problem Solving [9.2]

Applications of our work in algebra may lead to equations involving rational expressions. The five-step problem-solving algorithm is used along with the methods of Section 9.1 to solve the resulting equation.

Equations Involving Radicals [9.3]

To solve an equation involving radicals, apply the power property of equality.

Power Property of Equality Given any two expressions a and b and any positive integer n,

If $a = b$, then $a^n = b^n$.

That property is used in the following algorithm.

Solving Equations Involving Radicals

Step 1. Isolate a radical on one side of the equation.

Step 2. Raise each side of the equation to the smallest power that will eliminate the radical.

Step 3. If any radicals remain in the equation derived in step 2, return to step 1 and continue the solution process.

Step 4. Solve the resulting equation to determine any possible solutions.

Step 5. Check all solutions to determine whether extraneous solutions may have resulted from step 2.

To solve

$$\sqrt{2x - 3} + x = 9$$

$$\sqrt{2x - 3} = -x + 9$$

$$2x - 3 = x^2 - 18x + 81$$

$$0 = x^2 - 20x + 84$$

$$0 = (x - 6)(x - 14)$$

$$x = 6 \quad \text{or} \quad x = 14$$

By substitution, 6 is the only valid solution.

Equations That Are Quadratic in Form [9.4]

There are a variety of equations that are quadratic in form.

$2x - 11\sqrt{x} + 12 = 0$ is quadratic in $\sqrt{x}$.

$x^4 - 3x^2 - 10 = 0$ is quadratic in x^2.

$(x^2 - 1)^2 + 2(x^2 - 1) - 8 = 0$ is quadratic in $x^2 - 1$.

These equations can be solved by one of two techniques.

Solving by Substitution

To solve

$2x - 11\sqrt{x} + 12 = 0$

let $u = \sqrt{x}$. Then

$2u^2 - 11u + 12 = 0$

$(u - 4)(2u - 3) = 0$

$u = 4 \qquad \text{or} \qquad u = \dfrac{3}{2}$

Since $u = \sqrt{x}$,

$\sqrt{x} = 4 \qquad \text{or} \qquad \sqrt{x} = \dfrac{3}{2}$

$x = 16 \qquad\qquad x = \dfrac{9}{4}$

Both solutions are valid.

Step 1. Make an appropriate substitution so that the equation becomes quadratic.

Step 2. Solve the resulting equation for the intermediate variable.

Step 3. Use the intermediate values found in step 2 to find possible solutions for the original variable.

Step 4. Check the solutions of step 3 by substitution into the original equation.

Solving by Factoring

To solve

$x^4 - 3x^2 - 10 = 0$

$(x^2 - 5)(x^2 + 2) = 0$

$x^2 = 5 \qquad \text{or} \qquad x^2 = -2$

$x = \pm\sqrt{5} \qquad\qquad x = \pm i\sqrt{2}$

Step 1. Treat the original equation as quadratic in some power of the variable, and factor.

Step 2. Solve the resulting equations.

Step 3. Check the solutions of step 2 by substitution into the original equation.

Variation [9.5]

If y varies directly as x and $y = 64$ when $x = 4$, find the equation relating x and y.

$y = kx$

$64 = k \cdot 4 \qquad \text{so}$

$k = 16 \qquad \text{so}$

$y = 16x$

Direct Variation

If y *varies directly* as x (or y is *directly proportional* to x), we write

$y = kx$

where k is the *constant of variation*.

If r varies inversely as the square of s, the relating equation is

$r = \dfrac{k}{s^2}$

Inverse Variation

If y *varies inversely* as x (or y is *inversely proportional* to x), we write

$y = \dfrac{k}{x}$

Joint Variation

If y varies jointly as x and y, we write

$$y = kxy$$

If m varies jointly as n and the square root of p, the relating equation is

$$m = kn\sqrt{p}$$

Combined Variation

If z varies directly as x and inversely as y, we write

$$z = \frac{kx}{y}$$

If V varies directly as T and inversely as P, the relating equation is

$$V = \frac{kT}{P}$$

SUMMARY EXERCISES

This summary exercise set is provided to give you practice with each of the objectives of the chapter. Each exercise is keyed to the appropriate chapter section. The answers are provided in the instructor's manual that accompanies this text. Your instructor will provide guidelines on how best to use these exercises in your instructional program.

[9.1] Solve each of the following equations.

1. $\dfrac{1}{2x} + \dfrac{1}{3x} = \dfrac{1}{6}$

2. $\dfrac{5}{2x^2} - \dfrac{1}{4x} = \dfrac{1}{x}$

3. $\dfrac{x}{x-2} + 1 = \dfrac{x+4}{x-2}$

4. $\dfrac{2x-1}{x-3} - \dfrac{5}{x-3} = 1$

5. $\dfrac{2}{3x+1} = \dfrac{1}{x+2}$

6. $\dfrac{5}{x+1} + \dfrac{1}{x-2} = \dfrac{7}{x+1}$

7. $\dfrac{4}{x-1} - \dfrac{5}{3x-7} = \dfrac{3}{x-1}$

8. $\dfrac{7}{x} - \dfrac{1}{x-3} = \dfrac{9}{x^2-3x}$

9. $\dfrac{2}{x-3} - \dfrac{11}{x^2-9} = \dfrac{3}{x+3}$

10. $\dfrac{5}{x+3} + \dfrac{1}{x-5} = 1$

11. $\dfrac{2}{x-4} = \dfrac{x}{x-2} - \dfrac{x+4}{x^2-6x+8}$

12. $\dfrac{x}{x-5} = \dfrac{3x}{x^2-7x+10} + \dfrac{8}{x-2}$

[9.1] Solve each equation for the indicated variable.

13. $\dfrac{1}{T} + \dfrac{1}{T_1} = \dfrac{1}{T_2}$ for T_1

14. $R = \dfrac{R_1 R_2}{R_1 + R_2}$ for R_1

15. $\dfrac{1}{F} = \dfrac{1}{D_1} + \dfrac{1}{D_2}$ for D_2

16. $\dfrac{1}{x} = \dfrac{1}{a} + \dfrac{1}{b}$ for x

[9.1] Solve each of the following inequalities.

17. $\dfrac{x-2}{x+1} < 0$

18. $\dfrac{x+3}{x-2} > 0$

19. $\dfrac{x-4}{x-2} > 0$

20. $\dfrac{x+6}{x+3} < 0$

21. $\dfrac{x-5}{x+3} \le 0$

22. $\dfrac{x+3}{x-2} \le 0$

23. $\dfrac{2x-1}{x+3} \ge 0$

24. $\dfrac{3x-2}{x-4} \le 0$

25. $\dfrac{x}{x-3} + \dfrac{2}{x-3} \le 0$

26. $\dfrac{x}{x+5} - \dfrac{3}{x+5} > 0$

27. $\dfrac{x}{x+3} \le \dfrac{4}{x+3}$

28. $\dfrac{x}{x-5} \ge \dfrac{2}{x-5}$

29. $\dfrac{2x-5}{x-2} > 1$

30. $\dfrac{2x+3}{x+4} \ge 1$

[9.2] Solve each of the following applications.

31. One number is 4 times another number. If the sum of the reciprocals of the two numbers is $\dfrac{1}{4}$, what are the two numbers?

32. The sum of the reciprocals of two consecutive integers is equal to 9 times the reciprocal of the product of those integers. Find the two integers.

33. The same number is added to the numerator and denominator of $\dfrac{7}{10}$. If the resulting fraction is equal to $\dfrac{5}{6}$, what number was added to the numerator and denominator?

34. One integer is 2 more than another. If the sum of the reciprocals of the integers is $\dfrac{5}{12}$, find the two integers.

35. A yacht travels 25 mi/h in still water. If the yacht can travel 3 mi downstream on a river in the same amount of time it takes to travel 2 mi upstream, what is the rate of the current?

36. A light plane flew 800 mi against a steady 20-mi/h headwind. The plane then returned to its starting point with the same wind. If the entire trip took 9 h, what was the plane's airspeed (its rate in still air)?

37. On the first day of a business trip, Min Yeh drove 225 mi. On the second day, it took her 2 h longer to drive 315 mi. If her rate was the same each day, what was her driving time each day?

38. Brett made a trip of 240 mi using the freeway. Returning by a different route, he found that the distance was only 200 mi but that traffic slowed his speed by 8 mi/h. If the trip took the same time in both directions, what was Brett's rate each way?

39. A painter could paint an office complex in 10 h while it would take his helper 15 h. How long would it take to complete the job if the two worked together?

40. A water tank can be filled through an inlet pipe in 10 h. The tank will take 15 h to drain through an outlet pipe. The inlet pipe is opened to begin filling the tank, but the outlet valve is also inadvertently left open. How long will it take to fill the tank?

41. Salvatore and Susan can construct a fence in 6 h. If Susan could complete the same job by herself in 9 h, how long would it take Salvatore, working alone?

42. One model printer can print a company's monthly billings 3 times as fast as another model. If the two printers, working together, can complete the job in 9 h, how long will it take the faster model, working alone?

43. The sum of the reciprocal of an integer and the square of the reciprocal of that integer is $\dfrac{4}{9}$. Find the integer.

44. The difference between the reciprocal of an integer and the square of its reciprocal is $\dfrac{3}{16}$. What is the integer?

45. The difference of an integer and its square root is 20. What is the integer?

46. The sum of an integer and 3 times its square root is 28. Find the integer.

[9.3] Solve each of the following equations. Be sure to check your solutions.

47. $\sqrt{x-5} = 4$

48. $\sqrt{3x-2} + 2 = 5$

49. $\sqrt{y+7} = y - 5$

50. $\sqrt{2x-1} + x = 8$

51. $\sqrt[3]{5x+2} = 3$

52. $\sqrt[3]{x^2+2} - 3 = 0$

53. $\sqrt{z+7} = 1 + \sqrt{z}$

54. $\sqrt{4x+5} - \sqrt{x-1} = 3$

[9.3] Solve each of the following equations for the indicated variable.

55. $r = \sqrt{x^2 + y^2}$ for x

56. $t = 2\pi\sqrt{\dfrac{l}{10}}$ for l

[9.4] Solve each of the following equations either by factoring directly or by substitution.

57. $x^4 - 11x^2 + 18 = 0$

58. $x^4 + x^2 = 20$

59. $w^4 = 9w^2$

60. $p^4 - 16 = 0$

61. $m - \sqrt{m} - 12 = 0$

62. $(x - 3)^2 + 5(x - 3) = 14$

63. $(t^2 - 2t)^2 - 9(t^2 - 2t) + 8 = 0$

64. $x^{1/2} - 2x^{1/4} - 3 = 0$

65. $x^{2/3} + x^{1/3} = 2$

66. $\dfrac{10}{p^2} + \dfrac{3}{p} - 1 = 0$

67. $\dfrac{6}{(w + 1)^2} + \dfrac{1}{w + 1} - 1 = 0$

68. $\dfrac{2}{(x - 1)^2} - \dfrac{3}{x - 1} - 5 = 0$

69. $x^{1/2} - 5x^{1/4} + 6 = 0$

70. $x^{1/2} - 4x^{1/4} - 5 = 0$

[9.5] Translate each of the following statements of variation to an algebraic equation, using k as the constant of variation.

71. d varies directly as the square of t.

72. r varies inversely as the square root of s.

73. y is directly proportional to x and inversely proportional to w.

74. z varies jointly as the cube of x and the square root of y.

[9.5] Find k, the constant of variation, given each of the following sets of conditions.

75. y varies directly as the cube root of x; $y = 12$ when $x = 8$

76. p is inversely proportional to the square of q; $p = 3$ when $q = 4$

77. r varies jointly as s and the square of t; $r = 150$ when $s = 2$ and $t = 5$

78. t varies directly as the square of u and inversely as v; $t = 36$ when $u = 3$ and $v = 2$

[9.5] Solve each of the following variation problems.

79. Let z vary inversely as the square of w. If $z = 3$ when $w = 4$, find the value of z when $w = 2$.

80. Suppose that s varies directly as the square of t and that $s = 90$ when $t = 3$. Find the value of s when $t = 5$.

81. The variable m varies jointly as p and the square of n. If $m = 144$ when $n = 2$ and $p = 3$, find the value of m when $n = 3$ and $p = 4$.

82. Let p be directly proportional to the square of q and inversely proportional to r. If $p = 2$ when $q = 2$ and $r = 12$, find the value of p when $q = 4$ and $r = 24$.

83. The distance that a ball will fall (neglecting air resistance) is directly proportional to the square of time. If the ball falls 64 ft in 2 s, how far will it fall in 5 s?

84. If the temperature of a gas is held constant, the volume occupied by that gas varies inversely as the pressure. A gas has volume 200 ft^3 when it is subjected to a pressure of 20 lb/in^2. What will its volume be under a pressure of 25 lb/in^2?

SELF-TEST

The purpose of this self-test is to help you check your progress and to review for a chapter test in class. Allow yourself about an hour to take the test. When you are done, check your answers in the back of the book. If you missed any problems, be sure to go back and review the appropriate sections in the chapter and the exercises that are provided.

Solve each of the following equations.

1. $\dfrac{x}{x + 3} + 1 = \dfrac{3x - 6}{x + 3}$

2. $\dfrac{2x}{x + 1} = \dfrac{3}{x - 2} + \dfrac{1}{x^2 - x - 2}$

Solve and graph the solution for each of the following.

3. $\dfrac{x + 4}{x - 3} \le 0$

4. $\dfrac{x + 7}{x + 3} \ge 3$

Solve each of the following applications.

5. One positive number is 3 more than another. If the sum of the reciprocals of the two numbers is $\dfrac{1}{2}$, find the two numbers.

6. Stephen drove 250 mi to visit Jovita. Returning by a shorter route, he found that the trip was only 225 mi, but traffic slowed his speed by 5 mi/h. If the two trips took exactly the same time, what was his rate each way?

Solve the following equations. Be sure to check your solutions.

7. $\sqrt{x-7} - 2 = 0$

8. $\sqrt{3w+4} + w = 8$

9. $\sqrt{3z+3} - \sqrt{z-2} = 3$

10. $5 - \sqrt{5x-1} = x$

Solve the following equations either by factoring directly or by substitution.

11. $x^4 - 12x^2 + 27 = 0$

12. $y - 11\sqrt{y} + 18 = 0$

13. $(m-2)^2 + 2(m-2) = 15$

14. $\dfrac{6}{x^2} + \dfrac{1}{x} - 1 = 0$

Solve each equation for the indicated variable.

15. $\dfrac{1}{i} = \dfrac{1}{p} + \dfrac{1}{q}$ for q

16. $\dfrac{3}{x} - 5 = \dfrac{2}{q}$ for q

17. Let s vary inversely as the cube of t. If $s = 16$ when $t = 2$, find the value of s when $t = 4$.

18. Variable p is jointly proportional to r and the square root of s. If $p = 80$ when $r = 4$ and $s = 16$, find the value of p when $r = 3$ and $s = 25$.

19. Suppose that z is directly proportional to x and inversely proportional to the square of y. If $z = 32$ when $x = 4$ and $y = 2$, find the value of z when $x = 6$ and $y = 4$.

20. The pressure at a point under water varies directly as the depth. If the pressure at a point 4 ft below the surface of the water is 250 lb/ft^2, find the pressure at a depth of 8 ft.

CUMULATIVE REVIEW EXERCISES

Here is a review of selected topics from the first nine chapters.

Solve each of the following:

1. $2x - 3(x + 2) = 4(5 - x) + 7$

2. $|3x - 7| \geq 5$

Graph each of the following.

3. $5x - 7y = 35$

4. $-8(2 - x) \geq y$

5. Find the equation of the line that passes through the points $(2, -1)$ and $(-3, 5)$.

6. Graph the linear inequality $4x - 5y > 20$.

7. Solve the following system of equations:

$11x - 7y = 9$

$8x + 6y = 62$

Simplify each of the following.

8. $4x^2 - 3x + 8 - 2(x^2 + 5) - 3(x - 1)$

9. $(3x + 1)(2x - 5)$

Factor each of the following completely.

10. $2x^2 - x - 10$

11. $25x^3 - 16xy^2$

Perform the indicated operations.

12. $\dfrac{2}{x-4} - \dfrac{3}{x-5}$

13. $\dfrac{x^2 - x - 6}{x^2 + 2x - 15} \div \dfrac{x - 2}{x + 5}$

Simplify each of the following radical expressions.

14. $\sqrt{18} + \sqrt{50} - 3\sqrt{32}$

15. $(3\sqrt{2} + 2)(3\sqrt{2} + 2)$

16. $\dfrac{5}{\sqrt{5} - \sqrt{2}}$

17. Find three consecutive odd integers whose sum is 237.

Solve each of the following equations.

18. $x^2 + x - 2 = 0$

19. $2x^2 - 6x - 5 = 0$

20. Solve the following inequality:

$2x^2 + x - 3 \leq 0$

GRAPHS OF QUADRATIC FUNC-TIONS AND CONIC SECTIONS

CHAPTER TEN

RENEWABLE ENERGY RE-SOURCES—PART 1
Renewable energy resources include solar heating and power generation, hydropower generation, wind generators, biomass, and geothermal energy. The renewable energy resources will be the primary sources of energy in the future because even with high levels of conservation, all other energy resources will eventually be exhausted.

Aside from the burning of wood (biomass) for fuel in the underdeveloped world and hydroelectric power generation in North America and Europe, most potential renewable energy resources have gone undeveloped. It is projected that by the middle of the twenty-first century, 30 to 45 percent of the world's energy needs could be met by renewable sources.

Solar energy is energy derived from the sun. Sunlight can be used directly or indirectly to produce heat or electricity. By designing buildings to capture as much light as possible, the sun can be used as a heat source. These designs use the greenhouse effect to heat buildings. Solar heating can also be done with collectors which concentrate and store solar energy as heat for space or water heating.

Solar energy is also used to generate electricity. Installations in the United States and France use large arrays of mirrors to concentrate solar energy for producing steam used to generate electricity. Photovoltaic cell technology developed for the space program is also being developed for commercial use.

Hydroelectricity uses the energy stored in falling water to generate electricity. Although much of the hydropower potential in the developed nations has been developed, few developing countries have developed their hydroelectric resources.

Large-scale hydroelectric dams are not without their environmental costs, however. They cause the flooding of usually fertile lands, they disrupt the environment necessary for some species of fish and wildlife, and they sometimes cause the displacement of large groups of people. Dams are expensive to build and can cause economic problems for poorer countries.

Solar energy and hydropower are two types of energy that can be developed with current technology. With increased use of these renewable sources, we can begin to relieve some of the pressure exerted on the environment by current energy systems.

511

10.1 Graphing Functions of the Form $f(x) = a(x - h)^2 + k$

TAPE IN22

OBJECTIVES: Given functions of the form $f(x) = (x - h)^2 + k$,
1. To find the axis of symmetry
2. To find the vertex
3. To graph the parabola by the method of translation

In Chapter 3 we dealt with the graphs of linear equations in two variables of the form

$$ax + by = c \qquad \text{where } a \text{ and } b \text{ cannot both be } 0$$

The graphs of all the linear equations were straight lines. Suppose that we now allow the terms in x and/or y to be quadratic; that is, we will allow squares in one or both of those terms. The graphs of such equations will form a family of curves called the *conic sections*. Conic sections are curves formed when a plane cuts through, or forms a section of, a cone. The conic sections comprise four curves—the *parabola, circle, ellipse,* and *hyperbola.* Examples of how these curves are formed are shown below.

The inclination of the plane determines which of the sections is formed.

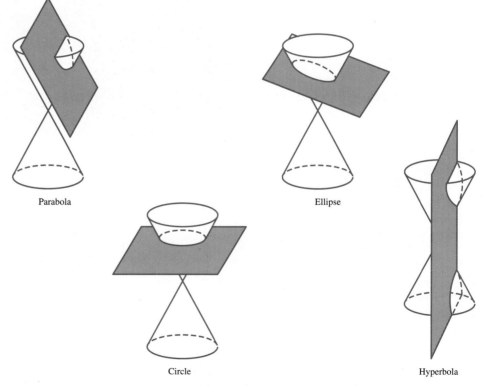

Parabola

Ellipse

Circle

Hyperbola

The names "ellipse," "parabola," and "hyperbola" are attributed to Apollonius, a third-century B.C. Greek mathematician and astronomer.

Our attention in this section is focused on the first of these sections, the parabola. Consider the function $f(x) = x^2$. This function is called a *quadratic function in x.* Its graph is a parabola. Let's look at that graph.

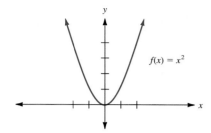

We could plot points, then connect them.

x	$f(x) = x^2$
-2	4
-1	1
0	0
1	1
2	4

Note that the parabola is rounded and is not pointed at the bottom.

There are three elements of the graph that should be noted.

1. The graph opens upward.

2. The y axis cuts the graph into two equal parts. A line that does this is called an *axis of symmetry*.

3. The graph has a minimum point, called the *vertex*.

Let's compare that graph to the graph of $f(x) = -x^2$.

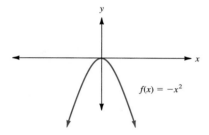

Looking at the three elements we examined earlier, we make three observations:

1. The graph opens downward.

2. The y axis is the *axis of symmetry*.

3. The graph has a maximum point, called the *vertex*.

It will always be the case that the sign of the coefficient of the x^2 term will determine which way the parabola opens. It will also be the case that a parabola opening upward has a minimum, and one opening downward has a maximum.
   For every function that is quadratic in x, we will look for three things:

1. Does the graph open upward or downward?

2. Where is the axis of symmetry?

3. What are the coordinates of the vertex?

EXAMPLE 1 Graphing a Parabola

Graph $f(x) = x^2 - 3$.

To better see the translation, graph both $y = x^2$ and $y = x^2 - 3$ on your calculator

Solution The difference between this graph and that of $f(x) = x^2$ is that each value of $f(x)$ has been decreased by 3. This results in a *translation* of 3 units in the negative direction on the y axis.

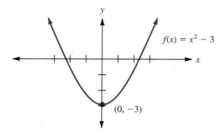

$f(x) = x^2 - 3$

$(0, -3)$

Note that the curve opens upward, the axis of symmetry is $x = 0$ (the y axis), and the vertex is $(0, -3)$.

Check Yourself 1

Graph $f(x) = -x^2 + 2$.

The graph of $f(x) = (x - h)^2$ will be translated along the x axis with the axis of symmetry at $x = h$.

EXAMPLE 2 Graphing a Parabola

Graph $f(x) = -(x - 3)^2$.

Solution Since the coefficient of the x^2 term is negative, the parabola opens downward and has a maximum point. Notice that when $x = 3$, y is 0 that is, $f(3) = 0$. If x is more than 3, $f(x)$ will be negative. Try to evaluate $f(x)$ for different values of x. When x is less than 3, $f(x)$ is also negative. Thus, the vertex is at $(3, 0)$, and the axis of symmetry is at $x = 3$.

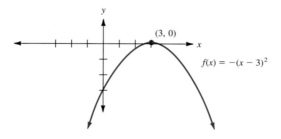

$(3, 0)$

$f(x) = -(x - 3)^2$

Check Yourself 2

Graph $f(x) = (x + 2)^2$.

Combining the lessons of the last two examples, we see that the graph of functions of the form $f(x) = (x - h)^2 + k$ is simply the parabola $f(x) = x^2$ translated horizontally h units and vertically k units.

EXAMPLE 3 Graphing a Parabola

Graph $f(x) = (x + 3)^2 + 1$.

Solution The parabola will be translated to the left 3 units and up 1 unit. The parabola opens upward, with an axis of symmetry at $x = -3$ and a vertex at $(-3, 1)$.

Note that the equation $y = a(x - h)^2 + k$ also defines quadratic functions since y is a function of x or $y = f(x)$.

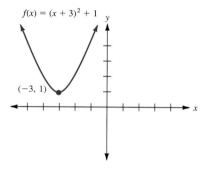

Check Yourself 3

Graph $f(x) = -(x - 2)^2 - 3$.

The rate at which the sides of a parabola rise (or fall) is determined by the coefficient of the x term.

EXAMPLE 4 Graphing a Parabola

Graph the parabolas $f(x) = 2x^2$, $f(x) = x^2$, and $f(x) = \dfrac{1}{2}x^2$ on the same axes.

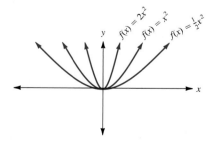

Notice that the larger the coefficient of x^2, the faster the parabola rises, thus the thinner the parabola appears.

Check Yourself 4

Graph the parabolas $f(x) = -2x^2$, $f(x) = -x^2$, and $f(x) = -\dfrac{1}{2}x^2$ on the same axes.

We can now graph any function of the form $f(x) = a(x - h)^2 + k$.

EXAMPLE 5 Graphing a Parabola

Graph $f(x) = -2(x - 3)^2 - 4$.

Solution This parabola will open downward, the axis of symmetry is at $x = 3$, the vertex is at $(3, -4)$, and it has the shape of $y = 2x^2$.

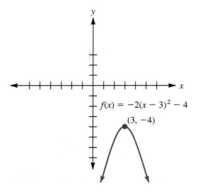

$$f(x) = -2(x - 3)^2 - 4$$

$(3, -4)$

Check Yourself 5

Graph $f(x) = \dfrac{1}{4}(x + 3)^2 - 1$.

Thus far we have looked only at parabolas that open upward or downward. An equation that is quadratic in y will open to the left or right. Equations of these parabolas are not functions of x.

EXAMPLE 6 Graphing a "Sideways" Parabola

Graph the equation $x = 2(y - 3)^2 + 2$.

Solution The parabola will open to the right [the coefficient of $(y - 3)^2$ is positive]. The axis of symmetry is the horizontal line $y = 3$. The vertex is at $(2, 3)$.

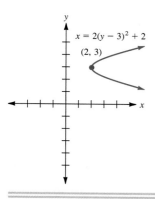

$x = 2(y - 3)^2 + 2$

$(2, 3)$

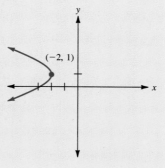

Note that this is not a function (it does not pass the vertical line test). Your calculator is designed to graph only functions. To graph this on the calculator, we would first solve for y

$x = 2(y - 3)^2 + 2$

$x - 2 = 2(y - 3)^2$

$\dfrac{1}{2x} - 1 = (y - 3)^2$

$y - 3 = \pm\sqrt{\left(\dfrac{1}{2x} - 1\right)}$

$y = 3 \pm \sqrt{\left(\dfrac{1}{2x} - 1\right)}$

This can then be separated into two functions

$y = 3 + \sqrt{\left(\dfrac{1}{2x} - 1\right)}$

and $y = 3 - \sqrt{\left(\dfrac{1}{2x} - 1\right)}$

If both of these are graphed simultaneously, the result will be the graph of the relation

$x = 2(y - 3)^2 + 2$

Check Yourself 6

Graph the equation $x = -\dfrac{1}{2}(y - 1)^2 - 2$.

CHECK YOURSELF ANSWERS

1.

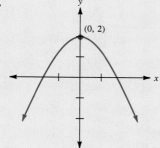

$(0, 2)$

2.

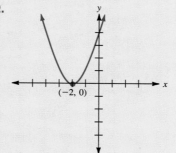

$(-2, 0)$

3.

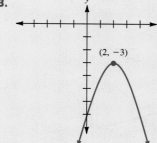

$(2, -3)$

4.

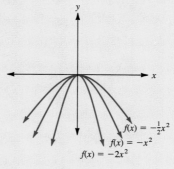

$f(x) = -\dfrac{1}{2}x^2$

$f(x) = -x^2$

$f(x) = -2x^2$

5.

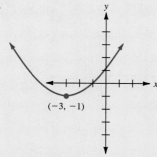

$(-3, -1)$

6.

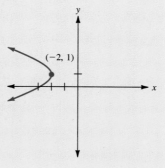

$(-2, 1)$

10.1 EXERCISES

Build Your Skills

Match the graphs to the equations.

(a) $y = (x - 3)^2 + 1$ (b) $y = (x + 3)^2 - 1$
(c) $y = (x - 3)^2 - 1$ (d) $y = (x + 3)^2 + 1$
(e) $x = (y - 3)^2 + 1$ (f) $x = (y + 3)^2 - 1$
(g) $x = (y - 3)^2 - 1$ (h) $x = (y + 3)^2 + 1$

1.

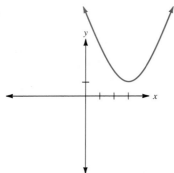

2.

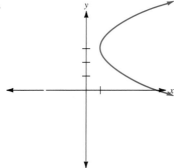

3.

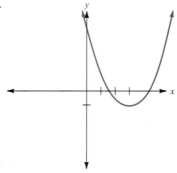

4.

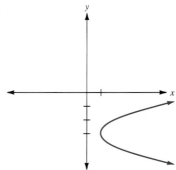

5.

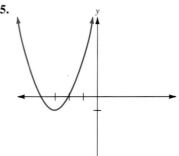

6.

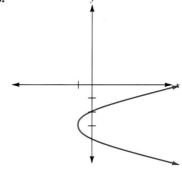

7.

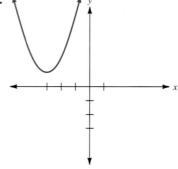

8.

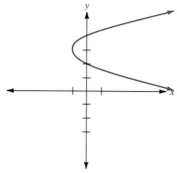

Determine whether the graph of each quadratic function is translated from the origin to the left, to the right, upward, or downward.

9. $f(x) = x^2 - 3$

10. $f(x) = (x - 3)^2$

11. $f(x) = (x + 5)^2$

12. $f(x) = x^2 + 7$

13. $f(x) = (x - 2)^2$

14. $f(x) = (x + 5)^2$

Find the equation of the axis of symmetry and the coordinates of the vertex; then graph each function or equation.

15. $f(x) = 2x^2 - 5$

16. $f(x) = -2(x - 5)^2$

17. $f(x) = -\dfrac{1}{2}(x + 3)^2$

18. $f(x) = \dfrac{1}{4}x^2 + 3$

19. $f(x) = (x - 3)^2 + 2$

20. $f(x) = -(x + 2)^2 - 1$

21. $f(x) = -2(x + 3)^2 - 5$

22. $f(x) = \dfrac{1}{2}(x - 3)^2 + 2$

23. $x = y^2 - 4$

24. $x = (y + 2)^2$

25. $x = y^2 + 1$

26. $x = (y - 7)^2$

27. $x = 2y^2 + 3$

28. $x = -2(y + 3)^2$

29. $x = -3(y - 3)^2 + 1$

30. $x = \dfrac{1}{2}(y + 2)^2 + 3$

Transcribe Your Skills

31. Explain why the equation $x = y^2$ does not represent a function.

32. What values of a will cause the graph of $y = ax^2$ to be narrower than the graph of $y = x^2$? Wider than the graph of $y = x^2$?

Think About These

33. A parabola can be defined as the set of points that are an equal distance from a point (called the *focal point*) and a line (called the *directrix*). In the drawing below, those distances are represented by d_1 and d_2. Use the distance formula to express d_1 and d_2 in terms of x, y, and p. Then derive the equation for the parabola.

$$y^2 = 4px$$

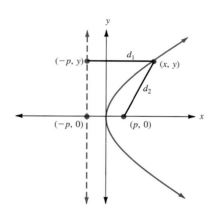

34. In the drawing below, derive the equation $x^2 = 4py$.

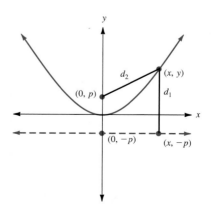

Graphing Utility

Describe a viewing window that would include the vertex and all intercepts for the graph of each of the following functions.

35. $f(x) = 3(x - 2)^2 - 5$

36. $f(x) = 9(x + 4)^2 - 7$

37. $f(x) = -2(x + 3)^2 + 5$

38. $f(x) = -5(x - 7)^2 + 12$

Skillscan (Section 8.2)

Determine the constant to add to each of the following expressions to form a perfect-square trinomial.

a. $x^2 + 4x$ **b.** $y^2 - 6y$ **c.** $y^2 - 10y$ **d.** $x^2 + 8x$
e. $x^2 - 20x$ **f.** $y^2 + 12y$ **g.** $y^2 + 16y$
h. $x^2 - 24x$

10.2   Graphing Functions of the Form $f(x) = ax^2 + bx + c$

TAPE IN22

OBJECTIVES: 1. To graph parabolas
                     2. To solve applications that result in quadratic equations

In the last section we graphed functions of the form $f(x) = a(x - h)^2 + k$. The standard form for a quadratic function in two variables is $f(x) = ax^2 + bx + c$. Any quadratic function can be written in either of these forms. The following example illustrates.

EXAMPLE 1 Graphing a Parabola by Completing the Square

Graph

$$f(x) = x^2 - 2x - 3$$

Solution Using the techniques of Section 8.2, we first complete the square.

One-half the middle term squared will complete the square.

$$f(x) = x^2 - 2x - 3$$
$$= x^2 - 2x + 1 - 3 - 1$$
$$= (x - 1)^2 - 4$$

Check the result by graphing both $f(x) = x^2 - 2x - 3$ and $f(x) = (x - 1)^2 - 4$ on your graphing utility. Confirm that they have the same graph.

From our work in the previous section, we know the parabola opens upward, the axis of symmetry is $x = 1$, and the vertex is at $(1, -4)$.

To improve the sketch, we can find the x intercepts. These are the x values for which $f(x) = 0$.

$$0 = (x - 1)^2 - 4$$
$$4 = (x - 1)^2$$
$$\pm 2 = x - 1$$

$$x = 1 + 2 \qquad \text{or} \qquad x = 1 - 2$$
$$= 3 \qquad\qquad\qquad\qquad = -1$$

The x intercepts are $(3, 0)$ and $(-1, 0)$.

Now draw a smooth curve connecting the vertex and the x intercepts.

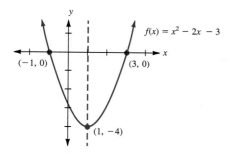

$f(x) = x^2 - 2x - 3$

$(-1, 0)$

$(3, 0)$

$(1, -4)$

Check Yourself 1

Graph $f(x) = -x^2 + 6x - 5$

Hint: Since the coefficient of x^2 is negative, the parabola opens downward.

A similar process will work if the quadratic member of the given equation is *not* factorable. In that case, one of two things happens:

1. The x intercepts are irrational. In this case, a calculator can be used to estimate the intercepts.

2. The x intercepts do not exist.

Consider the following example.

EXAMPLE 2 Graphing a Parabola by Completing the Square

Graph

$$f(x) = x^2 - 4x + 2$$
$$= x^2 - 4x + 2$$
$$= x^2 - 4x + 4 + 2 - 4$$
$$= (x - 2)^2 - 2$$

To keep the equation balanced, we both add and subtract 4.

The parabola opens upward, the axis of symmetry is $x = 2$, and the vertex is $(2, -2)$.

Again, we can improve the sketch if we find two symmetric points. Here the quadratic member is not factorable, and the x intercepts are irrational, so we prefer to find another pair of symmetric points.

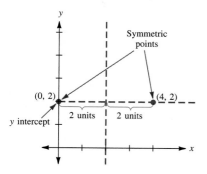

Symmetric points

$(0, 2)$

$(4, 2)$

2 units   2 units

y intercept

Note that $(0, 2)$ is the y intercept of the parabola. We found the axis of symmetry at $x = 2$ earlier. Note that the symmetric point to $(0, 2)$ lies along the horizontal line through the y intercept at the same distance (2 units) from the axis of symmetry, or at $x = 4$. Hence, $(4, 2)$ is the symmetric point.

Draw a smooth curve connecting the points found above to form the parabola.

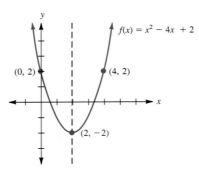

Check Yourself 2

Graph $f(x) = x^2 + 2x + 3$.

The coefficient of x^2 was 1 or -1 in our previous example and exercises. The following example shows the effect of different coefficients of the x^2 term.

EXAMPLE 3 Graphing a Parabola by Completing the Square

Graph

$$f(x) = 2x^2 - 4x + 3$$

Step 1 Complete the square.

$$f(x) = 2(x^2 - 2x) + 3$$
$$= 2(x^2 - 2x + 1) + 3 - 2$$
$$= 2(x - 1)^2 + 1$$

We have added 2 times 1, so we must also subtract 2.

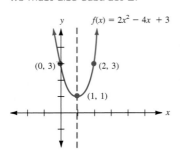

Step 2 The axis of symmetry is $x = 1$, and the vertex is at $(1, 1)$.

Step 3 Find symmetric points. Again, the quadratic member is not factorable, and we use the y intercept $(0, 3)$ and its symmetric point $(2, 3)$.

Step 4 Connect the points with a smooth curve to form the parabola.

Compare this curve to those in previous examples. Note that the parabola is "tighter" about the axis of symmetry. That is the effect of the larger x^2 coefficient.

Check Yourself 3

Graph $f(x) = \dfrac{1}{2}x^2 - 2x - 2$.

The following algorithm summarizes our work thus far in this section.

> **GRAPHING A QUADRATIC EQUATION**
> 1. **Complete the square for the quadratic variable.**
> 2. **Find the axis of symmetry and the vertex.**
> 3. **Determine two symmetric points.**
>
> **Note** You can use the x intercepts if the quadratic member of the given equation is factorable. Otherwise use the y intercept and its symmetric point.
>
> 4. **Draw a smooth curve connecting the points found above, to form the parabola. You may choose to find additional pairs of symmetric points at this time.**

If we use the algorithm to find the vertex of $f(x) = ax^2 + bx + c$, we get a useful result. The axis of symmetry will always occur at $x = -b/(2a)$.

EXAMPLE 4 A Business Application

A software company sells a word processing program for personal computers. It has found that the monthly profit in dollars P from selling x copies of the program is approximated by

$$P(x) = -0.2x^2 + 80x - 1200$$

Find the number of copies of the program that should be sold in order to maximize the profit.

Solution Since the relating equation is quadratic, the graph must be a parabola. Also, since the coefficient of x^2 is negative, the parabola must open downward, and thus the vertex will give the maximum value for the profit P. To find the vertex,

$$x = -\frac{b}{2a} = \frac{-80}{2(-0.2)} = \frac{-80}{-0.4} = 200$$

The maximum profit must then occur when $x = 200$, and we substitute that value into the original equation:

$$P(x) = -0.2(200)^2 + (80)(200) - 1200$$
$$= \$6800$$

The maximum profit will occur when 200 copies are sold per month, and that profit will be $6800.

Check Yourself 4

A company which sells portable radios finds that its weekly profit in dollars P and the number of radios sold x are related by

$$P(x) = -0.1x^2 + 20x - 200$$

Find the number of radios that should be sold to have the largest weekly profit and the amount of that profit.

EXAMPLE 5 A Fencing Application

A farmer has 1000 ft of fence and wishes to enclose the largest possible rectangular area with that fencing. Find the length and width of the largest possible area that can be enclosed.

As usual, when dealing with geometric figures, we start by drawing a sketch of the problem.

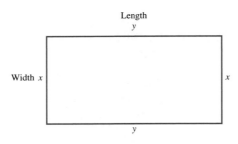

First, we can write the area A as

Area = length × width

$$A = xy \tag{1}$$

Also, since 1000 ft of fence is to be used, we know that

The perimeter of the region is

$2x + 2y$

$$2x + 2y = 1000$$
$$2y = 1000 - 2x$$
$$y = 500 - x \tag{2}$$

Substituting for y in Equation (1), we have

$$A = x(500 - x) = 500x - x^2$$
$$= -x^2 + 500x \tag{3}$$

Again, the graph for A is a parabola opening downward, and the largest possible area will occur at the vertex. As before,

The width x is 250 ft. From (2)

$y = 500 - 250$

$\quad = 250$ ft

The length is also 250 ft. The desired region is a square.

$$x = \frac{-500}{2(-1)} = \frac{-500}{-2} = 250$$

and the largest possible area is

$$A = -(250)^2 + 500(250) = 62{,}500 \text{ ft}^2$$

Check Yourself 5

We want to enclose the largest possible rectangular area by using 400 ft of fence. The fence will be connected to the house, so only three sides of fence will be needed. What should be the dimensions of the area?

We conclude this section with a formal geometric description of the parabola. As is true of all the conic sections, this geometric definition can be combined with the distance formula, developed earlier in this chapter, to write the corresponding algebraic equation for the parabola.

A *parabola* is the set of all points in the plane equidistant from a fixed point, called the *focus* of the parabola, and a fixed line, called the *directrix* of the parabola.

For any point (x, y) on the parabola,

$d_1 = d_2$

The parabola:

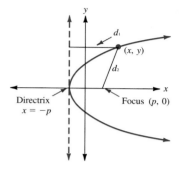

There are many physical models involving a parabolic path or design. The fact that a projectile follows a parabolic path was known in the time of Galileo. The reflective properties of the parabola also are important. If light rays enter a parabola along lines parallel to the axis of symmetry, the rays will reflect off the parabola and will pass through the focus of the parabola. That fact accounts for the design of parabolic radar antennas and solar collectors.

The reflective surface of an automobile headlight is also a parabolic surface. The light source in the headlight is placed approximately at the focus, so that the light rays will reflect off the surface as nearly parallel rays.

So far we have dealt with functions of the form

$$f(x) = ax^2 + bx + c$$

Suppose we now consider an equation of the form

$$x = ay^2 + by + c$$

which is quadratic in y but not in x. The graph of such an equation is once again a parabola, but this time the parabola will be horizontally oriented—opening to the right or left as a is positive or negative, respectively. This equation does not represent a function.

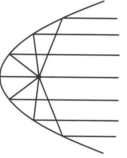

For $x = ay^2 + by + c$,

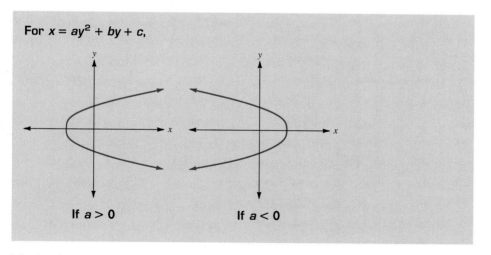

If $a > 0$ If $a < 0$

Much of what we did earlier is easily extended to this new case. The following example will illustrate the changes in the process.

EXAMPLE 6 Graphing a "Sideways" Parabola

Graph the equation

$$x = y^2 + 2y - 3$$

$$a = 1, b = 2$$

Step 1 Find the axis of symmetry.

The axis of symmetry is a horizontal line.

$$y = -\frac{b}{2a} = \frac{-2}{2 \cdot 1} = -1$$

Step 2 Find the vertex.
    If $y = -1$,

Since the axis of symmetry intersects the vertex, the y coordinate of the vertex is -1.

$$x = (-1)^2 + 2(-1) - 3 = -4$$

So $(-4, -1)$ is the vertex.

Step 3 Find two symmetric points.
    Here the quadratic member is factorable, so set $x = 0$ in the original equation. This gives the y intercepts:

$$0 = y^2 + 2y - 3$$

$$= (y + 3)(y - 1)$$

$$y + 3 = 0 \qquad \text{or} \qquad y - 1 = 0$$

$$y = -3 \qquad\qquad\qquad y = 1$$

The y intercepts then are at $(0, -3)$ and $(0, 1)$.

Step 4 Draw a smooth curve through the points found above, to form the parabola.

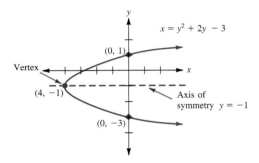

Check Yourself 6

Graph the equation $x = -y^2 + 5y - 4$.

From graphs of equations of the form $f(x) = ax^2 + bx + c$, we know that if $a > 0$, then the vertex is the lowest point on the graph (the minimum value). Also, if $a < 0$, then the vertex is the highest point on the graph (the maximum value). We can use this result to solve a variety of problems in which we want to find the maximum or minimum value of a variable.

CHECK YOURSELF ANSWERS

1. $f(x) = -x^2 + 6x - 5$

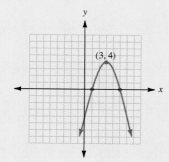

2. $f(x) = x^2 + 2x + 3$

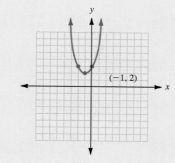

3. $f(x) = \dfrac{1}{2}x^2 - 2x - 2$

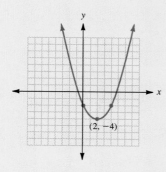

4. 100 radios, $800

5. Width 100 ft, length 200 ft

6. $x = -y^2 + 5y - 4$

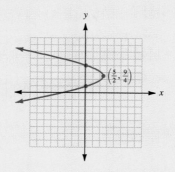

10.2 EXERCISES

Build Your Skills

Match each graph with one of the equations.

(a) $y = x^2 + 2$ (b) $y = 2x^2 - 1$
(c) $y = 2x + 1$ (d) $y = x^2 - 3x$
(e) $y = -x^2 - 4x$ (f) $y = -2x + 1$
(g) $y = x^2 + 2x - 3$ (h) $y = -x^2 + 6x - 8$

1.

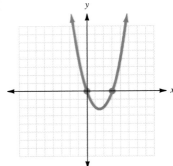

2.

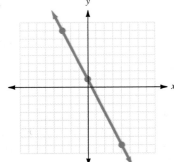

3.

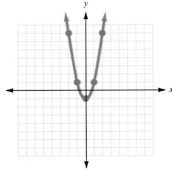

4.

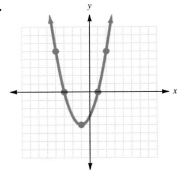

5.

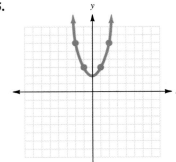

6.

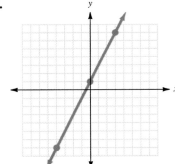

7.

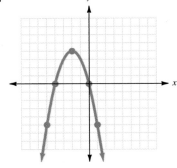

8.

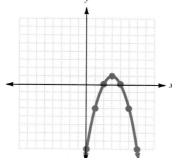

Which of the given conditions apply to the graphs of each of the following functions? Note that more than one condition will apply.

(*a*) The parabola opens upward.
(*b*) The parabola opens downward.
(*c*) The parabola has two x intercepts.
(*d*) The parabola has one x intercept.
(*e*) The parabola has no x intercept.

9. $f(x) = x^2 - 2$

10. $f(x) = -x^2 + 2x$

11. $f(x) = x^2 - 2x - 3$

12. $f(x) = x^2 - 4x + 5$

13. $f(x) = -x^2 - 4x + 12$

14. $f(x) = x^2 - 6x + 9$

Find the equation of the axis of symmetry, the coordinates of the vertex, and the x intercepts. Sketch the graph of each.

15. $f(x) = x^2 - 2x$

16. $f(x) = x^2 - 4$

17. $f(x) = -x^2 + 1$

18. $f(x) = x^2 + 4x$

19. $f(x) = -x^2 - 4x$

20. $f(x) = -x^2 + 3x$

21. $f(x) = x^2 - 2x - 3$

22. $f(x) = x^2 - x - 6$

23. $f(x) = x^2 - 5x + 4$

24. $f(x) = x^2 + 3x + 2$

25. $f(x) = -x^2 + 3x + 4$

26. $f(x) = x^2 - 6x + 8$

27. $f(x) = -x^2 + 6x - 5$

28. $f(x) = -x^2 + 4x - 3$

Find the equation of the axis of symmetry, the coordinates of the vertex, and at least two symmetric points. Sketch the graph of each function or equation.

29. $f(x) = x^2 - 2x - 2$

30. $f(x) = x^2 - 4x - 3$

31. $f(x) = x^2 + 4x + 5$

32. $f(x) = -x^2 + 3x - 3$

33. $f(x) = -x^2 + 6x - 2$

34. $f(x) = x^2 + 5x + 2$

35. $f(x) = 2x^2 - 4x + 1$

36. $f(x) = \dfrac{1}{2}x^2 + x - 2$

37. $f(x) = -\dfrac{1}{3}x^2 + x - 2$

38. $f(x) = -2x^2 + 4x - 3$

39. $f(x) = 3x^2 + 6x - 1$

40. $f(x) = -3x^2 + 12x - 5$

41. $x = y^2 - 4y$

42. $x = y^2 + 3y$

43. $x = y^2 - 3y - 4$

44. $x = -y^2 - y + 6$

45. A company's weekly profit P is related to the number of items sold by $P(x) = -0.2x^2 + 40x - 500$. Find the number of items that should be sold each week in order to maximize the profit. Then find the amount of that weekly profit.

46. A company's monthly profit P is related to the number of items sold by $P(x) = -0.1x^2 + 30x - 1000$. How many items should be sold each month to make the largest possible profit? What is that profit?

47. A builder wants to enclose the largest possible rectangular area with 1600 ft of fencing. What should the dimensions of the rectangle be, and what will its area be?

48. A farmer wants to enclose a field along a river on three sides. If 1200 ft of fencing is to be used, what dimensions will give the maximum enclosed area? Find that maximum area.

49. A ball is thrown upward into the air with an initial velocity of 64 ft/s. If h gives the height of the ball at time t, then the equation relating h and t is

$$h(t) = -16t^2 + 64t$$

Find the maximum height that the ball will attain.

50. A ball is thrown upward into the air with an initial velocity of 32 ft/s. If h gives the height of the ball at time t, then the equation relating h and t is

$$h(t) = -16t^2 + 32t$$

Find the maximum height that the ball will attain.

Transcribe Your Skills

51. Explain how to determine the domain and range of the function $f(x) = a(x - h)^2 + k$.

52. Give some specific practical situations where you might encounter a parabolic shape.

Think About These

Graph each inequality.

53. $y \geq -2(x - 3)^2 + 2$

54. $y \leq -\dfrac{1}{2}(x + 2)^2 + 5$

 55. The depth d, in centimeters, of a parabolic solar collector is related to its radius r, in centimeters, by the following equation:

$$d(r) = -0.008r^2 + 0.8r$$

What is the depth of the collector at its deepest point?

 56. The depth d, in meters, of a parabolic solar collector is related to its radius r, in meters, by the following equation:

$$d(r) = -r^2 + r$$

What is the depth of the collector at its deepest point?

 57. Determine the width of the collector in Exercise 55. (*Hint:* Determine the x intercepts.)

 58. Determine the width of the collector in Exercise 56. (*Hint:* Determine the x intercepts.)

▱ Graphing Utility

Describe a viewing window that would include the vertex and all intercepts for the graph of each of the following functions.

59. $f(x) = 3x^2 - 25$

60. $f(x) = 9x^2 - 5x - 7$

61. $f(x) = -2x^2 + 5x - 7$

62. $f(x) = -5x^2 + 2x + 12$

Skillscan (Section 3.4)

Find the distance between each of the following pairs of points.

a. $(2, 3)$ and $(1, 5)$ **b.** $(1, 4)$ and $(2, 5)$ **c.** $(-1, 2)$ and $(3, 2)$ **d.** $(2, -1)$ and $(2, 3)$ **e.** $(-2, -3)$ and $(-1, 1)$ **f.** $(-1, 4)$ and $(4, -1)$

 10.3   **The Circle**

TAPE IN22

OBJECTIVES: 1. To identify the graph of an equation as a line, parabola, or circle
2. To write the equation of a circle in standard form and graph the circle

In the first two sections of this chapter, we examined the parabola. In this section we will turn our attention to another conic section, the circle.

The distance formula is central to any discussion of conic sections, and we restate it here for reference.

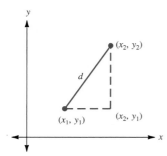

> **THE DISTANCE FORMULA**
>
> The *distance d* between two points (x_1, y_1) and (x_2, y_2) is given by
>
> $$d = \sqrt{(x_2 - x_1)^2 + (y_2 - y_1)^2}$$

Let's consider the circle. We offer the following definition.

A *circle* is the set of all points in the plane equidistant from a fixed point, called the *center* of the circle. The distance between the center of the circle and any point on the circle is called the *radius* of the circle.

From this definition we can use the distance formula to derive the algebraic equation of a circle, given its center and its radius.

Suppose a circle has its center at a point with coordinates (h, k) and radius r. If (x, y) represents any point on the circle, then by the definition the distance from (h, k) to (x, y) is r. Applying the distance formula, we have

$$r = \sqrt{(x - h)^2 + (y - k)^2}$$

Squaring both sides of the equation gives the equation of the circle

$$r^2 = (x - h)^2 + (y - k)^2$$

In general, we can write the following:

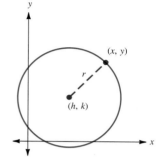

A special case is the circle centered at the origin with radius r. Then $(h, k) = (0, 0)$, and its equation is

$$x^2 + y^2 = r^2$$

EQUATION OF A CIRCLE

The equation of a circle with center (h, k) and radius r is

$$(x - h)^2 + (y - k)^2 = r^2 \qquad (1)$$

Equation (1) can be used in two ways. Given the center and radius of the circle, we can write its equation; or given its equation, we can find the center and radius of a circle.

EXAMPLE 1 Finding the Equation of a Circle

Find the equation of a circle with center at $(2, -1)$ and radius 3. Sketch the circle.

Solution Let $(h, k) = (2, -1)$ and $r = 3$. Applying Equation (1) yields

$$(x - 2)^2 + [y - (-1)]^2 = 3^2$$
$$(x - 2)^2 + (y + 1)^2 = 9$$

To sketch the circle, we locate the center of the circle. Then we determine four points which are 3 units, to the right and left and up and down, from the center of the circle.

Drawing a smooth curve through those four points completes the graph.

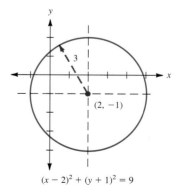

$(x - 2)^2 + (y + 1)^2 = 9$

Check Yourself 1

Find the equation of the circle with center at $(-2, 1)$ and radius 5. Sketch the circle.

Now, given an equation for a circle, we can also find the radius and center and then sketch the circle. We start with an equation in the special form of Equation (1).

EXAMPLE 2 Finding the Center and Radius of a Circle

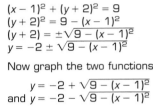
Just as was true of the parabola in Section 10.1, the circle can be graphed on the calculator by solving for y, then graphing both the upper half and lower half of the circle. In this case

$(x - 1)^2 + (y + 2)^2 = 9$
$(y + 2)^2 = 9 - (x - 1)^2$
$(y + 2) = \pm\sqrt{9 - (x - 1)^2}$
$y = -2 \pm \sqrt{9 - (x - 1)^2}$

Now graph the two functions

$y = -2 + \sqrt{9 - (x - 1)^2}$
and $y = -2 - \sqrt{9 - (x - 1)^2}$

on your calculator. (The display screen may need to be squared to obtain the shape of a circle.)

Find the center and radius of the circle with equation

$$(x - 1)^2 + (y + 2)^2 = 9$$

Solution Remember, the general form is

$$(x - h)^2 + (y - k)^2 = r^2$$

Our equation "fits" this form when it is written as

Note: $y + 2 = y - (-2)$

$$(x - 1)^2 + [y - (-2)]^2 = 3^2$$

So the center is at $(1, -2)$, and the radius is 3. The graph is shown.

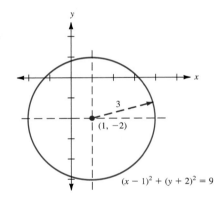

$(x - 1)^2 + (y + 2)^2 = 9$

Check Yourself 2

Find the center and radius of the circle with equation

$$(x + 3)^2 + (y - 2)^2 = 16$$

Sketch the circle.

To graph the equation of a circle that is not in standard form, we *complete the square.*

Let's see how completing the square can be used in graphing the equation of a circle.

To recognize the equation as having the form of a circle, note that the coefficients of x^2 and y^2 are equal.

The linear terms in x and y show a translation of the center away from the origin.

EXAMPLE 3 Finding the Center and Radius of a Circle

Find the center and radius of the circle with equation

$$x^2 + 2x + y^2 - 6y = -1$$

Then sketch the circle.

Solution We could, of course, simply substitute values of x and try to find the corresponding values for y. A much better approach is to rewrite the original equation so that it matches the standard form.

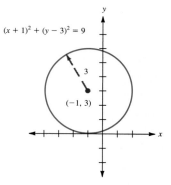

$(x + 1)^2 + (y - 3)^2 = 9$

First, add 1 to both sides to complete the square in x:

$$x^2 + 2x + 1 + y^2 - 6y = -1 + 1$$

Then add 9 to both sides to complete the square in y:

$$x^2 + 2x + 1 + y^2 - 6y + 9 = -1 + 1 + 9$$

We can factor the two trinomials on the left (they are both perfect squares) and simplify on the right:

$$(x + 1)^2 + (y - 3)^2 = 9$$

The equation is now in standard form, and we can see that the center is at $(-1, 3)$ and the radius is 3. The sketch of the circle is shown. Note the "translation" of the center to $(-1, 3)$.

Check Yourself 3

Find the center and radius of the circle with equation

$$x^2 - 4x + y^2 + 2y = -1$$

Sketch the circle.

CHECK YOURSELF ANSWERS

1. $(x + 2)^2 + (y - 1)^2 = 25$

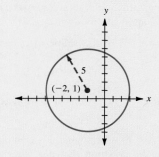

2. $(x + 3)^2 + (y - 2)^2 = 16$

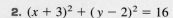
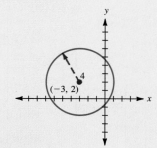

3. $(x - 2)^2 + (y + 1)^2 = 4$

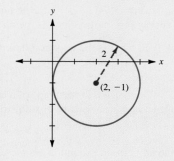

10.3 EXERCISES

Build Your Skills

Decide whether each of the following equations has as its graph a line, a parabola, a circle, or none of these.

1. $y = x^2 - 2x + 5$

2. $y^2 + x^2 = 64$

3. $y = 3x - 2$

4. $2y - 3x = 12$

5. $(x - 3)^2 + (y + 2)^2 = 10$

6. $y + 2(x - 3)^2 = 5$

7. $x^2 + 4x + y^2 - 6y = 3$

8. $4x = 3$

9. $y^2 - 4x^2 = 36$

10. $x^2 + (y - 3)^2 = 9$

11. $y = -2x^2 + 8x - 3$

12. $2x^2 - 3y^2 + 6y = 13$

Find the center and the radius for each circle.

13. $x^2 + y^2 = 25$

14. $x^2 + y^2 = 72$

15. $(x - 3)^2 + (y + 1)^2 = 16$

16. $(x + 3)^2 + y^2 = 81$

17. $x^2 + 2x + y^2 = 15$

18. $x^2 + y^2 - 6y = 72$

19. $x^2 - 6x + y^2 + 8y = 16$

20. $x^2 - 5x + y^2 - 3y = 8$

Graph each of the following circles by finding the center and the radius.

21. $x^2 + y^2 = 4$

22. $x^2 + y^2 = 25$

23. $4x^2 + 4y^2 = 36$

24. $9x^2 + 9y^2 = 144$

25. $(x - 1)^2 + y^2 = 9$

26. $x^2 + (y + 2)^2 = 16$

27. $(x - 4)^2 + (y + 1)^2 = 16$

28. $(x + 3)^2 + (y + 2)^2 = 25$

29. $x^2 + y^2 - 4y = 12$

30. $x^2 - 6x + y^2 = 0$

31. $x^2 - 4x + y^2 + 2y = -1$

32. $x^2 - 2x + y^2 - 6y = 6$

Transcribe Your Skills

33. Describe the graph of $x^2 + y^2 - 2x - 4y + 5 = 0$.

34. Describe how completing the square is used in graphing circles.

Think About These

Graph each inequality.

35. $y^2 + (x - 1)^2 \geq 25$

36. $(y + 3)^2 + (x - 2)^2 \leq 16$

37. $(y - 5)^2 + (x + 4)^2 > 9$

38. $y^2 + x^2 - 2y + 8x \leq 19$

Although parabolic mirrors are the most efficient solar collectors, they are also the most difficult to construct. For this reason, most commercial solar collectors are constructed in the shape of hemispheres (half spheres) or half cylinders. These shapes follow the principles of circles for their focusing properties.

While the efficiency of the circle is not as good as that of the parabola as a focusing tool, the circle's relatively high efficiency and its ease of construction combine to make it a more practical commercial application.

39. A solar oven is constructed in the shape of a hemisphere. If the equation

$$x^2 + y^2 + 500 = 1000$$

describes the circumference of the oven in centimeters, what is its radius?

40. A solar oven in the shape of a hemisphere is to have a diameter of 80 cm. Write the equation that describes the circumference of this oven.

41. A solar water heater is constructed in the shape of a half cylinder with the water supply pipe at its center. If the water heater has a diameter of $\frac{4}{3}$ m, what is the equation describing its circumference?

42. A solar water heater is constructed in the shape of a half cylinder having a circumference described by the equation

$$9x^2 + 9y^2 - 16 = 0$$

What is its diameter if the units for the equation are meters?

Graphing Utility

A circle can be graphed on any graphing utility by plotting the upper and lower semicircles on the same axes. For example, to graph $x^2 + y^2 = 16$, we solve for y:

$$y = \pm\sqrt{16 - x^2}$$

This is then graphed as two separate functions,

$$y = \sqrt{16 - x^2} \qquad \text{and} \qquad y = -\sqrt{16 - x^2}$$

Use that technique to graph each of the following circles.

43. $x^2 + y^2 = 36$

44. $(x - 3)^2 + y^2 = 9$

45. $(x + 5)^2 + y^2 = 36$

46. $(x - 2)^2 + (y + 1)^2 = 25$

Skillscan

In each of the following, find the value of x when $y = 0$.

a. $\dfrac{x^2}{25} + \dfrac{y^2}{16} = 1$ **b.** $\dfrac{x^2}{9} - \dfrac{y^2}{36} = 1$

c. $4y^2 + 16x^2 = 64$ **d.** $25x^2 - 9y^2 = 225$

e. $x^2 + y^2 = 16$ **f.** $\dfrac{x^2}{49} - \dfrac{y^2}{64} = 1$

The Ellipse and the Hyperbola   10.4

OBJECTIVES:
1. To identify the graph of an equation as a conic section
2. To graph an ellipse
3. To graph a hyperbola

TAPE IN22

Let's turn now to the third conic section, the ellipse. It can be described as an "oval-shaped" curve, and it has the following geometric description.

> An *ellipse* is the set of all points (x, y) such that the sum of the distances from (x, y) to two fixed points, called the *foci* of the ellipse, is constant.

Ellipses occur frequently in nature. The planets have elliptical orbits with the sun at one focus.

The reflecting properties of the ellipse are also interesting. Rays from one focus are reflected by the ellipse in such a way that they always pass through the other focus.

The following sketch illustrates the definition in two particular cases:

1. When the foci are located on the x axis and are symmetric about the origin
2. When the foci are located on the y axis and are symmetric about the origin

In either case, $d_1 + d_2$ is constant.

AN ELLIPSE

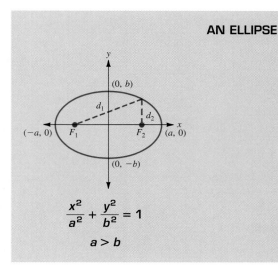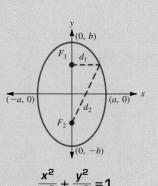

To sketch ellipses of the forms above quickly, we need to determine only *four points,* the points where the ellipse intercepts the coordinate axes.

Fortunately those points are easily found when the ellipse is written in standard form:

$$\frac{x^2}{a^2} + \frac{y^2}{b^2} = 1 \tag{1}$$

The *x* intercepts are at *a* and at $-a$. The *y* intercepts are at *b* and at $-b$. Let's use this information to sketch an ellipse.

EXAMPLE 1 Graphing an Ellipse

Sketch the ellipse

$$\frac{x^2}{9} + \frac{y^2}{4} = 1$$

Step 1 The equation is in standard form.

Step 2 Find the *x* intercepts. From Equation (1), $a^2 = 9$, and the *x* intercepts are at 3 and at -3.

Step 4 Find the *y* intercepts. From Equation (1), $b^2 = 4$, and the *y* intercepts are at 2 and at -2.

Step 4 Plot the intercepts found above, and draw a smooth curve to form the desired ellipse.

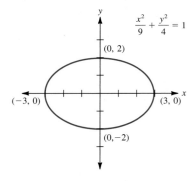

Check Yourself 1

Sketch the ellipse.

$$\frac{x^2}{16} + \frac{y^2}{9} = 1$$

EXAMPLE 2 Graphing an Ellipse

Sketch the ellipse with equation

$$9x^2 + 4y^2 = 36$$

Step 1 Since this equation is *not* in standard form (the right side is *not* 1), we divide both sides of the equation by the constant 36:

$$\frac{9x^2}{36} + \frac{4y^2}{36} = \frac{36}{36}$$

$$\frac{x^2}{4} + \frac{y^2}{9} = 1$$

We can now proceed as before. Comparing the derived equation with that in standard form, we deduce steps 2 and 3.

Step 2 The x intercepts are 2 and -2.

Step 3 The y intercepts are 3 and -3.

Step 4 We connect the intercepts with a smooth curve to complete the sketch of the ellipse.

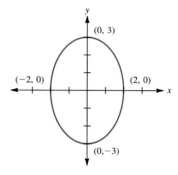

$9x^2 + 4y^2 = 36$

Check Yourself 2

Sketch the ellipse

$$25x^2 + 4y^2 = 100$$

Hint: First write the equation in standard form by dividing both sides of the equation by 100.

The following algorithm summarizes our work with graphing ellipses.

GRAPHING THE ELLIPSE

STEP 1   Write the given equation in standard form.

STEP 2   From that standard form, determine the x intercepts.

STEP 3   Also determine the y intercepts.

STEP 4   Plot the four intercepts and connect the points with a smooth curve, to complete the sketch.

Our discussion will now turn to the last of the conic sections, the hyperbola. As you will see, the geometric description of the hyperbola (and hence the corresponding standard form) is quite similar to that of the ellipse.

A *hyperbola* is the set of all points (x, y) such that the absolute value of the differences from (x, y) to each of two fixed points, called the *foci* of the hyperbola, is constant.

The following sketch illustrates the definition in the case where the foci are located on the x axis and are symmetric about the origin.

This is the first of two special cases we will investigate in this section.

The difference $|d_1 - d_2|$ remains constant for any point on the hyperbola.

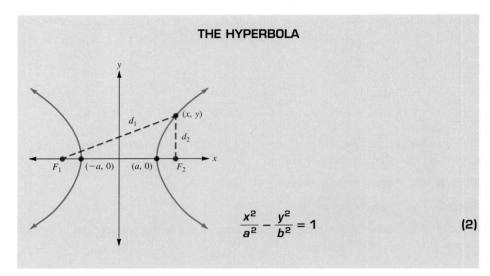

THE HYPERBOLA

$$\frac{x^2}{a^2} - \frac{y^2}{b^2} = 1 \qquad (2)$$

Before we try to sketch a hyperbola from its equation, let's examine the standard form more carefully. For

$$\frac{x^2}{a^2} - \frac{y^2}{b^2} = 1 \qquad (2)$$

the graph is a hyperbola which opens to the right and left and is symmetric about the x axis. The points where this hyperbola intercepts the x axis are called the *vertices* of the hyperbola. The vertices of the hyperbola are located at $(a, 0)$ and at $(-a, 0)$.

As we move away from the center of the hyperbola, the *branches* of the hyperbola will approach two straight lines called the *asymptotes* of the hyperbola. The equations of the two asymptotes of the hyperbola are

Although we show these equations, you will see an easier method for finding the asymptotes in our first example.

The equation of the hyperbola also has both x^2 and y^2 terms. Here the coefficients of those terms have *opposite* signs. If the x^2 coefficient is *positive,* the hyperbola will open *horizontally.*

$$y = \frac{b}{a}x \qquad \text{and} \qquad y = -\frac{b}{a}x$$

These asymptotes will prove to be extremely useful aids in sketching the hyperbola. In fact, for most purposes, the vertices and the asymptotes will be the only tools that we will need. The following example illustrates.

EXAMPLE 3 Graphing a Hyperbola

Sketch the hyperbola

$$\frac{x^2}{9} - \frac{y^2}{4} = 1$$

Step 1 The equation is in standard form.

Step 2 Find and plot the vertices.

From the standard form we can see that $a^2 = 9$ and so $a = 3$ or $a = -3$. The vertices of the hyperbola then occur at $(3, 0)$ and $(-3, 0)$.

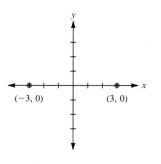

Step 3 Sketch the asymptotes.

Here is an easy way to sketch the asymptotes. Note from the standard form (2) that $b^2 = 4$, so $b = 2$ or $b = -2$. Plot the points $(0, 2)$ and $(0, -2)$ on the y axis.

Draw (using dashed lines) the rectangle whose sides are parallel to the x and y axes and which pass through the points determined in step 2 and step 3.

Draw the diagonals of the rectangle (again using dashed lines), and then extend those diagonals to form the desired asymptotes.

Step 4 Sketch the hyperbola.

We now complete our task by sketching the hyperbola as two smooth curves, passing through the vertices and approaching the asymptotes.

It is important to remember that the asymptotes are *not* a part of the graph. They are simply used as aids in sketching the graph as the branches get "closer and closer" to the lines.

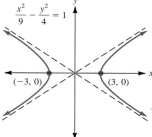

Check Yourself 3

Sketch the hyperbola.

$$\frac{x^2}{16} - \frac{y^2}{9} = 1$$

We now want to consider a second case of the hyperbola and its standard form. Suppose that the foci of the hyperbola are now on the y axis and symmetric about the origin. A sketch of such a hyperbola, and the equation that corresponds, follows.

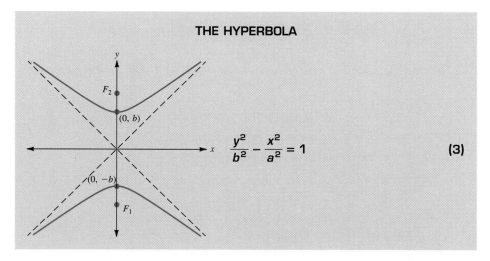

THE HYPERBOLA

$$\frac{y^2}{b^2} - \frac{x^2}{a^2} = 1 \qquad \textbf{(3)}$$

Some observations about this case are in order:

1. The vertices of the hyperbola are now at $(0, b)$ and $(0, -b)$.

2. The asymptotes of the hyperbola have the equations

$$y = \frac{b}{a}x \qquad \text{and} \qquad y = -\frac{b}{a}x$$

The following example illustrates sketching a hyperbola in this case.

Here the vertices are on the y axis.

The asymptotes are the same as before.

EXAMPLE 4 Graphing a Hyperbola

Sketch the hyperbola

$$4y^2 - 25x^2 = 100$$

You can recognize this equation as corresponding to a hyperbola since the coefficients of the squared terms are *opposite* in sign. Since the y^2 coefficient is *positive*, the hyperbola will open *vertically*.

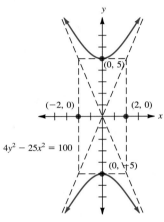

Step 1 Write the equation in the standard form (3) by dividing both sides by 100:

$$\frac{4y^2}{100} - \frac{25x^2}{100} = \frac{100}{100}$$

$$\frac{y^2}{25} - \frac{x^2}{4} = 1$$

Step 2 Find the vertices.

From the standard form (3), since $b^2 = 25$, we see that $b = 5$ or $b = -5$, so the vertices are at $(0, 5)$ and $(0, -5)$.

Step 3 Sketch the asymptotes.

Also from standard form (3), since $a^2 = 4$, we see that $a = 2$ or $a = -2$.

Plot $(2, 0)$ and $(-2, 0)$ on the x axis, and complete the dashed rectangle as before. The diagonals once again extend to form the asymptotes.

Step 4 Sketch the hyperbola.

Draw smooth curves, through the intercepts, that approach the asymptotes to complete the graph.

Check Yourself 4

Sketch the hyperbola $9y^2 - 4x^2 = 36$.

The following algorithm summarizes our work with sketching hyperbolas.

GRAPHING THE HYPERBOLA

STEP 1 Write the given equation in standard form.

STEP 2 Determine the vertices of the hyperbola.

If the x^2 coefficient is positive, the vertices are at $(a, 0)$ and $(-a, 0)$ on the x axis.

If the y^2 coefficient is positive, the vertices are at $(0, b)$ and $(0, -b)$ on the y axis.

STEP 3 Sketch the asymptotes of the hyperbola.

Plot points $(a, 0)$, $(-a, 0)$, $(0, b)$, and $(0, -b)$. Form a rectangle from these points. The diagonals (extended) are the asymptotes of the hyperbola.

STEP 4 Sketch the hyperbola.

Draw smooth curves, through the intercepts and approaching the asymptotes.

The following chart shows all the equation forms considered in this chapter.

Curve	Example	Recognizing the Curve
Straight line	$4x - 3y = 12$	The equation involves x and/or y to the first power only.
Parabola	$y = x^2 - 3x$ or $x = y^2 - 2y + 3$	Only one term, in x or in y, may be squared. The other variable appears to the first power.
Circle	$x^2 + 4x + y^2 = 5$	The equation has both x^2 and y^2 terms. The coefficients of those terms are equal.
Ellipse	$4x^2 + 9y^2 = 36$	The equation has both x^2 and y^2 terms. The coefficients of those terms have the same algebraic sign but different values.
Hyperbola	$4x^2 - 9y^2 = 36$ or $9y^2 - 16x^2 = 144$	The equation has both x^2 and y^2 terms. The coefficients of those terms have different algebraic signs.

CHECK YOURSELF ANSWERS

1. $\dfrac{x^2}{16} + \dfrac{y^2}{9} = 1$

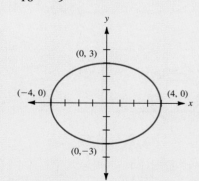

2. $\dfrac{x^2}{4} + \dfrac{y^2}{25} = 1$

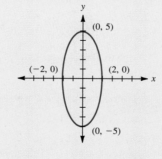

3. $\dfrac{x^2}{16} - \dfrac{y^2}{9} = 1$

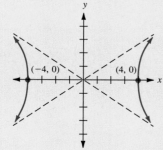

4. $\dfrac{y^2}{4} - \dfrac{x^2}{9} = 1$

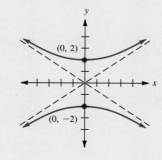

10.4 EXERCISES

Build Your Skills

Match each of the curves shown with the appropriate equation or function.

(*a*) $4x^2 + 25y^2 = 100$

(*b*) $f(x) = x^2 - 2x - 3$

(*c*) $x = \dfrac{1}{2}y^2 - 2y$

(*d*) $\dfrac{x^2}{9} + \dfrac{y^2}{16} = 1$

(*e*) $\dfrac{y^2}{25} - \dfrac{x^2}{4} = 1$

(*f*) $16x^2 - 9y^2 = 144$

(*g*) $(x - 2)^2 + (y - 2)^2 = 9$

(*h*) $x^2 + y^2 = 16$

4.

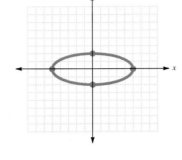

1.

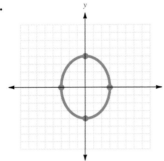

5.

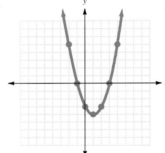

2.

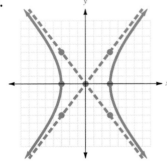

6.

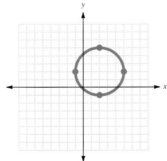

3.

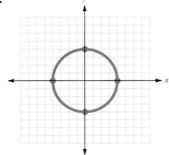

7.

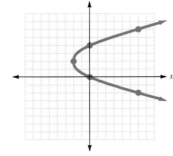

8.

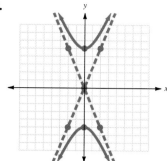

Identify the graph of each of the following equations as one of the conic sections (the parabola, circle, ellipse, or hyperbola).

9. $x^2 + y^2 = 16$

10. $\dfrac{x^2}{4} - \dfrac{y^2}{16} = 1$

11. $y = x^2 - 4$

12. $\dfrac{x^2}{16} + \dfrac{y^2}{9} = 1$

13. $9x^2 - 4y^2 = 36$

14. $x^2 = 4y$

15. $y^2 - 4x^2 = 4$

16. $x = y^2 - 2y + 1$

17. $x^2 - 6x + y^2 + 2x = 2$

18. $4x^2 + 25y^2 = 100$

19. $9y^2 - 16x^2 = 144$

20. $y = x^2 - 6x + 8$

Graph the following ellipses by finding the x and y intercepts. If necessary, write the equation in standard form.

21. $\dfrac{x^2}{4} + \dfrac{y^2}{9} = 1$

22. $\dfrac{x^2}{16} + \dfrac{y^2}{9} = 1$

23. $\dfrac{x^2}{9} + \dfrac{y^2}{25} = 1$

24. $\dfrac{x^2}{36} + \dfrac{y^2}{16} = 1$

25. $x^2 + 9y^2 = 36$

26. $4x^2 + y^2 = 16$

27. $4x^2 + 9y^2 = 36$

28. $25x^2 + 4y^2 = 100$

29. $4x^2 + 25y^2 = 100$

30. $9x^2 + 16y^2 = 144$

31. $25x^2 + 9y^2 = 225$

32. $16x^2 + 9y^2 = 144$

Graph the following hyperbolas by finding the vertices and asymptotes. If necessary, write the equation in standard form.

33. $\dfrac{x^2}{9} - \dfrac{y^2}{9} = 1$

34. $\dfrac{y^2}{9} - \dfrac{x^2}{4} = 1$

35. $\dfrac{y^2}{16} - \dfrac{x^2}{9} = 1$

36. $\dfrac{x^2}{25} - \dfrac{y^2}{16} = 1$

37. $\dfrac{x^2}{36} - \dfrac{y^2}{9} = 1$

38. $\dfrac{y^2}{25} - \dfrac{x^2}{9} = 1$

39. $x^2 - 9y^2 = 36$

40. $y^2 - 4x^2 = 36$

41. $9x^2 - 4y^2 = 36$

42. $9y^2 - 4x^2 = 36$

43. $16y^2 - 9x^2 = 144$

44. $4x^2 - 9y^2 = 36$

Transcribe Your Skills

45. How can you identify the graph of one of the conic sections (parabola, circle, ellipse, hyperbola) by simply looking at the equation?

46. Why does the equation that represents the hyperbola not describe a function?

Graphing Utility

The techniques used in the graphing utility exercises from Section 10.3 can also be used to graph ellipses and hyperbolas. Show the two functions you would graph to represent each equation on a graphing utility.

47.

48. $\dfrac{x^2}{25} + \dfrac{y^2}{9} = 1$

49. $4x^2 + 25y^2 = 100$

50. $\dfrac{x^2}{4} - \dfrac{y^2}{9} = 1$

Skillscan (Section 2.2)

Solve the following equations for x.

a. $y = 3x - 5$ **b.** $y = 2x + 7$ **c.** $y = \dfrac{1}{4}x - 3$

d. $y = \dfrac{1}{3}x + 2$ **e.** $y = x^2 + 2$ **f.** $y = 2x^2 - 1$

SUMMARY

Graphing Functions of the Form $f(x) = a(x - h)^2 + k$ [10.1]

The *conic sections* are the curves formed when a plane cuts through a cone. These include the *parabola, circle, ellipse,* and *hyperbola.*

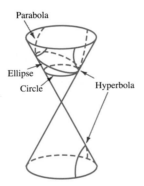

The graph of an equation quadratic in one variable and linear in the other is a parabola.

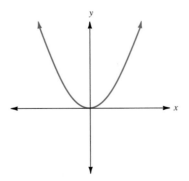

If the equation describes a function and is in the form

$$f(x) = a(x - h)^2 + k \tag{1}$$

The vertex of the parabola is at (h, k).

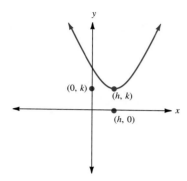

Graphing Functions of the Form $f(x) = ax^2 + bx + c$ [10.2]

The graph of

$$f(x) = ax^2 + bx + c \qquad a \neq 0 \qquad (2)$$

is a parabola.

The parabola opens *upward* if $a > 0$.

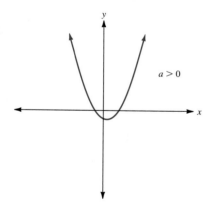

The parabola opens *downward* if $a < 0$.

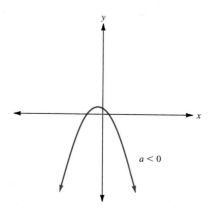

The *vertex* of the parabola (either the highest or the lowest point on the graph) is on the *axis of symmetry* with the equation

$$x = -\frac{b}{2a} \tag{3}$$

To graph $y = x^2 - 4x + 3$

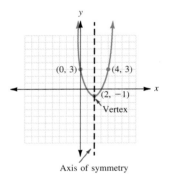

To graph a parabola:

Step 1. Find the axis of symmetry.

Step 2. Find the vertex.

Step 3. Determine two symmetric points.

> **Note** You can use the x intercepts if the quadratic member of the given equation is factorable and the x intercepts are distinct. Otherwise, use the y intercept and its symmetric point.

Step 4. Draw a smooth curve connecting the points found above, to form the parabola. You may choose to find additional pairs of symmetric points at this time.

The Circle [10.3]

The standard form for the circle with center (h, k) and radius r is

Given the equation

$(x - 2)^2 + (y + 3)^2 = 4$

we see that the center is at $(2, -3)$ and the radius is 2.

$$(x - h)^2 + (y - k)^2 = r^2 \tag{4}$$

Determining the center and radius of the circle from its equation allows us to easily graph the circle. *Note:* Completing the square may be used to derive an equivalent equation in standard form if the original equation is not in this form.

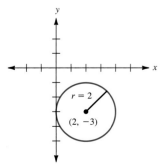

The Ellipse and the Hyperbola [10.4]

Ellipse

The standard form for the *ellipse,* whose foci are located on either the x or y axis and are symmetric about the origin, is

$$\frac{x^2}{a^2} + \frac{y^2}{b^2} = 1 \tag{5}$$

The x intercepts for the ellipse are at a and at $-a$.

The y intercepts for the ellipse are at b and at $-b$.

Determining the four intercepts of an ellipse allows us to sketch its graph. *Note:* If the given equation is not in standard form, we can divide both sides of the equation by the appropriate constant to derive the standard form.

Graph the equation

$$\frac{x^2}{9} + \frac{y^2}{4} = 1$$

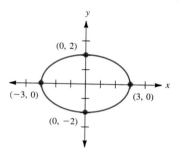

Hyperbola

The standard form for the *hyperbola* whose foci are located on the x axis and are symmetric about the origin is

$$\frac{x^2}{a^2} - \frac{y^2}{b^2} = 1 \tag{6}$$

The intercepts for this hyperbola are on the x axis, at a and at $-a$.

The asymptotes of the hyperbola have the equations

$$y = \pm\frac{b}{a}x \tag{7}$$

Determining and sketching the intercepts and asymptotes of a hyperbola will allow us to sketch its graph quickly. *Note:* Again, divide both sides of the given equation by the appropriate constant if the equation is not in standard form.

If the foci of the hyperbola are located on the y axis and are symmetric about the origin, the standard form is

$$\frac{y^2}{b^2} - \frac{x^2}{a^2} = 1 \tag{8}$$

The intercepts for this hyperbola are on the y axis at b and at $-b$. The equations for the asymptotes of the hyperbola remain the same as before.

Sketch the hyperbola

$$\frac{x^2}{4} - \frac{y^2}{9} = 1$$

The intercepts are at $(2, 0)$ and $(-2, 0)$. The asymptotes are $y = \frac{3}{2}x$ and $y = -\frac{3}{2}x$.

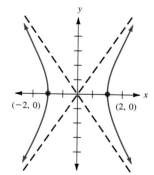

SUMMARY EXERCISES

This summary exercise set is provided to give you practice with each of the objectives of the chapter. Each exercise is keyed to the appropriate chapter section. The answers are provided in the instructor's manual that accompanies this text. Your instructor will provide guidelines on how best to use these exercises in your instructional program.

[10.1] Find the equation of the axis of symmetry and the coordinates for the vertex of each of the following.

1. $f(x) = x^2$

2. $f(x) = x^2 + 2$

3. $f(x) = x^2 - 5$

4. $f(x) = (x - 3)^2$

5. $f(x) = (x + 2)^2$

6. $f(x) = -(x - 3)^2$

7. $f(x) = (x + 3)^2 + 1$

8. $f(x) = -(x + 2)^2 - 3$

9. $f(x) = -(x - 5)^2 - 2$

10. $f(x) = 2(x - 2)^2 - 5$

11. $f(x) = -3(x + 4)^2 + 7$

12. $f(x) = -2(x + 2)^2 + 6$

[10.1] Graph each of the following.

13. $f(x) = x^2$

14. $f(x) = x^2 + 2$

15. $f(x) = x^2 - 5$

16. $f(x) = (x - 3)^2$

17. $f(x) = (x + 2)^2$

18. $f(x) = -(x - 3)^2$

19. $f(x) = (x + 3)^2 + 1$

20. $f(x) = -(x + 2)^2 - 3$

21. $f(x) = -(x - 5)^2 - 2$

22. $f(x) = 2(x - 2)^2 - 5$

23. $f(x) = -3(x + 4)^2 + 7$

24. $f(x) = -2(x + 2)^2 + 6$

[10.2] Find the equation of the axis of symmetry and the coordinates for the vertex of each of the following.

25. $f(x) = x^2 + 3$

26. $f(x) = x^2 - 4x$

27. $f(x) = -x^2 + 2x$

28. $f(x) = x^2 - 4x + 3$

29. $f(x) = -x^2 - x + 6$

30. $f(x) = x^2 + 4x + 5$

31. $f(x) = -x^2 - 6x + 4$

32. $f(x) = -x^2 + 2x - 2$

33. $f(x) = 2x^2 - 4x + 1$

34. $x = y^2 - 4y$

35. $x = -y^2 + 4y$

36. $x = y^2 - 5y + 6$

[10.2] Sketch the graph of each of the following functions.

37. $f(x) = x^2 + 3$

38. $f(x) = x^2 - 2$

39. $f(x) = x^2 - 4x$

40. $f(x) = -x^2 + 2x$

41. $f(x) = x^2 + 2x - 3$

42. $f(x) = x^2 - 4x + 3$

43. $f(x) = -x^2 - x + 6$

44. $f(x) = -x^2 + 3x + 4$

45. $f(x) = x^2 + 4x + 5$

46. $f(x) = x^2 - 6x + 4$

47. $f(x) = x^2 - 2x + 4$

48. $f(x) = -x^2 + 2x - 2$

49. $f(x) = 2x^2 - 4x + 1$

50. $f(x) = \frac{1}{2}x^2 - 4x$

[10.2] Sketch the graph of each of the following equations.

51. $x = y^2 - 4y$

52. $x = -y^2 + 4y$

53. $x = -y^2 + 2y + 3$

54. $x = y^2 - 5y + 6$

[10.3] Find the center and the radius of the graph of each equation.

55. $x^2 + y^2 = 16$

56. $x^2 + y^2 = 50$

57. $4x^2 + 4y^2 = 36$

58. $3x^2 + 3y^2 = 36$

59. $(x - 3)^2 + y^2 = 36$

60. $(x - 2)^2 + y^2 = 9$

61. $(x - 1)^2 + (y - 2)^2 = 16$

62. $x^2 + 6x + y^2 + 4y = 12$

63. $x^2 + 8x + y^2 + 10y = 23$

64. $x^2 - 6x + y^2 + 6y = 18$

65. $x^2 + y^2 - 4y - 5 = 0$

66. $x^2 - 2x + y^2 - 6y = 6$

[10.3] Graph each of the following.

67. $x^2 + y^2 = 16$

68. $4x^2 + 4y^2 = 36$

69. $x^2 + (y + 3)^2 = 25$

70. $(x - 2)^2 + y^2 = 9$

71. $(x - 1)^2 + (y - 2)^2 = 16$

72. $(x + 3)^2 + (y + 2)^2 = 25$

73. $x^2 + y^2 - 4y - 5 = 0$

74. $x^2 - 2x + y^2 - 6y = 6$

[10.4] For each of the following equations decide whether its graph is a line, parabola, circle, ellipse, or hyperbola.

75. $x + y = 16$

76. $x + y^2 = 5$

77. $4x^2 + 4y^2 = 36$

78. $3x + 3y = 36$

79. $y = (x - 3)^2$

80. $(x - 2)^2 + y^2 = 9$

81. $y = (x - 1)^2 + 1$

82. $x = y^2 + 4y + 4$

83. $\dfrac{x^2}{4} + \dfrac{y^2}{25} = 1$

84. $x^2 - 6x + y^2 + 6y = 18$

85. $\dfrac{x^2}{9} - \dfrac{y^2}{25} = 1$

86. $9x^2 - 4y^2 = 36$

87. $16x^2 + 4y^2 = 64$

88. $x^2 = -y^2 + 18$

89. $\dfrac{x^2}{4} - \dfrac{y^2}{9} = 1$

90. $4x^2 + 4y^2 = 36$

91. $4x - 6y = 12$

92. $3x^2 - y + 4 = 0$

[10.4] Graph each of the following equations.

93. $\dfrac{x^2}{25} + \dfrac{y^2}{9} = 1$

94. $\dfrac{x^2}{4} + \dfrac{y^2}{16} = 1$

95. $9x^2 + 4y^2 = 36$

96. $16x^2 + 9y^2 = 144$

97. $\dfrac{x^2}{9} - \dfrac{y^2}{4} = 1$

98. $\dfrac{y^2}{16} - \dfrac{x^2}{4} = 1$

99. $4x^2 - 9y^2 = 36$

100. $16x^2 - 9y^2 = 144$

SELF-TEST

The purpose of this self-test is to help you check your progress and to review for a chapter test in class. Allow yourself about an hour to take the test. When you are done, check your answers in the back of the book. If you missed any problems, be sure to go back and review the appropriate sections in the chapter and the exercises that are provided.

Find the equation of the axis of symmetry and the coordinates of the vertex of each of the following.

1. $y = -2x^2$

2. $y = (x - 3)^2$

3. $y = (x + 2)^2 - 5$

4. $y = -3(x + 2)^2 + 1$

5. $y = x^2 - 4x - 5$

6. $y = -2x^2 + 6x - 3$

7. $x = (y - 3)^2 - 2$

8. $x = y^2 - 6y + 2$

Graph each of the following functions.

9. $f(x) = (x - 5)^2$

10. $f(x) = (x + 2)^2 - 3$

11. $f(x) = -2(x - 3)^2 - 1$

12. $f(x) = 3x^2 + 9x + 2$

Graph each of the following equations.

13. $x = \dfrac{1}{2}(y - 4)^2 + 2$

14. $x = 2y^2 + 10y + 3$

Find the coordinates for the center and the radius of the graph of each equation.

15. $(x - 3)^2 + (y + 2)^2 = 36$

16. $x^2 + 2x + y^2 - 4y - 21 = 0$

Sketch the graph of each of the following equations.

17. $(x - 2)^2 + (y + 3)^2 = 9$

18. $\dfrac{x^2}{25} + \dfrac{y^2}{9} = 1$

19. $\dfrac{x^2}{9} - \dfrac{y^2}{16} = 1$

20. $4y^2 - 25x^2 = 100$

CUMULATIVE REVIEW EXERCISES

Here is a review of selected topics from the first 10 chapters.

Graph each of the following.

1. $f(x) = \dfrac{3}{2}x - 2$

2. $f(x) = -3x^2 + 2$

3. $f(x) = 2x^2 - 8x + 5$

4. $x^2 + y^2 = 64$

Solve each of the following equations.

5. $2x - 3 = 5x + 4$

6. $0 = (x - 3)(x + 1)$

7. $0 = x^2 - 5x - 24$

8. $2x^2 - 3x = 5$

9. $\sqrt{2x - 1} = 9$

Simplify each expression.

10. $(2x^2 - 3x + 1) - (4x^2 + 2x - 3)$

11. $\dfrac{x^2 - 3x - 10}{x^2 - 25}$

12. $\sqrt{4x^4 - 8x^2}$

Solve each of the following applications.

13. A newspaper stand has 300 coins at the end of the day, all nickels and quarters. If the total value of the coins is $34.80, how many of each coin are there?

14. Corina can drive to work in 45 min. If she decides to take the bus, the same trip takes 1 h 15 min. If the average rate for the car is 16 mi/h faster than the average rate for the bus, how far does she travel to work?

15. The sum of the squares of two consecutive odd integers is 202. What are the integers?

16. The length of a rectangle is 3 cm more than its width. If the area of the rectangle is 180 cm^2, find the dimensions.

17. A boat travels 25 mi/h in still water. If the boat can travel 3 mi downstream in the same time that it takes to travel 2 mi upstream, what is the rate of the current?

18. Ahmed and Joanne can build a deck in 36 h. If Ahmed could complete the same job by himself in 54 h, how long would it take Joanne, working alone?

If $f(x) = 3x + 2$ and $g(x) = x^2 - x + 1$, find each of the following.

19. $f(-2)$

20. $f(3) - g(-1)$

RENEWABLE ENERGY RESOURCES— PART 2

Wind, biomass, and geothermal energy are renewable energy resources which, while not as developed as the solar and hydropower resources discussed in the previous chapter, are also being developed. These resources are available worldwide and have not yet begun to be developed to their potential.

In 1990, there were approximately 20,000 wind turbines generating electricity worldwide. Electricity generated by these windfarms is cheaper than that from nuclear plants and may be as cheap as electricity from coal-fired plants by as early as 1995. Many regions of the world have sufficient winds to power turbines, and these machines emit no pollution and disrupt the environment very little.

Biomass, or plant material, has been a principal energy source since the discovery of fire. It supplies about 15 percent of the world's commercial energy and approximately half of the energy used in less developed countries. It is a renewable resource as long as trees and plants are replanted to replace the biomass which is harvested.

Biomass does have some disadvantages, however. It

takes a large amount of land to grow the necessary trees and plants. It can cause soil depletion, erosion, and flooding problems, and growth of biomass may displace large areas of food crops and thus disrupt regional food distribution and availability.

Geothermal energy is heat obtained from the earth's interior. It consists of dry steam, wet steam (steam and water droplets), and hot water trapped in porous rock in the earth's crust.

Dry steam is the easiest and least expensive geothermal resource to use to generate electricity, but it is also the rarest. Hot water geothermal sites are the most common but are not very effective for generating electricity. They do, however, provide highly effective space heating.

While the geothermal resource is potentially vast, easily accessible regions near the earth's surface are limited. Geothermal resources can be depleted if they are withdrawn from the earth too quickly, and they can cause air pollution from hydrogen sulfide gas. There is also some danger of water pollution from the minerals contained in the wet steam and hot water brought to the surface at geothermal plants. These problems, however, are no greater than the risks from fossil fuels and nuclear generators and should not deter from the promise that geothermal offers for future development.

11.1 Inverse Relations and Functions

TAPE IN6

OBJECTIVE: To write the inverse of a relation or a function

We want now to consider an extension of the concepts of relations and functions discussed in Chapter Two.

Suppose we are given the relation

$$\{(1, 2), (2, 4), (3, 6)\} \tag{1}$$

If we *interchange* the first and second components (the x and y values) of each of the ordered pairs in relation (1), we have

$$\{(2, 1), (4, 2), (6, 3)\} \tag{2}$$

which is another relation. Relations (1) and (2) are called *inverse relations,* and in general:

INVERSE OF A RELATION

The *inverse* of a relation is formed by interchanging the components of each of the ordered pairs in the given relation.

Since we know that relations are often specified by equations, it is natural for us to want to work with the concept of the inverse relation in that setting. We form the inverse relation by interchanging the roles of x and y in the defining equation. We then have the inverse relation.

EXAMPLE 1 Finding the Inverse of a Relation

Find the inverse of the relation

$$f = \{(x, y) \,|\, y = 2x - 4\} \tag{3}$$

Solution First interchange variables x and y to obtain

Note that x and y have been interchanged from the original equation.

$$x = 2y - 4$$

We now solve the defining equation for y.

$$2y = x + 4$$

$$y = \frac{1}{2}x + 2$$

and we rewrite the relation in the equivalent form:

$$f^{-1} = \left\{(x, y) \,\Big|\, y = \frac{1}{2}x + 2\right\} \tag{4}$$

Note The notation f^{-1} has a *different meaning* from the negative exponent, as in x^{-1} or $\frac{1}{x}$.

The inverse of the original relation (3) is now shown in (4) with the defining equation "solved for y." That inverse is denoted f^{-1} (this is read as "f inverse").

We use the notation f^{-1} to indicate the inverse of f when that inverse is *also a function.*

Check Yourself 1

Write the inverse relation for $g = \{(x, y) \mid y = 3x + 6\}$.

The graphs of relations and their inverses are related in an interesting way. First, note that the graphs of the ordered pairs (a, b) and (b, a) always have symmetry about the line $y = x$.

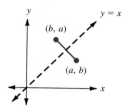

Now, with the above symmetry in mind, let's consider the following example.

EXAMPLE 2 Graphing a Relation and Its Inverse

Graph the relation f from Example 1 along with its inverse.

Solution Recall that

$$f = \{(x, y) \mid y = 2x - 4\}$$

and

$$f^{-1} = \left\{(x, y) \mid y = \frac{1}{2}x + 2\right\}$$

The graphs of f and f^{-1} are shown below.

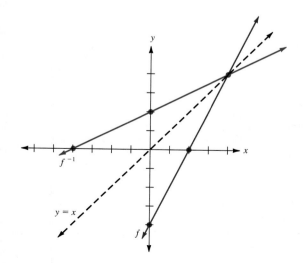

Note that the graphs of f and f^{-1} are symmetric about the line $y = x$. That symmetry follows from our earlier observation about the pairs (a, b) and (b, a) since we simply reversed the roles of x and y in forming the inverse relation.

Check Yourself 2

Graph the relation g from the Check Yourself 1 exercise along with its inverse.

From our work thus far, it should be apparent that every relation has an inverse. However, that inverse may or may not be a function.

EXAMPLE 3 Finding the Inverse of a Function

Find the inverses of the following functions.

(a) $f = \{(1, 3), (2, 4), (3, 9)\}$

Its inverse is

The elements of the ordered pairs have been interchanged.

$\{(3, 1), (4, 2), (9, 3)\}$

which is also a function.

(b) $g = \{(1, 3), (2, 6), (3, 6)\}$

Its inverse is

It's not a function because 6 is mapped to both 2 and 3.

$\{(3, 1), (6, 2), (6, 3)\}$

which is *not* a function.

Check Yourself 3

Write the inverses for each of the following relations. Which of the inverses are also functions?

1. $\{(-1, 2), (0, 3), (1, 4)\}$ **2.** $\{(2, 5), (3, 7), (4, 5)\}$

Can we predict in advance whether the inverse of a function will also be a function? The answer is yes.

We already know that for a relation to be a function, no element in its domain can be associated with more than one element in its range.

In addition, if the inverse of a function is to be a function, no element in the range can be associated with more than one element in the domain—that is, no two distinct ordered pairs in the function can have the same second component. A function that satisfies this additional restriction is called a *one-to-one function*.

Let's refer to the functions of Example 3:

$f = \{(1, 3), (2, 4), (3, 9)\}$

was a one-to-one function. Its inverse was also a function while

$g = \{(1, 3), (2, 6), (3, 6)\}$

was *not* a one-to-one function, and its inverse was *not* a function.

From those observations we can state the following general result.

INVERSE OF A FUNCTION

A function f has an inverse f^{-1}, which is also a function, if and only if f is a one-to-one function.

Because the statement is an "if and only if" statement, it can be turned around without changing the meaning. Here we use the same statement as a definition for a one-to-one function.

ONE-TO-ONE FUNCTION

A function f is a *one-to-one function* if and only if it has an inverse f^{-1} which is also a function.

Our result above regarding a one-to-one function and its inverse also has a convenient graphical interpretation, as the following example illustrates.

EXAMPLE 4 Graphing a Function and Its Inverse

Graph each function and its inverse. State which inverses are functions.

(a) $f = \{(x, y) \mid y = 4x - 8\}$

Since f is a one-to-one function (no value for y can be associated with more than one value for x), its inverse is also a function. Here

$$f^{-1} = \left\{ (x, y) \,\middle|\, y = \frac{1}{4}x + 2 \right\}$$

This is a *linear function* of the form $f = \{(x, y) \mid y = mx + b\}$. Its graph is a straight line. A linear function, where $m \neq 0$, is always one-to-one.

The graphs of f and f^{-1} are shown below.

The vertical-line test tells us that *both* f and f^{-1} are functions.

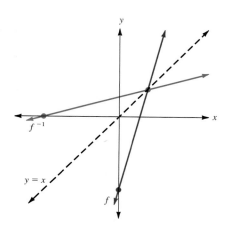

(b) $a = \{(x, y)\,|\,y = x^2\}$

This is a *quadratic function* of the form

$g = \{(x, y)\,|\,y = ax^2 + bx + c\}$ where $a \neq 0$

Its graph is always a parabola, and a quadratic function is *not* a one-to-one function.

For instance, 4 in the range is associated with both 2 and -2 from the domain. It follows that the inverse of g

$\{(x, y)\,|\,x = y^2\}$

or

$\{(x, y)\,|\,y = \pm\sqrt{x}\}$

is *not* a function.

The graphs of g and its inverse are shown below.

By the vertical-line test, we see that the inverse of g is *not* a function. Of course, that is because g was *not* one-to-one.

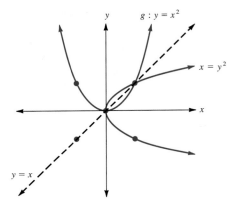

Note When a function is not one-to-one, as in Example 4*b*, we can restrict the domain of the function so that it will be one-to-one. In this case, if we redefine function g as

The domain is now restricted to nonnegative values for *x*.

$g = \{(x, y)\,|\,y = x^2,\ x \geq 0\}$

it will be one-to-one and its inverse

$g^{-1} = \{(x, y)\,|\,y = \sqrt{x}\}$

will be a function, as shown in the graphs below.

The function g is now one-to-one; its inverse g^{-1} is also a function.

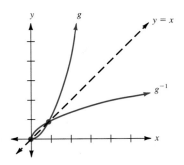

Check Yourself 4

Graph each function and its inverse. Which inverses are functions?

1. $f = \{(x, y) \mid y = 2x - 2\}$ **2.** $g = \{(x, y) \mid y = 2x^2\}$

It is easy to tell from the graph of a function whether that function is one-to-one. If any horizontal line can meet the graph of a function in at most one point, the function is one-to-one. The following examples illustrate.

EXAMPLE 5 Identifying a One-to-One Function

Which of the following graphs represent one-to-one functions?

(*a*)

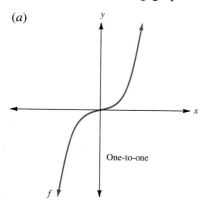

One-to-one

Since no horizontal line passes through any two points of the graph, *f* is one-to-one.

(*b*)

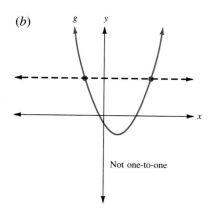

Not one-to-one

Since a horizontal line can meet the graph of function *g* at two points, *g* is *not* a one-to-one function.

Check Yourself 5

Consider the graphs of the functions of the Check Yourself 4 exercise. Which functions are one-to-one?

The following algorithm summarizes our work in this section.

FINDING INVERSE RELATIONS AND FUNCTIONS

1. Interchange the *x* and *y* components of the ordered pairs of the given relation or the roles of *x* and *y* in the defining equation.
2. If the relation was described in equation form, solve the defining equation of the inverse for *y*.
3. If desired, graph the relation and its inverse on the same set of axes. The two graphs will be symmetric about the line *y = x*.

CHECK YOURSELF ANSWERS

1. $g^{-1} = \left\{ (x, y) \,\middle|\, y = \dfrac{1}{3}x - 2 \right\}$ **2.**

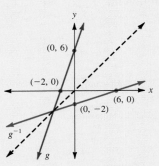

3. (1) $\{(2, -1), (3, 0), (4, 1)\}$—a function; (2) $\{(5, 2), (7, 3), (5, 4)\}$—*not* a function   **4.** (1) $f = \{(x, y) \mid y = 2x - 2\}$; (2) $g = \{(x, y) \mid y = 2x^2\}$

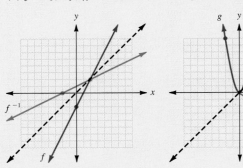

5. (1) is one-to one;   (2) is not one-to-one

Build Your Skills

Write the inverse relation for each of the following functions. In each case decide whether the inverse relation is also a function.

1. $\{(2, 3), (3, 4), (4, 5)\}$

2. $\{(2, 3), (3, 4), (4, 3)\}$

3. $\{(1, 2), (2, 2), (3, 2)\}$

4. $\{(5, 9), (3, 7), (7, 5)\}$

5. $\{(2, 4), (3, 9), (4, 16)\}$

6. $\{(-1, 2), (0, 3), (1, 2)\}$

Write an equation for the inverse of the relation defined by each of the following equations.

7. $y = 2x + 8$

8. $y = -2x - 4$

9. $y = \dfrac{x - 1}{2}$

10. $y = \dfrac{x + 1}{3}$

11. $y = x^2 - 1$

12. $y = -x^2 + 2$

13. $x^2 + 4y^2 = 36$

14. $4x^2 + y^2 = 36$

15. $x^2 - y^2 = 9$

16. $4y^2 - x^2 = 4$

Write an equation for the inverse of the relation defined by each of the following, and graph the relation and its inverse on the same set of axes. Determine which inverse relations are also functions.

17. $y = 3x - 9$

18. $y = 4x + 8$

19. $2x - 3y = 6$

20. $y = 3$

21. $y = x^2 + 1$

22. $y = -x^2 + 1$

Transcribe Your Skills

23. An inverse process is an operation that undoes a procedure. If the procedure is wrapping a present, describe in detail the inverse process.

24. If the procedure is the series of steps that take you from home to your classroom, describe the inverse process.

Think About These

If $f(x) = 3x - 6$, then $f^{-1}(x) = \dfrac{1}{3}x + 2$. Given these two functions, find each of the following.

25. $f(6)$

26. $f^{-1}(6)$

27. $f(f^{-1}(6))$

28. $f^{-1}(f(6))$

29. $f(f^{-1}(x))$

30. $f^{-1}(f(x))$

If $g(x) = \dfrac{x + 1}{2}$, then $g^{-1}(x) = 2x - 1$. Given these two functions, find each of the following.

31. $g(3)$

32. $g^{-1}(3)$

33. $g(g^{-1}(3))$

34. $g^{-1}(g(3))$

35. $g(g^{-1}(x))$

36. $g^{-1}(g(x))$

Given $h(x) = 2x + 8$, find each of the following.

37. $h(4)$

38. $h^{-1}(4)$

39. $h(h^{-1}(4))$

40. $h^{-1}(h(4))$

41. $h(h^{-1}(x))$

42. $h^{-1}(h(x))$

Skillscan (Section 2.3)

Write each of the following, using symbols.

a. y is 2 less than x.   **b.** The sum of s and t is 5.   **c.** x is 5 times as large as y.   **d.** w is one-fifth as long as l. **e.** C is pi times d.   **f.** Area A is length l times width w.   **g.** Rate r is distance d divided by time t.   **h.** Rate R is interest I divided by the product of principal P and time t.

TAPE IN23

11.2 Exponential Functions

OBJECTIVES: 1. To graph exponential functions
2. To apply exponential functions
3. To solve certain exponential equations

Up to this point in the text, we have worked with polynomial functions and other functions in which the variable was used as a base. We now want to turn to a new classification of functions, the *exponential function*.

These are functions whose defining equations involve the variable as an *exponent*. The introduction of these functions will allow us to consider many further applications, including population growth and radioactive decay.

EXPONENTIAL FUNCTIONS

An *exponential function* is a function that can be expressed in the form

$f(x) = b^x$

where $b > 0$ and $b \neq 1$. We call b the *base* of the exponential function.

The following are examples.

$$f(x) = 2^x \qquad g(x) = 3^x \qquad h(x) = \left(\frac{1}{2}\right)^x$$

As we have done with other new functions, we begin by finding some function values. We then use that information to graph the function.

EXAMPLE 1 Graphing an Exponential Function

Graph the exponential function

$y = f(x) = 2^x$

Solution Choosing convenient values for x, we have

Note

$2^{-2} = \dfrac{1}{2^2} = \dfrac{1}{4}$

$$f(0) = 2^0 = 1 \qquad f(-1) = 2^{-1} = \frac{1}{2} \qquad f(1) = 2^1 = 2$$

$$f(-2) = 2^{-2} = \frac{1}{4} \qquad f(2) = 2^2 = 4 \qquad f(-3) = 2^{-3} = \frac{1}{8}$$

$$f(3) = 2^3 = 8$$

We form a table of values from the information above, plot the corresponding points, and connect them with a smooth curve for the desired graph.

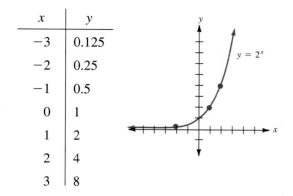

x	y
−3	0.125
−2	0.25
−1	0.5
0	1
1	2
2	4
3	8

Let's examine some of the characteristics of the graph of the exponential function. First, the vertical-line test shows that this is indeed the graph of a function. Also note that the horizontal-line test shows that the function is one-to-one.

The graph *approaches* the x axis on the left, but it does *not intersect* the x axis. The y intercept is 1 (this is because $2^0 = 1$ by definition). To the right the functional values get larger. We say that they "grow without bound."

A vertical line will cross the graph at one point at most. The same is true for a horizontal line.

There is no value for x such that

$2^x = 0$

so the graph never touches the x axis.

We call y = 0 (or the x axis) the *horizontal asymptote.*

Check Yourself 1

Sketch the graph of the exponential function

$y = g(x) = 3^x$

Let's look at an example in which the base of the function is less than 1.

EXAMPLE 2 Graphing an Exponential Function

Graph the exponential function

$$y = f(x) = \left(\frac{1}{2}\right)^x$$

Recall that

$$\left(\frac{1}{2}\right)^x = 2^{-x}$$

Solution We choose convenient values for x:

$$f(0) = \left(\frac{1}{2}\right)^0 = 1 \qquad f(-1) = \left(\frac{1}{2}\right)^{-1} = 2 \qquad f(1) = \left(\frac{1}{2}\right)^1 = \frac{1}{2}$$

$$f(-2) = \left(\frac{1}{2}\right)^{-2} = 4 \qquad f(2) = \left(\frac{1}{2}\right)^2 = \frac{1}{4} \qquad f(-3) = \left(\frac{1}{2}\right)^{-3} = 8$$

$$f(3) = \left(\frac{1}{2}\right)^3 = \frac{1}{8}$$

Again, we form a table of values and graph the desired function.

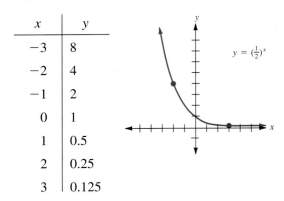

x	y
−3	8
−2	4
−1	2
0	1
1	0.5
2	0.25
3	0.125

Again, by the vertical- and horizontal-line tests, this is the graph of a one-to-one function.

Let's compare this graph and that of Example 1. Clearly the graph also represents a one-to-one function. As was true in the first example, the graph does not intersect the x axis but approaches that axis, here on the right. The values for the function again grow without bound, but this time on the left. The y intercept for both graphs occurs at 1.

The base of a growth function is greater than 1.

The base of a decay function is less than 1, but greater than 0.

Note that the graph of Example 1 was *increasing* (going up) as we moved from left to right. That function was an example of a *growth function*.

The graph of Example 2 was *decreasing* (going down) as we moved from left to right. It is an example of a *decay function*.

Check Yourself 2

Sketch the graph of the exponential function

$$y = \left(\frac{1}{3}\right)^x$$

The following algorithm summarizes our work thus far in this section.

GRAPHING AN EXPONENTIAL FUNCTION

STEP 1 Establish a table of values by considering the function in the form $y = b^x$

STEP 2 Plot points from that table of values and connect them with a smooth curve to form the graph.

STEP 3 If $b > 1$, the graph increases from left to right. If $0 < b < 1$, the graph decreases from left to right.

STEP 4 All graphs will have the following in common:
  a. The y intercept will be 1.
  b. The graphs will approach, but not touch, the x axis.
  c. The graphs will represent one-to-one functions.

We used bases of 2 and $\frac{1}{2}$ for the exponential functions of our examples because they provided convenient computations. A far more important base for an exponential function is an irrational number named e. In fact, when e is used as a base, the function defined by

$$f(x) = e^x$$

is called *the* exponential function.

The significance of this number will be made clear in later courses, particularly calculus. For our purposes e can be approximated as

$$e \approx 2.71828$$

The graph of $f(x) = e^x$ is shown below. Of course, it is very similar to the graphs seen earlier in this section.

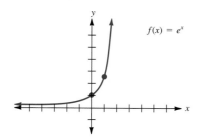

Exponential expressions involving base e occur frequently in real-world applications. The following are just two examples.

The use of the letter e as a base originated with Leonhard Euler (1707–1783), and e is sometimes called *Euler's number* for that reason.

Graph $y = e^x$ on your calculator. You may find the e^x key to be the 2nd (or inverse) function to the ln x key. Note that e^1 is approximately 2.71828.

EXAMPLE 3 A Population Application

(*a*) Suppose that presently the population of a city is 20,000 and that the population is expected to grow at a rate of 5 percent per year. The equation

$$P(5) = 20,000e^{(0.05)(5)} \approx 25,681$$

gives the town's population after t years. Find the population in 5 years.

Let $t = 5$ in the original equation to obtain

$$P(5) = 20,000e^{(0.05)(5)} \approx 25,681$$

which is the population expected 5 years from now.

Be certain that you enclose the multiplication (0.05×5) in parentheses. Otherwise the calculator will misinterpret your intended order of operation.

(*b*) Suppose that a principal of $1000 is invested at an annual rate of 8 percent, compounded continuously. The equation

$$A(t) = 1000e^{0.08t}$$

gives the amount in the account after t years. Find the amount in the account after 9 years.

Continuous compounding will give the highest accumulation of interest at any rate. However, daily compounding will result in an amount of interest that is only slightly less.

Note that in 9 years the amount in the account is a little more than *double* the original principal.

Let $t = 9$ in the original equation to obtain

$$A(9) = 1000e^{(0.08)(9)} \approx 2054$$

which is the amount in the account after 9 years.

Check Yourself 3

If a principal of $1000 is invested at an annual rate of 6 percent, compounded continuously, then the equation for the amount in the account after t years is

$$A(t) = 1000e^{0.06t}$$

If you have a scientific calculator available, find the amount in the account after 12 years.

As we have observed in this section, the exponential function is always one-to-one. This yields an important property that can be used to solve certain types of equations involving exponents.

If $b > 0$ and $b \neq 1$, then

$b^m = b^n$ if and only if $m = n$ (1)

where m and n are any real numbers.

The usefulness of this property is illustrated in the next example.

EXAMPLE 4 Solving an Exponential Equation

(*a*) Solve $2^x = 8$ for x.

We recognize that 8 is a power of 2, and we can write the equation as

$$2^x = 2^3 \qquad \text{Write with equal bases.}$$

Applying property (1) above, we have

$$x = 3 \qquad \text{Set exponents equal.}$$

and 3 is the solution.

(*b*) Solve $3^{2x} = 81$ for x.

Since $81 = 3^4$, we can write

$$3^{2x} = 3^4$$

$$2x = 4$$

The answer can easily be checked by substitution. Letting $x = 2$ gives

$$3^{2(2)} = 3^4 = 81$$

$$x = 2$$

We see that 2 is the solution for the equation.

(c) Solve $2^{x+1} = \dfrac{1}{16}$ for x.

Again we write $\dfrac{1}{16}$ as a power of 2, so that

$$2^{x+1} = 2^{-4}$$

Then

$$x + 1 = -4$$

$$x = -5$$

The solution is -5.

Check Yourself 4

Solve each of the following equations for x.

1. $2^x = 16$
2. $4^{x+1} = 64$
3. $3^{2x} = \dfrac{1}{81}$

CHECK YOURSELF ANSWERS

1. $y = g(x) = 3^x$

2. $y = \left(\dfrac{1}{3}\right)^x$

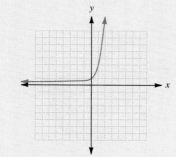

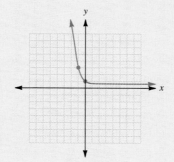

3. \$2054.43
4. (1) $x = 4$; (2) $x = 2$; (3) $x = -2$

11.2 EXERCISES

Build Your Skills

Match the graphs of Exercises 1 to 8 with the appropriate equation.

(a) $y = \left(\dfrac{1}{2}\right)^x$ (b) $y = 2x - 1$

(c) $y = 2^x$ (d) $y = x^2$

(e) $y = 1^x$ (f) $y = 5^x$

(g) $x = 2^y$ (h) $x = y^2$

1.

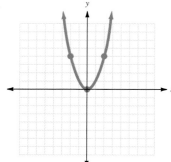

2.

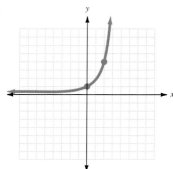

3.

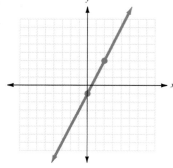

4.

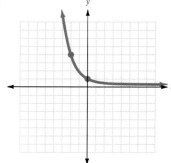

5.

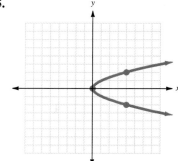

6.

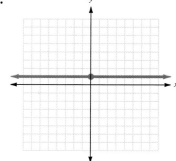

7.

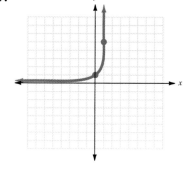

8.

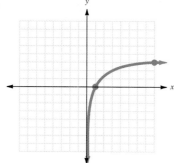

Let $f(x) = 4^x$ and find each of the following.

9. $f(0)$

10. $f(1)$

11. $f(2)$

12. $f(-2)$

Let $g(x) = 4^{x+1}$ and find each of the following.

13. $g(0)$

14. $g(1)$

15. $g(2)$

16. $g(-2)$

Let $h(x) = 4^x + 1$ and find each of the following.

17. $h(0)$

18. $h(1)$

19. $h(2)$

20. $h(-2)$

Let $f(x) = \left(\dfrac{1}{4}\right)^x$ and find each of the following.

21. $f(1)$

22. $f(-1)$

23. $f(-2)$

24. $f(2)$

Graph each of the following exponential functions.

25. $y = 4^x$

26. $y = \left(\dfrac{1}{4}\right)^x$

27. $y = \left(\dfrac{2}{3}\right)^x$

28. $y = \left(\dfrac{3}{2}\right)^x$

29. $y = 3 \cdot 2^x$

30. $y = 2 \cdot 3^x$

31. $y = 3^x$

32. $y = 2^{x-1}$

33. $y = 2^{2x}$

34. $y = \left(\dfrac{1}{2}\right)^{2x}$

35. $y = e^{-x}$

36. $y = e^{2x}$

Solve each of the following exponential equations for x.

37. $2^x = 32$

38. $4^x = 64$

39. $10^x = 10,000$

40. $5^x = 125$

41. $3^x = \dfrac{1}{9}$

42. $2^x = \dfrac{1}{16}$

43. $2^{2x} = 64$

44. $3^{2x} = 81$

45. $2^{x+1} = 64$

46. $4^{x-1} = 16$

47. $3^{x-1} = \dfrac{1}{27}$

48. $2^{x+2} = \dfrac{1}{8}$

Suppose that it takes 1 h for a certain bacterial culture to double by dividing in half. If there are 100 bacteria in the culture to start with, then the number of bacteria in the culture after x hours is given by $N(x) = 100 \cdot 2^x$. Use this function to find each of the following.

49. The number of bacteria in the culture after 2 h

50. The number of bacteria in the culture after 3 h

51. The number of bacteria in the culture after 5 h

52. Graph the relationship between the number of bacteria in the culture and the number of hours. Be sure to choose an appropriate scale for the N axis.

The half-life of radium is 1690 years. That is, after a 1690-year period, one-half of the original amount of radium will have decayed into another substance. If the original amount of radium was 64 grams (g), the formula relating the amount of radium left after time t is given by $R(t) = 64 \cdot 2^{-t/1690}$. Use that formula to find each of the following.

53. The amount of radium left after 1690 years

54. The amount of radium left after 3380 years

55. The amount of radium left after 5070 years

56. Graph the relationship between the amount of radium remaining and time. Be sure to use appropriate scales for the R and t axes.

If $1000 is invested in a savings account with an interest rate of 8 percent, compounded annually, the amount in the account after t years is given by $A(t) = 1000(1 + 0.08)^t$. Use a calculator to find each of the following.

57. The amount in the account after 2 years

58. The amount in the account after 5 years

59. The amount in the account after 9 years

60. Graph the relationship between the amount in the account and time. Be sure to choose appropriate scales for the A and t axes.

The so-called learning curve in psychology applies to learning a skill, such as typing, in which the performance level progresses rapidly at first and then levels off with time. One can approximate N, the number of words per minute that a person can type after t weeks of training, with the equation $N = 80(1 - e^{-0.06t})$. Use a calculator to find the following.

61. (a) N after 10 weeks, (b) N after 20 weeks, (c) N after 30 weeks

62. Graph the relationship between the number of words per minute N and the number of weeks of training t.

Transcribe Your Skills

63. Find two different calculators that have $\boxed{e^x}$ keys. Describe how to use the function on each of the calculators.

64. Are there any values of x for which e^x produces an exact answer on the calculator? Why are other answers not exact?

Think About These

A possible calculator sequence for evaluating the expression

$$\left(1 + \frac{1}{n}\right)^n$$

where $n = 10$ is

$\boxed{(}\ \boxed{1}\ \boxed{+}\ \boxed{1}\ \boxed{\div}\ \boxed{10}\ \boxed{)}\ \boxed{y^x}\ \boxed{10}\ \boxed{=}$

Use that sequence to find $\left(1 + \dfrac{1}{n}\right)^n$ for each of the following values of n.

65. $n = 100$

66. $n = 1000$

67. $n = 10,000$

68. $n = 100,000$

69. $n = 1,000,000$

70. What did you observe from the experiment above?

71. Graph the exponential function defined by $y = 2^x$.

72. Graph the function defined by $x = 2^y$ on the same set of axes as the previous graph. What do you observe? *Hint:* To graph $x = 2^y$, choose convenient values for y and then the corresponding values for x.

Skillscan (Section 11.1)

a. If $f = \{(x, y)|y = 2x + 1\}$, find f^{-1}. **b.** Graph the function f. **c.** Graph the function f^{-1} on the same set of axes. **d.** If $g = \{(x, y)|y = 3x - 2\}$, find g^{-1}. **e.** Graph the function g. **f.** Graph the function g^{-1} on the same set of axes.

11.3 Logarithmic Functions

TAPE IN23

OBJECTIVES: 1. To graph logarithmic functions
2. To convert between logarithmic and exponential equations
3. To solve certain logarithmic equations

Given our experience with the exponential function in Section 11.2 and our earlier work with the inverse of a function, we now can introduce the logarithmic function.

John Napier (1550–1617), a Scotsman, is credited with the invention of logarithms. The development of the logarithm grew out of a desire to ease the work involved in numerical computations, particularly in the field of astronomy. Today

Napier also coined the word "logarithm" from the Greek words "logos"—a ratio—and "arithmos"—a number.

the availability of inexpensive scientific calculators has made the use of logarithms as a computational tool unnecessary.

However, the concept of the logarithm and the properties of the logarithmic function that we describe in a later section still are very important in the solutions of particular equations, in calculus, and in the applied sciences.

Again, the applications for this new function are numerous. The Richter scale for measuring the intensity of an earthquake and the decibel scale for measuring the intensity of sound both make use of logarithms.

To develop the idea of a logarithmic function, we must return to the exponential function

$$f = \{(x, y) \mid y = b^x, \, b > 0, \, b \neq 1\} \tag{1}$$

Interchanging the roles of x and y, we have the inverse function

$$f^{-1} = \{(x, y) \mid x = b^y\} \tag{2}$$

Recall that f is a one-to-one function, so its inverse is also a function.

Presently we have no way to solve the equation $x = b^y$ for y. So, to write the inverse (2) in a more useful form, we offer the following definition.

> The *logarithm of x to base b* is denoted
>
> $\log_b x$
>
> and
>
> $y = \log_b x$ if and only if $x = b^y$

We can now write our inverse function, using this new notation, as

$$f^{-1} = \{(x, y) \mid y = \log_b x, \, b > 0, \, b \neq 1\} \tag{3}$$

Note that the restrictions on the base are the same as those used for the exponential function.

In general, any function defined in this form is called a *logarithmic function*.

At this point we should stress the meaning of this new relationship. Consider the equivalent forms:

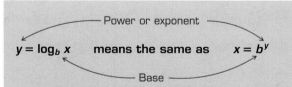

$y = \log_b x$ means the same as $x = b^y$

The logarithm y is the power to which we must raise b to get x. In other words, *a logarithm is simply a power or an exponent*. We return to this thought later when using the exponential and logarithmic forms of equivalent equations.

We begin our work by graphing a typical logarithmic function.

EXAMPLE 1 Graphing a Logarithmic Function

Graph the logarithmic function

$y = \log_2 x$

Solution Since $y = \log_2 x$ is equivalent to the exponential form

The base is 2, and the logarithm or power is y.

$$x = 2^y$$

we can find ordered pairs satisfying this equation by choosing convenient values for y and calculating the corresponding values for x.

Letting y take on values from -3 to 3 yields the table of values shown below. As before, we plot points from the ordered pairs and connect them with a smooth curve to form the graph of the function.

What do the vertical- and horizontal-line tests tell you about this graph?

x	y
$\dfrac{1}{8}$	-3
$\dfrac{1}{4}$	-2
$\dfrac{1}{2}$	-1
1	0
2	1
4	2
8	3

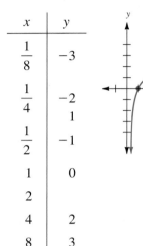

Use your calculator to compare the graphs of $y = 2^x$ and $y = \log_2 x$. Are they inverse functions? How can you tell?

We observe that the graph represents a one-to-one function whose domain is $\{x \mid x > 0\}$ and whose range is the set of all real numbers.

For base 2 (or for any base greater than 1) the function will always be increasing over its domain.

Recall from Section 11.1 that the graphs of a function and its inverse are always reflections of each other about the line $y = x$. Since we have defined the logarithmic function as the inverse of an exponential function, we can anticipate the same relationship.

The graphs of

$$f(x) = 2^x \qquad \text{and} \qquad f^{-1}(x) = \log_2 x$$

are shown below.

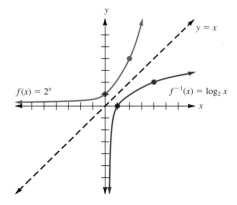

We see that the graphs of f and f^{-1} are indeed reflections of each other about the line $y = x$. In fact, this relationship provides an alternate method of sketching $y = \log_b x$. We can sketch the graph of $y = b^x$ and then reflect that graph about the line $y = x$ to form the graph of the logarithmic function.

Check Yourself 1

Graph the logarithmic function defined by

$$y = \log_3 x$$

Hint: Consider the equivalent form $x = 3^y$.

For our later work in this chapter it will be necessary to be able to convert back and forth between exponential and logarithmic forms.

The conversion is straightforward. You need only keep in mind the basic relationship

$y = \log_b x$ **means the same as**   **$x = b^y$**

Again, this tells us that a logarithm is an exponent or a power.

Look at the following examples.

EXAMPLE 2 Writing Equations in Logarithmic Form

Convert to logarithmic form.

(a) $3^4 = 81$ is equivalent to $\log_3 81 = 4$.

The base is 3; the exponent or power is 4.

(b) $10^3 = 1000$ is equivalent to $\log_{10} 1000 = 3$.

(c) $2^{-3} = \dfrac{1}{8}$ is equivalent to $\log_2 \dfrac{1}{8} = -3$.

(d) $9^{1/2} = 3$ is equivalent to $\log_9 3 = \dfrac{1}{2}$.

Check Yourself 2

Convert each statement to logarithmic form.

1. $4^3 = 64$ **2.** $10^{-2} = 0.01$ **3.** $3^{-3} = \dfrac{1}{27}$ **4.** $27^{1/3} = 3$

The next example shows how to write a logarithmic expression in exponential form.

EXAMPLE 3 Writing Equations in Exponential Form

Convert to exponential form.

Here the base is 2; the logarithm, which is the power, is 3.

(a) $\log_2 8 = 3$ is equivalent to $2^3 = 8$.

(b) $\log_{10} 100 = 2$ is equivalent to $10^2 = 100$.

(c) $\log_3 \frac{1}{9} = -2$ is equivalent to $3^{-2} = \frac{1}{9}$.

(d) $\log_{25} 5 = \frac{1}{2}$ is equivalent to $25^{1/2} = 5$.

Check Yourself 3

Convert to exponential form.

1. $\log_2 32 = 5$ **2.** $\log_{10} 1000 = 3$ **3.** $\log_4 \frac{1}{16} = -2$ **4.** $\log_{27} 3 = \frac{1}{3}$

Certain logarithms can be directly calculated by changing an expression to the equivalent exponential form. Two examples follow.

EXAMPLE 4 Evaluating Logarithmic Expressions

(a) Evaluate $\log_3 27$.

If $x = \log_3 27$, in exponential form we have

Recall that $b^m = b^n$ if and only if $m = n$.

$3^x = 27$

$3^x = 3^3$

$x = 3$

We then have $\log_3 27 = 3$.

(b) Evaluate $\log_{10} \frac{1}{10}$.

If $x = \log_{10} \frac{1}{10}$, we can write

$10^x = \frac{1}{10}$

Write each side as a power of the same base.

$= 10^{-1}$

We then have $x = -1$ and

$\log_{10} \frac{1}{10} = -1$

Check Yourself 4

Evaluate each logarithm.

1. $\log_2 64$

2. $\log_3 \dfrac{1}{27}$

The relationship between exponents and logarithms also allows us to solve certain equations involving logarithms where two of the quantities in the equation $y = \log_b x$ are known. The next example illustrates.

EXAMPLE 5 Solving Logarithmic Equations

(a) Solve $\log_5 x = 3$ for x.

Since $\log_5 x = 3$, in exponential form we have

$$x = 5^3$$
$$= 125$$

(b) Solve $y = \log_4 \dfrac{1}{16}$ for y.

The original equation is equivalent to

$$4^y = \dfrac{1}{16}$$
$$= 4^{-2}$$

We then have $y = -2$ as the solution.

(c) Solve $\log_b 81 = 4$ for b.

In exponential form the equation becomes

$$b^4 = 81$$
$$b = 3$$

Keep in mind that the base must be *positive*, so we do not consider the possible solution $b = -3$.

Check Yourself 5

Solve each of the following equations for the variable cited.

1. $\log_4 x = 4$ for x **2.** $\log_b \dfrac{1}{8} = -3$ for b **3.** $y = \log_9 3$ for y

To conclude this section, we turn to two common applications of the logarithmic function. The *decibel scale* is used in measuring the loudness of various sounds.

Loudness can be measured in *bels* (B), a unit named for Alexander Graham Bell. This unit is rather large, so a more practical unit is the *decibel* (dB), a unit one-tenth as large.

Variable I_0 is the intensity of the minimum sound level detectable by the human ear.

If I represents the intensity of a given sound and I_0 represents the intensity of a "threshold sound," then the decibel (dB) rating of the given sound is given by

$$L = 10 \log_{10} \frac{I}{I_0}$$

where $I_0 = 10^{-16}$ watt per square centimeter (W/cm^2). Consider the following examples.

EXAMPLE 6 A Decibel Application

(*a*) A whisper has intensity $I = 10^{-14}$. Its decibel rating is

$$L = 10 \log_{10} \frac{10^{-14}}{10^{-16}}$$

$$= 10 \log_{10} 10^2$$

$$= 10 \cdot 2$$

$$= 20$$

(*b*) A rock concert has intensity $I = 10^{-4}$. Find its decibel rating.

$$L = 10 \log_{10} \frac{10^{-4}}{10^{-16}}$$

$$= 10 \log_{10} 10^{12}$$

$$= 10 \cdot 12$$

$$= 120$$

Check Yourself 6

Ordinary conversation has intensity $I = 10^{-12}$. Find its rating on the decibel scale.

The scale was named after Charles Richter, a U.S. geologist.

Another commonly used logarithmic scale is the *Richter scale*. Geologists use that scale to convert seismographic readings, which give the intensity of the shock waves of an earthquake, to a measure of the magnitude of that earthquake.

The magnitude M of an earthquake is given by

$$M = \log_{10} \frac{a}{a_0}$$

A "zero-level" earthquake is the quake of least intensity that is measurable by a seismograph.

where a is the intensity of its shock waves and a_0 is the intensity of the shock wave of a zero-level earthquake.

EXAMPLE 7 A Richter Scale Application

How many times stronger is an earthquake measuring 5 on the Richter scale than one measuring 4 on the Richter scale?

Solution Suppose a_1 is the intensity of the earthquake with magnitude 5 and a_2 is the intensity of the earthquake with magnitude 4. Then

$$5 = \log_{10} \frac{a_1}{a_0} \quad \text{and} \quad 4 = \log_{10} \frac{a_2}{a_0}$$

We convert these logarithmic expressions to exponential form.

$$10^5 = \frac{a_1}{a_0} \quad \text{and} \quad 10^4 = \frac{a_2}{a_0}$$

or

$$a_1 = a_0 \cdot 10^5 \quad \text{and} \quad a_2 = a_0 \cdot 10^4$$

We want the ratio of the intensities of the two earthquakes, so

$$\frac{a_1}{a_2} = \frac{a_0 \cdot 10^5}{a_0 \cdot 10^4} = 10^1 = 10$$

The earthquake of magnitude 5 is *10 times stronger* than the earthquake of magnitude 4.

On your calculator, the $\boxed{\text{log}}$ key is actually $\log_{10} x$.

The ratio of a_1 to a_2 is

$\dfrac{a_1}{a_2}$

Check Yourself 7

How many times stronger is an earthquake of magnitude 6 than one of magnitude 4?

CHECK YOURSELF ANSWERS

1. $y = \log_3 x$.

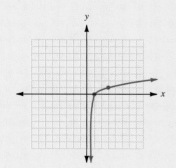

2. (1) $\log_4 64 = 3$; (2) $\log_{10} 0.01 = -2$; (3) $\log_3 \dfrac{1}{27} = -3$; (4) $\log_{27} 3 = \dfrac{1}{3}$

3. (1) $2^5 = 32$; (2) $10^3 = 1000$; (3) $4^{-2} = \dfrac{1}{16}$; (4) $27^{1/3} = 3$

4. (1) $\log_2 64 = 6$; (2) $\log_3 \dfrac{1}{27} = -3$ **5.** (1) $x = 256$; (2) $b = 2$; (3) $y = \dfrac{1}{2}$

6. 40 dB   **7.** 100 times

11.3 EXERCISES

Build Your Skills

Sketch the graph of the function defined by each of the following equations.

1. $y = \log_4 x$

2. $y = \log_{10} x$

3. $y = \log_2 (x - 1)$

4. $y = \log_3 (x + 1)$

5. $y = \log_e x$ (Use your calculator.)

6. $y = \log_3 x + 1$

Convert each of the following statements to logarithmic form.

7. $2^4 = 16$

8. $3^5 = 243$

9. $10^2 = 100$

10. $4^3 = 64$

11. $3^0 = 1$

12. $10^0 = 1$

13. $4^{-2} = \dfrac{1}{16}$

14. $3^{-4} = \dfrac{1}{81}$

15. $10^{-3} = \dfrac{1}{1000}$

16. $2^{-5} = \dfrac{1}{32}$

17. $16^{1/2} = 4$

18. $125^{1/3} = 5$

19. $64^{-1/3} = \dfrac{1}{4}$

20. $36^{-1/2} = \dfrac{1}{6}$

21. $8^{2/3} = 4$

22. $9^{3/2} = 27$

23. $27^{-2/3} = \dfrac{1}{9}$

24. $16^{-3/2} = \dfrac{1}{64}$

Convert each of the following statements to exponential form.

25. $\log_2 16 = 4$

26. $\log_3 3 = 1$

27. $\log_5 1 = 0$

28. $\log_3 27 = 3$

29. $\log_{10} 10 = 1$

30. $\log_2 32 = 5$

31. $\log_5 125 = 3$

32. $\log_{10} 1 = 0$

33. $\log_3 \dfrac{1}{27} = -3$

34. $\log_5 \dfrac{1}{25} = -2$

35. $\log_{10} 0.01 = -2$

36. $\log_{10} \dfrac{1}{1000} = -3$

37. $\log_{16} 4 = \dfrac{1}{2}$

38. $\log_{125} 5 = \dfrac{1}{3}$

39. $\log_8 4 = \dfrac{2}{3}$

40. $\log_9 27 = \dfrac{3}{2}$

41. $\log_{25} \dfrac{1}{5} = -\dfrac{1}{2}$

42. $\log_{64} \dfrac{1}{16} = -\dfrac{2}{3}$

Evaluate each of the following logarithms.

43. $\log_2 32$

44. $\log_3 81$

45. $\log_4 64$

46. $\log_{10} 1000$

47. $\log_3 \dfrac{1}{81}$

48. $\log_4 \dfrac{1}{64}$

49. $\log_{10} \dfrac{1}{100}$

50. $\log_5 \dfrac{1}{25}$

51. $\log_{25} 5$

52. $\log_{27} 3$

Solve each of the following equations for the unknown variable.

53. $y = \log_5 25$

54. $\log_2 x = 4$

55. $\log_b 64 = 3$

56. $y = \log_3 1$

57. $\log_{10} x = 2$

58. $\log_b 125 = 3$

59. $y = \log_5 5$

60. $y = \log_3 81$

61. $\log_{3/2} x = 3$

62. $\log_b \dfrac{4}{9} = 2$

63. $\log_b \dfrac{1}{25} = -2$

64. $\log_3 x = -3$

65. $\log_{10} x = -3$

66. $y = \log_2 \dfrac{1}{16}$

67. $y = \log_8 \dfrac{1}{64}$

68. $\log_b \dfrac{1}{100} = -2$

69. $\log_{27} x = \dfrac{1}{3}$

70. $y = \log_{100} 10$

71. $\log_b 5 = \dfrac{1}{2}$

72. $\log_{64} x = \dfrac{2}{3}$

73. $y = \log_{27} \dfrac{1}{9}$

74. $\log_b \dfrac{1}{8} = -\dfrac{3}{4}$

See Example 6 for the decibel formula to use in each of the following.

75. A television commercial has a volume with intensity $I = 10^{-11}$ W/cm^2. Find its rating in decibels.

76. The sound of a jet plane on takeoff has an intensity $I = 10^{-2}$ W/cm^2. Find its rating in decibels.

77. The sound of a computer printer has an intensity of $I = 10^{-9}$ W/cm^2. Find its rating in decibels.

78. The sound of a busy street has an intensity of $I = 10^{-8}$ W/cm^2. Find its rating in decibels.

The formula for the decibel rating L can be solved for the intensity of the sound as $I = I_0 \cdot 10^{L/10}$.

79. Find the intensity of the sound in an airport waiting area if the decibel rating is 80.

80. Find the intensity of the sound of conversation in a crowded room if the decibel rating is 70.

81. What is the ratio of intensity of the sound of 80 dB to that of 70 dB?

82. What is the ratio of intensity of a sound of 60 dB to one measuring 40 dB?

83. What is the ratio of intensity of a sound of 70 dB to one measuring 40 dB?

84. Derive the formula for intensity provided above. *Hint:* First divide both sides of the decibel formula by 10. Then write the equation in exponential form.

See Example 7 for the formula for magnitude on the Richter scale.

85. An earthquake has an intensity a of $10^6 \cdot a_0$, where a_0 is the intensity of the zero-level earthquake. What was its magnitude?

86. The great San Francisco earthquake of 1906 had an intensity of $10^{8.3} \cdot a_0$. What was its magnitude?

87. An earthquake can begin causing damage to buildings with a magnitude of 5 on the Richter scale. Find its intensity in terms of a_0.

88. An earthquake may cause moderate building damage with a magnitude of 6 on the Richter scale. Find its intensity in terms of a_0.

Transcribe Your Skills

89. The *learning curve* describes the relationship between learning and time. Its graph is a logarithmic curve in the first quadrant. Describe that curve as it relates to learning.

90. In which scientific fields would you expect to again encounter a discussion of logarithms?

The *half-life* of a radioactive substance is the time it takes for half of the original amount of the substance to decay to a nonradioactive element. The half-life of radioactive waste is very important in figuring how long the waste must be kept isolated from the environment in some sort of storage facility. Half-lives of various radioactive waste products vary from a few seconds to millions of years. It usually takes at least 10 half-lives for a radioactive waste product to be considered safe.

The half-life of a radioactive substance can be determined by the following formula.

$$\ln \frac{1}{2} = -\lambda x$$

where λ is the radioactive decay constant and x is the half-life.

Find the half-lives of the following important radioactive waste products given the radioactive decay constant (RDC).

 91. Plutonium 239, RDC = 0.000029

 92. Strontium 90, RDC = 0.024755

 93. Thorium 230, RDC = 0.000009

 94. Cesium 135, RDC = 0.00000035

 95. How many years will it be before each of these waste products will be considered safe?

Skillscan (Section 5.1)

Simplify each expression.

a. $a^3 \cdot a^2$ **b.** $b^5 \cdot b^3$ **c.** $x^5 \cdot x^0$ **d.** $r^0 \cdot r^6$

e. $\dfrac{y^8}{y^5}$ **f.** $\dfrac{m^7}{m}$ **g.** $(b^4)^2$ **h.** $(x^5)^4$

11.4 Properties of Logarithms

TAPE IN23

OBJECTIVE: To apply the properties of logarithms

As we mentioned earlier, logarithms were developed as aids to numerical computations. The early utility of the logarithm was due to the properties that we will discuss in this section. Even with the advent of the scientific calculator, that utility remains important today. We can apply these same properties to applications in a variety of areas that lead to exponential or logarithmic equations.

Since a logarithm is, by definition, an exponent, it seems reasonable that our knowledge of the properties of exponents should lead to useful properties for logarithms. That is, in fact, the case.

We start with two basic facts that follow immediately from the definition of the logarithm.

The properties follow from the facts that

$$b^1 = b \quad \text{and} \quad b^0 = 1$$

For $b > 0$ and $b \neq 1$,

1. $\log_b b = 1$
2. $\log_b 1 = 0$

We know that the logarithmic function $y = \log_b x$ and the exponential function $y = b^x$ are inverses of each other. So for $f(x) = b^x$, we have $f^{-1}(x) = \log_b x$.

Also recall from our work in Section 11.1 that for any one-to-one function f,

$f^{-1}(f(x)) = x$ for any x in domain of f

and

$f(f^{-1}(x)) = x$ for any x in domain of f^{-1}

The inverse has "undone" whatever f did to x.

Since $f(x) = b^x$ is a one-to-one function, we can apply the above to the case where

$f(x) = b^x$ and $f^{-1}(x) = \log_b x$

to derive the following.

---

3. $\log_b b^x = x$

4. $b^{\log_b x} = x$ for $x > 0$

For Property 3,

$f^{-1}(f(x)) = f^{-1}(b^x) = \log_b b^x$

But in general, for any one-to-one function f,

$f^{-1}(f(x)) = x$

Since logarithms are exponents, we can again turn to the familiar exponent rules to derive some further properties of logarithms. Consider the following.
We know that

$\log_b M = x$ if and only if $M = b^x$

and

$\log_b N = y$ if and only if $N = b^y$

Then

$$M \cdot N = b^x \cdot b^y = b^{x+y} \tag{1}$$

From Equation (1) we see that $x + y$ is the power to which we must raise b to get the product MN. In logarithmic form that becomes

$$\log_b MN = x + y \tag{2}$$

Now since $x = \log_b M$ and $y = \log_b N$, we can substitute in (2) to write

$$\log_b MN = \log_b M + \log_b N \tag{3}$$

This is the first of the basic logarithmic properties which are written below. The remaining properties may all be proved by arguments similar to that above.

PROPERTIES OF LOGARITHMS

PRODUCT PROPERTY

$$\log_b MN = \log_b M + \log_b N$$

QUOTIENT PROPERTY

$$\log_b \frac{M}{N} = \log_b M - \log_b N$$

POWER PROPERTY

$$\log_b M^p = p \log_b M$$

In all cases, $M, N > 0$, $b > 0$, $b \ne 1$, and p is any real number.

Many applications of logarithms require using the above properties to write a single logarithmic expression as the sum or difference of simpler expressions. The following example illustrates.

EXAMPLE 1 Using the Properties of Logarithms

Expand, using the properties of logarithms.

(a) $\log_b xy = \log_b x + \log_b y$ Product property

(b) $\log_b \dfrac{xy}{z} = \log_b xy - \log_b z$ Quotient property

        $= \log_b x + \log_b y - \log_b z$ Product property

(c) $\log_{10} x^2 y^3 = \log_{10} x^2 + \log_{10} y^3$ Product property

        $= 2 \log_{10} x + 3 \log_{10} y$ Power property

Recall $\sqrt{a} = a^{1/2}$.

(d) $\log_b \sqrt{\dfrac{x}{y}} = \log_b \left(\dfrac{x}{y}\right)^{1/2}$ Definition of exponent

        $= \dfrac{1}{2} \log_b \dfrac{x}{y}$ Power property

        $= \dfrac{1}{2} (\log_b x - \log_b y)$ Quotient property

---

Check Yourself 1

Expand each expression, using the properties of logarithms.

1. $\log_b x^2 y^3 z$ **2.** $\log_{10} \sqrt{\dfrac{xy}{z}}$

In some cases we will reverse the process and use the properties to write a single logarithm, given a sum or difference of logarithmic expressions.

EXAMPLE 2 Rewriting Logarithmic Expressions

Write each expression as a single logarithm with coefficient 1.

(a) $2 \log_b x + 3 \log_b y$

    $= \log_b x^2 + \log_b y^3$ Power property

    $= \log_b x^2 y^3$ Product property

(b) $5 \log_{10} x + 2 \log_{10} y - \log_{10} z$

$\quad = \log_{10} x^5 y^2 - \log_{10} z$

$\quad = \log_{10} \dfrac{x^5 y^2}{z}$ $\qquad$ Quotient property

(c) $\dfrac{1}{2}(\log_2 x - \log_2 y)$

$\quad = \dfrac{1}{2}\left(\log_2 \dfrac{x}{y}\right)$

$\quad = \log_2 \left(\dfrac{x}{y}\right)^{1/2}$ $\qquad$ Power property

$\quad = \log_2 \sqrt{\dfrac{x}{y}}$

Check Yourself 2

Write each expression as a single logarithm with coefficient 1.

1. $3 \log_b x + 2 \log_b y - 2 \log_b z$ $\qquad$ **2.** $\dfrac{1}{3}(2 \log_2 x - \log_2 y)$

The following example illustrates the basic concept of the use of logarithms as a computational aid.

EXAMPLE 3 Evaluating Logarithmic Expressions

Suppose that $\log_{10} 2 = 0.301$ and that $\log_{10} 3 = 0.477$. Given these values, find:

(a) $\log_{10} 6$

$\qquad$ Since $6 = 2 \cdot 3$,

$\log_{10} 6 = \log_{10}(2 \cdot 3)$

$\qquad = \log_{10} 2 + \log_{10} 3$

$\qquad = 0.301 + 0.477$

$\qquad = 0.778$

(b) $\log_{10} 18$

$\qquad$ Since $18 = 2 \cdot 3 \cdot 3$,

$\log_{10} 18 = \log_{10}(2 \cdot 3 \cdot 3)$

$\qquad = \log_{10} 2 + \log_{10} 3 + \log_{10} 3$

$\qquad = 1.255$

We have written the logarithms correct to three decimal places and will follow this practice throughout the remainder of this chapter.

 Keep in mind, however, that this is an approximation and that $10^{0.301}$ will only approximate 2. Verify this with your calculator.

We have extended the product rule for logarithms.

(c) $\log_{10} \dfrac{1}{9}$

Since $\dfrac{1}{9} = \dfrac{1}{3^2}$,

$\log_{10} \dfrac{1}{9} = \log_{10} \dfrac{1}{3^2}$

Note that $\log_b 1 = 0$ for any base b.

$= \log_{10} 1 - \log_{10} 3^2$

$= 0 - 2 \log_{10} 3$

$= -0.954$

(d) $\log_{10} 16$

Since $16 = 2^4$,

$\log_{10} 16 = \log_{10} 2^4 = 4 \log_{10} 2$

$= 1.204$

Verify each answer with your calculator.

(e) $\log_{10} \sqrt{3}$

Since $\sqrt{3} = 3^{1/2}$,

$\log_{10} \sqrt{3} = \log_{10} 3^{1/2} = \dfrac{1}{2} \log_{10} 3$

$= 0.239$

Check Yourself 3

Given the values above for $\log_{10} 2$ and $\log_{10} 3$, find each of the following.

1. $\log_{10} 12$ **2.** $\log_{10} 27$ **3.** $\log_{10} \sqrt[3]{2}$

LOGARITHMS TO PARTICULAR BASES

You can easily check the results in the example and exercise above by using the $\boxed{\log}$ key on your calculator. For instance, in Example 3d, to find $\log_{10} 16$, enter

16 $\boxed{\log}$

and the result (to three decimal places) will be 1.204. As you can see, the $\boxed{\log}$ key on your calculator provides logarithms to base 10. This is one of two types of logarithms which are used most frequently in mathematics.

Logarithms to base 10

Logarithms to base e

Of course, the use of logarithms to base 10 is convenient because our number system has base 10. We call logarithms to base 10 *common logarithms,* and it is customary to omit the base in writing a common (or base-10) logarithm. So

log *N*   means $\log_{10} N$

Note When no base for "log" is written, it is assumed to be 10.

The following table shows the common logarithms for various powers of 10.

Exponential Form	Logarithmic Form
$10^3 = 1000$	$\log 1000 = 3$
$10^2 = 100$	$\log 100 = 2$
$10^1 = 10$	$\log 10 = 1$
$10 = 1$	$\log 1 = 0$
$10^{-1} = 0.1$	$\log 0.1 = -1$
$10^{-2} = 0.01$	$\log 0.01 = -2$
$10^{-3} = 0.001$	$\log 0.001 = -3$

EXAMPLE 4 Approximating Logarithms with a Calculator

Verify each of the following with a calculator.

(*a*) $\log 4.8 = 0.681$
(*b*) $\log 48 = 1.681$
(*c*) $\log 480 = 2.681$
(*d*) $\log 4800 = 3.681$
(*e*) $\log 0.48 = -0.319$

The number 4.8 lies between 1 and 10, so log 4.8 lies between 0 and 1.

Note that

$480 = 4.8 \times 10^2$

and

$\log (4.8 \times 10^2)$

$= \log 4.8 + \log 10^2$

$= \log 4.8 + 2$

$= 2 + \log 4.8$

The value of log 0.48 is really $-1 + 0.681$. Your calculator will combine the signed numbers.

Check Yourself 4

Use your calculator to find each of the following logarithms, correct to three decimal places.

1. $\log 2.3$ **2.** $\log 23$ **3.** $\log 230$

4. $\log 2300$ **5.** $\log 0.23$ **6.** $\log 0.023$

Let's look at an application of common logarithms from chemistry. Common logarithms are used to define the pH of a solution. This is a scale which measures whether the solution is acidic or basic.

The pH of a solution is defined as

$$pH = -\log [H^+]$$

where $[H^+]$ is the hydrogen ion concentration, in moles per liter (mol/L), in the solution.

Note A solution is *neutral* with pH = 7, *acidic* if the pH is less than 7, and *basic* if the pH is greater than 7.

EXAMPLE 5 A pH Application

Find the pH of each of the following.

(*a*) Rainwater: $[H^+] = 1.6 \times 10^{-7}$

From the definition,

$$
\begin{aligned}
\text{pH} &= -\log [H^+] \\
&= -\log (1.6 \times 10^{-7}) \\
&= -(\log 1.6 + \log 10^{-7}) \\
&= -[0.204 + (-7)] \\
&= -(-6.796) = 6.796
\end{aligned}
$$

Note the use of the product rule here.

Also, in general, $\log_b b^x = x$, so $\log 10^{-7} = -7$.

The rain is just slightly acidic.

(*b*) Household ammonia: $[H^+] = 2.3 \times 10^{-8}$

$$
\begin{aligned}
\text{pH} &= -\log (2.3 \times 10^{-8}) \\
&= -(\log 2.3 + \log 10^{-8}) \\
&= -[0.362 + (-8)] \\
&= 7.638
\end{aligned}
$$

The ammonia is slightly basic.

(*c*) Vinegar: $[H^+] = 2.9 \times 10^{-3}$

$$
\begin{aligned}
\text{pH} &= -\log (2.9 \times 10^{-3}) \\
&= -(\log 2.9 + \log 10^{-3}) \\
&= 2.538
\end{aligned}
$$

The vinegar is very acidic.

Check Yourself 5

Find the pH for the following solutions. Are they acidic or basic?

1. Orange juice: $[H^+] = 6.8 \times 10^{-5}$

2. Drain cleaner: $[H^+] = 5.2 \times 10^{-13}$

Many applications require reversing the process. That is, given the logarithm of a number, we must be able to find that number. The process is straightforward.

EXAMPLE 6 Using a Calculator to Estimate Antilogarithms

Suppose that $\log x = 2.1567$. We want to find a number x whose logarithm is 2.1567. Using a calculator requires one of the following sequences:

Because it is a one-to-one function, the logarithmic function has an inverse.

2.1567 $\boxed{10^x}$ or   2.1567 $\boxed{\text{INV}}$ $\boxed{\log}$

Both give the result 143.45, often called the *antilogarithm* of 2.1567.

Check Yourself 6

Find the value of the antilogarithm of x.

1. $\log x = 0.828$

2. $\log x = 1.828$

3. $\log x = 2.828$

4. $\log x = -0.172$

Let's return to the application from chemistry for an example requiring the use of the antilogarithm.

EXAMPLE 7 A pH Application

Suppose that the pH for tomato juice is 6.2. Find the hydrogen ion concentration $[H^+]$.

Solution Recall from our earlier formula that

$$pH = -\log [H^+]$$

In this case we have

$$6.2 = -\log [H^+]$$

or

$$\log [H^+] = -6.2$$

The desired value for $[H^+]$ is then the antilogarithm of -6.2, and we use the following calculator sequence:

6.2 $\boxed{+/-}$ $\boxed{INV}$ $\boxed{\log}$

The result is 0.00000063, and we can write

$$[H^+] = 6.3 \times 10^{-7}$$

Check Yourself 7

The pH for eggs is 7.8. Find $[H^+]$ for the same solution.

As we mentioned, there are two systems of logarithms in common use. The second type of logarithm uses the number e as a base, and we call logarithms to base e the *natural logarithms*.

As with common logarithms, a convenient notation has developed, as the following definition shows.

> The *natural logarithm* is a logarithm to base e, and it is denoted ln x, where
>
> ln $x = \log_e x$

Natural logarithms are also called *napierian logarithms* after Napier. The importance of this system of logarithms was not fully understood until later developments in the calculus.

The restrictions on the domain of the natural logarithmic function are the same as before. The function is defined only if $x > 0$.

By the general definition of a logarithm,

$y = \ln x$ means the same as $x = e^y$

and this leads us directly to the following:

$\ln 1 = 0$ since $e^0 = 1$

$\ln e = 1$ since $e^1 = e$

Also

$\ln e^2 = 2$ and $\ln e^{-3} = -3$

In general,

$\log_b b^x = x$ $b \neq 1$

EXAMPLE 8 Estimating Natural Logarithms

To find other natural logarithms, we can again turn to a scientific calculator. To find the value of ln 2, use the sequence

2 $\boxed{\ln}$

The result is 0.693 (to three decimal places).

Check Yourself 8

Use a calculator to find each of the following.

1. ln 3   **2.** ln 6   **3.** ln 4   **4.** $\ln \sqrt{3}$

Of course, the properties of logarithms are applied in an identical fashion, no matter what the base.

EXAMPLE 9 Evaluating Logarithms

If $\ln 2 = 0.693$ and $\ln 3 = 1.099$, find the following.

Recall that

$\log_b MN = \log_b M + \log_b N$

$\log_b M^p = p \log_b M$

(a) $\ln 6 = \ln (2 \cdot 3) = \ln 2 + \ln 3 = 1.792$

(b) $\ln 4 = \ln 2^2 = 2 \ln 2 = 1.386$

(c) $\ln \sqrt{3} = \ln 3^{1/2} = \dfrac{1}{2} \ln 3 = 0.549$

Again, verify these results with your calculator.

Check Yourself 9

Use $\ln 2 = 0.693$ and $\ln 3 = 1.099$ to find the following.

1. ln 12   **2.** ln 27

The natural logarithm function plays an important role in both theoretical and applied mathematics. The following example illustrates just one of the many applications of this function.

EXAMPLE 10 A Learning Curve Application

A class of students took a final mathematics examination and received an average score of 76.

In a psychological experiment, the students are retested at weekly intervals over the same material. If t is measured in weeks, then the new average score after t weeks is given by

$$S(t) = 76 - 5 \ln (t + 1)$$

(*a*) Find the score after 10 weeks.

$$S(t) = 76 - 5 \ln (10 + 1)$$
$$= 76 - 5 \ln 11$$
$$\approx 64$$

(*b*) Find the score after 20 weeks.

$$S(t) = 76 - 5 \ln (20 + 1)$$
$$\approx 61$$

(*c*) Find the score after 30 weeks.

$$S(t) = 76 - 5 \ln (30 + 1)$$
$$\approx 59$$

Recall that we read $S(t)$ as "S of t," which means that S is a function of t.

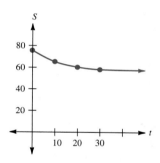

This is an example of a "forgetting" curve. Note how it drops more rapidly at first.

Compare this curve to the learning curve drawn in Section 11.2, Exercise 62.

Check Yourself 10

The average score for a group of biology students, retested after time t (in months), is given by

$$S(t) = 83 - 9 \ln (t + 1)$$

Find the average score after

1. 3 months

2. 6 months

We conclude this section with one final property of logarithms. This property will allow us to quickly find the logarithm of a number to any base. Although work with logarithms with bases other than 10 or e is relatively infrequent, the relationship between logarithms of different bases is in itself interesting. Consider the following argument.

Suppose that

$$x = \log_2 5$$

or

$$2^x = 5 \tag{4}$$

Taking the logarithm to base 10 of both sides of Equation (4) yields

$$\log 2^x = \log 5$$

or

$$x \log 2 = \log 5 \qquad \text{Use the power property of logarithms.} \tag{5}$$

(Note that we omit the 10 for the base and write log 2, for example.) Now, dividing both sides of Equation (5) by log 2, we have

$$x = \frac{\log 5}{\log 2}$$

We can now find a value for x with the calculator. Dividing with the calculator log 5 by log 2, we get an approximate answer of 2.3219.

Since $x = \log_2 5$ and $x = \dfrac{\log 5}{\log 2}$, then

$$\log_2 5 = \frac{\log 5}{\log 2}$$

Generalizing our result, we find the following:

CHANGE-OF-BASE FORMULA

For the positive real numbers a and x,

$$\log_a x = \frac{\log x}{\log a}$$

Note that the logarithm on the left side has base a while the logarithms on the right side have base 10. This allows us to calculate the logarithm to base a of any positive number, given the corresponding logarithms to base 10 (or any other base). The following example illustrates.

EXAMPLE 11 Evaluating Logarithms

Find $\log_5 15$.

Solution From the change-of-base formula with $a = 5$ and $b = 10$,

$$\log_5 15 = \frac{\log_{10} 15}{\log_{10} 5}$$

$$= 1.683$$

We have written $\log_{10} 15$ rather than $\log 15$ to emphasize the change-of-base formula.

The calculator sequence for the above computation is

15 $\boxed{\log}$ $\boxed{\div}$ 5 $\boxed{\log}$ $\boxed{=}$

Note $\log_5 5 = 1$ and $\log_5 25 = 2$, so the result for $\log_5 15$ must be between 1 and 2.

Check Yourself 11

Use the change-of-base formula to find $\log_8 32$.

CAUTION
Be Careful! A *common error* is to write

$$\frac{\log 15}{\log 5} \stackrel{?}{=} \log 15 - \log 5$$

This is *not* a logarithmic property. A true statement would be

$$\log \frac{15}{5} = \log 15 - \log 5$$

but

$$\log \frac{15}{5} \quad \text{and} \quad \frac{\log 15}{\log 5}$$

are *not* the same.

Note A special case of the change-of-base formula allows us to find natural logarithms in terms of common logarithms:

$$\ln x = \frac{\log x}{\log e}$$

so

$$\ln x \approx \frac{\log x}{0.434} \quad \text{or, since} \quad \frac{1}{0.434} \approx 2.304 \text{ then } \ln x \approx 2.304 \log x$$

The $\log_e x$ is called the *natural log* of x. We use "ln x" to designate the natural log of x.

CHECK YOURSELF ANSWERS

1. (1) $2 \log_b x + 3 \log_b y + \log_b z$;   (2) $\dfrac{1}{2}(\log_{10} x + \log_{10} y - \log_{10} z)$

2. (1) $\log_b \dfrac{x^3 y^2}{z^2}$; (2) $\log_2 \sqrt[3]{\dfrac{x^2}{y}}$ **3.** (1) 1.079; (2) 1.431; (3) 0.100

4. (1) 0.362; (2) 1.362; (3) 2.362; (4) 3.362; (5) -0.638; (6) -1.638

5. (1) 4.17, acidic; (2) 12.28, basic   **6.** (1) 6.73; (2) 67.3; (3) 673; (4) 0.673

7. $[H^+] = 1.6 \times 10^{-8}$ **8.** (1) 1.099; (2) 1.792; (3) 1.386; (4) 0.549

9. (1) 2.485; (2) 3.297   **10.** (1) 70.5; (2) 65.5

11. $\log_8 32 = \dfrac{\log 32}{\log 8} \approx 1.667$

11.4 EXERCISES

Build Your Skills

Use the properties of logarithms to expand each of the following expressions.

1. $\log_b 5x$

2. $\log_3 7x$

3. $\log_4 \dfrac{x}{3}$

4. $\log_b \dfrac{2}{y}$

5. $\log_3 a^2$

6. $\log_5 y^4$

7. $\log_5 \sqrt{x}$

8. $\log \sqrt[3]{z}$

9. $\log_b x^3 y^2$

10. $\log_5 x^2 z^4$

11. $\log_4 y^2 \sqrt{x}$

12. $\log_b x^3 \sqrt[3]{z}$

13. $\log_b \dfrac{x^2 y}{z}$

14. $\log_5 \dfrac{3}{xy}$

15. $\log \dfrac{xy^2}{\sqrt{z}}$

16. $\log_4 \dfrac{x^3 \sqrt{y}}{z^2}$

17. $\log_5 \sqrt[3]{\dfrac{xy}{z^2}}$

18. $\log_b \sqrt[4]{\dfrac{x^2 y}{z^3}}$

Write each of the following expressions as a single logarithm.

19. $\log_b x + \log_b y$

20. $\log_5 x - \log_5 y$

21. $2 \log_2 x - \log_2 y$

22. $3 \log_b x + \log_b z$

23. $\log_b x + \dfrac{1}{2} \log_b y$

24. $\dfrac{1}{3} \log_b x - 2 \log_b z$

25. $\log_b x + 2 \log_b y - \log_b z$

26. $2 \log_5 x - (3 \log_5 y + \log_5 z)$

27. $\dfrac{1}{2} \log_6 x + 2 \log_6 y - 3 \log_6 z$

28. $\log_b x - \dfrac{1}{3} \log_b y - 4 \log_b z$

29. $\dfrac{1}{3} (2 \log_b x + \log_b y - \log_b z)$

30. $\dfrac{1}{5} (2 \log_4 x - \log_4 y + 3 \log_4 z)$

Given that $\log 2 = 0.301$ and $\log 3 = 0.477$, find each of the following logarithms.

31. $\log 24$

32. $\log 36$

33. $\log 8$

34. $\log 81$

35. $\log \sqrt{2}$

36. $\log \sqrt[3]{3}$

37. $\log \dfrac{1}{4}$

38. $\log \dfrac{1}{27}$

Using your calculator, find each of the following logarithms.

39. $\log 6.8$

40. $\log 68$

41. $\log 680$

42. $\log 6800$

43. $\log 0.68$

44. $\log 0.068$

Find the pH of each of the following, given the hydrogen ion concentration [H^+] for each solution. Are the solutions acidic or basic? See Example 5 for the appropriate formula.

45. Blood: $[H^+] = 3.8 \times 10^{-8}$

46. Lemon juice: $[H^+] = 6.4 \times 10^{-3}$

Using your calculator, find the antilogarithm for each of the following logarithms.

47. 0.749

48. 1.749

49. 3.749

50. −0.251

Given the pH of the following solutions, find the hydrogen ion concentration [H⁺]. See Example 7.

51. Wine: pH = 4.7

52. Household ammonia: pH = 7.8

Using your calculator, find each of the following logarithms.

53. ln 2

54. ln 3

55. ln 10

56. ln 30

The average score on a final examination for a group of psychology students, retested after time t (in weeks), is given by

$$S = 85 - 8 \ln (t + 1)$$

Find the average score on the retests:

57. After 3 weeks

58. After 12 weeks

Use the change-of-base formula to find each of the following logarithms.

59. $\log_3 25$

60. $\log_5 30$

 The amount of a radioactive substance remaining after a given amount of time is given by the following formula:

$$\ln \frac{A}{A_0} = -\lambda t$$

where A is the amount remaining after time t, variable A_0 is the original amount of the substance, and λ is the radioactive decay constant.

 61. How much plutonium 239 will remain after 50,000 years if 24 kg was originally stored? Plutonium 239 has a radioactive decay constant of 0.000029.

 62. How much plutonium 241 will remain after 100 years if 52 kg was originally stored? Plutonium 241 has a radioactive decay constant of 0.053319.

 63. How much strontium 90 was originally stored if after 56 years it is discovered that 15 kg still remains? Strontium 90 has a radioactive decay constant of 0.024755.

 64. How much cesium 137 was originally stored if after 90 years it is discovered that 20 kg still remains? Cesium 137 has a radioactive decay constant of 0.023105.

Transcribe Your Skills

65. Which keys on your calculator are function keys and which are operation keys? What is the difference?

66. How is the pH factor relevant to your selection of a hair care product?

Skillscan (Sections 8.1 and 9.1)

Solve each of the following equations for x.

a. $x(x - 4) = 5$ **b.** $x(x + 5) = 36$
c. $(x - 4)(x + 2) = 7$ **d.** $(x + 7)(x - 1) = 20$
e. $\dfrac{x + 1}{x} = 10$ **f.** $\dfrac{x - 2}{x + 3} = 16$

Logarithmic and Exponential Equations   11.5

OBJECTIVE: To solve logarithmic and exponential equations

 TAPE IN23

Much of the importance of the properties of logarithms developed in the previous section lies in the application of those properties to the solution of equations involving logarithms and exponentials. Our work in this section will consider solution techniques for both types of equations. Let's start with a definition.

A *logarithmic equation* is an equation that contains a logarithmic expression. We solved some simple examples in Section 11.3. Let's review for a moment.

To solve $\log_3 x = 4$ for x, recall that we simply convert the logarithmic equation to exponential form. Here

$$x = 3^4$$

so

$$x = 81$$

and 81 is the solution to the given equation.

Now, what if the logarithmic equation involves more than one logarithmic term? The following example illustrates how the properties of logarithms must then be applied.

EXAMPLE 1 Solving a Logarithmic Equation

Solve each logarithmic equation.

(*a*) $\log_5 x + \log_5 3 = 2$

The original equation can be written as

We apply the product rule for logarithms:

$\log_b M + \log_b N = \log_b MN$

$$\log_5 3x = 2$$

Now, since only a single logarithm is involved, we can write the equation in the equivalent exponential form:

$$3x = 5^2$$

$$3x = 25$$

$$x = \frac{25}{3}$$

(*b*) $\log x + \log (x - 3) = 1$

Write the equation as

Since no base is written, it is assumed to be 10.

$$\log x(x - 3) = 1$$

or

Given the base of 10, this is the equivalent exponential form.

$$x(x - 3) = 10^1$$

We now have

$$x^2 - 3x = 10$$

$$x^2 - 3x - 10 = 0$$

$$(x - 5)(x + 2) = 0$$

Possible solutions are $x = 5$ or $x = -2$.

Checking possible solutions is particularly important here.

Note that substitution of -2 into the original equation gives

$$\log (-2) + \log (-5) = 1$$

Since logarithms of negative numbers are *not* defined, -2 is an extraneous solution and we must reject it. The only solution for the original equation is 5.

Check Yourself 1

Solve $\log_2 x + \log_2 (x + 2) = 3$ for x.

The quotient property is used in a similar fashion for solving logarithmic equations. Consider the following example.

EXAMPLE 2 Solving a Logarithmic Equation

Solve each equation for x.

(a) $\log_5 x - \log_5 2 = 2$

Rewrite the original equation as

$$\log_5 \frac{x}{2} = 2$$

Now

$$\frac{x}{2} = 5^2$$

$$\frac{x}{2} = 25$$

$$x = 50$$

We apply the quotient rule for logarithms:

$$\log_b M - \log_b N = \log_b \frac{M}{N}$$

(b) $\log_3 (x + 1) - \log_3 x = 3$

$$\log_3 \frac{x + 1}{x} = 3$$

$$\frac{x + 1}{x} = 27$$

$$x + 1 = 27x$$

$$1 = 26x$$

$$x = \frac{1}{26}$$

Again you should verify that substituting $\frac{1}{26}$ for x leads to a positive value in each of the original logarithms.

Check Yourself 2

Solve $\log_5 (x + 3) - \log_5 x = 2$ for x.

The solution of certain types of logarithmic equations calls for the one-to-one property of the logarithmic function.

If $\qquad \log_b M = \log_b N$

then $\qquad\qquad M = N$

EXAMPLE 3 Solving a Logarithmic Equation

Solve the following equation for x:

$\log (x + 2) - \log 2 = \log x$

Solution Again we rewrite the left-hand side of the equation. So

$$\log \frac{x + 2}{2} = \log x$$

Since the logarithmic function is one-to-one, this is equivalent to

$$\frac{x + 2}{2} = x$$

or

$$x = 2$$

Check Yourself 3

Solve for x.

$\log (x + 3) - \log 3 = \log x$

The following algorithm summarizes our work in solving logarithmic equations.

SOLVING LOGARITHMIC EQUATIONS

STEP 1 Use the properties of logarithms to combine terms containing logarithmic expressions into a single term.

STEP 2 Write the equation formed in step 1 in exponential form.

STEP 3 Solve for the indicated variable.

STEP 4 Check your solutions to make sure that possible solutions do not result in the logarithms of negative numbers.

Let's look now at *exponential equations*. These are equations in which the variable appears as an exponent.

We solved some particular exponential equations in Section 11.2. In solving an equation like

$$3^x = 81$$

we wrote the right-hand member as a power of 3, so that

$$3^x = 3^4$$

or

$$x = 4$$

Again we want to write both sides as a power of the same base, here 3.

The technique above will work only when both sides of the equation can be conveniently expressed as powers of the same base. If that is not the case, we must use logarithms for the solution of the equation, as illustrated in the following examples.

EXAMPLE 4 Solving an Exponential Equation

Solve $3^x = 5$ for x.

Solution We begin by taking the common logarithm of both sides of the original equation:

$$\log 3^x = \log 5$$

Now we apply the power property so that the variable becomes a coefficient on the left:

$$x \log 3 = \log 5$$

Again:

If $M = N$, then

$$\log_b M = \log_b N$$

Dividing both sides of the equation by $\log 3$ will isolate x, and we have

$$x = \frac{\log 5}{\log 3}$$

$$= 1.465 \qquad \text{(to three decimal places)}$$

CAUTION

Be Careful! This is *not* log 5 − log 3, a common error.

Note You can verify the approximate solution by using the $\boxed{y^x}$ key on your calculator. Raise 3 to power 1.465.

Check Yourself 4

Solve $2^x = 10$ for x.

The next example shows how to solve an equation with a more complicated exponent.

EXAMPLE 5 Solving an Exponential Equation

Solve $5^{2x+1} = 8$ for x.

On the left, we apply

$\log_b M^p = p \log_b M$

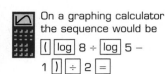

On a graphing calculator the sequence would be

(log 8 ÷ log 5 –

1) ÷ 2 =

Another calculator sequence to find x would be

8 log ÷ 5 log – 1 =

÷ 2 =

Solution The solution begins as in Example 4:

$$\log 5^{2x+1} = \log 8$$

$$(2x + 1) \log 5 = \log 8$$

$$2x + 1 = \frac{\log 8}{\log 5}$$

$$2x = \frac{\log 8}{\log 5} - 1$$

$$x = \frac{1}{2}\left(\frac{\log 8}{\log 5} - 1\right)$$

$$x \approx 0.146$$

Check Yourself 5

Solve $3^{2x-1} = 7$ for x.

The procedure is similar if the variable appears as an exponent in more than one term of the equation.

EXAMPLE 6 Solving an Exponential Equation

Solve $3^x = 2^{x+1}$ for x.

$$\log 3^x = \log 2^{x+1}$$

$$x \log 3 = (x + 1) \log 2$$

$$x \log 3 = x \log 2 + \log 2$$

Use the power property to write the variables as coefficients.

$$x \log 3 - x \log 2 = \log 2$$

We now isolate x on the left.

$$x(\log 3 - \log 2) = \log 2$$

$$x = \frac{\log 2}{\log 3 - \log 2}$$

$$\approx 1.710$$

Check Yourself 6

Solve $5^{x+1} = 3^{x+2}$ for x.

To check the reasonableness of this result, use your calculator to verify that

$3^{1.710} \approx 2^{2.710}$

The following algorithm summarizes our work with solving exponential equations.

SOLVING EXPONENTIAL EQUATIONS

STEP 1 Try to write each side of the equation as a power of the same base. Then equate the exponents to form an equation.

STEP 2 If the above procedure is not applicable, take the common logarithm of both sides of the original equation.

STEP 3 Use the power rule for logarithms to write an equivalent equation with the variables as coefficients.

STEP 4 Solve the resulting equation.

There are many applications of our work with exponential equations. Consider the following.

If an investment of P dollars earns interest at an annual interest rate r and the interest is compounded n times per year, then the amount in the account after t years is given by

$$A = P\left(1 + \frac{r}{n}\right)^{nt} \tag{1}$$

EXAMPLE 7 An Interest Application

If \$1000 is placed in an account with an annual interest rate of 6 percent, find out how long it will take the money to double when:

(*a*) The interest is compounded annually.

Using Equation (1) with $A = 2000$ (we want the original 1000 to double), $P = 1000$, $r = 0.06$, and $n = 1$, we have

$$2000 = 1000(1 + 0.06)^t$$

Dividing both sides by 1000 yields

$$2 = (1.06)^t$$

We now have an exponential equation which can be solved by our earlier techniques.

$$\log 2 = \log (1.06)^t$$
$$= t \log 1.06$$

or

$$t = \frac{\log 2}{\log 1.06}$$
$$\approx 11.9 \text{ years}$$

It takes just a little less than 12 years for the money to double.

> Since the interest is compounded *once* a year, $n = 1$.

> From accounting, we have the "rule of 72." It states that the doubling time is approximately 72 divided by the interest rate as a percentage. Here $\frac{72}{6} = 12$ years.

(b) The interest is compounded quarterly.

Now $n = 4$ in Equation (1), so

Since the interest is compounded 4 times per year, $n = 4$.

$$2000 = 1000\left(1 + \frac{0.06}{4}\right)^{4t}$$

$$2 = (1.015)^{4t}$$

$$\log 2 = \log (1.015)^{4t}$$

$$\log 2 = 4t \log 1.015$$

$$\frac{\log 2}{4 \log 1.015} = t$$

$$t \approx 11.6 \text{ years}$$

Note that the doubling time is reduced by approximately 3 months by the more frequent compounding.

Check Yourself 7

Find the doubling time in Example 7 if the interest is compounded monthly.

Problems involving rates of growth or decay can also be solved by using exponential equations.

EXAMPLE 8 A Population Application

A town's population is presently 10,000. Given a projected growth rate of 7 percent per year, t years from now the population P will be given by

$$P = 10,000e^{0.07t}$$

In how many years will the town's population double?

Solution We want the time t when P will be 20,000 (doubled in size). So

$$20,000 = 10,000e^{0.07t}$$

or

Divide both sides by 10,000. $2 = e^{0.07t}$

In this case we take the *natural logarithm* of both sides of the equation. This is because e is involved in the equation.

$$\ln 2 = \ln e^{0.07t}$$

Apply the power property. $\ln 2 = 0.07t \ln e$

Note $\ln e = 1$ $\ln 2 = 0.07t$

$$\frac{\ln 2}{0.07} = t$$

$$t \approx 9.9 \text{ years}$$

The population will double in approximately 9.9 years.

Check Yourself 8

If \$1000 is invested in an account with an annual interest rate of 6 percent, compounded continuously, the amount A in the account after t years is given by

$$A = 1000e^{0.06t}$$

Find the time t that it will take for the amount to double ($A = 2000$). Compare this time with the result of the Check Yourself 7 exercise.

CHECK YOURSELF ANSWERS

1. 2 years   **2.** $\dfrac{1}{8}$ **3.** $\dfrac{3}{2}$ **4.** 3.322   **5.** 1.386   **6.** 1.151   **7.** 11.58   **8.** 11.55 years

Build Your Skills

Solve each of the following logarithmic equations for x.

1. $\log_4 x = 3$

2. $\log_3 x = -2$

3. $\log (x + 1) = 2$

4. $\log_5 (2x - 1) = 2$

5. $\log_2 x + \log_2 8 = 6$

6. $\log 5 + \log x = 2$

7. $\log_3 x - \log_3 6 = 3$

8. $\log_4 x - \log_4 8 = 3$

9. $\log_2 x + \log_2 (x + 2) = 3$

10. $\log_3 x + \log_3 (2x + 3) = 2$

11. $\log_7 (x + 1) + \log_7 (x - 5) = 1$

12. $\log_2 (x + 2) + \log_2 (x - 5) = 3$

13. $\log x - \log (x - 2) = 1$

14. $\log_5 (x + 5) - \log_5 x = 2$

15. $\log_3 (x + 1) - \log_3 (x - 2) = 2$

16. $\log (x + 2) - \log (2x - 1) = 1$

17. $\log (x + 5) - \log (x - 2) = \log 5$

18. $\log_3 (x + 12) - \log_3 (x - 3) = \log_3 6$

19. $\log_2 (x^2 - 1) - \log_2 (x - 2) = 3$

20. $\log (x^2 + 1) - \log (x - 2) = 1$

Solve each of the following exponential equations for x. Give your solutions in decimal form, correct to three decimal places.

21. $5^x = 625$

22. $4^x = 64$

23. $2^{x+1} = \dfrac{1}{8}$

24. $9^x = 3$

25. $8^x = 2$

26. $3^{2x-1} = 27$

27. $3^x = 7$

28. $5^x = 30$

29. $4^{x+1} = 12$

30. $3^{2x} = 5$

31. $7^{3x} = 50$

32. $6^{x-3} = 21$

33. $5^{3x-1} = 15$

34. $8^{2x+1} = 20$

35. $4^x = 3^{x+1}$

36. $5^x = 2^{x+2}$

37. $2^{x+1} = 3^{x-1}$

38. $3^{2x+1} = 5^{x+1}$

See Example 7 for the appropriate formula for the following exercises.

39. If $5000 is placed in an account with an annual interest rate of 9 percent, how long will it take the amount to double if the interest is compounded annually?

40. Repeat Exercise 39 if the interest is compounded semi-annually.

41. Repeat Exercise 39 if the interest is compounded quarterly.

42. Repeat Exercise 39 if the interest is compounded monthly.

Suppose that the number of bacteria present in a culture after t hours is given by $N(t) = N_0 \cdot 2^{t/2}$, where N_0 is the initial number of bacteria.

43. How long will it take the bacteria to increase from 12,000 to 20,000?

44. How long will it take the bacteria to increase from 12,000 to 50,000?

45. How long will it take the bacteria to triple? *Hint:* Let $N = 3N_0$.

46. How long will it take the culture to increase to 5 times its original size? *Hint:* Let $N = 5N_0$.

The radioactive element strontium 90 has a half-life of approximately 28 years. That is, in a 28-year period, one-half of the initial amount will have decayed into another substance. If A_0 is the initial amount of the element, then the amount A remaining after t years is given by

$$A(t) = A_0\left(\frac{1}{2}\right)^{t/28}$$

47. If the initial amount of the element is 100 g, in how many years will 60 g remain?

48. If the initial amount of the element is 100 g, in how many years will 20 g remain?

49. In how many years will 75 percent of the original amount remain? *Hint:* Let $A = 0.75A_0$.

50. In how many years will 10 percent of the original amount remain? *Hint:* Let $A = 0.1A_0$.

Given projected growth, t years from now a city's population P can be approximated by $P(t) = 25{,}000e^{0.045t}$.

51. How long will it take the city's population to reach 35,000?

52. How long will it take the population to double?

The number of bacteria in a culture after t hours can be given by $N(t) = N_0e^{0.03t}$, where N_0 is the initial number of bacteria in the culture.

53. In how many hours will the size of the culture double?

54. In how many hours will the culture grow to 4 times its original population?

The atmospheric pressure P, in inches of mercury (inHg), at an altitude h feet above sea level is approximated by $P(t) = 30e^{-0.00004h}$.

55. Find the altitude if the pressure at that altitude is 25 inHg.

56. Find the altitude if the pressure at that altitude is 20 inHg.

Carbon 14 dating is used to measure the age of specimens and is based on the radioactive decay of the element carbon 14. This decay begins once a plant or animal dies. If A_0 is the initial amount of carbon 14, then the amount remaining after t years is $A(t) = A_0e^{-0.000124t}$.

57. Estimate the age of a specimen if 70 percent of the original amount of carbon 14 remains.

58. Estimate the age of a specimen if 20 percent of the original amount of carbon 14 remains.

Transcribe Your Skills

59. In some of the earlier exercises, we talked about bacteria cultures that double in size every few minutes. Can that go on forever? Explain.

60. The population of the United States has been doubling every 45 years. Is it reasonable to assume that this rate will continue? What factors will start to limit that growth?

Graphing Utility

Use your graphing utility to find the graph for each equation, then explain the result.

61. $y = \log 10^x$

62. $y = 10^{\log x}$

63. $y = \ln e^x$

64. $y = e^{\ln x}$

Skillscan (Section 5.3)

Multiply the following.

a. $(x + y)^2$ **b.** $(3x + 2)^2$ **c.** $(x + y)^3$

d. $(2x - 3)^3$ **e.** $(x - y)^2$ **f.** $(x + y)^4$

SUMMARY

Inverse Relations and Functions [11.1]

The inverse of a relation is formed by interchanging the components of each of the ordered pairs in the given relation.

    If a relation (or function) is specified by an equation, interchange the roles of x and y in the defining equation to form the inverse.

The inverse of the relation

$\{(1, 2), (2, 3), (4, 3)\}$ is

$\{(2, 1), (3, 2), (3, 4)\}$.

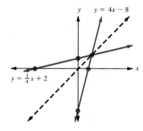

The inverse of a function f may or may not be a function. If the inverse *is* also a function, we denote that inverse as f^{-1}, read "the inverse of f."

    A function f has an inverse f^{-1} which is also a function if and only if f is a *one-to-one* function. That is, no two ordered pairs in the function have the same second component.

    The *horizontal-line test* can be used to determine whether a function is one-to-one.

If $f(x) = 4x - 8$, then

$$f^{-1}(x) = \frac{1}{4}x + 2$$

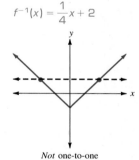

Not one-to-one

Exponential Functions [11.2]

An *exponential function* is any function defined by an equation of the form

$$y = f(x) = b^x \qquad b > 0,\ b \neq 1$$

If b is greater than 1, the function is always increasing (a *growth function*). If b is less than 1, the function is always decreasing (a *decay function*).

    In both cases the exponential function is one-to-one. The domain is the set of all real numbers, and the range is the set of positive real numbers.

    The function defined by $y = e^x$, where e is an irrational number (approximately 2.71828), is called *the* exponential function.

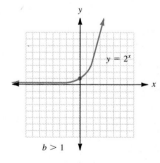

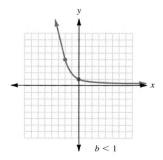

Logarithmic Functions [11.3]

In the expression

$$y = \log_b x$$

log₃ 9 = 2 is in logarithmic form.

$3^2 = 9$ is the exponential form.

log₃ 9 = 2 is equivalent to

$3^2 = 9$

2 is the power to which we must raise 3 in order to get 9.

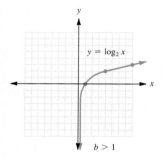

y is called the *logarithm of x to base b*, where $b > 0$ and $b \neq 1$.
An expression such as $y = \log_b x$ is said to be in *logarithmic form.*
An expression such as $x = b^y$ is said to be in *exponential form.*

$$y = \log_b x \quad \text{means the same as} \quad x = b^y$$

A logarithm is an exponent or a power. The logarithm of x to base b is the power to which we must raise b in order to get x.

A *logarithmic function* is any function defined by an equation of the form

$$y = f(x) = \log_b x \qquad b > 0,\ b \neq 1$$

The logarithm function is the inverse of the corresponding exponential function. The function is one-to-one with domain $\{x \mid x > 0\}$ and range composed of the set of all real numbers.

Properties of Logarithms [11.4]

If M, N, and b are positive real numbers with $b \neq 1$ and if p is any real number, then we can state the following properties of logarithms:

log₂ 1 = 0

$$\log_b 1 = 0$$

log 10 = 1

$$\log_b b = 1$$

$3^{\log_3 2} = 2$

$$b^{\log_b x} = x$$

log₅ 5ˣ = x

$$\log_b b^x = x$$

log₃ x + log₃ y = log₃ xy

Product Property $\quad \log_b MN = \log_b M + \log_b N$

log₅ 8 − log₅ 3 = log₅ $\frac{8}{3}$

Quotient Property $\log_b \dfrac{M}{N} = \log_b M - \log_b N$

log 3² = 2 log 3

Power Property $\quad \log_b M^p = p \log_b M$

Common logarithms are logarithms to base 10. For convenience, we omit the base in writing common logarithms:

log₁₀ 1000 = log 1000

 = log 10³ = 3

$$\log M = \log_{10} M$$

Natural logarithms are logarithms to base e. By custom we also omit the base in writing natural logarithms:

ln 3 = log_e 3

$$\ln M = \log_e M$$

Logarithmic and Exponential Equations [11.5]

A *logarithmic equation* is an equation that contains a logarithmic expression.

$$\log_2 x = 5$$

is a logarithmic equation.

To solve $\log_2 x = 5$:

Write the equation in the equivalent exponential form to solve

$$x = 2^5 \quad \text{or} \quad x = 32$$

Solving Logarithmic Equations

Step 1. Use the properties of logarithms to combine terms involving logarithmic expressions into a single term.

Step 2. Write the equation formed in step 1 in exponential form.

Step 3. Solve for the indicated variable.

Step 4. Check your solutions to make sure that possible solutions do not result in the logarithms of negative numbers.

To solve:

$$\log_4 x + \log_4 (x - 6) = 2$$
$$\log_4 x(x - 6) = 2$$
$$x(x - 6) = 4^2$$
$$x^2 - 6x - 16 = 0$$
$$(x - 8)(x + 2) = 0$$
$$x = 8 \quad \text{or} \quad x = -2$$

Since substituting -2 for x in the original equation results in the logarithm of a negative number, we reject that answer. The only solution is 8.

An *exponential equation* is an equation in which the variable appears as an exponent.

The following algorithm summarizes the steps in solving any exponential equation.

To solve $4^x = 64$:

Since $64 = 4^3$, write

$$4^x = 4^3 \quad \text{or} \quad x = 3$$

Solving Exponential Equations

Step 1. Try to write each side of the equation as a power of the same base. Then equate the exponents to form an equation.

Step 2. If the above procedure is not applicable, take the logarithm of both sides of the original equation.

Step 3. Use the power rule for logarithms to write an equivalent equation with the variables as coefficients.

Step 4. Solve the resulting equation.

$$2^{x+3} = 5^x$$
$$\log 2^{x+3} = \log 5^x$$
$$(x + 3) \log 2 = x \log 5$$
$$x \log 2 + 3 \log 2 = x \log 5$$
$$x \log 2 - x \log 5 = -3 \log 2$$
$$x(\log 2 - \log 5) = -3 \log 2$$
$$x = \frac{-3 \log 2}{\log 2 - \log 5} \approx 2.269$$

SUMMARY EXERCISES

This summary exercise set is provided to give you practice with each of the objectives of the chapter. Each exercise is keyed to the appropriate chapter section. The answers are provided in the instructor's manual that accompanies this text. Your instructor will provide guidelines on how best to use these exercises in your instructional program.

[11.1] Write the inverse relation for each of the following functions. Which inverses are also functions?

1. $\{(1, 5), (2, 7), (3, 9)\}$

2. $\{(3, 1), (5, 1), (7, 1)\}$

3. $\{(2, 4), (4, 3), (6, 4)\}$

[11.1] Write an equation for the inverse of the relation defined by each of the following equations.

4. $y = 3x - 6$

5. $y = \dfrac{x + 1}{2}$

6. $y = x^2 - 2$

[11.1] Write an equation for the inverse of the relation defined by each of the following equations. Then graph the relation and its inverse on the same set of axes. Which inverses are also functions?

7. $y = 3x + 6$

8. $y = -x^2 + 3$

9. $4x^2 + 9y^2 = 36$

[11.2] Graph the exponential functions defined by each of the following equations.

10. $y = 3^x$

11. $y = \left(\dfrac{3}{4}\right)^x$

[11.3] Solve each of the following exponential equations for x.

12. $5^x = 125$

13. $2^{2x+1} = 32$

14. $3^{x-1} = \dfrac{1}{9}$

[11.3] If it takes 2 h for the population of a certain bacteria culture to double (by dividing in half), then the number N of bacteria in the culture after t hours is given by $N = 1000 \cdot 2^{t/2}$, where the initial popula-

tion of the culture was 1000. Using this formula, find the number in the culture:

15. After 4 h

16. After 12 h

17. After 15 h

[11.3] Graph the logarithmic functions defined by each of the following equations.

18. $y = \log_3 x$

19. $y = \log_2 (x - 1)$

[11.3] Convert each of the following statements to logarithmic form.

20. $3^4 = 81$

21. $10^3 = 1000$

22. $5^0 = 1$

23. $5^{-2} = \dfrac{1}{25}$

24. $25^{1/2} = 5$

25. $16^{3/4} = 8$

[11.3] Convert each of the following statements to exponential form.

26. $\log_3 81 = 4$

27. $\log 1 = 0$

28. $\log_{81} 9 = \dfrac{1}{2}$

29. $\log_5 25 = 2$

30. $\log 0.001 = -3$

31. $\log_{32} \dfrac{1}{2} = -\dfrac{1}{5}$

[11.3] Solve each of the following equations for the unknown variable.

32. $y = \log_5 125$

33. $\log_b \dfrac{1}{9} = -2$

34. $\log_8 x = 2$

35. $y = \log_5 1$

36. $\log_b 3 = \dfrac{1}{2}$

37. $y = \log_{16} 2$

38. $y = \log_8 2$

[11.3] The decibel (dB) rating for the loudness of a sound is given by

$$L = 10 \log \frac{I}{I_0}$$

where I is the intensity of that sound in watts per square centimeter and I_0 is the intensity of the "threshold" sound $I_0 = 10^{-16}$ W/cm². Find the decibel rating of each of the given sounds.

39. A table saw in operation with intensity $I = 10^{-6}$ W/cm²

40. The sound of a passing car horn with intensity $I = 10^{-8}$ W/cm²

[11.3] The formula for the decibel rating of a sound can be solved for the intensity of the sound as

$$I = I_0 \cdot 10^{L/10}$$

where L is the decibel rating of the given sound.

41. What is the ratio of intensity of a 60-dB sound to one of 50 dB?

42. What is the ratio of intensity of a 60-dB sound to one of 40 dB?

The magnitude of an earthquake on the Richter scale is given by

$$M = \log \frac{a}{a_0}$$

where a is the intensity of the shock wave of the given earthquake and a_0 is the intensity of the shock wave of a zero-level earthquake. Use that formula to solve the following.

43. The Alaskan earthquake of 1964 had an intensity of $10^{8.4}a_0$. What was its magnitude on the Richter scale?

44. Find the ratio of intensity of an earthquake of magnitude 7 to an earthquake of magnitude 6.

[11.4] Use the properties of logarithms to expand each of the following expressions.

45. $\log_b x^2 y$

46. $\log_4 \dfrac{y^3}{5}$

47. $\log_3 \dfrac{xy^2}{z}$

48. $\log_5 x^3 yz^2$

49. $\log \dfrac{xy}{\sqrt{z}}$

50. $\log_b \sqrt[3]{\dfrac{x^2 y}{z}}$

[11.4] Use the properties of logarithms to write each of the following expressions as a single logarithm.

51. $\log x + 2 \log y$

52. $3 \log_b x - 2 \log_b z$

53. $\log_b x + \log_b y - \log_b z$

54. $2 \log_5 x - 3 \log_5 y - \log_5 z$

55. $\log x - \dfrac{1}{2} \log y$

56. $\dfrac{1}{3} (\log_b x - 2 \log_b y)$

[11.4] Given that $\log 2 = 0.301$ and $\log 3 = 0.477$, find each of the following logarithms. Verify your results with a calculator.

57. $\log 18$

58. $\log 16$

59. $\log \dfrac{1}{8}$

60. $\log \sqrt{3}$

Use your calculator to find the pH of each of the following solutions, given the hydrogen ion concentration [H⁺] for each solution, where

$$\text{pH} = -\log [\text{H}^+]$$

Are the solutions acidic or basic?

61. Coffee: $[\text{H}^+] = 5 \times 10^{-6}$

62. Household detergent: $[\text{H}^+] = 3.2 \times 10^{-10}$

Given the pH of the following solutions, find the hydrogen ion concentration [H⁺].

63. Lemonade: pH = 3.5

64. Ammonia: pH = 10.2

[11.4] The average score on a final examination for a group of chemistry students, retested after time t (in weeks), is given by

$$S(t) = 81 - 6 \ln (t + 1)$$

Find the average score on the retests after the given times.

65. After 5 weeks

66. After 10 weeks

67. After 15 weeks

68. Graph these results.

[11.4] The formula for converting from a logarithm with base b to a logarithm with base a is

$$\log_a x = \frac{\log_b x}{\log_b a}$$

Use that formula to find each of the following logarithms.

69. $\log_4 20$

70. $\log_8 60$

[11.5] Solve each of the following logarithmic equations for x.

71. $\log_3 x + \log_3 5 = 3$

72. $\log_5 x - \log_5 10 = 2$

73. $\log_3 x + \log_3 (x + 6) = 3$

74. $\log_5 (x + 3) + \log_5 (x - 1) = 1$

75. $\log x - \log (x - 1) = 1$

76. $\log_2 (x + 3) - \log_2 (x - 1) = \log_2 3$

[11.5] Solve each of the following exponential equations for x. Give your results correct to three decimal places.

77. $3^x = 243$

78. $5^x = \dfrac{1}{25}$

79. $5^x = 10$

80. $4^{x-1} = 8$

81. $6^x = 2^{2x+1}$

82. $2^{x+1} = 3^{x-1}$

[11.5] If an investment of P dollars earns interest at an annual rate of 12 percent and the interest is com-

pounded n times per year, then the amount A in the account after t years is

$$A(t) = P\left(1 + \frac{0.12}{n}\right)^{nt}$$

Use that formula to solve each of the following.

83. If $1000 is invested and the interest is compounded quarterly, how long will it take the amount in the account to double?

84. If $3000 is invested and the interest is compounded monthly, how long will it take the amount in the account to reach $8000?

[11.5] A certain radioactive element has a half-life of 50 years. The amount A of the substance remaining after t years is given by

$$A(t) = A_0 \cdot 2^{-t/50}$$

where A_0 is the initial amount of the substance. Use this formula to solve each of the following.

85. If the initial amount of the substance is 100 milligrams (mg), after how long will 40 mg remain?

86. After how long will only 10 percent of the original amount of the substance remain?

[11.5] A city's population is presently 50,000. Given the projected growth, t years from now the population P will be given by $P(t) = 50,000e^{0.08t}$. Use this formula to solve each of the following.

87. How long will it take the population to reach 70,000?

88. How long will it take the population to double?

The atmospheric pressure, in inches of mercury, at an altitude h miles above the surface of the earth, is approximated by $P(h) = 30e^{-0.021h}$. Use this formula to solve the following exercises.

89. Find the altitude at the top of Mt. McKinley in Alaska if the pressure is 27.7 inHg.

90. Find the altitude outside an airliner in flight if the pressure is 26.1 inHg.

SELF-TEST

The purpose of this self-test is to help you check your progress and to review for a chapter test in class. Allow yourself about an hour to take the test. When you are done, check your answers in the back of the book. If you missed any problems, be sure to go back and review the appropriate sections in the chapter and the exercises that are provided.

1. Use $f(x) = 4x - 2$ and $g(x) = x^2 + 1$ in each of the following.

(a) Find the inverse of f. Is the inverse also a function?

(b) Find the inverse of g. Is the inverse also a function?

(c) Graph f and its inverse on the same set of axes.

Graph the exponential functions defined by each of the following equations.

2. $y = 4^x$

3. $y = \left(\dfrac{2}{3}\right)^x$

4. Solve each of the following exponential equations for x.

(a) $5^x = \dfrac{1}{25}$

(b) $3^{2x-1} = 81$

5. Graph the logarithmic function defined by the following equation.

$y = \log_4 x$

Convert each of the following statements to logarithmic form.

6. $10^4 = 10{,}000$

7. $27^{2/3} = 9$

Convert each of the following statements to exponential form.

8. $\log_5 125 = 3$

9. $\log 0.01 = -2$

Solve each of the following equations for the unknown variable.

10. $y = \log_2 64$

11. $\log_b \dfrac{1}{16} = -2$

12. $\log_{25} x = \dfrac{1}{2}$

Use the properties of logarithms to expand each of the following expressions.

13. $\log_b x^2 y z^3$

14. $\log_5 \sqrt{\dfrac{xy^2}{z}}$

Use the properties of logarithms to write each of the following expressions as a single logarithm.

15. $\log x + 3 \log y$

16. $\dfrac{1}{3}(\log_b x - 2\log_b z)$

Solve each of the following logarithmic equations for x.

17. $\log_6 (x + 1) + \log_6 (x - 4) = 2$

18. $\log (2x + 1) - \log (x - 1) = 1$

Solve each of the following exponential equations for x. Give your results correct to three decimal places.

19. $3^{x+1} = 4$

20. $5^x = 3^{x+1}$

CUMULATIVE REVIEW EXERCISES

Find the equation of the axis of symmetry and the coordinates for the vertex of each of the following.

1. $y = x^2 - 3$

2. $y = -2(x + 3)^2 + 2$

3. $y = 2x^2 - 4x - 3$

4. $x = y^2 - 3y + 2$

Find the center and the radius of the graph of each equation.

5. $x^2 + y^2 = 49$

6. $(x - 3)^2 + y^2 = 9$

7. $(x - 2)^2 + (y + 5)^2 = 16$

8. $x^2 + y^2 - 8x + 2y = 53$

Graph each of the following.

9. $y = -2x^2 + 3$

10. $4x^2 - 9y^2 = 36$

11. $4x^2 + 25y^2 = 100$

12. $(x + 3)^2 + (y - 1)^2 = 25$

For each of the following relations, give the domain and range. Which relations are also functions?

13. $\{(1, 2), (2, 3), (1, 4)\}$

14. $\{(x, y) \mid y = x^2\}$

Graph each of the following relations. Use the graph to find the domain and range of the relation. Tell whether the given relation is also a function.

15. $y = x^2 + 4x$

16. $4x^2 + 25y^2 = 100$

If $f(x) = 2x - 5$ and $g(x) = -x^2 + 2$, find each of the following.

17. $f(0)$

18. $f(x + h)$

19. $f(2) + g(2)$

20. $g(f(2))$

CHAPTER TWELVE

SEQUENCES AND SERIES

NUCLEAR POWER In the 1950s, nuclear energy was touted as the energy source of the future. It was clean, abundant, and inexpensive. It was claimed by some proponents to be "too cheap to meter." Nuclear power was projected to replace most other energy resources by the middle of the twenty-first century. What happened that changed these optimistic forecasts into the much smaller figures of the last decade of the twentieth century?

People realized that nuclear power is not clean, cheap, or plentiful. Although it doesn't pollute the air like coal and oil, nuclear power could cause catastrophic damage from radiation contamination if there was an accident. It is not as abundant as projected. The uranium used to create nuclear fuel will probably be exhausted as a resource sometime in the next century. And nuclear power is not cheap. It is nearly twice as expensive to use as coal in some regions of the world.

Also nuclear waste from commercial generating plants has become a major political problem. The used fuel from the nuclear energy generating process is highly radioactive. This radioactive material must be kept isolated from the surrounding environment until the amount of radiation reaches a safe level.

Another problem for the nuclear industry is the fear of major accidents. On March 29, 1979, the nuclear plant at Three Mile Island near Harrisburg, Pennsylvania, experienced a partial meltdown, or melting of the reactor core. No one was killed, but it is projected that there may be several premature deaths to nearby residents as a result of radiation exposure.

People near Kiev were not so fortunate. On April 26, 1986, two big explosions ripped off the top of the Chernobyl nuclear plant, set the core on fire, and sent radioactive debris thousands of meters into the air. As of 1989, this accident had killed 36 workers, firefighters, and rescuers, with another 237 people hospitalized with acute radiation sickness. By 1988, the Soviet Union had spent $14.4 billion on the accident.

Even though nuclear energy has lost much of its appeal in recent years, it will still be a necessary part of energy resources in the immediate future. New nuclear technologies are showing promise for less dangerous reactors. If the waste disposal problems can be solved, nuclear energy may become acceptable once again. ■

Binomial Expansion 12.1

OBJECTIVES: 1. To generate Pascal's triangle
2. To compute factorials
3. To expand binomials with the binomial formula

Expressions of the form

$$(x + y)^n$$

where n is a positive integer, occur frequently in mathematics, particularly in the areas of probability and statistics. Of course, you can always find such an expansion by repeated multiplication; however, with large values of n, the calculation can be very tedious. Consider the following expansions (they can be verified by multiplying). The patterns involved will allow us to find larger powers directly.

Suppose that you want $(x + y)^8$. The binomial $x + y$ is a factor 8 times in the product.

$$(x + y)^1 = x + y$$
$$(x + y)^2 = x^2 + 2xy + y^2$$
$$(x + y)^3 = x^3 + 3x^2y + 3xy^2 + y^3$$
$$(x + y)^4 = x^4 + 4x^3y + 6x^2y^2 + 4xy^3 + y^4$$
$$(x + y)^5 = x^5 + 5x^4y + 10x^3y^2 + 10x^2y^3 + 5xy^4 + y^5$$

Some useful observations can be made by looking at the patterns common to the five expansions: In the expansion of $(x + y)^n$,

1. There are always $n + 1$ terms.

2. The exponents of x begin with the original exponent of the binomial and then decrease by 1, term by term.

3. The exponents of y begin with 0 ($y^0 = 1$) in the first term and then increase by 1, term by term.

In general, the variable factors in the terms are then

$$x^n,\ x^{n-1}y,\ x^{n-2}y^2,\ \ldots,\ xy^{n-1},\ y^n$$

As a computational aid, note that the sum of the exponents in each term is always n (the original power).

EXAMPLE 1 Expanding a Binomial: The Variable Factors

In the expansion of

$$(x + y)^6$$

there are seven terms ($7 = 6 + 1$). The variable factors in the terms are

$$x^6 \qquad x^5y \qquad x^4y^2 \qquad x^3y^3 \qquad x^2y^4 \qquad xy^5 \qquad y^6$$

Check Yourself 1

What are the variable factors in the expansion of $(x + y)^4$?

Now if you want to write an expansion directly, the only problem remaining is how to determine the coefficients. Arranging the binomial coefficients in a certain way yields the following pattern.

PASCAL'S TRIANGLE

This triangular array of numbers was named after Blaise Pascal (1623–1662), who called the array the *arithmetic triangle*. However, it was known much earlier, and it is due to Pascal's application of the triangle to the field of probability and his development of its theory that the triangle bears his name today.

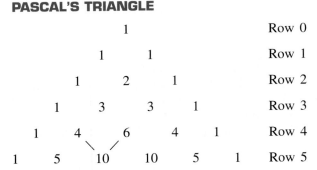

Comparing the pattern to our earlier expansions, you may have noticed that each row of Pascal's triangle gives the desired coefficients. To build this pattern:

1. Each row starts and ends with 1.

2. Each other entry is the sum of the two entries from the row immediately above, to the left and to the right. (The 10 in row 5 is the sum of 4 and 6 from row 4, and so on.)

The use of these observations is illustrated in the example below.

EXAMPLE 2 Generating Pascal's Triangle

Generate row 6 of Pascal's triangle.

Row 5		1		5		10		10		5		1	
Row 6	1		6		15		20		15		6		1

Check Yourself 2

Generate row 4 of Pascal's triangle.

Now we know that each row of Pascal's triangle gives the desired coefficients of the binomial expansion. Since we also know the variable patterns, we are ready to combine these observations to write an expansion.

EXAMPLE 3 Expanding a Binomial

Expand $(x + y)^6$.

Solution In row 6 the entries are

1   6   15   20   15   6   1

and the desired expansion is

$$(x + y)^6 = x^6 + 6x^5y + 15x^4y^2 + 20x^3y^3 + 15x^2y^4 + 6xy^5 + y^6$$

Check Yourself 3

Expand $(x + y)^4$.

Observing algebraic sign patterns will also allow us to expand a binomial expression of the form $(x - y)^n$ very easily.

EXAMPLE 4 Expanding a Binomial

Expand $(a - b)^6$.

Solution Following the same coefficient patterns as in Example 3, we have

$$(a - b)^6 = a^6 + 6a^5(-b) + 15a^4(-b)^2 + 20a^3(-b)^3 + 15a^2(-b)^4$$
$$+ 6a(-b)^5 + (-b)^6$$

or $= a^6 - 6a^5b + 15a^4b^2 - 20a^3b^3 + 15a^2b^4 - 6ab^5 + b^6$

Note that $(-b)^2 = b^2$, $(-b)^4 = b^4$, and so on; $(-b)^3 = -b^3$, $(-b)^5 = -b^5$, and so on.

What do you notice? In expanding a binomial that is a *difference* of two terms, the algebraic signs of the expansion *alternate*.

Check Yourself 4

Expand $(x - y)^4$.

If coefficients are involved in a binomial, the process is much the same. Consider the following.

EXAMPLE 5 Expanding a Binomial

Expand $(2x + y)^7$.

Solution Using row 7 of Pascal's triangle, we have

$$(2x + y)^7 = (2x)^7 + 7(2x)^6y + 21(2x)^5y^2$$
$$+ 35(2x)^4y^3 + 35(2x)^3y^4 + 21(2x)^2y^5$$
$$+ 7(2x)y^6 + y^7$$

We will leave it to you to verify that row 7 in the triangle is
1   7   21   35   35   21   7

CAUTION
Note that since the first term of the binomial is $2x$, the first term of the expansion is $(2x)^7$, *not* $2x^7$, a common error.

Expanding the powers and multiplying, we then have

$$(2x + y)^7 = 128x^7 + 448x^6y + 672x^5y^2$$
$$+ 560x^4y^3 + 280x^3y^4 + 84x^2y^5$$
$$+ 14xy^6 + y^7$$

Check Yourself 5

Expand $(x - 2y)^4$.

While we certainly can always use Pascal's triangle to expand any positive integral power of a binomial, the process may get a bit cumbersome, so it is worthwhile to develop an alternate strategy. First a new definition will be needed.

We write $n!$ (read "n factorial") to indicate the product of the consecutive integers from n down to 1.

> **DEFINITION**
>
> n **factorial is given by**
>
> $n! = n(n - 1)(n - 2) \cdots 2 \cdot 1$
>
> **We define two special cases:**
>
> $1! = 1$ and $0! = 1$

The last part of the definition may look a bit strange, but it is necessary for consistency. We know that

$n! = n(n - 1)!$

If $n = 1$, then

$1! = 1(1 - 1)!$

$1 = 1 \cdot 0!$

and 0! must be 1.

EXAMPLE 6 Evaluating Factorials

(a) $6! = 6 \cdot 5 \cdot 4 \cdot 3 \cdot 2 \cdot 1 = 720$

(b) $1! = 1$

(c) $(8 - 2)! = 6! = 720$

(d) $\dfrac{7!}{(7 - 3)!} = \dfrac{7!}{4!}$

$$= \frac{7 \cdot 6 \cdot 5 \cdot 4!}{4!}$$

$$= 7 \cdot 6 \cdot 5 = 210$$

The form shown in (e) will have special significance for our work in this section.

(e) $\dfrac{6!}{2!(6 - 2)!} = \dfrac{6!}{2! \, 4!}$

$$= \frac{6 \cdot 5 \cdot 4!}{2! \, 4!} = \frac{6 \cdot 5}{2 \cdot 1} = 15$$

Check Yourself 6

Evaluate.

1. $5!$

2. $(12 - 8)!$

3. $\dfrac{7!}{(7 - 5)!}$

In Example 6, we mentioned a special significance for a quotient of the form

$$\frac{n!}{r!(n-r)!}$$

We are now ready to make a definition involving that form and to introduce some notation.

$$\binom{n}{r} = \frac{n!}{r!(n-r)!} \qquad n \geq r$$

The notation $\binom{n}{r}$ is frequently used in probability and statistics and is read "the number of combinations of n elements taken r at a time."

EXAMPLE 7 Evaluating Combinations

Evaluate.

(a) $\binom{5}{2} = \frac{5!}{2!(5-2)!} = \frac{5!}{2!\,3!}$

Here $n = 5$ and $r = 2$.

$$= \frac{5 \cdot 4 \cdot 3!}{2!\,3!} = \frac{5 \cdot 4}{2 \cdot 1} = 10$$

(b) $\binom{5}{5} = \frac{5!}{5!(5-5)!}$

Now $n = r = 5$.

$$= \frac{5!}{5!\,0!}$$

Remember that $0! = 1$.

$$= 1$$

Continuing the work of Example 7, we could write

$$\binom{5}{0} \qquad \binom{5}{1} \qquad \binom{5}{2} \qquad \binom{5}{3} \qquad \binom{5}{4} \qquad \text{and} \qquad \binom{5}{5}$$

as   1   5   10   10   5   and   1

You should recognize these values. They are exactly the same as row 5 of Pascal's triangle and are therefore the binomial coefficients in the expansion of $(x + y)^5$. This leads us to the following form for what is called the *binomial formula*.

For this reason $\binom{n}{r}$ is also called a *binomial coefficient*.

Check Yourself 7

Evaluate.

1. $\binom{8}{3}$

2. $\binom{12}{9}$

THE BINOMIAL FORMULA

$$(x + y)^n = \binom{n}{0}x^n + \binom{n}{1}x^{n-1}y + \binom{n}{2}x^{n-2}y^2$$

$$+ \cdots + \binom{n}{r}x^{n-r}y^r + \cdots + \binom{n}{n}y^n$$

EXAMPLE 8 Using the Binomial Formula

(*a*) Expand $(x + y)^6$, using the binomial formula.

$$(x + y)^6 = \binom{6}{0}x^6 + \binom{6}{1}x^5y + \binom{6}{2}x^4y^2 + \cdots + \binom{6}{6}y^6$$

$$= x^6 + 6x^5y + 15x^4y^2 + 20x^3y^3 + 15x^2y^4 + 6xy^5 + y^6$$

You have no doubt noticed the symmetry of the binomial coefficients. That will save you considerable effort. You need only evaluate the coefficient until you reach the fourth term in this case. The pattern then repeats back down to 1.

(*b*) Find the first three terms in the expansion of $(x^2 - 2y)^{12}$.

$$(x^2 - 2y)^{12} = \binom{12}{0}(x^2)^{12} + \binom{12}{1}(x^2)^{11}(-2y) + \binom{12}{2}(x^2)^{10}(-2y)^2 + \cdots$$

$$= x^{24} + 12(x^{22})(-2y) + 66(x^{20})(4y^2) + \cdots$$

$$= x^{24} - 24x^{22}y + 264x^{20}y^2 + \cdots$$

Check Yourself 8

Expand $(x + y)^4$, using the binomial formula.

CHECK YOURSELF ANSWERS

1. x^4, x^3y, x^2y^2, xy^3, y^4 **2.** 1 4 6 4 1 **3.** $x^4 + 4x^3y + 6x^2y^2 + 4xy^3 + y^4$ **4.** $x^4 - 4x^3y + 6x^2y^2 - 4xy^3 + y^4$ **5.** $x^4 - 8x^3y + 24x^2y^2 - 32xy^3 + 16y^4$ **6.** (1) 120; (2) 24; (3) 2520 **7.** (1) 56; (2) 220
8. $x^4 + 4x^3y + 6x^2y^2 + 4xy^3 + y^4$

12.1 EXERCISES

Build Your Skills

Using Pascal's triangle, expand each of the following.

1. $(x + y)^5$

2. $(m - n)^6$

3. $(2a + b)^6$

4. $(c - 2d)^4$

5. $(x^2 + y)^7$

6. $(a - b^2)^5$

7. $(a - 3)^5$

8. $(x + 2)^6$

Evaluate each of the following.

9. $5!$

10. $7!$

11. $(7 - 2)!$

12. $7! - 2!$

13. $\dfrac{8!}{5!}$

14. $\dfrac{7!}{3!}$

15. $\dfrac{8!}{6!\,2!}$

16. $\dfrac{10!}{7!\,3!}$

17. $\dfrac{9!}{3!\,(9 - 3)!}$

18. $\dfrac{7!}{2!\,(7 - 2)!}$

19. $\binom{6}{2}$

20. $\binom{5}{3}$

21. $\binom{8}{0}$

22. $\binom{10}{10}$

Use the binomial formula to find the first four terms in each of the following expansions.

23. $(x + y)^7$

24. $(a - b)^9$

25. $(m - 3)^8$

26. $(c + 4)^6$

27. $(x + 2y)^{10}$

28. $(2m - n)^8$

29. $(a^2 - b)^5$

30. $(r + s^3)^7$

 Note that

$$(1.01)^6 = (1 + 0.01)^6$$
$$= 1^6 + 6(1)^5(0.01) + 15(1)^4(0.01)^2$$
$$\quad + 20(1)^3(0.01)^3 + \cdots$$
$$= 1 + 0.06 + 0.0015 + 0.000020 + \cdots$$
$$\approx 1.061520$$

and we have an approximation of $(1.01)^6$ accurate to six decimal places. You can easily verify this with your calculator. Approximate each of the following, using the first four terms of a binomial expansion. You should verify your results with a calculator.

31. $(1.02)^8$

32. $(0.99)^7$

In Pascal's triangle, the sum of the entries in row 0 is 1; the sum of the entries in row 1 is 2; in row 2, the sum is 4; and so on. Using that pattern, find the following.

33. The sum of the entries in row 3

34. The sum of the entries in row 5

35. The sum of the entries in row 6

36. The sum of the entries in row n

Think About These

The probability of getting exactly two heads in flipping a fair coin 4 times is given by the third term of the expansion:
$$\left(\frac{1}{2} + \frac{1}{2}\right)^4.$$

37. Find the probability of getting exactly three heads.

38. Find the probability of getting exactly four heads.

The greatest number of intersection points for three lines in a plane is three. For four lines, the greatest number of intersection points is six, and for five lines, ten points are possible.

39. Can you give the greatest number of intersection points for six lines?

40. For n lines?

A handy computational pattern is available for generating the binomial coefficients in any expansion. It is illustrated below.

$$(x + y)^7 = x^7 + 7x^6y + 21x^5y^2 + 35x^4y^3 + \cdots$$

with the pattern $\dfrac{7 \cdot 6}{2}$ and $\dfrac{21 \cdot 5}{3}$ shown over terms (1) (2) (3) (4)

In each case the coefficient of any term can be found by dividing the product of the coefficient and the power of the first factor in the preceding term by the number of that term. Use the pattern above to expand each of the following.

41. $(x + y)^6$

42. $(2a - b)^5$

43. $(x + 3)^4$

44. $(m - n^2)^7$

Skillscan

Evaluate each expression for the given values of the variables.

a. $3x + z$; $x = 1$, $z = 2$ **b.** $\dfrac{x}{2}(y + z)$; $x = 4$, $y = 3$, $z = 2$ **c.** $\dfrac{x}{3}(2y - w)$; $x = 6$, $y = 3$, $w = -1$

d. $\dfrac{2x}{5} - (w - z)$; $x = 10$, $w = -2$, $z = -1$

e. $\dfrac{3w}{4}[2x - 3(y + 1)]$; $w = 8$, $x = 4$, $y = -2$

f. $\dfrac{2x}{7}[y - (w - x)]$; $x = 14$, $y = 3$, $w = 6$

12.2 Arithmetic Sequences and Series

OBJECTIVES: 1. To find the nth term of an arithmetic sequence
2. To find the sum of an arithmetic sequence

Sequences and series have wide applications in both applied and theoretical mathematics. Our work in this and the following section will deal with the basic definitions, notation, and formulas used when we work with two particular types of sequences and series. Let's begin by looking at two applications that lead to sequences of numbers.

Suppose that you take a position which pays $20,000 per year. At the end of each year, you will receive a $1000 raise. Your salary will then look like this:

Year 1	Year 2	Year 3	Year 4
$20,000	$21,000	$22,000	$23,000

and these salary amounts form a sequence of numbers.

Again, you start a job paying $20,000 per year, but this time you will receive a 5 percent raise each year. Now your salary will be

Year 1	Year 2	Year 3	Year 4
$20,000	$21,000	$22,050	$23,152.50

and again these salary figures give a sequence of numbers. Note that in both cases a salary figure was associated with a number (1 for the first year, 2 for the second year, etc.), which is the essential idea for sequences of numbers.

Let's return to our first salary example. Note that in the sequence $20,000, $21,000, $22,000, $23,000, . . . , each term differs from the preceding term by a fixed amount (here $1000). We will see that this type of sequence occurs frequently enough in applications that it is worth developing special terminology and formulas to deal with it.

Consider

2, 5, 8, 11, 14 . . .

For this sequence, every term after the first can be found by adding the same fixed number to the preceding term. We call that fixed number the *common difference*, and it is always denoted by d (of course in the sequence above $d = 3$). This type of sequence is called an *arithmetic sequence* or *progression*.

The words "sequence" and "progression" are used interchangeably in this case.

The sequence *increases* as d is *positive*. The sequence *decreases* as d is *negative*.

EXAMPLE 1 Identifying an Arithmetic Sequence

(*a*) 3, 7, 11, 15, . . . is an arithmetic sequence with a common difference 4.
(*b*) 12, 9, 6, 3, . . . is also an arithmetic sequence. This time the common difference is -3.
(*c*) 1, 4, 9, 16, . . . is *not* an arithmetic sequence.

The differences between the terms are *not fixed* in this case. The difference $4 - 1 = 3$ and $9 - 4 = 5$.

Check Yourself 1

Find the common difference for each sequence.

1. $-3, 1, 5, 9, \ldots$

2. $-7, -1, 5, 11, \ldots$

The terms of a sequence are usually written as

$$a_1, a_2, a_3, \ldots, a_n, \ldots$$

This is illustrated in our next example.

EXAMPLE 2 Finding Terms for an Arithmetic Sequence

Find the first six terms of the arithmetic sequence with a first term of 5 and a common difference of 4.

$a_1 = 5$

$a_2 = 5 + 4 = 9$

$a_3 = 9 + 4 = 13$

$a_4 = 13 + 4 = 17$

$a_5 = 17 + 4 = 21$

$a_6 = 21 + 4 = 25$

In each case we add 4 to the preceding term to generate the next term.

The first six terms are 5, 9, 13, 17, 21, and 25.

Check Yourself 2

Find the first six terms for a sequence with a first term of -2 and a common difference of 3.

Suppose that you are asked to find the 50th term of an arithmetic sequence. Writing out 50 terms of the sequence is possible, but not very practical!

As should be clear from the work above, it would be very convenient to have a formula to find specific terms of an arithmetic sequence.

Fortunately, in the case of arithmetic sequences, it is always possible to represent the general or nth term in terms of the first term and the common difference. Consider the following:

In any arithmetic sequence:

First term: a_1

Second term: $a_2 = a_1 + d$

Third term: $a_3 = a_1 + 2d$

Fourth term: $a_4 = a_1 + 3d$

$$\vdots$$

nth term: $a_n = a_1 + (n-1)d$

Since $a_3 = a_2 + d$
$\qquad = (a_1 + d) + d$
$\qquad = a_1 + 2d$
Since $a_4 = a_3 + d$
$\qquad = (a_1 + 2d) + d$
$\qquad = a_1 + 3d$

Note that in each case the coefficient of d is *1 less than* the term number.

In general, we see that the nth term of an arithmetic sequence can be found by adding $n - 1$ times the common difference to the first term. This gives our first formula for arithmetic sequences.

THE nTH TERM OF AN ARITHMETIC SEQUENCE

In any arithmetic sequence,

$$a_n = a_1 + (n - 1)d \qquad\qquad (1)$$

where a_1 is the first term and d is the common difference.

This formula will allow us to solve a variety of problems involving arithmetic sequences. To start, once the beginning terms of a sequence are known, we can always find a specific term of that sequence.

EXAMPLE 3 Finding the nth Term of an Arithmetic Sequence

(a) Find the 20th term of the arithmetic sequence with $a_1 = -3$ and $d = 8$.

Let $a_1 = -3$, $d = 8$, and $n = 20$ in Equation (1).

$$\begin{aligned} a_{20} &= -3 + (20 - 1)8 \\ &= -3 + (19)8 \\ &= -3 + 152 \\ &= 149 \end{aligned}$$

(b) Find the 30th term of the arithmetic sequence

$$12, 8, 4, 0, -4, \ldots$$

First note that $a_1 = 12$ and $d = -4$. Now apply Equation (1) with $n = 30$.

$$\begin{aligned} a_{30} &= 12 + (30 - 1)(-4) \\ &= 12 + (29)(-4) \\ &= 12 - 116 \\ &= -104 \end{aligned}$$

Check Yourself 3

Find the 15th term of the arithmetic sequence with $a_1 = -2$ and $d = 4$.

Given any two specific terms of an arithmetic sequence, we can always completely describe the sequence. Our next two examples illustrate.

EXAMPLE 4 Finding Terms for an Arithmetic Sequence

Suppose that in an arithmetic sequence we have $a_1 = 3$ and $a_{10} = 48$. Find the first 10 terms.

Solution Again using Equation (1), we let $a_1 = 3$ and $n = 10$ so that a_n or $a_{10} = 48$. We can then solve the resulting equation for d.

$$48 = 3 + (10 - 1)d$$

$$48 = 3 + 9d$$

$$45 = 9d$$

$$d = 5$$

Once d has been found, we can write the first 10 terms of the sequence:

3, 8, 13, 18, 23, 28, 33, 38, 43, 48

The desired terms are found by adding 5 to each preceding term.

Check Yourself 4

Given an arithmetic sequence with $a_1 = 1$ and $a_{12} = 56$, find the first 12 terms.

EXAMPLE 5 Finding the Difference

Suppose we are given $a_3 = 12$ and $a_{10} = 61$ in an arithmetic sequence. Find a_1 and d.

Solution Since a_1 is *not* known, we will need two applications of Equation (1):

For a_3: $a_3 = a_1 + 2d$ or $12 = a_1 + 2d$

For a_{10}: $a_{10} = a_1 + 9d$ or $61 = a_1 + 9d$

Now consider the system of two equations in the two unknowns a_1 and d.

$$a_1 + 2d = 12$$

$$a_1 + 9d = 61$$

The system is easily solved by elimination to find a_1 and d.

Solving, we have $a_1 = -2$ and $d = 7$, and the sequence is

$-2, 5, 12, \ldots, 61, \ldots$

Check Yourself 5

Suppose you are given $a_3 = 6$ and $a_{12} = 60$. Find a_1 and d.

As we mentioned earlier in this section, many applications in algebra lead to arithmetic sequences. Consider the following example.

EXAMPLE 6 A Salary Application

Sandra will start in a new position in 1995 with an annual salary of $21,000. She will receive an annual raise of $1500 each year. Find her salary in 2005.

Solution Consider her salary amounts as an arithmetic sequence with $a_1 = 21{,}000$ and $d = 1500$. Her salary in the year 2005 is a_{11} in the sequence, so

$$a_{11} = 21{,}000 + (11 - 1)(1500)$$
$$= 21{,}000 + (10)(1500)$$
$$= 21{,}000 + 15{,}000$$
$$= 36{,}000$$

Sandra's salary in the year 2005 will be $36,000.

Check Yourself 6

Dien charges $15 to type a 1-page paper and $3 for every page after that. What will he charge for a 14-page paper?

Let's turn our attention now to a series associated with arithmetic sequences. An *arithmetic series* is the sum of the terms of an arithmetic sequence, or progression. For instance, given the sequence

5, 8, 11, 14, 17, 20

The associated series is

We use S_n to denote the *sum* of *n* terms of the sequence. Here S_6 is the sum of the first six terms.

$5 + 8 + 11 + 14 + 17 + 20$

and we can easily add to find that $S_6 = 75$.

Suppose that we want to find S_{20} or S_{50}. We could continue generating terms of the sequence and then find the sum. But again, a convenient formula is available that will allow us to find S_n in the case of an arithmetic sequence. Consider the following. We know that

$$S_n = a_1 + a_2 + a_3 + \cdots + a_n$$

or

$$S_n = a_1 + (a_1 + d) + (a_1 + 2d) + \cdots + [a_1 + (n - 1)d] \qquad (2)$$

Now we reverse the order, writing S_n as the sum from a_n *down to* a_1, this time subtracting d each time.

Mathematical historians pass on the story that the German mathematician Karl Friedrich Gauss (1777–1855) used a method similar to this when assigned (in elementary school) the problem of finding the sum of the first 100 natural numbers.

$$S_n = a_n + (a_n - d) + (a_n - 2d) + \cdots + [a_n - (n - 1)d] \qquad (3)$$

Now adding the two equations gives

$$2S_n = \underbrace{(a_1 + a_n) + (a_1 + a_n) + \cdots + (a_1 + a_n)}_{n \text{ terms}}$$

or

$$2S_n = n(a_1 + a_n)$$

Now dividing by 2, we have a general formula for S_n.

> **THE SUM OF *n* TERMS OF AN ARITHMETIC SEQUENCE**
>
> **In any arithmetic sequence,**
>
> $$S_n = \frac{n}{2}(a_1 + a_n)$$ **(4)**
>
> **where a_1 is the first term and a_n is the *n*th term.**

Thinking of the formula in the equivalent form

$$S_n = n\left(\frac{a_1 + a_n}{2}\right)$$

gives a helpful memory aid. "The sum of *n* terms is *n* times the *average* of the first and last terms."

    This formula allows us to solve a new type of problem involving arithmetic sequences.

EXAMPLE 7 Finding the Sum of an Arithmetic Sequence

Find S_{20} in the arithmetic sequence

3, 7, 11, 15, . . .

Solution Here $a_1 = 3$ and $d = 4$. First we find a_{20} as before, using Equation (1).

$$a_{20} = 3 + (20 - 1)4$$
$$= 3 + (19)(4)$$
$$= 3 + 76$$
$$= 79$$

Now from Equation (4), where $n = 20$, $a_1 = 3$, and $a_n = 79$,

$$S_{20} = \frac{20}{2}(3 + 79)$$
$$= 10(82)$$
$$= 820$$

So 820 is the sum of the first 20 terms.

Check Yourself 7

Find S_{20} in the arithmetic sequence 1, 4, 7, 10, ...

EXAMPLE 8 Finding the Sum of an Arithmetic Sequence

Find the sum of the first 50 *even* natural numbers.

Solution The associated sequence is

2, 4, 6, 8, . . . , 100

Note If $a_n = 100$ where $a_1 = 2$ and $d = 2$, then $100 = 2 + (n - 1)2$, and solving for n, we have $n = 50$. So 100 is the 50th even natural number.

Do you see that this is an arithmetic sequence with $a_1 = 2$ and $a_{50} = 100$? Therefore by Equation (4) the desired sum is

$$S_{50} = \frac{50}{2}(2 + 100)$$

$$= (25)(102)$$

$$= 2550$$

Check Yourself 8

Find the sum of the first 100 odd positive integers.

In the examples above, it was necessary to find the last term a_n of the sequence before calculating S_n by Equation (4). A third formula exists that will allow us to find S_n in a single step. We restate Equations (1) and (4):

$$a_n = a_1 + (n - 1)d \tag{1}$$

$$S_n = \frac{n}{2}(a_1 + a_n) \tag{4}$$

Substituting for a_n in Equation (4) yields

$$S_n = \frac{n}{2}[a_1 + a_1 + (n - 1)d]$$

$$= \frac{n}{2}[2a_1 + (n - 1)d]$$

And as a result, we have an alternate formula for S_n.

THE SUM OF n TERMS OF AN ARITHMETIC SEQUENCE

In any arithmetic sequence,

$$S_n = \frac{n}{2}[2a_1 + (n - 1)d] \tag{5}$$

EXAMPLE 9 Finding the Sum of an Arithmetic Sequence

Find S_{50} in the arithmetic sequence

50, 46, 42, 38, . . .

Solution Here $a_1 = 50$, $d = -4$, and $n = 50$. Using Equation (5), we have

$$S_{50} = \frac{50}{2}[2 \cdot 50 + 49(-4)]$$

$$= 25(100 - 196)$$

$$= 25(-96)$$

$$= -2400$$

Should you use Equation (4) or (5)? Both give the same result, but (5) allows a single computation for S_n in problems such as this. The big disadvantage is that there is another formula to learn! You will have to be the judge.

Check Yourself 9

Find S_{30} in the arithmetic sequence 50, 47, 44, 41, ...

Formula (5) is also useful in applications which involve arithmetic series, as the following example illustrates.

EXAMPLE 10 A Savings Application

Marcia decides to start a savings plan in which she will deposit $100 the first month and increase that deposit by $25 each month. What will she have saved in the first year of the plan?

Solution The savings amounts per month form an arithmetic sequence

100, 125, 150, 175, . . .

with $a_1 = 100$ and $d = 25$. To find the total savings for the first year, we want S_{12}, so

Variable S_{12} represents the sum saved after 12 months (1 year).

$$S_{12} = \frac{12}{2}[2 \cdot 100 + (12 - 1)(25)]$$

$$= 6(200 + 11 \cdot 25)$$

$$= 6(200 + 275)$$

$$= 6(475)$$

$$= 2850$$

She will have saved $2850 in the first year.

Check Yourself 10

Marcia decides to change her savings plan. She will deposit $200 the first month and increase that deposit by $10 each month. What will she have saved in the first year of this plan?

CHECK YOURSELF ANSWERS

1. (1) 4; (2) 6   **2.** -2, 1, 4, 7, 10, 13   **3.** 54   **4.** 1, 6, 11, 16, 21, 26, 31, 36, 41, 46, 51, 56   **5.** $a_1 = -6$, $d = 6$ **6.** $54   **7.** 590   **8.** 10,000   **9.** 195   **10.** $3060

12.2 EXERCISES

Build Your Skills

Which of the following sequences are arithmetic? For those that are arithmetic, find d, the common difference.

1. 2, 4, 6, 8, . . .

2. 15, 12, 9, 6, . . .

3. 2, 4, 8, 16, . . .

4. 5, 13, 21, 29, . . .

5. 2, $\dfrac{8}{3}$, $\dfrac{10}{3}$, 4, . . .

6. 1, 4, 9, 16, . . .

Write the first six terms of the arithmetic sequences given the following.

7. $a_1 = 3$, $d = 2$

8. $a_1 = -2$, $d = -2$

9. $a_1 = 3$, $d = \dfrac{3}{4}$

10. $a_1 = 8$, $d = -0.5$

Find the specified term for each of the following arithmetic sequences.

11. With $a_1 = 4$ and $d = 3$; a_{20}

12. With $a_1 = 50$ and $d = -5$; a_{12}

13. With $a_1 = -10$ and $d = 4$; a_{25}

14. With $a_1 = 2$ and $d = \dfrac{2}{3}$; a_{10}

15. 2, 9, 16, 23, . . . ; a_{20}

16. 20, 17, 14, 11, . . . ; a_{30}

17. 2, $\dfrac{7}{2}$, 5, $\dfrac{13}{2}$, . . . ; a_{10}

18. 1, $1 + \sqrt{2}$, $1 + \sqrt{8}$, $1 + \sqrt{18}$, . . . ; a_8

Each of the following exercises refers to an arithmetic sequence.

19. If $a_1 = 3$ and $a_{10} = 39$, find d.

20. If $a_1 = 28$ and $a_{12} = 6$, find d.

21. If $a_3 = 15$ and $a_{15} = 75$, find a_1 and d.

22. If $a_5 = -8$ and $a_{12} = 13$, find a_1 and d.

Find the indicated sum of each of the following arithmetic sequences.

23. With $a_1 = 5$ and $a_{20} = 43$; S_{20}

24. With $a_1 = 12$ and $a_{15} = -30$; S_{15}

25. With $a_1 = 3$ and $d = 3$; S_{30}

26. With $a_1 = -5$ and $d = 4$; S_{25}

27. 3, 10, 17, 24, . . . ; S_{20}

28. $-2, -4, -6, -8, . . .$; S_{10}

Solve each of the following applications.

29. Find the sum of the first 100 positive integers.

30. Find the sum of the first 100 positive odd integers.

31. Find the sum of the multiples of 5 between 1 and 201.

32. Find the sum of the multiples of 3 between 100 and 200.

33. In the first year of a new job, Beau saves $500. If he increases that amount by $250 each year, what will he have saved in 5 years?

34. A rock dropped from a cliff falls 16 ft in the first second, 48 ft during the second, 80 ft during the third, and so on. Give that same pattern, how far will the ball fall in 8 s?

35. A certain job pays $15,000 per year with an annual raise of $800. How much would a person be paid during the eighth year in the position? What would be the total salary paid for the first 8 years?

36. A $15,000 automobile will depreciate $2500 in the first year, $2100 in the second, $1700 in the third, and so on. Given the same pattern, find the value of the car after 6 years.

37. A woman's salary in 1985 was $18,000. Each year she receives a $1200 raise. Find her 1998 salary. What would be her total salary over that period?

38. A contest offers 30 prizes with first prize being $20,000, second prize $19,500, third prize $19,000, and so on. Find the value of the 30th prize and the total prize money for the first 30 places.

39. A bricklayer has stacked bricks so that there are 25 bricks in the bottom row of the pile, 23 bricks in the next-to-the-bottom row, and so on, up to 1 brick in the top row. How many bricks are in the pile?

40. If bricks are arranged in the pattern of Exercise 39 and there are 400 bricks in the pile, how many bricks are there in the bottom row?

41. Find a formula for the sum of the first n positive odd integers.

42. Find a formula for the sum of the first n positive even integers.

Magic squares have appeared in the mathematical histories of many cultures. These are squares in which the values along each row, each column, and each diagonal all have the same sum. Given that the values along each diagonal, the middle row, and the middle column also form arithmetic sequences, complete the following squares.

43.

	3	8
	5	
2		

44.

	2	
	10	
4		

45.

	15	20
	25	
		10

46.

$x + 1$		
	$x + 4$	
		$x + 7$

Skillscan (Section 1.3)

Evaluate the following.

a. $(2)^3$ **b.** $(-3)^2$ **c.** $\left(\dfrac{1}{2}\right)^3$ **d.** $(3)^{-2}$

e. $(-2)^{-3}$ **f.** $\left(-\dfrac{1}{2}\right)^2$

Geometric Sequences and Series   12.3

OBJECTIVES: 1. To find the nth term of a geometric sequence
2. To find the sum of n terms of a geometric sequence
3. To find the sum of an infinite geometric sequence

In the preceding section, we looked at a salary that was increased a flat amount each year. In this section we'll look at a salary that is increased a certain percentage each year. Suppose you begin a job with an initial annual salary of $20,000 and receive a 5 percent raise each year. The salary amounts are then

Year 1	Year 2	Year 3	Year 4
$20,000	$21,000	$22,050	$23,152.50

Let's examine the sequence formed by writing the amounts in a different manner.

Year 1 $20,000

Year 2   ($20,000)(1.05) or $21,000

Year 3   ($21,000)(1.05) or $22,050

Year 4   ($22,050)(1.05) or $23,152.50

> In year 2 the salary will be $20,000 + $20,000(0.05), or $20,000(1.05).

Note that each term after the first is obtained by multiplying the preceding term by 1.05.

This leads us to a second major classification of sequences, the geometric sequence.

In general, a *geometric sequence* or *progression* is a sequence in which each term after the first is obtained by multiplying the preceding term by the same fixed number. In the case of the geometric sequence, that number is called the *common ratio* and is denoted by r. You can always determine the common ratio by dividing a given term by the term which immediately precedes it.

In our salary example, we can find the common ratio as $\frac{21,000}{20,000}$, or 1.05.

EXAMPLE 1 Identifying a Geometric Sequence

(*a*) Show that 2, 6, 18, 54, . . . is a geometric sequence.

To determine r, find the ratio of a_2 to a_1.

Of course, you could use the ratio of a_3 to a_2 with the same result for r.

$$r = \frac{a_2}{a_1} = \frac{6}{2} = 3$$

You can easily verify that a_3 and a_4 are generated by multiplying the preceding terms by 3.

(*b*) Show that 27, 9, 3, 1, . . . is a geometric sequence.

Here

$$r = \frac{9}{27} = \frac{1}{3}$$

The subsequent terms are found by multiplying by $\frac{1}{3}$.

Check Yourself 1

Find r in the geometric sequence 16, 4, 1, $\frac{1}{4}$, ...

Given the first term and the common ratio, one can always generate a geometric sequence term by term.

EXAMPLE 2 Finding Terms for a Geometric Sequence

(*a*) Write the first five terms of the geometric sequence with $a_1 = 4$ and $r = -2$.

Starting with $a_1 = 4$, multiply a_1 by -2 to find a_2, then a_2 by -2 to find a_3, and so on. The desired terms are

4, -8, 16, -32, 64

(b) Write the first six terms of the geometric sequence with $a_1 = 16$ and $r = \dfrac{1}{4}$.

The desired terms are

$$16, \ 4, \ 1, \ \frac{1}{4}, \ \frac{1}{16}, \ \frac{1}{64}$$

Once again we can generate terms of a geometric sequence by simply multiplying by the common ratio. However, we can develop a useful formula that will allow us to write specific terms directly. Look at the following pattern.

First term: a_1

Second term: $a_2 = a_1 r$

Third term: $a_3 = a_1 r^2$

Fourth term: $a_4 = a_1 r^3$

$$\vdots$$

nth term: $a_n = a_1 r^{n-1}$

In general, we have

Since $a_3 = a_2 r$
$\quad = (a_1 r)r$
$\quad = a_1 r^2$
Since $a_4 = a_3 r$
$\quad = (a_1 r^2)r$
$\quad = a_1 r^3$

THE nTH TERM OF A GEOMETRIC SEQUENCE

For any geometric sequence,

$$a_n = a_1 r^{n-1} \tag{1}$$

where a_1 is the first term, r is the common ratio, and $r \neq 0$.

Check Yourself 2

Given a geometric sequence with $a_1 = 3$ and $r = -4$, find the first five terms.

EXAMPLE 3 Finding the nth Term of a Geometric Sequence

Find the general or nth term of the geometric progression

3, 12, 48, 192, . . .

Solution First we see that $a_1 = 3$ and $r = 4$. From Equation (1),

$$a_n = 3 \cdot 4^{n-1}$$

Check Yourself 3

Find the general, or nth, term of the geometric sequence $-2, -6, -18, -54, \ldots$.

We can also use Equation (1) to find specific terms of a geometric sequence. This is shown in the next example.

EXAMPLE 4 Finding Terms for a Geometric Sequence

(a) Find a_6 if $a_1 = 5$ and $r = 2$.

By Equation (1),

$$a_6 = 5 \cdot 2^{6-1}$$
$$= 5 \cdot 2^5$$
$$= 5 \cdot 32$$
$$= 160$$

(b) Find a_8 in the geometric sequence

$$64, \ -32, \ 16, \ -8, \ \ldots$$

Note that $r = -\dfrac{32}{64} = -\dfrac{1}{2}$.

Here we see that $a_1 = 64$ and $r = -\dfrac{1}{2}$. Again using Equation (1), we have

$$a_8 = 64\left(-\frac{1}{2}\right)^{8-1}$$
$$= 64\left(-\frac{1}{2}\right)^{7}$$
$$= 64\left(-\frac{1}{128}\right)$$
$$= -\frac{1}{2}$$

Check Yourself 4

Find a_5 if $a_1 = 3$ and $r = -2$.

EXAMPLE 5 Finding Terms for a Geometric Sequence

Find the second and third terms of the geometric sequence

$$2, \ a_2, \ a_3, \ 250$$

Solution Here $a_1 = 2$ and $a_4 = 250$. Also by Equation (1) we know that $a_4 = a_1 r^3$, so

$$250 = 2r^3$$

or

$$125 = r^3 \quad \text{and} \quad 5 = r$$

Using 5 as the common ratio, we get

$$a_2 = a_1 \cdot 5 = 2 \cdot 5 = 10 \quad \text{and} \quad a_3 = a_2 \cdot 5 = 10 \cdot 5 = 50$$

Note 10 and 50 are called the two *geometric means* between 2 and 250.

Check Yourself 5

Find the second and third terms in the geometric sequence 1, a_2, a_3, 125.

The sum of the terms of a geometric sequence is called a *geometric series*. For instance,

$$3 + 6 + 12 + 24 + 48$$

is the geometric series associated with the geometric sequence

3, 6, 12, 24, 48

Of course, to find the sum of any finite sequence, you can just add the terms. However, in the case of a geometric sequence, there is again a convenient formula for the sum of the first n terms. Consider the following:

We know that

$$S_n = a_1 + a_2 + a_3 + \cdots + a_n$$

Again, S_n denotes the sum of n terms of the sequence.

or

$$S_n = a_1 + a_1 r + a_1 r^2 + \cdots + a_1 r^{n-1}$$

Multiplying both sides of this equation by r, we have

$$rS_n = a_1 r + a_1 r^2 + \cdots + a_1 r^n$$

Subtracting the second equation from the first gives

$$S_n - rS_n = a_1 - a_1 r^n$$

Be certain that you go back and see what happened to all the other terms on the right side.

Factoring both members gives

$$S_n(1 - r) = a_1(1 - r^n) \quad \text{or} \quad S_n = \frac{a_1(1 - r^n)}{1 - r}$$

and we can write the general formula:

THE SUM OF n TERMS OF A GEOMETRIC SEQUENCE

In any geometric sequence,

$$S_n = \frac{a_1(1 - r^n)}{1 - r} \tag{2}$$

where a_1 is the first term, r is the common ratio, and $r \neq 1$.

EXAMPLE 6 Finding the Sum of a Geometric Sequence

(a) Find the sum of the first six terms of the geometric progression

3, 6, 12, 24, . . .

First we see that $a_1 = 3$ and $r = 2$. Applying Equation (2) with $n = 6$, we have

$$S_6 = \frac{3(1 - 2^6)}{1 - 2}$$

$$= \frac{3(1 - 64)}{-1}$$

$$= \frac{3(-63)}{-1}$$

$$= 189$$

(b) Find the sum of the first 10 terms of the geometric progression

8, 4, 2, 1, . . .

Here $a_1 = 8$ and $r = \dfrac{1}{2}$, so again by Equation (2) with $n = 10$, we have

$$S_{10} = \frac{8\left[1 - \left(\dfrac{1}{2}\right)^{10}\right]}{1 - \dfrac{1}{2}}$$

$$= \frac{8\left(1 - \dfrac{1}{1024}\right)}{\dfrac{1}{2}}$$

$$= \frac{8\left(\dfrac{1023}{1024}\right)}{\dfrac{1}{2}}$$

$$= 8\left(\frac{1023}{1024}\right)\left(\frac{2}{1}\right) = \frac{1023}{64} = 15\frac{63}{64}$$

Note This sum is approximately 15.98. We will comment on this later.

Check Yourself 6

Find the sum of the first six terms of the geometric sequence 2, 8, 32, 128, ...

Let's return to the sequence of Example 6b. The terms of the sequence are

$$8, 4, 2, 1, \frac{1}{2}, \frac{1}{4}, \frac{1}{8}, \frac{1}{16}, \ldots$$

Since $a_1 = 8$ and $r = \dfrac{1}{2}$, by Equation (2) the sum of n terms will be

$$S_n = \frac{8\left[1 - \left(\dfrac{1}{2}\right)^n\right]}{\dfrac{1}{2}}$$

Consider the term $\left(\dfrac{1}{2}\right)^n$ and its effect on the sum.

$$\left(\frac{1}{2}\right)^2 = \frac{1}{4} \qquad \left(\frac{1}{2}\right)^3 = \frac{1}{8} \qquad \left(\frac{1}{2}\right)^4 = \frac{1}{16}$$

$$\left(\frac{1}{2}\right)^5 = \frac{1}{32} \qquad \left(\frac{1}{2}\right)^6 = \frac{1}{64} \qquad \text{and so on}$$

It is clear that the term $\left(\dfrac{1}{2}\right)^n$ is becoming smaller as the powers increase. In fact, we can make $\left(\dfrac{1}{2}\right)^n$ as small as we please by taking sufficiently large values of n. This is true for any value of r between -1 and 1. So, whenever $|r| < 1$ or $-1 < r < 1$, r^n approaches 0 as n becomes larger. Now consider the sum S_n. As n becomes larger,

$$S_n = \frac{a_1(1 - r^n)}{1 - r} \rightarrow \frac{a_1(1 - 0)}{1 - r} = \frac{a_1}{1 - r}$$

So the sum of the terms approaches

$$\frac{a_1}{1 - r}$$

as n becomes larger, and we write the sum of the *infinite geometric series,* denoted S, as follows:

THE SUM OF AN INFINITE GEOMETRIC SERIES

In any geometric sequence with common ratio r, such that $|r| < 1$,

$$S = \frac{a_1}{1 - r} \tag{3}$$

Note If $|r| \geq 1$, the infinite geometric series has no sum.

EXAMPLE 7 Finding the Sum of an Infinite Geometric Sequence

Find the sum of the infinite geometric sequence

$8, 4, 2, 1, \ldots .$

Solution Again, $a_1 = 8$ and $r = \dfrac{1}{2}$ and since $|r| < 1$, we can apply Equation (3).

When we found the sum of 10 terms earlier, that sum was approximately 15.98, very close to 16. The more terms we take, the closer the sum will be to 16.

$$S = \frac{8}{1 - \dfrac{1}{2}} = \frac{8}{\dfrac{1}{2}} = 8 \cdot \frac{2}{1} = 16$$

Thus the sum approaches 16 (called the *limit*), and this is the sum of the infinite geometric series.

Check Yourself 7

Find the sum of the infinite geometric sequence $3, 1, \dfrac{1}{3}, \dfrac{1}{9}, \ldots$.

Let's look at an increasing application of the sum of an infinite geometric series. Recall that any repeating decimal is a rational number. We can write that repeating decimal as an infinite geometric series to convert the decimal to a quotient of two integers.

EXAMPLE 8

Write 0.636363. . . as the quotient of two integers.

Solution First we rewrite the repeating decimal as

$$0.\overline{63} = 0.63 + 0.0063 + 0.000063 + \cdots$$

Note

$r = \dfrac{a_2}{a_1}$

$= \dfrac{0.0063}{0.63} = 0.01$

We now have a geometric sequence with $a_1 = 0.63$ and $r = 0.01$. From Equation (3), the sum of the infinite geometric series is then

$$S = \frac{0.63}{1 - 0.01} = \frac{0.63}{0.99}$$

Multiplying numerator and denominator by 100 and simplifying, we have

$$S = \frac{63}{99} = \frac{7}{11}$$

and

$$0.\overline{63} = \frac{7}{11}$$

Check Yourself 8

Write 0.2727272727. . . . as the quotient of two integers.

CHECK YOURSELF ANSWERS

1. $r = \dfrac{1}{4}$ **2.** 3, -12, 48, -192, 768   **3.** $-2 \cdot 3^{n-1}$ **4.** 48

5. 5, 25   **6.** 2730   **7.** 4.5   **8.** $\dfrac{3}{11}$

12.3 EXERCISES

Build Your Skills

Which of the following are geometric sequences? For those that are geometric, find the common ratio r.

1. 1, 4, 16, 64, . . .

2. 3, 6, 9, 12, . . .

3. 1, $\dfrac{1}{3}$, $\dfrac{1}{9}$, $\dfrac{1}{27}$, . . .

4. 2, -4, 8, -16, . . .

5. 1, 4, 9, 16, . . .

6. 1, 8, 27, 64, . . .

Find the specified term for each geometric sequence, given the following.

7. $a_1 = 2$, $r = 4$; a_5

8. $a_1 = 3$, $r = 2$; a_8

9. $a_1 = 4$, $r = \dfrac{3}{2}$, a_6

10. $a_1 = 32$, $r = \dfrac{1}{2}$; a_6

11. 2, 6, 18, 54, . . . ; a_6

12. 2, -4, 8, -16, . . . ; a_8

13. 27, 18, 12, 8, . . . ; a_7

14. 64, -32, 16, -8, . . . ; a_6

15. 1, $\sqrt{2}$, 2, $2\sqrt{2}$, . . . ; a_7

16. 1, x, x^2, x^3, . . . ; a_{10}

Find a_2 and a_3 in the given geometric sequences.

17. 3, a_2, a_3, 192

18. 32, a_2, a_3, 4

Find the indicated sum for each geometric sequence, given the following.

19. $a_1 = 3$, $r = 2$; S_6

20. $a_1 = 4$, $r = 4$; S_4

21. $a_1 = 5$, $r = -2$; S_6

22. $a_1 = 6$, $r = -3$, S_4

23. $a_1 = 4$, $r = \dfrac{3}{2}$; S_5

24. 3, 9, 27, . . . ; S_4

25. 18, 6, 2, . . . ; S_5

26. 32, -16, 8, . . . ; S_8

27. 27, -9, 3, . . . ; S_6

28. 4, 3, $\dfrac{9}{4}$, . . . ; S_5

Find the sum of the following series. If no sum exists, give the reason why.

29. $1 + \dfrac{1}{2} + \dfrac{1}{4} + \dfrac{1}{8} + \cdots$

30. $3 + 1 + \dfrac{1}{3} + \dfrac{1}{9} + \cdots$

31. $2 + 4 + 8 + 16 + \cdots$

32. $27 - 9 + 3 - 1 + \cdots$

33. $8 + 6 + \dfrac{9}{2} + \dfrac{27}{8} + \cdots$

34. $2 + 3 + \dfrac{9}{2} + \dfrac{27}{4} + \cdots$

Write each of the following repeating decimals as a quotient of two integers.

35. 0.777 . . .

36. 0.555 . . .

37. 0.373737 . . .

38. 0.494949 . . .

39. 0.123123 . . .

40. 0.513513 . . .

Solve each of the following. A calculator will be needed on many of the exercises.

41. An amount of $1000 is invested for 5 years at an interest rate of 6 percent. What will the balance be at the end of that period?

42. A bacteria culture doubles every 2 h. If 1000 bacteria were in the culture originally, how many are present after 10 h?

43. A radioactive substance has a half-life of 1600 years. If 640 g is present today, how much of the substance will be present in 6400 years?

44. A $20,000 investment in a land purchase should increase 10 percent in value each year. Find its value after 6 years. Round your answer to the nearest dollar.

45. A piece of machinery had a purchase price of $2500. Each year its value is four-fifths that of the preceding year. Find its value at the end of 5 years.

46. A ball rebounds one-half the distance from which it is dropped. If it is dropped from a height of 16 ft, how high will it rebound on the fifth bounce?

47. Your direct ancestors are your parents, grandparents, great-grandparents, and so on. How many ancestors did you have in the sixth generation back? How many ancestors did you have over those six generations?

48. The length of arc through which a pendulum swings is four-fifths that of the previous swing. If the length of the original swing of the pendulum is 25 cm, find the length of the fourth swing and the total distance the tip of the pendulum has traveled in those four swings.

49. Given the pendulum of Exercise 48, how far will the tip of the pendulum have traveled before the pendulum comes to rest?

50. Given the ball of Exercise 46, find the total distance traveled by the ball before it comes to rest. *Hint:* Consider the distance traveled *after* the ball first strikes the ground.

Magic squares may be designed so that all rows, columns, and diagonals have the same product. In a particular type of such a magic square, the diagonals, the middle row, and the middle column all form geometric sequences. Given those conditions, complete the following squares.

51.

	1	32
	8	
2		

52.

	10	
	100	5

53.

x		
	$4x$	1

54.

	1	
	x^3	
x		

SUMMARY

Binomial Expansion [12.1]

A *binomial* is an expression with two terms. When the expression is raised to a power, the binomial can be expanded. Pascal's triangle yields the coefficients for the expansion.

The variable terms in the expansion of $(x + y)^n$ are

$$x^n, x^{n-1}y, x^{n-2}y^2, \ldots, xy^{n-1}, y^n$$

By combining the coefficients from Pascal's triangle with the variable terms, we can expand any binomial.

A binomial can also be squared by using the binomial formula.

$$(x + y)^n = \binom{n}{0}x^n + \binom{n}{1}x^{n-1}y + \binom{n}{2}x^{n-2}y^2 + \cdots + \binom{n}{r}x^{n-r}y^r + \cdots + \binom{n}{n}y^n$$

Arithmetic Sequences and Series [12.2]

An *arithmetic sequence* is a sequence of numbers in which each pair of consecutive terms has the same difference. In general terms, we write the sequence as

$$a_1, a_2, a_3, \ldots, a_n, \ldots$$

In any arithmetic sequence, the value of the nth term a_n can be found with the formula

$$a_n = a_1 + (n - 1)d$$

where d is the common difference in terms.

The sum of the first n terms of an arithmetic sequence S_n can be found by using either the formula

$$S_n = \frac{n}{2}(a_1 + a_n)$$

or the formula

$$S_n = \frac{n}{2}[2a_1 + (n - 1)d]$$

Geometric Sequences and Series [12.3]

A *geometric sequence* is a sequence of numbers in which each pair of consecutive terms has a common ratio.

In any geometric sequence, the value of the nth term a_n can be found from the formula

$$a_n = a_1 r^{n-1}$$

where r is the common ratio of terms.

The sum of the first n terms of a geometric sequence S^n can be found by using the formula

$$S^n = \frac{a_1(1 - r^n)}{1 - r}$$

The sum of an infinite geometric series with $|r| < 1$ can be found with the formula

$$S = \frac{a_1}{1 - r}$$

SUMMARY EXERCISES

[12.1]   Evaluate each of the following.

1. $4!$

2. $8!$

3. $(8 - 4)!$

4. $8! - 4!$

5. $\dfrac{8!}{4!}$

6. $\dfrac{8!}{4!} \, 4!$

7. $\dfrac{9!}{5!} \, 4!$

8. $\dbinom{8}{5}$

[12.1]   Use Pascal's triangle to expand each of the following.

9. $(x + y)^4$

10. $(a - b)^5$

11. $(x + y^2)^3$

12. $(x + 3)^4$

[12.1]   Use the binomial theorem to find the first four terms in each expansion.

13. $(x - y)^9$

14. $(r + 3s)^7$

15. $(m - 2)^{12}$

16. $(a^2 - 2b)^6$

[12.2]   Write the first six terms of each arithmetic sequence.

17. $a_1 = 5$, $d = 4$

18. $a_1 = -5$, $d = 9$

19. $a_1 = 2$, $d = -3$

20. $a_1 = 56$, $d = -12$

[12.2]   Find the specified term in each arithmetic sequence.

21. With $a_1 = 3$ and $d = 5$; a_{20}

22. With $a_1 = -36$ and $d = 11$; a_{10}

23. $-4, 5, 14, 23, \ldots$; a_{15}

24. $2, 2 - \sqrt{2}, 2 - \sqrt{8}, 2 - \sqrt{18}, \ldots$; a_{20}

[12.2]   Find the indicated sum for each arithmetic sequence.

25. $a_1 = 2$ and $d = 5$; S_{12}

26. $a_1 = 53$ and $d = -6$; S_{20}

27. $3, 7, 11, 15, \ldots$; S_{16}

28. $-1, -2, -3, -4, \ldots$; S_{100}

[12.3]   Find the specified term for each geometric sequence.

29. $a_1 = 2$, $r = 3$; a_7

30. $a_1 = 3$, $r = -1$; a_{101}

31. $1, -3, 9, -27, \ldots$; a_7

32. $x^2, x^5, x^8, x^{11}, \ldots$; a_{20}

[12.3]   Find the indicated sum for each geometric sequence.

33. $a_1 = 4$, $r = 2$; S_8

34. $a_1 = 2$, $r = \dfrac{3}{2}$; S_{10}

35. $192, 96, 48, 24, \ldots$; S_{12}

36. $12, -6, 3, -\dfrac{3}{2}, \ldots$; S_8

[12.3]   Find the sum of each infinite geometric series.

37. $a_1 = 64$, $r = \dfrac{1}{2}$

38. $a_1 = 2$, $r = -\dfrac{1}{3}$

39. $54 - 18 + 6 - 2 + \cdots$

40. $4, \dfrac{8}{3}, \dfrac{16}{9}, 32, 37, \ldots$

1. Use Pascal's triangle to expand $(x - 2)^5$.

2. Evaluate $\dfrac{12!}{3!} \, 9!$.

3. Evaluate $\dbinom{15}{4}$.

4. Use the binomial theorem to find the first four terms in the expansion of $(2x - y^2)^5$.

5. Use the binomial theorem to find the first five terms in the expansion of $(a - 3b)^8$.

6. Find the common difference in the arithmetic sequence $5, 2, -1, -4, \ldots$.

7. Write the first five terms of the arithmetic sequence where $a_1 = -32$ and $d = 12$.

8. Find a_{20} if $a_1 = 8$ and $d = 5$.

9. Find a_{15} if $a_1 = 300$ and $d = -75$.

10. Find d if $a_1 = 10$ and $a_{10} = 46$.

11. Find a_1 if $a_{12} = 100$ and $d = 14$.

12. Find the sum of the first 10 terms of the arithmetic sequence with $a_1 = 250$ and $d = -40$.

13. Find the sum of the first 100 positive even integers.

14. Given $a_1 = -10$ and $d = 12$, find S_{20}.

15. Find the common ratio for the geometric sequence $3, -6, 12, -24, \ldots$.

16. Find the fifth term of the geometric sequence $3, 15, 75, 375, \ldots$.

17. Find a_2 and a_3 in the geometric sequence $2, a_2, a_3, 432$.

18. Find the sum of the first six terms of the geometric sequence in which $a_1 = 3$ and $r = -2$.

19. Find the sum of the first five terms of the geometric sequence $18, -6, 2, -\dfrac{2}{3}, \cdots$.

20. Find the sum of the infinite geometric sequence $16 + 4 + 1 + \dfrac{1}{4} + \cdots$.

This review covers selected topics from all 12 chapters.

Find the equation of the axis of symmetry and the coordinates for the vertex of each of the following.

1. $y = x^2 + 5$

2. $y = -3(x - 1)^2 - 1$

3. $y = x^2 - 3x - 4$

4. $x = y^2 - y - 2$

Find the center and the radius of the graph of each equation.

5. $x^2 + y^2 = 64$

6. $(x + 2)^2 + (y - 3)^2 = 49$

7. $x^2 + 2x + y^2 - 6y = -1$

8. $x^2 - 6x + y^2 = 0$

Graph each of the following.

9. $y = -3x^2 + 2x$

10. $3x^2 - 4y^2 = 12$

11. $5x^2 + 7y^2 = 35$

12. $(x - 3)^2 + (y + 2)^2 = 49$

For each of the following, give the domain and range. Which relations are also functions?

13. $\{(x, y) \,|\, y = x^2\}$

14. $\{(x, y) \,|\, x^2 + y^2 = 25\}$

15. $\{(x, y) \,|\, 3x + 2y = 6\}$

If $f(x) = 3x + 2$ and $g(x) = -2x^2 + 3$, find each of the following.

16. $f(3) + g(-2)$

17. $f[g(3)]$

18. In an arithmetic sequence, find a_{20} if $a_1 = 4$ and $d = 3$.

19. Find the sum of the series $3 + 1 + \dfrac{1}{3} + \dfrac{1}{9} + \cdots$

20. Find the seventh term in the geometric sequence $2, -4, 8, -16, \ldots, a_7$.

ANSWERS TO SECTION EXERCISES AND SELF-TESTS

Note Because answers to the Transcribe Your Skills exercises will vary, answers to those exercises are not included.

Section 1.1

1. Infinite **3.** Finite **5.** Finite **7.** Infinite
9. N, Z, Q, R **11.** Q, R **13.** Z, Q, R **15.** Q', R
17. Q', R **19.** N, Z, Q, R **21.** Q, R **23.** Q, R
25.

$$2 \quad 4 \quad 6$$
$$\longleftarrow\!\!+\!\!+\!\!\bullet\!\!+\!\!\bullet\!\!+\!\!\bullet\!\!\longrightarrow$$
$$0$$

27.

$$-\tfrac{1}{2} \qquad 2\tfrac{7}{3}$$
$$\longleftarrow\!\!+\!\!\bullet\!\!+\!\!+\!\!\bullet\!\!+\!\!+\!\!\longrightarrow$$
$$-1 \quad 0 \quad 1 \quad 2 \quad 3 \quad 4$$

29.

$$-\sqrt{2} \qquad \sqrt{3}$$
$$\longleftarrow\!\!\bullet\!\!+\!\!+\!\!\bullet\!\!+\!\!+\!\!\longrightarrow$$
$$-2 \quad -1 \quad 0 \quad 1 \quad 2 \quad 3$$

31. $5 \in Z$

33. $-4 \notin N$ **35.** $\sqrt{2} \notin Z$ **37.** $\dfrac{5}{7} \in Q$ **39.** $\sqrt{2} \in R$
41. $\sqrt{36} \in R$ **43.** $0 \notin N$ **45.** 6 **47.** $\sqrt{5}, 2\pi$
49. $-5, 0, 6$ **55.** True **57.** True **59.** False
61. False **63.** True **65.** True **67.** False **a.** True
b. True **c.** False **d.** False

Section 1.2

1. $a + 9$ **3.** $13 + m$ **5.** $2x + 18$ **7.** $m + 15$ **9.** b
11. $21x$ **13.** x **15.** x **17.** $15x + 20$
19. $3w^2 + 9w$ **21.** $3x + 5$ **23.** $10x + 15y + 25$
25. $10a$ **27.** $6x$ **29.** $2x$ **31.** $\dfrac{5}{6}z$ **33.** $9y + 3$
35. $10x + 5$ **37.** $5n + 16$ **39.** $10a + 4$ **41.** $9x$
43. $8x + 5$ **45.** $5w^2 + 7w$ **47.** 2040 **49.** 4975
51. 24,072 **53.** Commutative property of addition
55. Distributive property
57. Associative property of addition
59. Commutative property of addition
61. Distributive property **63.** Additive inverse
69. False, $2 + 4x + 20$ **71.** True **73.** False, $6x + 6$
75. True **77.** True **79.** False, $7x$ **a.** -3 **b.** -4
c. $\sqrt{2}$ **d.** -6 **e.** 2.6 **f.** $\dfrac{1}{6}$

Section 1.3

1. Symmetric property **3.** Transitive property
5. Transitive property or substitution principle
7. Reflexive property **9.** Substitution principle **11.** $7 > 5$
13. $-2 < -1$ **15.** $-3.8 > -3.9$ **17.** $-\dfrac{5}{4} > -\dfrac{4}{3}$
19. $\sqrt{2} > 1.41$ **21.** $1.25 = \dfrac{5}{4}$ **23.** $|-2| > -2$

25. $-|4| < |-4|$ **27.** x is greater than 5.
29. t is less than or equal to 3.
31. x is greater than or equal to y.
33. m is less than 0 or m is negative.
35. -2 is less than x, and x is less than 5; or x is between -2 and 5. **37.**

$$x < 5$$
$$\longleftarrow\!\!\!\longleftarrow\!\!+\!\!\circ\!\!\longrightarrow$$
$$0 \quad 5$$

39.

$$x > -5$$
$$\longleftarrow\!\!\circ\!\!+\!\!\longrightarrow$$
$$-5 \quad 0$$

41.

$$-2 \geq x$$
$$\longleftarrow\!\!\bullet\!\!+\!\!\longrightarrow$$
$$-2 \quad 0$$

43.

$$x \geq 2$$
$$\longleftarrow\!\!+\!\!\bullet\!\!\longrightarrow$$
$$0 \quad 2$$

45.

$$2 < x < 3$$
$$\longleftarrow\!\!+\!\!\circ\!\!\circ\!\!\longrightarrow$$
$$0 \quad 2 \quad 3$$

47.

$$x < 2$$
$$\longleftarrow\!\!\!\longleftarrow\!\!\circ\!\!\longrightarrow$$
$$0 \quad 2$$

49.

$$x \geq -1$$
$$\longleftarrow\!\!\bullet\!\!+\!\!\longrightarrow$$
$$-1 \quad 0$$

51.

$$x > 5$$
$$\longleftarrow\!\!+\!\!\circ\!\!\longrightarrow$$
$$0 \quad 5$$

53.

$$x \leq -1$$
$$\longleftarrow\!\!\bullet\!\!+\!\!\longrightarrow$$
$$-1 \quad 0$$

55.

$$2 < x < 5$$
$$\longleftarrow\!\!+\!\!\circ\!\!\circ\!\!\longrightarrow$$
$$0 \quad 2 \quad 5$$

  57. 4 **59.** 3.5

61. $\dfrac{7}{8}$ **63.** -1.5 **65.** -2 **67.** 10 **69.** -5
75. True **77.** True **79.** True
81. False, this is not true for negative values of a. **83.** True
85. False, every real number has an absolute value.
87. $N \geq 8$, $A > 300$, $D > 40$ **89.** $75 \leq E \leq 1000$ **a.** 15
b. 6.9 **c.** $\dfrac{7}{12}$ **d.** 30 **e.** 16.32 **f.** $\dfrac{2}{15}$

Section 1.4

1. 2 **3.** -3 **5.** 2 **7.** -7 **9.** -5 **11.** -6
13. 6 **15.** 2 **17.** -2 **19.** 6 **21.** -2 **23.** 2

25. −10 **27.** −10 **29.** −20 **31.** −8 **33.** 6
35. 9 **37.** −3 **39.** 0 **41.** 0 **43.** 28th floor
45. $170 **47.** $716 **49.** 5 **51.** −3 **53.** 6
55. 0 **57.** −3x **59.** −19m **61.** −8x **63.** 9x + 2y
65. 2x − 4y **67.** −10 **69.** −15 **71.** 21 **73.** 30

75. −60 **77.** 1 **79.** $\frac{3}{2}$ **81.** 36 **83.** −14

85. −21 **87.** −26 **89.** 0 **91.** 80 **93.** −3
95. 7 **97.** 5 **99.** 9 **101.** −22 **103.** 4
105. 14 **107.** 2 **109.** 20 **111.** 20 **113.** −56
115. −56 **121.** Always **123.** Always
125. Answers may vary. **127.** Always
129. Always **131.** Sometimes **133.** Always
135. −350, or 350-hectare loss **137.** 261-hectare gain **a.** 3
b. 6 **c.** 7 **d.** 5 **e.** 15 **f.** −5

Chapter 1 Self-Test
1. Associative property of multiplication **2.** Additive inverse
3. Associative property of addition **4.** Distributive property
5.
6.
7.
8.
9. 7 − 5 = 2 **10.** 2 **11.** −15 **12.** 7.6 **13.** −30
14. 8 **15.** 35 **16.** −54 **17.** −9a − 4b
18. 13x + 12 **19.** −1 **20.** −16

Section 2.1

1. 5 **3.** 7 **5.** 11 **7.** −4 **9.** 6 **11.** $\frac{2}{3}$ **13.** 9

15. 7 **17.** $\frac{5}{2}$ **19.** −9 **21.** 5 **23.** 5 **25.** $\frac{5}{2}$

27. 5 **29.** $-\frac{4}{3}$ **31.** −13 **33.** 7 **35.** 30 **37.** 6

39. 15 **41.** 3 **43.** $\frac{3}{2}$ **45.** 20 **47.** Conditional

49. Contradiction **51.** Identity **53.** Contradiction
55. Identity **61.** True **63.** False **65.** True
a. $-\frac{1}{2}$ **b.** $-\frac{3}{2}$ **c.** $\frac{1}{2}$ **d.** $-\frac{1}{3}$ **e.** −2 **f.** $\frac{13}{2}$

g. −8 **h.** $\frac{1}{2}$

Section 2.2

1. $h = \frac{V}{B}$ **3.** $r = \frac{C}{2\pi}$ **5.** $H = \frac{V}{LW}$ **7.** $h = \frac{V}{\pi r^2}$

9. $B = \frac{3V}{h}$ **11.** $R = \frac{E}{I}$ **13.** $x = -\frac{b}{a}$

15. $W = \frac{P - 2L}{2}$ **17.** $S = C - nD$ **19.** $r = \frac{R - C}{C}$

21. $b = \frac{2A - hB}{h}$ **23.** $C = \frac{5}{9}(F - 32)$ or $C = \frac{5F - 160}{9}$

27. 3 cm **29.** 6 percent **31.** 8 cm **33.** 12 ft

35. 5 years **37.** (V * P)/K **39.** $15 \le C \le 30$ **41.** 500 m
a. x + y **b.** w − 3 **c.** 4b **d.** $\frac{1}{2}x$ **e.** m − 6
f. 2s + 5 **g.** 7p **h.** 3x − 2

Section 2.3
1. 7, 31 **3.** 13, 16 **5.** 5, 6 **7.** 18, 20, 22
9. Length 75 ft, width 40 ft **11.** Length 20 cm, width 15 cm
13. 50 dimes, 150 quarters **15.** 300 main floor, 200 balcony
17. 105 lb, $4; 45 lb, $6.50
19. 100 mL of 15% solution, 200 mL of 45% solution
21. 50 cm³ **23.** $7000 at 8 percent, $5000 at 9 percent
25. $5000 at 7 percent, $10,500 at 10 percent
27. 56 mi/h going, 48 mi/h returning
29. 45 mi/h for 3 h and 55 mi/h for 2 h **31.** 3 P.M.
33. 6 h going, 8 h returning, 288 mi **35.** 800 flashlights
37. 175 calculators **39.** $21 **41.** 250 m
43. $35 per ton cardboard, $45 per ton white ledger
45. 7 plastic recyclers, 10 cardboard recyclers, 20 newspaper recyclers **a.**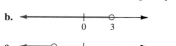

b.
c.
d.
e.
f.
g.
h.

Section 2.4
1. x < 7
3. x ≥ −2
5. x > 5
7. x ≥ 5
9. $x < \frac{7}{2}$
11. x ≤ −6
13. x > −3
15. x > 9

17. $x \le \dfrac{4}{3}$

19. $x > -2$

21. $x \ge -4$

23. $x < \dfrac{2}{5}$

25. $x \le 3$ **27.** $x > 1$ **29.** $x > -3$ **31.** $x < 19$
33. $x \ge -11$ **35.** $x \ge 12$ **37.** $x > 15$ **39.** $x \le 10$
41. $2 \le x \le 4$

43. $-4 < x < 2$

45. $2 \le x \le \dfrac{9}{2}$

47. $-1 < x < 2$

49. $x < -2$ or $x > 4$

51. $x < -3$ or $x > 4$

57. $x > 50$ **59.** $x \ge 92$ **61.** $x \ge 12$ **63.** $190 \le m \le 230$
65. $x \le 19 \times 10^6 \, \text{t}$ **a.** 3   **b.** 7   **c.** 6   **d.** 4   **e.** 0
f. 0   **g.** 7   **h.** 6

Section 2.5
1. $-5, 5$ **3.** $-1, 5$ **5.** -6 **7.** $-4, 10$ **9.** $-3, 6$
11. $1, \dfrac{3}{2}$ **13.** $-24, 4$ **15.** $-\dfrac{16}{3}, 16$ **17.** No solution
19. $-5, 5$ **21.** $0, 4$ **23.** $-2, 5$ **25.** $-20, 12$
27. $1, 7$ **29.** No solution   **31.** $-\dfrac{2}{3}, 4$ **33.** $-\dfrac{2}{7}, 2$
35. $\dfrac{1}{2}$ **37.** 2   **39.** All real numbers
41. $-5 < x < 5$

43. $x \le -7$ or $x \ge 7$

45. $x < 2$ or $x > 6$

47. $-10 \le x \le -2$

49. $x < -2$ or $x > 8$

51. No solution
53. $1 < x < 4$

55. $x \le -3$ or $x \ge \dfrac{1}{3}$

57. $x < -\dfrac{4}{5}$ or $x > 2$

59. $-3 < x < \dfrac{13}{3}$

61. $x \le -\dfrac{4}{5}$ or $x \ge 2$

63. $-\dfrac{8}{3} < x < 16$

69. $\text{abs}(x + 2)$ **71.** $\text{abs}(2 * x - 3)$ **73.** $\text{abs}(3 * x + 2) - 4$
75. $2 * \text{abs}(3 * x - 1)$ **a.** Not equivalent   **b.** Equivalent
c. Equivalent   **d.** Not equivalent   **e.** Equivalent
f. Equivalent

Section 2.6
1. Linear   **3.** Linear   **5.** Linear   **7.** Not linear
9. Linear   **11.** Not linear   **13.** 3   **15.** 20   **17.** $x \ge 7$
19. -3 **21.** 2   **25.** g **27.** b **29.** d **31.** a
a.

b.

c.

d.

e.

f.

Chapter 2 Self-Test
1. $\dfrac{4}{5}$ **2.** 2   **3.** 2   **4.** $\dfrac{3}{2}$ **5.** $r = \dfrac{A - P}{Pt}$
6. $h = \dfrac{2A}{B + b}$ **7.** $x \le 4$ **8.** $x > \dfrac{17}{2}$ **9.** $x > 1$
10. $-2 \le x \le 4$ **11.** $-\dfrac{2}{3}, 4$ **12.** $\dfrac{2}{3}, 4$ **13.** $-\dfrac{3}{2} < x < 3$
14. $x \le -2$ or $x \ge \dfrac{9}{}$ **15.** $-13, -11$
16. Length 44 ft; width 17 ft   **17.** 100 lb at \$3.40; 125 lb at \$1.60
18. 250 mL of 20% solution, 150 mL of 60% solution
19. \$15,000 at 6 percent, \$7,000 at 9 percent   **20.** 3:30 P.M.

Section 3.1
1. Q I   **3.** Q III   **5.** On the x axis   **7.** Q II

9. On the y axis   **11–19.** See figure below   **21.** $(3, 5)$ **5.** $x + y = 6$

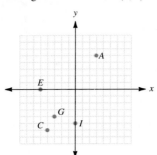

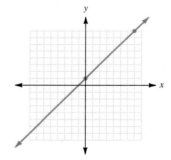

23. $(-6, 0)$ **25.** $(-5, -4)$ **27.** $(0, -4)$ **29.** $(6, -2)$
33. The points lie on a straight line; $(4, 5)$
35. The points lie on a straight line; $(2, -6)$ **a.** 5   **b.** -3 **7.** $y = x - 2$
c. 1   **d.** 5   **e.** 2   **f.** 2   **g.** 6   **h.** 6

Section 3.2
1. Domain $\{1, 3, 5\}$, range $\{1, 2\}$, function
3. Domain $\{2, 4\}$, range $\{1, 3\}$, not a function
5. Domain $\{-7, -5, 2\}$, range $\{1, 4\}$, function
7. Domain $\{-1, 1, 2\}$, range $\{5, 7\}$, function
9. Domain $\{-2, -1\}$, range $\{0, 1\}$, not a function
11. Function   **13.** Function   **15.** Not a function
17. Function   **23.** Domain $\{-1, 2, 4\}$; range $\{-1, 2, 6\}$
25. Domain $\{-4, -2, -1\}$; range $\{-2, -1, 1, 3\}$ **a.** 14
b. 5   **c.** 0   **d.** -1 **e.** 2   **f.** 9

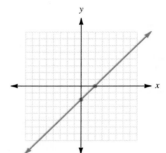

9. $y = x + 1$

Section 3.3
1. $y = x + 1$

x	y
-2	-1
-1	0
0	1
1	2
2	3

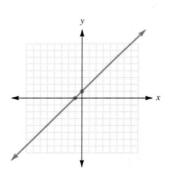

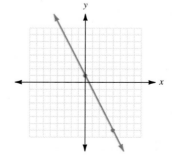

11. $y = -2x + 1$

3. $y = x - 4$

x	y
-2	-6
-1	-5
0	-4
1	-3
2	-2

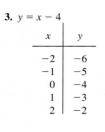

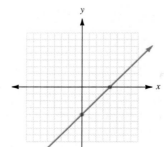

13. $y = \dfrac{1}{2}x - 3$

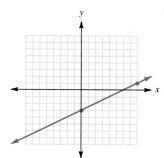

15. $y = -x - 3$

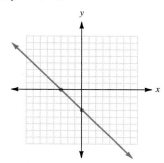

17. $x + 2y = 0$

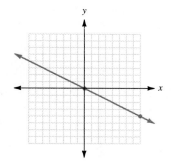

19. $x - 2y = 4$
Intercepts: $(4, 0)$, $(0, -2)$

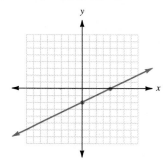

21. $2x - y = 6$
Intercepts: $(3, 0)$, $(0, -6)$

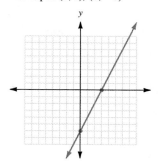

23. $2x + 5y = 10$
Intercepts: $(5, 0)$, $(0, 2)$

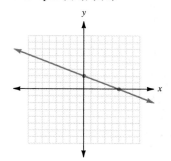

25. $x + 4y + 8 = 0$
Intercepts: $(-8, 0)$, $(0, -2)$

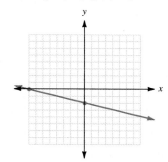

27. $x = 4$

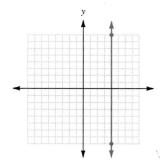

29. $y = 4$

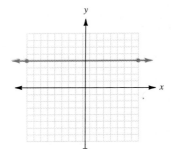

31. $C = 0.08s + 12$

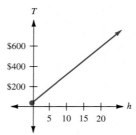

33. $T = 25h + 40$

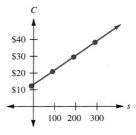

39. $y = 3x$ **41.** $y = 2x + 3$ **43.** $r + s = 7$
45. $x = 4y + 7$
47. $(5, 1)$

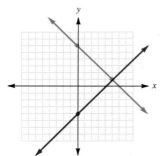

49. $(1, 2)$

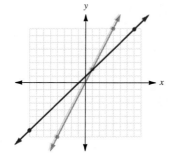

51. The line corresponding to $y = 2x$ is steeper than the line corresponding to $y = x$.

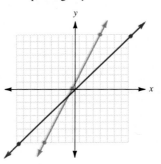

53. The two lines appear to be parallel.

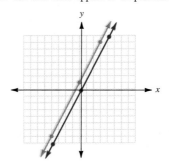

55. The lines appear to be perpendicular.

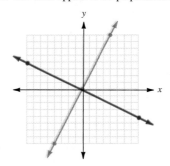

57.

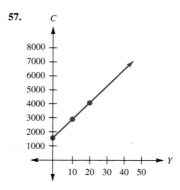

59.

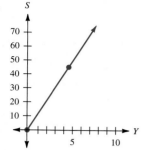

a. 1   **b.** 3   **c.** 2   **d.** 4   **e.** $\dfrac{7}{2}$ **f.** $-\dfrac{1}{4}$ **g.** 0

h. Undefined

Section 3.4

1. 2   **3.** $\dfrac{1}{2}$ **5.** 0   **7.** Undefined   **9.** $\dfrac{1}{5}$ **11.** $-\dfrac{2}{3}$

13. -3 **15.** $-\dfrac{1}{2}$ **17.** $\dfrac{2}{3}$ **19.** $-\dfrac{3}{4}$ **21.** Parallel

23. Neither   **25.** Perpendicular   **27.** $\dfrac{1}{3}$ **29.** 12

31. Collinear   **33.** Not collinear   **35.** Not collinear

37.

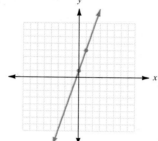

61.

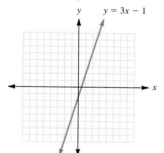

63.

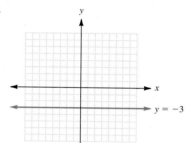

$y = 3x - 1$

39.

65.

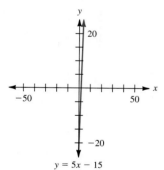

$y = 5x - 15$

41.

67.

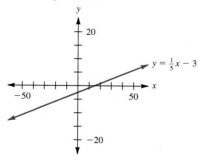

$y = \frac{1}{5}x - 3$

43.

45.

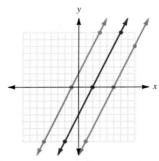

c. $y = -\dfrac{1}{3}x + 3$ **d.** $y = -\dfrac{2}{3}x + 2$ **e.** $y = 3x$

f. $y = -\dfrac{1}{2}x$ **g.** $y = \dfrac{2}{5}x - 2$ **h.** $y = \dfrac{3}{4}x - 3$

Section 3.7

1. (e)   **3.** (a)   **5.** (b)   **7.** (h)

9. $y = -x + 5$, $m = -1$, y intercept is 5

11. $y = 2x + 2$, $m = 2$, y intercept is 2

13. $y = -\dfrac{1}{3}x + 3$, $m = -\dfrac{1}{3}$, y intercept is 3

15. $y = \dfrac{2}{3}x - 2$, $m = \dfrac{2}{3}$, y intercept is -2

17. $y = 2x$, $m = 2$, y intercept is 0

19. $y = -3$, $m = 0$, y intercept is -3 **21.** $y = 3x + 2$

23. $y = \dfrac{3}{2}x + 2$ **25.** $y = 4$ **27.** $y = \dfrac{5}{4}x - 5$

29. $y = 3x - 1$ **31.** $y = -3x - 9$ **33.** $y = \dfrac{2}{5}x - 5$

35. $x = 2$ **37.** $y = -\dfrac{4}{5}x + 4$ **39.** $y = x + 1$

41. $y = \dfrac{3}{4}x - \dfrac{3}{2}$ **43.** $y = 2$ **45.** $y = \dfrac{3}{2}x - 3$

47. $y = \dfrac{5}{2}x + 4$ **49.** $y = 4x - 2$ **51.** $y = -\dfrac{1}{2}x + 2$

53. $y = 4$ **55.** $y = 5x - 13$ **57.** $y = 3x + 3$

59. $f(x) = \dfrac{1}{2}x + 4$ **61.** $f(x) = 3$ **63.** $f(x) = 2x + 8$

65. $f(x) = \dfrac{4}{3}x - 2$ **67.** $f(x) = \dfrac{1}{3}x - \dfrac{11}{3}$

69. $f(x) = -\dfrac{1}{2}x$ **73.** $f(x) = -\dfrac{4}{7}x + \dfrac{18}{7}$

75. $F = \dfrac{9}{5}C + 32$ **77.** $V = -1500t + 10{,}000$

79. $C = -\dfrac{1}{3}t + 9.9$ **81.** $I = -\dfrac{32}{9}t + 100$

49. Parallelogram, rectangle   **51.** Parallelogram, not a rectangle

53.

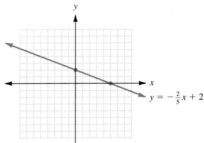

$y = -\frac{2}{5}x + 2$

55.

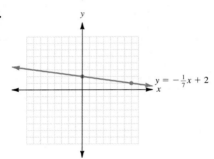

$y = -\frac{1}{7}x + 2$

a. 7   **b.** 5   **c.** 3   **d.** 5   **e.** $\sqrt{13}$ **f.** $\sqrt{5}$

g. $2x - 5y = 10$ **h.** $3x - 4y = 12$

Section 3.5

1. 2   **3.** 10   **5.** $\sqrt{45} = 3\sqrt{5}$ **7.** $\sqrt{26}$ **9.** $\sqrt{82}$

11. Show that $(AB)^2 + (BC)^2 = (AC)^2$.   **13.** $(-4, 4)$

15. $(5, 12)$ **17.** $\left(-\dfrac{7}{2}, -\dfrac{3}{2}\right)$ **19.** $\left(\dfrac{5}{4}, -\dfrac{13}{4}\right)$ **21.** 36

23. Square diagonals are perpendicular.   **25.** Not collinear

27. Not collinear   **29.** Collinear   **33.** 3   **35.** 2.1

a. 5   **b.** 9   **c.** 1   **d.** -2 **e.** -4 **f.** -2

Section 3.6

1. 17   **3.** -3 **5.** -19 **7.** 5   **9.** -13 **11.** -4

13. -1 **15.** 3   **17.** 1   **19.** 1   **21.** 3   **23.** 1

25. No real solution   **27.** $5a - 1$ **29.** $5x + 4$

31. $5x + 5h - 1$ **33.** $m^2 + 2$ **35.** $x^2 + 4x + 6$ **39.** 5

41. $(1, 5), (3, 9)$ **43.** Yes   **45.** 5   **47.** 5   **49.** 13

51. 41   **53.** 41   **55.** 364.5   **57.** 629.856   **59.** -3

61. 69   **a.** $y = -x + 3$ **b.** $y = 2x - 5$

83.

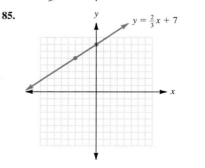

$y = \frac{3}{5}x - 6$

85.

$y = \frac{2}{3}x + 7$

87.

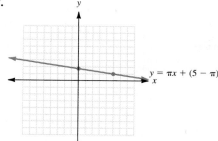

$y = \pi x + (5 - \pi)$

a. $x < 3$

b. $x \geq -2$

c. $2x \leq 8$

d. $3x > -9$

e. $-3x < 12$

f. $-2x \leq 10$

g. $\dfrac{2}{3}x \leq 4$

h. $-\dfrac{3}{4}x \geq 6$

Section 3.8

1. $x + y < 4$

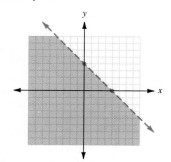

3. $x - y \geq 3$

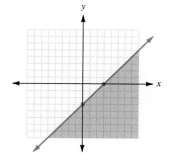

5. $y \geq 2x + 1$

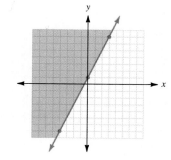

7. $2x + 3y < 6$

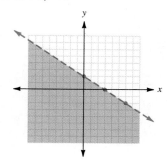

9. $x - 4y > 8$

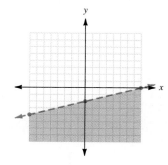

11. $y \geq 3x$

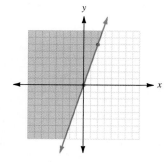

13. $x - 2y > 0$

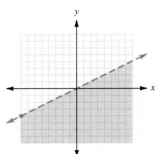

15. $x < 3$

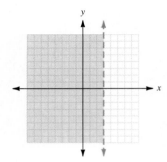

17. $y > 3$

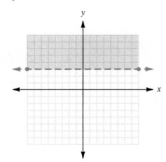

19. $3x - 6 \leq 0$

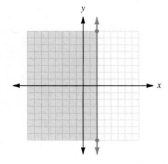

21. $0 < x < 1$

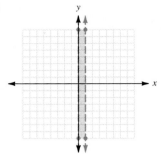

23. $1 \leq x \leq 3$

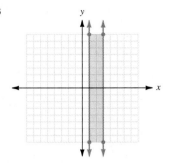

25. $0 \leq x \leq 3$
$2 \leq y \leq 4$

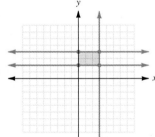

27. $x + 2y \leq 4$
$x \geq 0$
$y \geq 0$

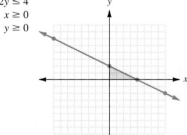

31. $\qquad x \geq 0$
$y \geq 0$
$12x + 18y \leq 360$

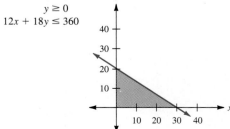

33. $x \geq 1000$
$y \geq 500$
$x + y \leq 3000$

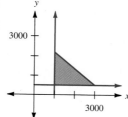

35. The solution is the region in which the graph of $y = 2x + 1$ is above the graph of $y = 5$
a. -1 **b.** -4 **c.** 9

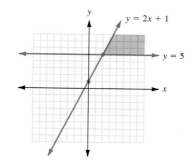

d. 4 **e.** -4 **f.** $-\dfrac{1}{4}$

g. $\dfrac{9}{4}$ **h.** 0

Chapter 3 Self-Test
1.

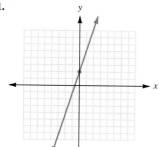

2.

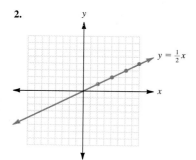

3.

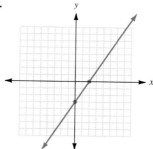

4.

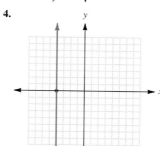

5. $D = \{-1, 0, 2, 4\}$
$R = \{2, 4, 8, 16\}$

6. $D = \left\{ -\dfrac{7}{2}, -\dfrac{3}{2}, -\dfrac{1}{2} \right\}$
$R = \left\{ \dfrac{1}{3}, -\dfrac{1}{3} \right\}$

7. Perpendicular   **8.** Parallel   **9.** $\dfrac{2}{3}$ **10.** $-\dfrac{5}{2}$

11. $\sqrt{34}$ **12.** $\sqrt{29}$ **13.** $(3, -8)$ **14.** $\left(\dfrac{1}{2}, -\dfrac{1}{2} \right)$

15. $f(-1) = -5, f(3) = -13$ **16.** $f(-1) = -1, f(3) = \dfrac{}{3}$

17. $y - \dfrac{4}{3}x + \dfrac{25}{3}$ **18.** $y = -\dfrac{3}{2}x + \dfrac{5}{2}$

19.

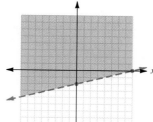

20.

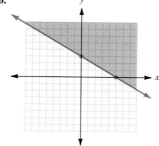

Section 4.1

1. $(5, 1)$

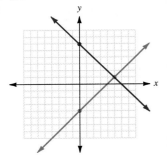

3. $(2, 1)$

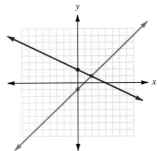

5. Inconsistent

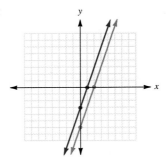

7. $(6, 2)$

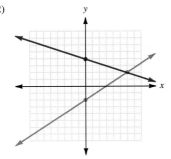

9. $(2, 3)$ **11.** $\left(-5, \dfrac{3}{2}\right)$ **13.** $(2, 1)$ **15.** Dependent

17. $(5, -3)$ **19.** Inconsistent **21.** $(-8, -2)$

23. $(5, -2)$ **25.** $(-4, -3)$ **27.** Dependent

29. $\left(\dfrac{1}{3}, 2\right)$ **31.** $(10, 1)$ **33.** Inconsistent

35. $\left(-2, -\dfrac{8}{3}\right)$ **37.** $(3, 4)$ **39.** $(4, -5)$ **41.** $(12, -6)$

43. $(9, -15)$ **47.** $\left(\dfrac{2}{3}, \dfrac{2}{5}\right)$ **49.** $\left(\dfrac{1}{3}, -\dfrac{3}{2}\right)$

51. $y = \dfrac{3}{2}x - 2$ **53.** 4 per car, 43 per bus

55. 3 per car, 45 per bus

57.

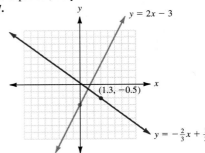

59.

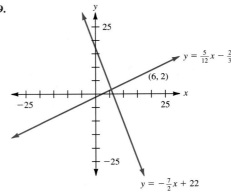

a. $8x - 2z$ **b.** $2x - 8y$ **c.** $7x + 5y$ **d.** $5x + 5z$
e. $5x + z$ **f.** $10x - 5y$ **g.** $-4x - 5z$ **h.** $-7y - 11z$

Section 4.2

1. $(1, 2, 4)$ **3.** $(-2, 1, 2)$ **5.** $(-4, 3, 2)$

7. An infinite number of solutions **9.** $\left(3, \dfrac{1}{2}, -\dfrac{7}{2}\right)$

11. $(3, 2, -5)$ **13.** $(2, 0, -3)$ **15.** $\left(4, -\dfrac{1}{2}, \dfrac{3}{2}\right)$

17. Inconsistent **19.** $\left(2, \dfrac{5}{2}, -\dfrac{3}{2}\right)$ **23.** $(1, 2, -1, -2)$

25. $(7, 2)$ **27.** $T = 20$, $C = 25$, $B = 40$ **a.** 3 **b.** -8
c. 4 **d.** 3 **e.** -3 **f.** -10 **g.** 6 **h.** -6

Section 4.3

1. (d) **3.** (g) **5.** (h) **7.** (c) **9.** 8, 21
11. 13 nickels, 15 quarters **13.** 550 adult, 200 student
15. Length 27 in, width 15 in **17.** \$1.80 mulch, \$3.20 fertilizer
19. 105 lb of \$4 bean, 45 lb of \$6.50 bean
21. \$7000 time deposit, \$5000 bond
23. 100 mL of 10% solution, 300 mL of 50% solution
25. 15 mi/h boat, 3 mi/h current **27.** 26
29. 15 battery-powered models, 20 solar models **31.** 15
33. 100 mi **35.** 3, 5, 8 **37.** 3 nickels, 5 dimes, 17 quarters
39. 4 cm, 7 cm, 8 cm
41. \$3000 savings, \$9000 bond, \$5000 money market **43.** 243
45. 30 **47.** 110 single, 230 carpool
49. Roy 8 mi, Sally 16 mi, Jeff 26 mi

a. $x + y > 8$

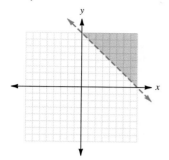

b. $2x - y \leq 6$

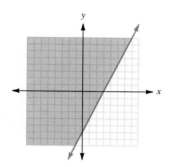

c. $3x + 4y \geq 12$

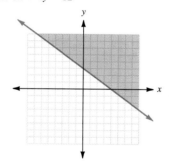

d. $y > 2x$

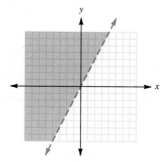

e. $y \leq -3$

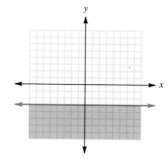

f. $x > 5$

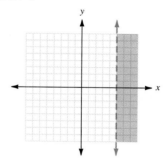

Section 4.4

1. $x + 2y \leq 4$
   $x - y \geq 1$

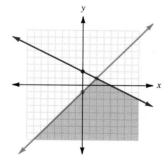

3. $3x + y < 6$
   $x + y > 4$

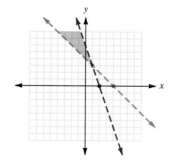

5. $x + 3y \le 12$
$2x - 3y \le 6$

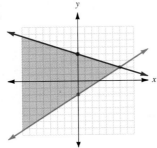

7. $3x + 2y \le 12$
$x \ge 2$

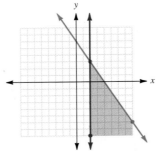

9. $2x + y < 8$
$x > 1$
$y > 2$

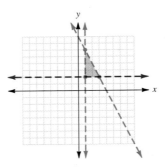

11. $x + 2y \le 8$
$2 \le x \le 6$
$y \ge 0$

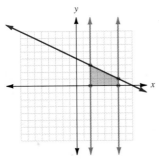

13. $3x + y \le 6$
$x + y \le 4$
$x \ge 0$
$y \ge 0$

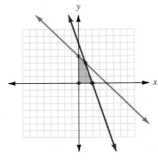

15. $4x + 3y \le 12$
$x + 4y \le 8$
$x \ge 0$
$y \ge 0$

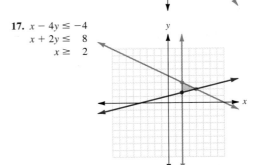

17. $x - 4y \le -4$
$x + 2y \le 8$
$x \ge 2$

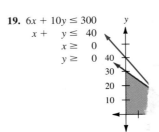

19. $6x + 10y \le 300$
$x + y \le 40$
$x \ge 0$
$y \ge 0$

a. $-4x - 6y = -8$ **b.** -3
d. $-4x + 15x = 1$ **e.** $4x -$
f. $12x - 4y + 4z = 8$

Section 4.5

1. $\begin{bmatrix} 2 & -3 & \vdots & 5 \\ 1 & 4 & \vdots & 2 \end{bmatrix}$ **3.** $\begin{bmatrix} 1 & -5 \\ 0 & 1 \end{bmatrix}$

5. $\begin{bmatrix} 1 & 2 & -1 & \vdots & 3 \\ 1 & 0 & 3 & \vdots & 1 \\ 0 & 1 & -2 & \vdots & 4 \end{bmatrix}$ **7.** $x + 2y =$
$x + 5y = -6$

9. $x + 3y = 5$ **11.** $x + 2y = 4$ **13.** $(1, -2)$
$y = 2$ $y + 5z = 3$
$x + y + z = 1$

15. $(5, -3)$ **17.** $(2, -4)$ **19.** $(4, -3)$ **21.** Inconsistent

23. $(1, -2, 0)$ **25.** $(-1, -8, -3)$ **27.** $\left(\dfrac{5}{2}, -3, -\dfrac{7}{2} \right)$

29. $(-1, 1, 0)$ **a.** 10   **b.** 2   **c.** 10   **d.** -26 **e.** -3
f. 3

Section 4.6

1. 2   **3.** 0   **5.** 79   **7.** 3   **9.** $(5, 2)$ **11.** $(0, -1)$
13. $\left(3, \dfrac{2}{3} \right)$ **15.** $(0, 0)$ **17.** $(3, 7)$ **19.** 11   **21.** -1

23. 43　**25.** 12　**27.** 44　**29.** -13　**31.** $(2, -1, -5)$

33. $\left(2, \dfrac{3}{2}, -2\right)$　**35.** 0　**37.** 24　**39.** 0

41. (a) -8, (b) -8　**43.** -36　**45.** $2x + 3y - 18 = 0$

47. -203　**a.** 8　**b.** 9　**c.** 9　**d.** -16　**e.** -64

f. $\dfrac{4}{9}$

Chapter 4 Self-Test

1. $(-3, 4)$　**2.** Dependent　**3.** Inconsistent　**4.** $(-2, -5)$

5. $(5, 0)$　**6.** $\left(3, -\dfrac{5}{3}\right)$　**7.** $(-1, 2, 4)$　**8.** $\left(2, -3, -\dfrac{1}{2}\right)$

9. Disks \$2.50, ribbons \$6

10. 60 lb jawbreakers, 40 lb licorice

11. Four 5-in sets, six 12-in sets

12. \$8000 savings, \$4000 bond, \$2000 mutual fund

13. 50 by 80 ft

14.

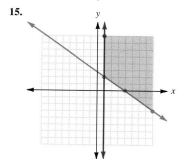

15.

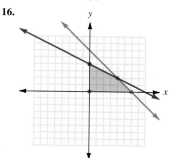

16.

17. $(10, -3)$　**18.** Inconsistent, no solution　**19.** $(3, -2, 2)$

20. Inconsistent, no solution

Section 5.1

1. x^9　**3.** y^{12}　**5.** 3^5　**7.** $(-5)^{10}$　**9.** $3s^{11}$　**11.** $\dfrac{1}{x^5}$

13. $\dfrac{1}{25}$　**15.** $\dfrac{1}{25}$　**17.** $-\dfrac{1}{8}$　**19.** $\dfrac{27}{8}$　**21.** $\dfrac{3}{x^2}$

23. $\dfrac{-5}{a^4}$　**25.** $\dfrac{1}{9x^2}$　**27.** x^3　**29.** $\dfrac{2x^3}{5}$　**31.** $\dfrac{y^4}{x^3}$

33. a^6b^5　**35.** x^9y^7　**37.** $-24x^{10}$　**39.** $-30a^8$

41. $30a^4b^5$　**43.** $a^8b^8c^3$　**45.** x^2　**47.** $\dfrac{1}{a^3}$　**49.** $\dfrac{1}{z^{10}}$

51. 1　**53.** x^3　**55.** x^3y^8　**57.** $a^4b^2c^2$　**59.** $3s^3t^3$

61. $\dfrac{1}{x^3}$　**63.** x^{15}　**65.** $2x^{11}$　**67.** $3a^9$　**69.** $x^{10}y$

71. $a^{17}b^8c^{13}$　**73.** $\dfrac{1}{w^{15}}$　**75.** b^8　**77.** $\dfrac{x^{10}}{y^6}$　**79.** $m^{12}n^6$

81. $\dfrac{r^{15}}{32}$　**83.** $\dfrac{z^{12}}{w^6}$　**85.** $\dfrac{m^8}{n^4}$　**87.** $\dfrac{48}{y^8}$　**89.** $27b^{16}$

91. 4×10^{20}　**93.** $x^{42}y^{33}z^{25}$　**95.** $75a^6$　**97.** $144a^{22}$

99. $\dfrac{27a^{18}}{8b^{27}}$　**101.** $-567a^{22}b^{25}$　**103.** $\dfrac{6}{a^2b^5}$　**105.** y

107. $\dfrac{3xy^5}{4z^6}$　**109.** $\dfrac{1}{x^{15}y^9}$　**111.** a^6　**113.** 9.3×10^7

115. 1.3×10^{11}　**117.** 28　**119.** 18　**121.** 0.0000075

123. 0.000521　**125.** 3×10^{-6}　**127.** 5.1×10^{-5}

129. 6×10^{-4}　**131.** 5×10^{-6}　**133.** 8×10^9

135. 2×10^2　**137.** 6×10^{16}　**139.** 66 years

141. 1.55×10^{23}　**143.** $\approx 4.66 \times 10^{15}$　**149.** x^{5n}

151. w^2　**153.** y^{3n^2}　**155.** a^2　**a.** $6x + 10$

b. $12x - 3$　**c.** $4x - 6y$　**d.** $11x$　**e.** $6y^2$　**f.** $7x + 10$

Section 5.2

1. Binomial, 1　**3.** Monomial, 3　**5.** Trinomial, 6

7. Binomial, 1　**9.** Binomial, 5　**11.** $9a + 4$

13. $5m^2 + 7mn + 2n^2$　**15.** $17x^2 - 8xy - 3y^2$

17. $10x - 1$　**19.** $2x + 13$　**21.** $12y^2 + 5y$

23. $3x^2 - 2x - 5$　**25.** $-3a^2 + 7a + 5$　**27.** $4x - 7$

29. $-x^2 - 7x + 3$　**31.** $13x^2 + 5$　**33.** $-3x - 3$

35. $3x + 4$　**37.** $x - 12$　**39.** 23　**41.** -3　**43.** -8

45. -41　**49.** $50x - 2000$　**51.** $-3x^2 + 50x - 2000$

53. $6ah + 3h^2 - 2h$　**55.** $x - 11$　**57.** $-3x + 5$

59. $6w + 2$　**61.** -15　**63.** $-2kC - 3x$　**a.** $6x^5$

b. $-6y^8$　**c.** $32m^3n^5$　**d.** $21a^5b^8$　**e.** $3w^2$　**f.** $-8x$

g. $-5xy^4$　**h.** $7pq$

Section 5.3

1. $20xy$　**3.** $-18x^5$　**5.** $30r^5s^5$　**7.** $6x^3 - 9x^2$

9. $5a^4 - 15a^3 + 5a^2$　**11.** $5x^4y^3 - 15x^3y^4 + 5x^2y^5$

13. $-2a^3b + 2a^2b^2 - 6a^2b^3 - 2ab^4$　**15.** $x^2 + 4xy + 3y^2$

17. $m^2 + 5mn - 14n^2$　**19.** $25a^2 - 80ab + 63b^2$

21. $21c^2 - 43cd + 20d^2$　**23.** $15x^3 + 10x^2y^2 - 6xy - 4y^3$

25. $x^2 + 10x + 25$　**27.** $4a^2 - 12a + 9$

29. $16c^2 - 24cd + 9d^2$　**31.** $16x^2 + 24xy^2 + 9y^4$

33. $x^2 - 9y^2$　**35.** $4p^2 - 9q^2$　**37.** $16a^4 - 9b^2$

39. $3x^3 + 8x^2y - 6xy^2 + y^3$　**41.** $x^3 - 8y^3$

43. $x^3 - 2x^2 - 3x$　**45.** $2x^3 + 2x^2y - 24xy^2$

51. $2x^3 + 5x^2 - 3x$　**53.** $100x - 0.2x^2$, 4500　**55.** 2499

57. 884　**59.** 3575　**61.** $\pi R^2 - \pi r^2$　**a.** 1, 2, 5, 10

b. 1, 2, 3, 6, 9, 18　**c.** 1, 5, 25　**d.** 1, 23

e. 1, 2, 4, 7, 14, 28　**f.** 1, 2, 4, 13, 26, 52

g. 1, 2, 4, 5, 10, 20, 25, 50, 100

h. 1, 2, 3, 4, 6, 8, 9, 12, 18, 24, 36, 72

Section 5.4

1. $3(2x + 3y)$　**3.** $4x(x - 3)$　**5.** $9mn(2m + 3n)$

7. $3(3x^2 - 2x + 1)$　**9.** $5x(x^2 - 3x + 5)$

11. $6mn(2m^2 - 1 + 3n)$　**13.** $4ab(a^2b - 2a + 3b - 1)$

15. $(y - z)(x + 3)$ **17.** $(m - n)(3 + 5m - 5n)$
19. $5(x - y)(x^2 - 2x + 3)$ **21.** $(b - c)(a + b)$
23. $(r + 2s)(6r - 1)$ **25.** $(a - 2)(b^2 + 3)$
27. $(4a - b^2)(4a^2 - b)$ **29.** $(3x - 2y)(x + 1)$
31. $(x + 2)(x - 5y)$ **33.** $(m - 3n)(m + 2n^2)$
35. $2\pi r(h + r)$ **41.** $(x + y)(x^2 - x + 3)$
43. $(a - b)(a^2 - 3a + 3)$ **a.** $x^2 - 4$ **b.** $4a^2 - 9$
c. $25m^2 - n^2$ **d.** $9c^2 - 25d^4$ **e.** $x^3 + 1$ **f.** $w^3 - 8$
g. $a^3 - b^3$ **h.** $8p^3 + q^3$

Section 5.5

1. $(x + 7)(x - 7)$ **3.** $(a + 9)(a - 9)$
5. $(3p + 1)(3p - 1)$ **7.** $(5a + 4)(5a - 4)$
9. $(xy + 5)(xy - 5)$ **11.** $(2c + 5d)(2c - 5d)$
13. $(7p + 8q)(7p - 8q)$ **15.** $(x^2 + 4y)(x^2 - 4y)$
17. $a(a + 2b)(a - 2b)$ **19.** $2c(b + 3c)(b - 3c)$
21. $(a^2 + 4b^2)(a + 2b)(a - 2b)$ **23.** $(x + 4)(x^2 - 4x + 16)$
25. $(m - 5)(m^2 + 5m + 25)$ **27.** $(ab - 3)(a^2b^2 + 3ab + 9)$
29. $(2w + z)(4w^2 - 2wz + z^2)$
31. $(r - 4s)(r^2 + 4rs + 16s^2)$
33. $(2x - 3y)(4x^2 + 6xy + 9y^2)$
35. $(2x + y^2)(4x^2 - 2xy^2 + y^4)$
37. $4(x - 2y)(x^2 + 2xy + 4y^2)$
39. $5m(m + 2n)(m^2 - 2mn + 4n^2)$
41. $(x + 3)(x + 2)(x - 2)$ **43.** $(2x + 3)(2x + 1)(2x - 1)$
45. $(a + 3)(3a + 2)(3a - 2)$
47. $(x - y)(x + y)(x^2 - xy + y^2)$ **49.** $a^3 + b^3$
53. $(x + y)(x^2 - xy + y^2)(x - y)(x^2 + xy + y^2)$
55. The factors are $x^2 - xy + y^2$ and $x^2 + xy + y^2$
57. Represent the area by the sum of the "inner square" plus 4
times the area of each of the "tabs." **a.** $x^2 + 2x - 15$
b. $m^2 + 13m + 40$ **c.** $2a^2 - 13a + 15$
d. $6d^2 - 11d - 10$ **e.** $5w^2 - 2wz - 3z^2$
f. $21c^2 + 11cd - 2d^2$ **g.** $3m^3 + m^2 - 2m$
h. $4y^3 - 26y^2 + 42y$

Section 5.6

1. $(x + 3)(x + 4)$ **3.** $(b - 8)(b - 1)$ **5.** $(y - 10)(y - 5)$
7. $(m + 10)(m - 3)$ **9.** $(x - 6)(x - 4)$
11. $(a - 11)(a + 4)$ **13.** $(x + 3y)(x + 5y)$
15. $(m - 11n)(m - 5n)$ **17.** $(3x - 4)(x + 5)$
19. $(5x - 2)(x + 4)$ **21.** $(3y + 5)(4y + 1)$ **23.** $(2x + 5)^2$
25. $(5a - 6)(a + 5)$ **27.** $(5b - 6)(b + 6)$
29. $(2x - 3)(5x + 4)$ **31.** $(4y + 5)^2$
33. $(7r - 3s)(r - 2s)$ **35.** $(4x - y)(2x - 7y)$
37. $3(x - 5)(x - 3)$ **39.** $2(n - 4)(n - 9)$
41. $x(6x - 1)(x - 5)$ **43.** $b(b + 4)(5b - 6)$
45. $3m(m - 6n)(m + n)$ **51.** $(x^2 + 1)(x^2 + 2)$
53. $(m^2 - 11)(m^2 + 3)$ **55.** $(y^3 - 5)(y^3 + 3)$
57. $x(x^2 + 2)(x^2 - 8)$ **59.** $(a + 3)(a - 3)(a^2 + 4)$
61. $(y - 2)(y^2 + 2y + 4)(y^3 + 2)$ **a.** $2x^4 + 6x^3 - 10x^2$
b. $10a^2 + 14a - 12$ **c.** $25m^2 - 9n^2$ **d.** $x^3 - 8y^3$
e. $10w^2 - 17wz + 3z^2$ **f.** $x^3 - 25xy^2$ **g.** $a^3 + 27b^3$
h. $12s^3 + 2s^2r - 2sr^2$

Section 5.7

1. GCF, $x(x - 3)$ **3.** Trial and error, $(x - 8)(x + 3)$
5. GCF, $(x - y)(x + 2)$ **7.** GCF, $2y(x^2 - 3x + 4y)$
9. Trial and error, $(y - 5)(y - 8)$
11. Trial and error, $(b + 7)(3b - 4)$
13. Trial and error, $(3x + 4y)(x - 6y)$
15. Trial and error, $(2a + 3)(a + 4)$

17. Sum of cubes, $(5r + s)(25r^2 - 5rs + s^2)$
19. GCF then trial and error, $3(x - 3)(x - 7)$
21. GCF, then sum of cubes, $5(2a + 1)(4a^2 - 2a + 1)$
23. GCF, then trial and error, $2(w - 9)(w + 2)$
25. GCF, then difference of squares, $3b(a + 4b)(a - 4b)$
29. $(x + 8)(x - 18)$ **31.** $(x + 2y + 4)(x + 2y - 4)$
33. $(2x - 5)(3x - 1)$ **35.** $\left(\dfrac{x}{2} - \dfrac{1}{3}\right)\left(\dfrac{x^2}{4} + \dfrac{x}{6} + \dfrac{1}{9}\right)$
a. x **b.** a^2 **c.** $2m^2$ **d.** $4y$ **e.** $2xy$ **f.** $3mn$
g. $4ab$ **h.** $4xy^2$

Section 5.8

1. $3b - 4$ **3.** $4a + 6$ **5.** $3y^2 + 2y - 1$
7. $-2a^4 + 4a^2 - 1$ **9.** $4x^2y - 3y^2 + 2x$ **11.** $x - 3$
13. $y + 1$ **15.** $a + 2b$ **17.** $x^2 - 3x + 9$
19. $(m + 2)(m^2 + 4)$ **21.** $x^2 + 3$ **23.** $3x - 4$
25. $x - 3$ **27.** $y + 13 + \dfrac{50}{y - 5}$ **29.** $2x - 4 + \dfrac{3}{2x + 7}$
31. $2m^2 - 4m + 5 + \dfrac{-10}{m + 5}$ **33.** $x^2 - 2x + 3 + \dfrac{1}{2x + 1}$
35. $x^2 + x + 2 + \dfrac{9}{x - 2}$ **37.** $5x^2 + 2x + 1 + \dfrac{2}{5x - 2}$
39. $x^2 + 4x + 5 + \dfrac{2}{x - 2}$ **41.** $y^3 - 2y^2 + 5y - 10 + \dfrac{4}{y + 2}$
43. $x^3 + 2x^2 + 4x + 8$ **45.** $w^2 - 3w + 9 + \dfrac{-22w - 5}{w^2 + 3w + 1}$
49. $(x + 2)(2x - 3)(x + 4)$ **51.** $(x - 5)(2x + 1)(2x - 3)$
53. 50 **55.** -10 **a.** 1 **b.** -1 **c.** 11 **d.** -20
e. 0 **f.** -19

Section 5.9

1. 14 **3.** -106 **5.** 5 **7.** 3 **9.** 70
11. Quotient: $x + 5$, remainder: 10
13. Quotient: $3x^2 - x - 4$, remainder: 6
15. Quotient: $4x^2 + 12x + 33$, remainder: 101
19. $P(-2) = 0$ **21.** $P(-3) = 0$ **a.** $(x + 3)(x - 3)$
b. $(x - 2)(x + 1)$ **c.** $(2x - 3)(x + 1)$ **d.** $4(x + 2)(x - 2)$
e. $(x + 2)(x^2 - 3)$ **f.** $x(3x + 1)(2x - 3)$

Chapter 5 Self-Test

1. $-6x^3y^4$ **2.** $\dfrac{16m^4n^{10}}{p^6}$ **3.** x^8y^{-10} **4.** $\dfrac{c^2d^7}{2}$
5. 4.23×10^9 **6.** 2.5×10^{-5} **7.** $9x^2 - 2y^2$
8. $-a^2 + 3ab + 7b^2$ **9.** $5x + 12$ **10.** 4
11. $6a^2 - ab - 35b^2$ **12.** $25m^2 - 9n^2$
13. $4a^2 + 12ab + 9b^2$ **14.** $2x^3 - 13x^2 + 26x - 15$
15. $7ab(2ab - 3a + 5b)$ **16.** $(x - 3y)(x + 5)$
17. $(5c - 8d)(5c + 8d)$ **18.** $(3x - 1)(9x^2 + 3x + 1)$
19. $2a(2a + b)(4a^2 - 2ab + b^2)$ **20.** $(x - 8)(x + 6)$
21. $(5x - 2)(2x - 7)$ **22.** $3x(x + 3)(2x - 5)$ **23.** $2x + 5$
24. $3x + 2 - \dfrac{10}{x - 2}$ **25.** $18x^2 + 13$ **26.** $12x + 5$
27. -51 **28.** 18 **29.** Quotient: $3x + 2$, remainder: -3
30. Quotient: $2x^2 + 4x - 7$, remainder: 16

Section 6.1

1. 5 **3.** Never undefined **5.** $\dfrac{1}{2}$ **7.** 0 **9.** -2
11. 0 **13.** 2 **15.** 3 **17.** $\dfrac{2}{3}$ **19.** $\dfrac{2x^3}{3}$ **21.** $\dfrac{2xy^3}{5}$

23. $\dfrac{-3x^2}{y^2}$ 25. $\dfrac{a^3b^2}{3c^2}$ 27. $\dfrac{6}{x+4}$ 29. $\dfrac{x+1}{6}$

31. $\dfrac{x+2}{x+7}$ 33. $3b+1$ 35. $\dfrac{y-z}{y+3z}$

37. $\dfrac{x^2+4x+16}{x+4}$ 39. $\dfrac{(a^2+9)(a-3)}{a+2}$ 41. $\dfrac{y-2}{x+5}$

43. $\dfrac{x+6}{x^2-2}$ 45. $\dfrac{-2}{m+5}$ 47. $\dfrac{-x-7}{2x+1}$ 53. 2

55. 3 57. $2x+h$ a. $\dfrac{8}{15}$ b. $\dfrac{10}{33}$ c. $\dfrac{5}{14}$

d. $\dfrac{3}{14}$ e. $\dfrac{2}{3}$ f. $\dfrac{1}{2}$ g. $\dfrac{9}{5}$ h. $\dfrac{5}{3}$

Section 6.2

1. $\dfrac{2}{x}$ 3. $\dfrac{3}{a^4}$ 5. $\dfrac{5}{12x}$ 7. $\dfrac{16b^3}{3a}$ 9. $5mn$

11. $\dfrac{15x}{2}$ 13. $\dfrac{9b}{8}$ 15. x^2+2x 17. $\dfrac{3(c-2)}{5}$

19. $\dfrac{5x}{2(x-2)}$ 21. $\dfrac{5d}{4(d-3)}$ 23. $\dfrac{x-5}{2x+3}$

25. $\dfrac{2a+1}{2a}$ 27. $\dfrac{-6}{w+2}$ 29. $\dfrac{-a}{6}$ 31. $\dfrac{2}{x}$ 33. $\dfrac{3}{m}$

35. $\dfrac{5}{x}$ 39. $\dfrac{9}{2}$ 41. $-\dfrac{3}{10}$ 43. π 45. 3

47. $\dfrac{2000}{x}$ a. $\dfrac{5}{8}$ b. $\dfrac{1}{3}$ c. $\dfrac{5}{4}$ d. $\dfrac{1}{10}$ e. $\dfrac{29}{24}$

f. $\dfrac{11}{40}$ g. $\dfrac{41}{48}$ h. $\dfrac{23}{30}$

Section 6.3

1. $\dfrac{6}{x^2}$ 3. $\dfrac{7}{3a+7}$ 5. 2 7. $\dfrac{y-1}{2}$ 9. 2

11. $\dfrac{3}{x+2}$ 13. $\dfrac{19}{6x}$ 15. $\dfrac{3(2a+1)}{a^2}$ 17. $\dfrac{2(n-m)}{mn}$

19. $\dfrac{9b+20}{12b^3}$ 21. $\dfrac{a-4}{a(a-2)}$ 23. $\dfrac{5x+7}{(x+1)(x+2)}$

25. $\dfrac{4(y+2)}{(y-3)(y+1)}$ 27. $\dfrac{w(3w-11)}{(w-7)(w-2)}$

29. $\dfrac{7x}{(3x-2)(2x+1)}$ 31. $\dfrac{4}{m-7}$ 33. $\dfrac{2x+11}{(x+4)(x-4)}$

35. $\dfrac{3m+1}{(m-1)(m-2)}$ 37. $\dfrac{15}{y-5}$ 39. $\dfrac{-12}{x-4}$

41. $\dfrac{5z+14}{(z+2)(z-2)(z+4)}$ 45. 1 47. $\dfrac{49}{25}$ 49. 2

51. 8 53. Undefined 55. $\dfrac{3t^2+6t+2}{t(t+1)(t+2)}$ a. 6

b. 14 c. 6 d. $9w$ e. $2n$ f. $3y^2$ g. $5rs$ h. $3b$

Section 6.4

1. $\dfrac{8}{9}$ 3. $\dfrac{14}{5}$ 5. $\dfrac{5}{6}$ 7. $\dfrac{1}{2x}$ 9. $\dfrac{m}{2}$

11. $\dfrac{2(y+1)}{y-1}$ 13. $\dfrac{3b}{a^2}$ 15. $\dfrac{x}{(x+2)(x-3)}$

17. $\dfrac{2x-1}{2x+1}$ 19. $y-x$ 21. $\dfrac{x-y}{y}$ 23. $\dfrac{a+4}{a+3}$

25. $\dfrac{x^2y(x+y)}{x-y}$ 27. $\dfrac{x}{x-2}$ 29. $\dfrac{y+2}{(y-1)(y+4)}$

31. $\dfrac{x}{3}$ 33. 1 35. $\dfrac{2a}{a^2+1}$ 37. $\dfrac{2x+1}{x+1}$

41. $\dfrac{3x+2}{2x+1}$ 43. $\dfrac{5x+3}{3x+2}$ 45. $44\frac{4}{9}$ mi/h

47. $54\frac{6}{11}$ mi/h 49. $\dfrac{x-3}{x(x+3)}$ a. 4 b. 27

c. -25 d. 25 e. 64 f. -32

Chapter 6 Self-Test

1. $\dfrac{-3x^4}{4y^2}$ 2. $\dfrac{w+1}{w-2}$ 3. $\dfrac{x-3}{x+2}$ 4. $\dfrac{x^2+xy+y^2}{x+y}$

5. $\dfrac{4a^2}{7b}$ 6. $\dfrac{m-4}{4}$ 7. $\dfrac{2}{x-1}$ 8. $\dfrac{2x+y}{x(x+3y)}$

9. -5 10. $\dfrac{x-9}{(2x+5)^2}$ 11. $\dfrac{2(2x+1)}{x(x-2)}$ 12. $\dfrac{2}{x-3}$

13. $\dfrac{4}{x-2}$ 14. $\dfrac{8x+17}{(x-4)(x+1)(x+4)}$

15. $\dfrac{3x^2-7}{(x+2)(x+1)(x-3)}$ 16. $\dfrac{-x-19}{(x+3)(x-3)(x-1)}$

17. $\dfrac{y}{3y+x}$ 18. $\dfrac{z-1}{2(z+3)}$ 19. $\dfrac{1}{xy(x-y)}$ 20. -1

Section 7.1

1. 7 3. -6 5. ±9 7. Not a real number 9. 3

11. -4 13. -6 15. 3 17. -2 19. -2

21. Not a real number 23. -3 25. $\dfrac{2}{3}$ 27. $\dfrac{2}{3}$ 29. 6

31. 3 33. 4 35. 3 37. $|x|$ 39. y 41. $3|x|$

43. $|a^2b^3|$ 45. $4x^2$ 47. $|y^5|$ 49. $|m^2n^3|$ 51. $5a$

53. $2xy^3$ 55. 3.873 57. 12.490

59. Not a real number 61. 4.362 63. 2.618 65. -2.466

69. False 71. True 73. False 75. False 77. 7 m

a. $18x^3$ b. $20a^5$ c. $75m^3$ d. $16r^4$ e. $128y^7$

f. $16w^7$ g. $98a^7$ h. $200s^5$

Section 7.2

1. $2\sqrt{3}$ 3. $5\sqrt{2}$ 5. $-6\sqrt{3}$ 7. $2\sqrt{13}$ 9. $2\sqrt{15}$

11. $-5\sqrt{5}$ 13. $12\sqrt{2}$ 15. $15\sqrt{2}$ 17. $2\sqrt[3]{2}$

19. $-2\sqrt[3]{6}$ 21. $3\sqrt[3]{5}$ 23. $2\sqrt[4]{2}$ 25. $3\sqrt[3]{3}$

27. $3z\sqrt{2}$ 29. $3x^2\sqrt{7}$ 31. $7m\sqrt{2m}$ 33. $4xy\sqrt{5y}$

35. $2b\sqrt[3]{5}$ 37. $2p^3\sqrt[3]{6}$ 39. $3m^2\sqrt[3]{2m}$

41. $-2ab\sqrt[3]{3a^2b}$ 43. $2x^2yz\sqrt[3]{7y^2z}$ 45. $2x^2\sqrt[4]{2}$

47. $3a^3\sqrt[4]{3a^3}$ 49. $2wz^3\sqrt[4]{6wz}$ 51. $2w^2\sqrt[5]{2}$ 53. $\dfrac{\sqrt{5}}{4}$

55. $\dfrac{x^2}{5}$ 57. $\dfrac{\sqrt{5}}{3y^2}$ 59. $\dfrac{\sqrt[3]{5}}{2}$ 61. $\dfrac{\sqrt[3]{4x^2}}{3}$ 63. $\dfrac{\sqrt[4]{3}}{3a^2}$

65. $\dfrac{2\sqrt{5}}{5}$ 67. $\dfrac{3\sqrt{10}}{10}$ 69. $\dfrac{\sqrt{10}}{4}$ 71. $\dfrac{\sqrt{42}}{7}$

73. $\dfrac{\sqrt{30}}{5}$ 75. $\dfrac{\sqrt[3]{14}}{2}$ 77. $\dfrac{5\sqrt[3]{4}}{4}$ 79. $\dfrac{\sqrt{3x}}{x}$

81. $\dfrac{2\sqrt{3w}}{w}$ 83. $\dfrac{2m\sqrt{10mn}}{5n}$ 85. $\dfrac{\sqrt[3]{5y^2}}{y}$ 87. $\dfrac{3\sqrt[3]{4x^2}}{2x}$

89. $\dfrac{\sqrt[3]{50x}}{5x}$ 91. $\dfrac{a\sqrt[3]{a^2b^2}}{b^3}$ 95. False 97. True

99. True 101. x^5y^5 103. $\dfrac{\sqrt{16\pi^2-k^2}}{4\pi}$ a. $13a$

b. $6x$ c. $-5y$ d. $20w$ e. x f. b g. $5m+5n$

h. $8r+3s$

Section 7.3
1. $7\sqrt{5}$ **3.** $3\sqrt{7}$ **5.** Cannot be simplified   **7.** $8\sqrt{x}$
9. $3\sqrt{3a}$ **11.** $5x\sqrt{5}$ **13.** Cannot be simplified
15. $9\sqrt{2}$ **17.** $9\sqrt{6}$ **19.** $3\sqrt{5}$ **21.** $8\sqrt{2}$ **23.** $\sqrt{6}$
25. $8\sqrt{3}$ **27.** $5\sqrt{7}$ **29.** $6\sqrt{2}$ **31.** $4\sqrt{3}$ **33.** $5\sqrt[3]{2}$
35. $3\sqrt{5x}$ **37.** $\sqrt{6w}$ **39.** $5x\sqrt{2x}$ **41.** $2a\sqrt{3a}$
43. $\sqrt[3]{3m^2}$ **45.** $\sqrt[4]{2x}$ **47.** $5x\sqrt[3]{2x}$ **49.** $3w\sqrt[3]{2w^2}$
51. $\dfrac{4}{3}\sqrt{3}$ **53.** $\dfrac{2}{3}\sqrt{6}$ **55.** $\dfrac{15}{7}\sqrt{21}$ **57.** $\dfrac{4}{5}\sqrt{5x}$
59. $\dfrac{11}{5}\sqrt{15x}$ **61.** $\dfrac{3}{2}\sqrt[3]{6}$ **63.** $\dfrac{1}{3}\sqrt{3}$
65. $\dfrac{b\sqrt{a}+a\sqrt{b}}{ab}$ **69.** $14x^2\sqrt[3]{x}$ **71.** $5a\sqrt[4]{a}$ **73.** False
75. True   **77.** False   **79.** $45\sqrt{10}$ **a.** $3a+18$
b. $5x-20$ **c.** y^2-7y **d.** m^2+9m **e.** s^2-9
f. x^2-25 **g.** $w^2-2wz+z^2$ **h.** $m^2+16m+64$

Section 7.4
1. $\sqrt{42}$ **3.** $\sqrt{11a}$ **5.** $\sqrt{91x}$ **7.** $\sqrt{33ab}$ **9.** $\sqrt{42}$
11. $\sqrt[3]{36}$ **13.** $\sqrt[3]{45a^2}$ **15.** $\sqrt[4]{56}$ **17.** 6   **19.** 7
21. $5x\sqrt{3}$ **23.** $4x\sqrt{2x}$ **25.** $5\sqrt{3}$ **27.** $2\sqrt[3]{3}$
29. $3p\sqrt[3]{2}$ **31.** $2xy\sqrt[3]{5y}$ **33.** $2\sqrt[4]{3}$ **35.** $3a\sqrt[4]{2a}$
37. $\sqrt{6}+5\sqrt{2}$ **39.** $\sqrt{15}-\sqrt{10}$ **41.** $3\sqrt{2}+\sqrt{15}$
43. $\sqrt{10}+2\sqrt{15}$ **45.** $2\sqrt{6}$ **47.** $4x\sqrt{3}$
49. $a+a\sqrt{b}$ **51.** $6\sqrt[3]{2}$ **53.** $x\sqrt[3]{y^2}-y\sqrt[3]{x^2}$
55. $-10-\sqrt{2}$ **57.** $13-7\sqrt{3}$
59. $5+\sqrt{10}-\sqrt{15}-\sqrt{6}$ **61.** $x-\sqrt{x}-6$
63. $-28+\sqrt{10}$ **65.** -4 **67.** 3   **69.** 3   **71.** $x-9$
73. $a-3$ **75.** $28-10\sqrt{3}$ **77.** $a+6\sqrt{a}+9$
79. $x+2\sqrt{xy}+y$ **81.** $\dfrac{\sqrt{21}}{7}$ **83.** $\dfrac{\sqrt{6ab}}{3b}$ **85.** $\dfrac{3\sqrt[3]{2}}{2}$
87. $\dfrac{\sqrt{3x}}{2x}$ **89.** $2-\sqrt{3}$ **91.** $6+2\sqrt{5}$
93. $3\sqrt{6}+3\sqrt{2}$ **95.** $\dfrac{5-\sqrt{15}}{2}$ **97.** $\dfrac{11-4\sqrt{7}}{3}$
99. $3+2\sqrt{2}$ **101.** $\dfrac{a-2\sqrt{a}}{a-4}$ **103.** $\dfrac{w+6\sqrt{w}+9}{w-9}$
105. $\dfrac{x-2\sqrt{xy}+y}{x-y}$ **109.** $x\sqrt{x+1}$ **111.** $x(x-2)$
113. $\sqrt{2x+3}$ **115.** $\dfrac{\sqrt{x^2+3x+2}}{x+1}$ **117.** $\sqrt{2x-1}$
119. $\dfrac{\sqrt{x^2+5x+6}}{x+2}$ **121.** $\dfrac{\sqrt{2xz}-\sqrt{2yz}}{2z}$ **a.** 4   **b.** 3
c. -5 **d.** 27   **e.** 16   **f.** 8   **g.** $\dfrac{1}{32}$ **h.** $\dfrac{1}{16}$

Section 7.5
1. 6   **3.** -5 **5.** Not a real number   **7.** 3   **9.** 3
11. $\dfrac{2}{3}$ **13.** 9   **15.** 16   **17.** 4   **19.** 729   **21.** $\dfrac{4}{9}$
23. $\dfrac{1}{5}$ **25.** $\dfrac{1}{3}$ **27.** $\dfrac{1}{27}$ **29.** $\dfrac{1}{32}$ **31.** $\dfrac{5}{2}$ **33.** x
35. $y^{4/5}$ **37.** $b^{13/6}$ **39.** $x^{1/3}$ **41.** s **43.** $w^{3/4}$
45. x **47.** $a^{3/5}$ **49.** $\dfrac{1}{y^6}$ **51.** a^4b^9 **53.** $32xy^3$
55. $st^{1/3}$ **57.** $4pq^{5/3}$ **59.** $x^{2/5}y^{1/2}z$ **61.** $a^{1/2}b^{1/4}$
63. $\dfrac{s^2}{r^4}$ **65.** $\dfrac{x^3}{y^2}$ **67.** $\dfrac{1}{mn^2}$ **69.** $\dfrac{s^3}{r^2t}$ **71.** $\dfrac{2xz^3}{y^2}$

73. $\dfrac{2n}{m^{1/5}}$ **75.** $xy^{3/4}$ **77.** $\sqrt[4]{a^3}$ **79.** $2\sqrt[3]{x^2}$
81. $3\sqrt[5]{x^2}$ **83.** $\sqrt[5]{9x^2}$ **85.** $(7a)^{1/2}$ **87.** $2m^2n^3$
89. 9.946   **91.** 0.370   **95.** $a^2+a^{5/4}$ **97.** $a-4$
99. $m-n$ **101.** $x+4x^{1/2}+4$ **103.** $r+2r^{1/2}s^{1/2}+s$
105. $(x^{1/3}+1)(x^{1/3}+3)$ **107.** $(a^{2/5}-3)(a^{2/5}-4)$
109. $(x^{2/3}-2)(x^{2/3}+2)$ **111.** x^{5n} **113.** y^{4n} **115.** r^2
117. $a^{6n}b^{4n}$ **119.** x **121.** $\sqrt[4]{x}$ **123.** $\sqrt[8]{y}$
125. 2×10^4 **127.** 2×10^{-3} **129.** 4×10^{-4}
131. 40   **a.** $6x+10x^2$ **b.** $-6x+15x^2$
c. $9+3x-2x^2$ **d.** $15-26w+8w^2$ **e.** $49-4a^2$
f. $25-9m^2$ **g.** $25+10y+y^2$ **h.** $16-24r+9r^2$

Section 7.6
1. $4i$ **3.** $-8i$ **5.** $i\sqrt{21}$ **7.** $2i\sqrt{3}$ **9.** $-6i\sqrt{3}$
11. $8+3i$ **13.** $1+5i$ **15.** $2+2i$ **17.** $5-7i$
19. $7+11i$ **21.** $3+11i$ **23.** $6-3i$ **25.** $0+0i$
27. $-15+9i$ **29.** $28+12i$ **31.** $-6-8i$
33. $-5+4i$ **35.** $13i$ **37.** $23+14i$ **39.** $18+i$
41. $21-20i$ **43.** $3+2i$, 13   **45.** $2-3i$, 13
47. $-3+2i$, 13   **49.** $-5i$, 25   **51.** $2-3i$
53. $-2-3i$ **55.** $\dfrac{6}{29}-\dfrac{15}{29}i$ **57.** $2-3i$
59. $\dfrac{17}{25}+\dfrac{6}{25}i$ **61.** $-\dfrac{7}{25}-\dfrac{24}{25}i$ **65.** $-\sqrt{35}$ **67.** -6
69. $-3\sqrt{10}$ **71.** -10 **73.** -1 **75.** 1   **77.** -1
79. $-i$ **a.** $x(x-3)$ **b.** $2x(x^2-2)$ **c.** $(2x-3)(x+1)$
d. $(x-3)(x-3)$ **e.** $(2x+1)(x-1)$ **f.** $(x+4)(x-4)$

Chapter 7 Self-Test
1. $7a^2$ **2.** $-3w^2z^3$ **3.** $p^2q\sqrt[3]{9pq^2}$ **4.** $\dfrac{7x}{8y}$
5. $\dfrac{\sqrt{10xy}}{4y}$ **6.** $\dfrac{\sqrt[3]{3x^2}}{x}$ **7.** $3x\sqrt{3x}$ **8.** $5m\sqrt[3]{2m}$
9. $4x\sqrt{3}$ **10.** $3-2\sqrt{2}$ **11.** $\dfrac{1}{64x^6}$ **12.** $\dfrac{9m}{n^4}$
13. $\dfrac{8s^3}{r}$ **14.** $a^2b\sqrt[5]{a^4b}$ **15.** $5p^3q^2$ **16.** $3+10i$
17. $-14+22i$ **18.** $5-5i$ **19.** $\dfrac{1}{125}$ **20.** 8

Section 8.1
1. $-1,-3$ **3.** $-3,5$ **5.** $5,6$ **7.** $-3,7$ **9.** $-5,10$
11. $-5,7$ **13.** $0,8$ **15.** $0,-10$ **17.** $0,5$
19. $-5,5$ **21.** $-8,8$ **23.** $-\dfrac{3}{2}$ **25.** $4,\dfrac{9}{2}$
27. $-1,\dfrac{4}{3}$ **29.** $\dfrac{1}{2},\dfrac{2}{3}$ **31.** $-3,9$ **33.** $0,6$
35. $0,3$ **37.** $-3,5$ **39.** $-\dfrac{3}{2},3$ **41.** $-\dfrac{7}{3},2$
43. $-2,6$ **45.** $-\dfrac{3}{2},5$ **47.** $2,6$ **49.** $-5,1$
53. $x^2+x-6=0$ **55.** $x^2-8x+12=0$ **57.** $0,-2,5$
59. $0,-3,3$ **61.** $-1,-2,2$ **63.** $1,-1,3,-3$
65. $0,100$ cm   **67.** $0,132$ cm   **69.** Too much water
71. The x intercepts are the same as the solutions to number 48.
a. 5   **b.** 11   **c.** $4\sqrt{2}$ **d.** $5\sqrt{2}$ **e.** $1+\sqrt{3}$
f. $1+\sqrt{5}$

Section 8.2

1. $-5, -1$ **3.** $-5, 7$ **5.** $-\dfrac{1}{2}, 3$ **7.** $-\dfrac{1}{2}, \dfrac{2}{3}$ **9.** $-6, 6$

11. $-\sqrt{7}, \sqrt{7}$ **13.** $-\sqrt{6}, \sqrt{6}$ **15.** $-2i, 2i$

17. $-1 \pm 2\sqrt{3}$ **19.** $\dfrac{-1 \pm \sqrt{3}}{2}$ **21.** 36 **23.** 16

25. $\dfrac{9}{4}$ **27.** $\dfrac{1}{4}$ **29.** $\dfrac{1}{16}$ **31.** a^2 **33.** $-6 \pm \sqrt{38}$

35. $-2, 4$ **37.** $1 \pm \sqrt{6}$ **39.** $-5 \pm 2\sqrt{3}$

41. $\dfrac{5 \pm \sqrt{53}}{2}$ **43.** $\dfrac{1 \pm \sqrt{13}}{2}$ **45.** $\dfrac{-1 \pm \sqrt{17}}{4}$

47. $\dfrac{-1 \pm \sqrt{3}}{2}$ **49.** $\dfrac{3 \pm \sqrt{15}}{3}$ **51.** $\dfrac{1 \pm i\sqrt{35}}{3}$

53. $-4 \pm 2i$ **57.** a^2 **59.** $\dfrac{9}{4}a^2$ **61.** 1

63. $-a \pm \sqrt{4 + a^2}$

65. The x intercepts match the solutions to number 33
67. The x intercepts match the solutions to number 35
a. -23 **b.** -23 **c.** 17 **d.** 0 **e.** 20 **f.** -7

Section 8.3

1. $-2, 7$ **3.** $-13, 5$ **5.** $-1, \dfrac{1}{5}$ **7.** $\dfrac{3}{4}$ **9.** $1 \pm \sqrt{6}$

11. $\dfrac{-3 \pm 3\sqrt{13}}{2}$ **13.** $\dfrac{3 \pm \sqrt{15}}{2}$ **15.** $\dfrac{2 \pm \sqrt{2}}{2}$ **17.** $-\dfrac{2}{3}, 1$

19. $\dfrac{1 \pm \sqrt{41}}{4}$ **21.** $1, 3$ **23.** 4 **25.** $\dfrac{1 \pm \sqrt{7}}{2}$

27. $\dfrac{1 \pm i\sqrt{2}}{3}$ **29.** $\dfrac{1 \pm \sqrt{161}}{8}$ **31.** $\dfrac{3 \pm \sqrt{39}}{15}$

33. $\dfrac{-1 \pm \sqrt{41}}{2}$ **35.** $\dfrac{1 \pm \sqrt{23}}{2}$ **37.** $\dfrac{5 \pm \sqrt{37}}{6}$

39. $\dfrac{3 \pm \sqrt{33}}{4}$ **41.** $1 \pm \sqrt{6}$ **43.** 25, two **45.** 0, one

47. 37, two **49.** -63, none **51.** 4 **53.** $\dfrac{7 \pm \sqrt{37}}{6}$

55. $0, \dfrac{2}{5}$ **57.** $\dfrac{-5 \pm \sqrt{33}}{4}$ **59.** $\pm 3i$ **61.** $-2, 5$

63. $\dfrac{7 \pm \sqrt{21}}{2}, \approx 1.2$ or 5.8 s **65.** $-1 + \sqrt{6}, \approx 1.4$ s

69. $\pm\sqrt{z^2 - y^2}$ **71.** $\pm 6a$ **73.** $\dfrac{a}{2}, -3a$

75. $\dfrac{-a \pm a\sqrt{17}}{4}$ **77.** $-1 \pm \sqrt{6}$ **79.** $-1, \dfrac{1 \pm i\sqrt{3}}{2}$

81. $-1, 1, \dfrac{-1 \pm i\sqrt{3}}{2}, \dfrac{1 \pm i\sqrt{3}}{2}$ **83.** 55.9 or 84.1 cm

85. $x \approx 3.4, x \approx -1.4$ **87.** $x \approx -6.9, x \approx 3.9$

a. $x > 2$

b. $x > 3$

c. $x < -4$

d. $x \le -5$

e. $1 < x < 5$

f. $-2 \le x \le 3$

g. $x < -3$ or $x > 2$

h. $x \le -4$ or $x \ge 5$

Section 8.4

1. $-4 < x < 3$

3. $x < -4$ or $x > 3$

5. $-4 \le x \le 3$

7. $x \le -4$ or $x \ge 3$

9. $x < -1$ or $x > 4$

11. $-4 \le x \le 3$

13. $x \le 2$ or $x \ge 3$

15. $-6 \le x \le 4$

17. $x < -9$ or $x > 3$

19. $-2 \le x \le \dfrac{3}{2}$

21. $-1 < x < \dfrac{3}{4}$

23. $-4 \le x \le 4$

25. $x \le -5$ or $x \ge 5$

27. $x < -2$ or $x > 2$

29. $0 \le x \le 4$

31. $x \le 0$ or $x \ge 6$

33. $0 < x < 4$

35. $x = 2$

37. $-4 \le x \le 7$

41. $x < -1$ or $0 < x < 2$

43. $-2 \le x \le 1$ or $x \ge 3$

45. $x \le -3$ or $0 \le x \le 5$

47. $50 \le x \le 60$ **49.** $2 \le t \le 3$ **a.** 10, 14   **b.** 7, 19
c. 31, 33   **d.** 22, 24, 26   **e.** 9 by 14 cm   **f.** 6 by 15 in

Section 8.5

1. 5, 13   **3.** 4, 6   **5.** -9, -8 or 8, 9   **7.** 5, 6   **9.** 4, 5
11. -5, -4, -3 or 3, 4, 5   **13.** 4   **15.** 7 by 10 ft
17. 5 by 17 cm   **19.** 5 by 6 cm
21. $\dfrac{6\sqrt{5}}{5}$, $\dfrac{12\sqrt{5}}{5}$, or 2.7, 5.4 m
23. $2 + 2\sqrt{3}$, $3 + 2\sqrt{3}$, $5 + 2\sqrt{3}$, or 5.5, 6.5, 8.5 in
25. $\dfrac{25 \pm 7\sqrt{7}}{2}$ or 3.2, 21.8 m
27. $15 + 5\sqrt{11}$, $30 + 10\sqrt{11}$, or 31.6 by 63.2 cm
29. 4 by 8 ft   **31.** $10 + 4\sqrt{5}$ or 18.9 cm
33. $\dfrac{-25 + 5\sqrt{41}}{2}$ or 3.5 ft   **35.** $-8 + \sqrt{104}$ or 2.2 ft
37. $\dfrac{-65 + 5\sqrt{193}}{4}$ or 1.1 m   **39.** $-25 + 5\sqrt{31}$ or 2.8 cm
41. $\dfrac{9 - \sqrt{61}}{2}$ or 0.6 m   **43.** $6 - \sqrt{29}$ or 0.6 m

45. 9, 12, 15 cm   **47.** 5 by 12 in   **49.** $\dfrac{5\sqrt{2}}{2}$ or 3.5 cm
51. $3 + 3\sqrt{2}$ or 7.2 in   **53.** (a) 4 s, (b) 1 s   **55.** 50
57. $2 + \sqrt{6}$ or 4.4 s   **59.** $3 + \sqrt{11}$ or 6.3 s
61. $h(t) = 100 + 20t - 5t^2$ **63.** 5 s
65. $h(t) = 100 + 20t - 16t^2$ **67.** $\dfrac{5 + \sqrt{105}}{8}$ or 1.9 s
69. $45 \pm 5\sqrt{13}$ or 63, 27   **71.** $9 - \sqrt{65}$ or $0.94
a. $2x + 12$ **b.** $3x + 2$ **c.** $2x^3 + 6x$ **d.** $x^2 + 6x$
e. $-7x + 21$ **f.** $2x + 2$

Chapter 8 Self-Test

1. -3, 12   **2.** $-\dfrac{5}{3}$, $\dfrac{5}{3}$ **3.** -3, $-\dfrac{1}{2}$ **4.** $-\dfrac{5}{2}$, $\dfrac{2}{3}$
5. 0, $+\dfrac{3}{2}$, $-\dfrac{3}{2}$ **6.** $\pm\sqrt{5}$ **7.** $1 \pm \sqrt{10}$ **8.** $\dfrac{2 \pm \sqrt{23}}{2}$
9. $\dfrac{-3 \pm \sqrt{13}}{2}$ **10.** $\dfrac{5 \pm \sqrt{19}}{2}$ **11.** $\dfrac{5 \pm \sqrt{37}}{2}$
12. $-2 \pm \sqrt{11}$ **13.** $-\dfrac{2}{3}$, 4   **14.** $-7 < x < 2$
15. $x \le -3$ or $x \ge 6$ **16.** $-2 \le x \le \dfrac{5}{3}$ **17.** 7, 9
18. Width 5 cm, length 17 cm   **19.** 4.4, 5.4 in   **20.** 2.7 s

Section 9.1

1. Equation, 36   **3.** Expression, $\dfrac{3x}{10}$ **5.** Equation, 5
7. Equation, 3   **9.** 5   **11.** 8   **13.** $\dfrac{3}{2}$ **15.** -1
17. $-\dfrac{9}{5}$ **19.** 2   **21.** No solution   **23.** -23 **25.** 6
27. 4   **29.** 8   **31.** 4   **33.** $\dfrac{3}{2}$ **35.** No solution
37. 7   **39.** 5   **41.** 2, $-\dfrac{3}{2}$ **43.** 5, 6   **45.** 4, 6

47. $-\dfrac{1}{2}$, 6   **49.** $-\dfrac{1}{2}$ **51.** $\dfrac{ab}{b - a}$ **53.** $\dfrac{RR_2}{R_2 - R}$
55. $\dfrac{y + 1}{y - 1}$ **57.** $\dfrac{A}{1 + rt}$ **59.** $-1 < x < 2$
61. $x < 2$ or $x > 4$ **63.** $-3 < x \le 5$ **65.** $x < -3$ or $x \ge \dfrac{1}{2}$
67. $-2 \le x < 3$ **69.** $-3 < x \le 4$ **71.** $x < 2$ or $x > 3$
73. 24   **75.** $\dfrac{25}{7}$ **77.** $\dfrac{3}{4}$ **79.** 3, 13   **83.** 60 mi/h
85. $\dfrac{FF_2 + FF_1 - F_1F_2}{F}$ **a.** 7, 12   **b.** -2, 1   **c.** 11, 12
d. 9, 11   **e.** 4 mi/h, 6 mi/h   **f.** 2:30 P.M.

Section 9.2

1. 8, 24   **3.** 5, 6   **5.** 10   **7.** 3   **9.** 5   **11.** 6, 8
13. 4, 5   **15.** 4 mi/h   **17.** 150 mi/h
19. Bicycling 25 mi/h, driving 55 mi/h   **21.** 3 h, 2 h
23. Bus 50 mi/h, train 70 mi/h   **25.** 6 h   **27.** 15 h   **29.** 30 h
31. 6 h   **33.** 12 min, 24 min   **35.** 4 h   **37.** 80 years
39. The car, 27 mpg; the truck, 21 mpg   **41.** 15 and 9 gal
a. 5   **b.** 6   **c.** 8   **d.** 5   **e.** 3   **f.** 13   **g.** 15
h. 4

Section 9.3

1. 4   **3.** $\dfrac{1}{4}$ **5.** 4   **7.** 6   **9.** No solution   **11.** 1
13. All real numbers   **15.** $\dfrac{7}{2}$ **17.** $\dfrac{7}{3}$ **19.** 3   **21.** 4
23. 6   **25.** 7   **27.** 0, $\dfrac{1}{2}$ **29.** 32   **31.** ± 3 **33.** 1
35. 1   **37.** $-\dfrac{7}{4}$ **39.** 3, 7   **41.** 0   **43.** 6
45. -15, 3   **47.** 7   **49.** 16   **51.** $x \ge 1$ **55.** $\dfrac{h^2}{p}$
57. $\dfrac{v^2}{2g}$ **59.** $2\pi r^2$ **61.** $\dfrac{\pi hr^2}{2}$
63. $1 \pm \sqrt{d^2 - (y - 2)^2}$ **65.** 1.88 s   **67.** $\dfrac{t^2g}{4\pi^2}$ **69.** 9
71. 16   **a.** $(x - 2)(x - 3)$ **b.** $(x - 5)(x + 3)$
c. $(x^2 - 2)(x^2 - 3)$ **d.** $(x^2 - 5)(x^2 + 3)$
e. $(x - 4)(x - 9)$ **f.** $(x - 1)(x + 5)$
g. $(x + 2)(x - 2)(x + 3)(x - 3)$ **h.** $(x + 1)(x - 1)(x^2 + 5)$

Section 9.4

1. $\pm\sqrt{5}$, ± 2 **3.** $\pm\sqrt{3}$, $\pm 2i$ **5.** $\dfrac{\pm\sqrt{2}}{2}$, ± 2 **7.** $\pm\sqrt{2}$
9. $\dfrac{\pm\sqrt{6}}{3}$, $\pm i\sqrt{6}$ **11.** $\pm\sqrt{5}$, $\pm i\sqrt{7}$ **13.** 0, $\pm\sqrt{5}$
15. ± 2, $\pm\sqrt{5}$ **17.** $\dfrac{\pm\sqrt{10}}{2}$, $\pm i\sqrt{3}$ **19.** 16, 256
21. 81   **23.** 7, 0   **25.** -1, $-1 \pm i$ **27.** 36   **29.** 0, 8
31. ± 2, $\dfrac{\pm i\sqrt{6}}{2}$ **35.** 1, 81   **37.** 16   **39.** -27, 1
41. $\dfrac{1}{4}$, $\dfrac{1}{2}$ **43.** -3, $\dfrac{1}{2}$ **45.** $\dfrac{5}{3}$, $\dfrac{5}{2}$ **47.** 16   **49.** 2
a. 6   **b.** 30   **c.** 12   **d.** 10   **e.** $\dfrac{48}{5}$ **f.** 72

Section 9.5

1. $s = kx^2$ **3.** $r = \dfrac{k}{s}$ **5.** $V = \dfrac{kT}{P}$ **7.** $V = khr^2$

9. $w = \dfrac{kxy}{z^2}$ **11.** 9   **13.** 80   **15.** 5   **17.** 20

19. 8   **21.** 96   **23.** 96   **25.** 6   **27.** 15 lb
29. 22 A   **31.** 32π cm^3 **33.** 180 ft   **35.** 75 mi
37. 1 Ω **a.** 5   **b.** -29 **c.** 0   **d.** 1   **e.** -9 **f.** 11

Chapter 9 Self-Test

1. 9   **2.** $-\dfrac{1}{2}$, 4

3. $-4 \le x < 3$

4. $-3 < x \le -1$

5. 3, 6   **6.** 50 mi/h, 45 mi/h   **7.** 11   **8.** 4   **9.** 2, 11
10. 2   **11.** $\pm\sqrt{3}$, ± 3 **12.** 4, 81   **13.** -3, 5   **14.** 3, -2
15. $\dfrac{ip}{p - i}$ **16.** $\dfrac{2x}{3 - 5x}$ **17.** 2   **18.** 75   **19.** 12
20. 500 lb/ft^2

Section 10.1

1. (a)   **3.** (c)   **5.** (b)   **7.** (d)   **9.** Down   **11.** Left
13. Right   **15.** $x = 0$, $(0, -5)$ **17.** $x = -3$, $(-3, 0)$
19. $x = 3$, $(3, 2)$ **21.** $x = -3$, $(-3, -5)$ **23.** $y = 0$, $(-4, 0)$
25. $y = 0$, $(1, 0)$ **27.** $y = 0$, $(3, 0)$ **29.** $y = 3$, $(1, 3)$
33. $d_1 = x + p$, $d_2 = \sqrt{(x - p)^2 + y^2}$, set $d_1 = d_2$, and simplify
35. The y values must include -5 and 7. The x values should
include -4 and 0.   **37.** The y values must include -13 and 5.
The x values should include -5 and 0.   **a.** 4   **b.** 9
c. 25   **d.** 16   **e.** 100   **f.** 36   **g.** 64   **h.** 144

Section 10.2

1. (d)   **3.** (b)   **5.** (a)   **7.** (e)   **9.** (a), (c)
11. (a), (c)   **13.** (b), (c)
15. $f(x) = x^2 - 2x$; $x = 1$

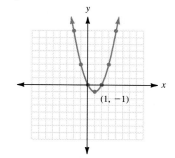

17. $f(x) = -x^2 + 1$; $x = 0$

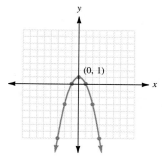

19. $f(x) = -x^2 - 4x$; $x = -2$

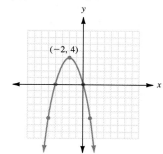

21. $f(x) = x^2 - 2x - 3$; $x = 1$

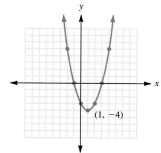

23. $f(x) = x^2 - 5x + 4$; $x = \dfrac{5}{2}$

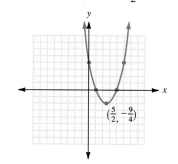

25. $f(x) = -x^2 + 3x + 4$; $x = \dfrac{3}{2}$

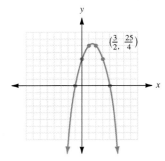

$\left(\dfrac{3}{2}, \dfrac{25}{4}\right)$

27. $f(x) = -x^2 + 6x - 5$; $x = 3$

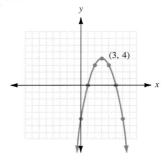

$(3, 4)$

29. $f(x) = x^2 - 2x - 2$; $x = 1$

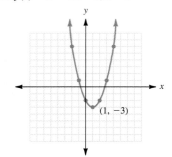

$(1, -3)$

31. $f(x) = x^2 + 4x + 5$; $x = -2$

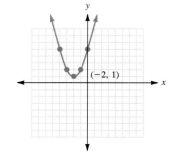

$(-2, 1)$

33. $f(x) = -x^2 + 6x - 2$; $x = 3$

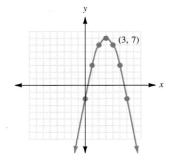

$(3, 7)$

35. $f(x) = 2x^2 - 4x + 1$; $x = 1$

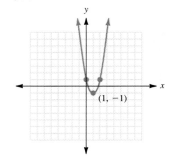

$(1, -1)$

37. $f(x) = -\dfrac{1}{3}x^2 + x - 2$; $x = \dfrac{3}{2}$

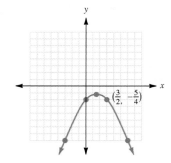

$\left(\dfrac{3}{2}, -\dfrac{5}{4}\right)$

39. $f(x) = 3x^2 + 6x - 1$; $x = -1$

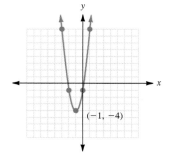

$(-1, -4)$

41. $x = y^2 - 4y$; $y = 2$

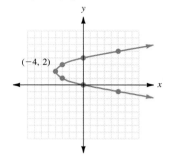

$(-4, 2)$

21. $x^2 + y^2 = 4$

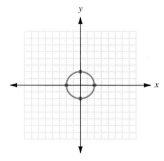

43. $x = y^2 - 3y - 4$; $y = \dfrac{3}{2}$

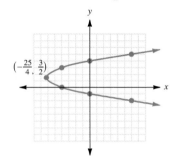

$\left(-\dfrac{25}{4}, \dfrac{3}{2}\right)$

23. $4x^2 + 4y^2 = 36$

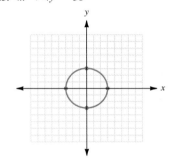

45. 100 items, \$1500   **47.** 400 by 400 ft, 160,000 ft^2
49. 64 ft
53. $y \geq -2(x - 3)^2 + 2$

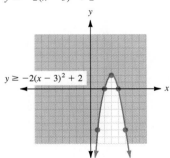

$y \geq -2(x - 3)^2 + 2$

25. $(x - 1)^2 + y^2 = 9$

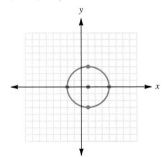

55. 20 cm   **57.** 100 cm   **59.** The y values must include -25 and 0. The x values should include -3 and 3.   **61.** The y values must include -7 and $-\dfrac{17}{4}$. The x values must include $\dfrac{5}{4}$.
a. $\sqrt{5}$ **b.** $\sqrt{2}$ **c.** 4   **d.** 4   **e.** $\sqrt{17}$ **f.** $5\sqrt{2}$

27. $(x - 4)^2 + (y + 1)^2 = 16$

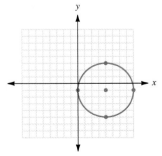

Section 10.3
1. Parabola   **3.** Line   **5.** Circle   **7.** Circle   **9.** None
11. Parabola   **13.** $(0, 0)$, $r = 5$ **15.** $(3, -1)$, $r = 4$
17. $(-1, 0)$, $r = 4$ **19.** $(3, -4)$, $r = \sqrt{41}$

29. $x^2 + y^2 - 4y = 12$

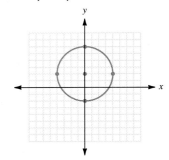

31. $x^2 - 4x + y^2 + 2y = -1$

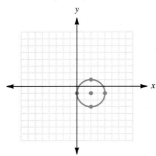

35.

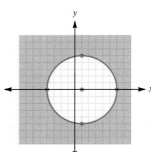

37.

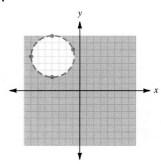

39. $10\sqrt{5} \approx 22.4$ cm   **41.** $x^2 + y^2 = \dfrac{4}{9}$

43.

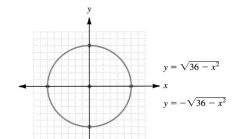

$y = \sqrt{36 - x^2}$

$y = -\sqrt{36 - x^2}$

45.

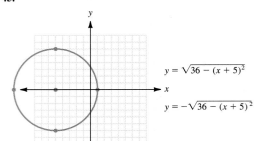

$y = \sqrt{36 - (x + 5)^2}$

$y = -\sqrt{36 - (x + 5)^2}$

a. ± 5 **b.** ± 3 **c.** ± 2 **d.** ± 3 **e.** ± 4 **f.** ± 7

Section 10.4
1. (d)   **3.** (h)   **5.** (b)   **7.** (c)   **9.** Circle
11. Parabola   **13.** Hyperbola   **15.** Hyperbola
17. Circle   **19.** Hyperbola
21. $\dfrac{x^2}{4} + \dfrac{y^2}{9} = 1$

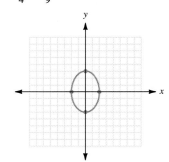

23. $\dfrac{x^2}{9} + \dfrac{y^2}{25} = 1$

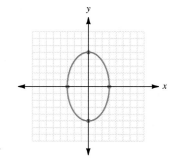

25. $x^2 + 9y^2 = 36$

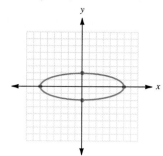

27. $4x^2 + 9y^2 = 36$

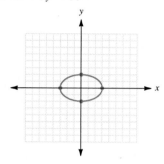

29. $4x^2 + 25y^2 = 100$

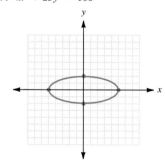

31. $25x^2 + 9y^2 = 225$

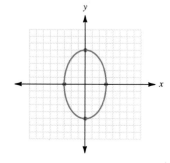

33. $\dfrac{x^2}{9} - \dfrac{y^2}{9} = 1$

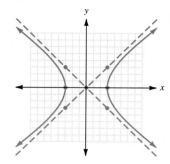

35. $\dfrac{y^2}{16} - \dfrac{x^2}{9} = 1$

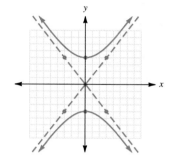

37. $\dfrac{x^2}{36} - \dfrac{y^2}{9} = 1$

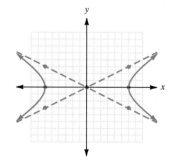

39. $x^2 - 9y^2 = 36$

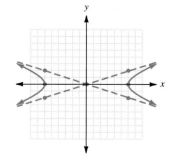

41. $9x^2 - 4y^2 = 36$

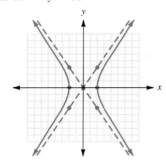

43. $16y^2 - 9x^2 = 144$

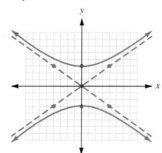

47. $y = \dfrac{1}{2}\sqrt{36 - 9x^2}$ **49.** $y = \dfrac{1}{5}\sqrt{100 - 4x^2}$

$y = -\dfrac{1}{2}\sqrt{36 - 9x^2}$ $y = -\dfrac{1}{5}\sqrt{100 - 4x^2}$

a. $\dfrac{y + 5}{3}$ **b.** $\dfrac{y - 7}{2}$ **c.** $4y + 12$ **d.** $3y - 6$

e. $\sqrt{y - 2}$ **f.** $\sqrt{\dfrac{y + 1}{2}}$

Chapter 10 Self-Test

1. $x = 0$, $(0, 0)$ **2.** $x = 3$, $(3, 0)$ **3.** $x = -2$, $(-2, -5)$

4. $x = -2$, $(-2, 1)$ **5.** $x = 2$, $(2, -9)$ **6.** $x = \dfrac{3}{2}$,

$\left(\dfrac{3}{2}, \dfrac{3}{2}\right)$ **7.** $y = 3$, $(-2, 3)$ **8.** $y = 3$, $(-7, 3)$

9. $f(x) = (x - 5)^2$

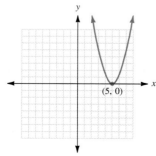

10. $f(x) = (x + 2)^2 - 3$

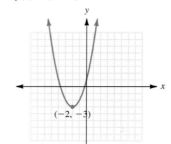

11.

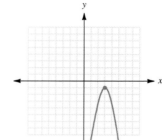

12.

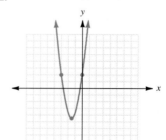

13.

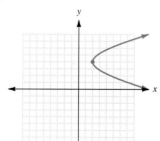

14.

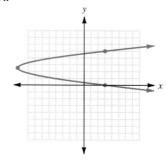

15. $(3, -2)$, $r = 6$ **16.** $(1, -2)$, $r = \sqrt{26}$

17.

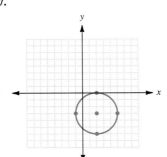

18.

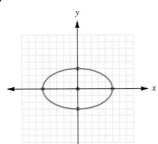

19.

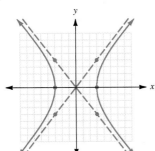

20.

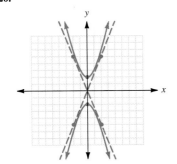

Section 11.1

1. $\{(3, 2), (4, 3), (5, 4)\}$, function

3. $\{(2, 1), (2, 2), (2, 3)\}$, not a function

5. $\{(4, 2), (9, 3), (16, 4)\}$, function   **7.** $y = \dfrac{1}{2}x - 4$

9. $y = 2x + 1$ **11.** $y = \pm\sqrt{x + 1}$ or $x = y^2 - 1$

13. $4x^2 + y^2 = 36$ **15.** $y^2 - x^2 = 9$

17.

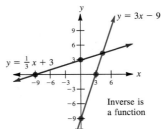

19.

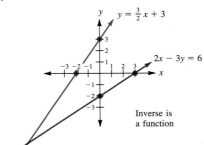

21.

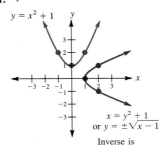

25. 12   **27.** 6   **29.** x **31.** 2   **33.** 3   **35.** x

37. 16   **39.** 4   **41.** x **a.** $y = x - 2$ **b.** $s + t = 5$

c. $x = 5y$ **d.** $w = \dfrac{1}{5}l$ **e.** $C = \pi d$ **f.** $A = lw$

g. $r = \dfrac{d}{t}$ **h.** $R = \dfrac{I}{Pt}$

Section 11.2

1. (c) **3.** (b) **5.** (h) **7.** (f) **9.** 1   **11.** 16

13. 4   **15.** 64   **17.** 2   **19.** 17   **21.** $\dfrac{1}{4}$ **23.** 16

25. $y = 4^x$

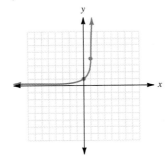

27. $y = \left(\dfrac{2}{3}\right)^x$

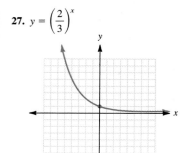

29. $y = 3 \cdot 2^x$

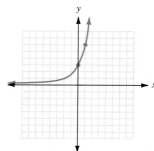

31. $y = 3^x$

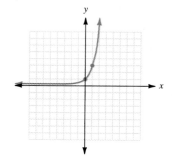

33. $y = 2^{2x}$

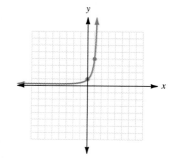

35. $y = e^{-x}$

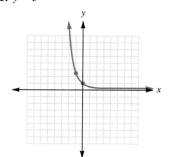

37. 5   **39.** 4   **41.** -2 **43.** 3   **45.** 5   **47.** -2
49. 400   **51.** 3200   **53.** 32 g   **55.** 8 g   **57.** \$1166.40
59. \$1999   **61.** (a) 36, (b) 56, (c) 67   **65.** 2.7048
67. 2.71815   **69.** 2.71828   **71.** See text for graph of $y = 2^x$.

a. $f^{-1} = \left\{(x, y) \,\middle|\, y = \dfrac{x}{2} - \dfrac{1}{2}\right\}$

b–c.

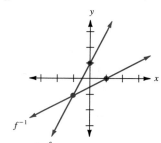

d. $g^{-1} = \left\{(x, y) \,\middle|\, y = \dfrac{x}{3} + \dfrac{2}{3}\right\}$

e–f.

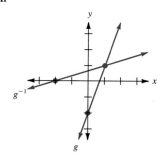

Section 11.3
1. $y = \log_4 x$

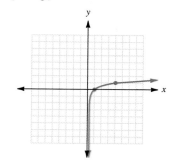

3. $y = \log_2 (x - 1)$

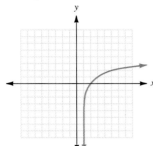

5. $y \log_e x$

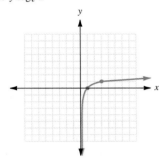

7. $\log_2 16 = 4$ **9.** $\log_{10} 100 = 2$ **11.** $\log_3 1 = 0$

13. $\log_4 \dfrac{1}{16} = -2$ **15.** $\log_{10} \dfrac{1}{1000} = -3$

17. $\log_{16} 4 = \dfrac{1}{2}$ **19.** $\log_{64} \dfrac{1}{4} = -\dfrac{1}{3}$ **21.** $\log_8 4 = \dfrac{2}{3}$

23. $\log_{27} \dfrac{1}{9} = -\dfrac{2}{3}$ **25.** $2^4 = 16$ **27.** $5^0 = 1$

29. $10^1 = 10$ **31.** $5^3 = 125$ **33.** $3^{-3} = \dfrac{1}{27}$

35. $10^{-2} = 0.01$ **37.** $16^{1/2} = 4$ **39.** $8^{2/3} = 4$

41. $25^{-1/2} = \dfrac{1}{5}$ **43.** 5 **45.** 3 **47.** -4 **49.** -2

51. $\dfrac{1}{2}$ **53.** 2 **55.** 4 **57.** 100 **59.** 1 **61.** $\dfrac{27}{8}$

63. 5 **65.** $\dfrac{1}{1000}$ **67.** -2 **69.** 3 **71.** 25

73. $-\dfrac{2}{3}$ **75.** 50 **77.** 70 **79.** 10^{-8} **81.** 10

83. 1000 **85.** 6 **87.** $10^5 \cdot a_0$ **91.** 24,000 yr
93. 77,000 yr **95.** Pu-239, 240,000 yr; Sr-90, 280 yr; Th-230,
770,000 yr; Cs-135, 20,000,000 yr **a.** a^5 **b.** b^8 **c.** x^5
d. r^6 **e.** y^3 **f.** m^6 **g.** b^8 **h.** x^{20}

Section 11.4
1. $\log_b 5 + \log_b x$ **3.** $\log_4 x - \log_4 3$ **5.** $2 \log_3 a$

7. $\dfrac{1}{2} \log_5 x$ **9.** $3 \log_b x + 2 \log_b y$

11. $2 \log_4 y + \dfrac{1}{2} \log_4 x$ **13.** $2 \log_b x + \log_b y - \log_b z$

15. $\log x + 2 \log y - \dfrac{1}{2} \log z$

17. $\dfrac{1}{3} (\log_5 x + \log_5 y - 2 \log_5 z)$ **19.** $\log_b xy$

21. $\log_2 \dfrac{x^2}{y}$ **23.** $\log_b x\sqrt{y}$ **25.** $\log_b \dfrac{xy^2}{z}$

27. $\log_6 \dfrac{\sqrt{xy^2}}{z^3}$ **29.** $\log_b \sqrt[3]{\dfrac{x^2 y}{z}}$ **31.** 1.380 **33.** 0.903

35. 0.151 **37.** -0.602 **39.** 0.833 **41.** 2.833
43. -0.167 **45.** 7.42, basic **47.** 5.61 **49.** 5610
51. 2×10^{-5} **53.** 0.693 **55.** 2.303 **57.** 74
59. 2.930 **61.** 5.6 kg **63.** 60 kg **a.** 5, -1
b. 4, -9 **c.** 5, -3 **d.** 3, -9 **e.** $\dfrac{1}{9}$ **f.** $-\dfrac{10}{3}$

Section 11.5
1. 64 **3.** 99 **5.** 8 **7.** 162 **9.** 2 **11.** 6
13. $\dfrac{20}{9}$ **15.** $\dfrac{19}{8}$ **17.** $\dfrac{15}{4}$ **19.** 3, 5 **21.** 4

23. -4 **25.** $\dfrac{1}{3}$ **27.** 1.771 **29.** 0.792 **31.** 0.670

33. 0.894 **35.** 3.819 **37.** 4.419 **39.** 8 years
41. 7.79 years **43.** 1.47 h **45.** 3.17 h **47.** 20.6 years
49. 11.6 years **51.** 7.5 years **53.** 23.1 h **55.** 4558 ft
57. 2876 years **a.** $x^2 + 2xy + y^2$ **b.** $9x^2 + 12x + 4$
c. $x^3 + 3x^2y + 3xy^2 + y^3$ **d.** $8x^3 - 36x^2 + 54x - 27$
e. $x^2 - 2xy + y^2$ **f.** $x^4 + 4x^3y + 6x^2y^2 + 4xy^3 + y^4$

Chapter 11 Self-Test

1. $(a)\ f^{-1} = \left\{ (x, y) \,\middle|\, y = \dfrac{1}{4}x + \dfrac{1}{2} \right\}$, a function

$(b)\ g^{-1} = \{(x, y) \mid y = \pm\sqrt{x - 1}\}$, not a function
(c)

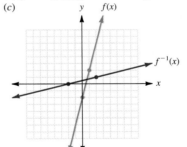

2.

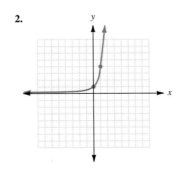

3.

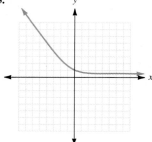

4. (a) -2 (b) $\dfrac{5}{2}$ **5.**

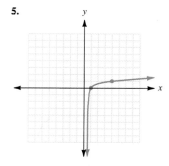

6. $\log 10.000 = 4$ **7.** $\log_{27} 9 = \dfrac{2}{3}$ **8.** $5^3 = 125$

9. $10^2 = 0.01$ **10.** 6   **11.** 4   **12.** 5
13. $2 \log_b x + \log_b y + 3 \log_b z$

14. $\dfrac{1}{2}(\log_b x + 2 \log_b y - \log_b z)$ **15.** $\log xy^2$

6. $\log_b \sqrt[3]{\dfrac{x}{z^2}}$ **17.** 8   **18.** $\dfrac{11}{8}$ **19.** 0.262   **20.** 2.151

Section 12.1

1. $x^5 + 5x^4y + 10x^3y^2 + 10x^2y^3 + 5xy^4 + y^5$
3. $64a^6 + 192a^5b + 240a^4b^2 + 160a^3b^3 + 60a^2b^4 + 12ab^5 + b^6$
5. $x^{14} + 7x^{12}y + 21x^{10}y^2 + 35x^8y^3 + 35x^6y^4 + 21x^4y^5 + 7x^2y^6 + y^7$
7. $a^5 - 15a^4 + 90a^3 - 270a^2 + 405a - 243$ **9.** 120
11. 120   **13.** 336   **15.** 28   **17.** 84   **19.** 15
21. 1   **23.** $x^7 + 7x^6y + 21x^5y^2 + 35x^4y^3$
25. $m^8 - 24m^7 + 252m^6 - 1512m^5$
27. $x^{10} + 20x^9y + 180x^8y^2 + 960x^7y^3$
29. $a^{10} - 5a^8b + 10a^6b^2 - 10a^4b^3$ **31.** 1.171648   **33.** 8

35. 64   **37.** $\dfrac{1}{4}$ **39.** $\dbinom{6}{2}$ or 15

41. $x^6 + 6x^5y + 15x^4y^2 + 20x^3y^3 + 15x^2y^4 + 6xy^5 + y^6$
43. $x^4 + 12x^3 + 54x^2 + 108x + 81$ **a.** 5   **b.** 10
c. 14   **d.** 5   **e.** 66   **f.** 44

Section 12.2

1. Arithmetic, 2   **3.** Not arithmetic   **5.** Arithmetic, $\dfrac{2}{3}$

7. 3, 5, 7, 9, 11, 13   **9.** $3, \dfrac{15}{4}, \dfrac{9}{2}, \dfrac{21}{4}, 6, \dfrac{27}{4}$ **11.** 61

13. 86   **15.** 135   **17.** $\dfrac{31}{2}$ **19.** $d = 4$

21. $a_1 = 5, d = 5$ **23.** 480   **25.** 1395   **27.** 1390
29. 5050   **31.** 4100   **33.** \$5000   **35.** \$20,600, \$142,400
37. \$33,600, \$361,200   **39.** 169 bricks   **41.** n^2
43.

4	3	8
9	5	1
2	7	6

45.

40	15	20
5	25	45
30	35	10

a. 8   **b.** 9   **c.** $\dfrac{1}{8}$ **d.** $\dfrac{1}{9}$ **e.** $-\dfrac{1}{8}$ **f.** $\dfrac{1}{4}$

Section 12.3

1. Geometric, $r = 4$ **3.** Geometric, $r = \dfrac{1}{3}$

5. Not geometric   **7.** 512   **9.** $\dfrac{243}{8}$ **11.** 486   **13.** $\dfrac{64}{27}$

15. 8   **17.** 12, 48   **19.** 189   **21.** -105 **23.** $\dfrac{211}{4}$

25. $\dfrac{242}{9}$ **27.** $\dfrac{182}{9}$ **29.** 2   **31.** No sum exists, $r = 2$

33. 32   **35.** $\dfrac{7}{9}$ **37.** $\dfrac{37}{99}$ **39.** $\dfrac{41}{333}$ **41.** \$1338.23

43. 40 g   **45.** \$819.20   **47.** 64,126   **49.** 125 cm
51.

16	1	32
16	8	4
2	64	4

53.

x	16	$4x^2$
$16x^2$	$4x$	1
4	x^2	$16x$

Chapter 12 Self-Test
1. $x^2 - 4x + 4$ **2.** 220   **3.** 1365
4. $32x^5 - 80x^4y^2 + 80x^3y^4 - 40x^2y^6$
5. $a^8 - 21a^7b + 189a^6b^2 - 945a^5b^3 + 315a^4b^2$ **6.** -3
7. $-32, -20, -8, 4, 16$ **8.** 103   **9.** -750 **10.** 4
11. -54 **12.** 700   **13.** 10,100   **14.** 2080   **15.** -2
16. 1875   **17.** 12, 72   **18.** -63 **19.** $\dfrac{122}{9}$ **20.** $\dfrac{64}{3}$

INDEX